Birkhäuser

Operator Theory: Advances and Applications
Volume 235

Founded in 1979 by Israel Gohberg

Yuli Eidelman
Israel Gohberg
Iulian Haimovici

Separable Type Representations of Matrices and Fast Algorithms

Volume 2
Eigenvalue Method

Birkhäuser

Yuli Eidelman
Israel Gohberg
Iulian Haimovici
School of Mathematical Sciences
Raymond and Beverly Sackler
Faculty of Exact Sciences
Tel Aviv University
Tel Aviv
Israel

ISSN 0255-0156 ISSN 2296-4878 (electronic)
ISBN 978-3-0348-0611-4 ISBN 978-3-0348-0612-1 (eBook)
DOI 10.1007/978-3-0348-0612-1
Springer Basel Heidelberg New York Dordrecht London

Library of Congress Control Number: 2013951156

Mathematics Subject Classification (2010): 15A18, 65F15, 65H04

Printed on acid-free paper

Springer Basel is part of Springer Science+Business Media (www.birkhauser-science.com)

Contents

Part VI Divide and Conquer Method for the Eigenproblem

Part VII Algorithms for QR Iterations and for Reduction to Hessenberg Form

Introduction

The book. This book contains a systematic theoretical and computational study of several types of generalizations of separable matrices. It is related to semiseparable, quasiseparable, band and companion representations of matrices. For them their natural parameters, called generators, are analyzed and algorithms are expressed in terms of generators.

In this volume we apply results and methods developed in Volume 1 to study in detail the eigenstructures of matrices with quasiseparable representations and to develop fast eigenvalue algorithms. A large number of illustrative examples are provided. The present volume contains Parts V–VIII.

Description of parts. The volume addresses the eigenvalue problem for matrices with quasiseparable structure. In Part V, for matrices with quasiseparable representation of the first order, we perform a detailed study of the properties of the characteristic polynomials of principal leading submatrices, the structure of eigenspaces, and the basic methods to compute eigenvalues. Part VI is devoted to the divide and conquer method; the main algorithms are derived also for matrices with quasiseparable representation of order one. In Part VII, for some classes of matrices with quasiseparable of any order representations, we study the QR iteration method. This method is used in Part VIII in order to get a fast solver for the polynomial root finding problem.

The first volume consists of Parts I–IV. The titles are as follows. Part I: Basics on separable, semiseparable and quasiseparable representations of matrices; Part II: Completion of matrices with specified band; Part III: Quasiseparable representations of matrices, descriptor systems with boundary conditions and first applications; Part IV: Factorization and inversion.

Part V

The Eigenvalue Structure of Order One Quasiseparable Matrices

Introduction to Part V

In this part the eigenstructure of scalar matrices with quasiseparable order one is studied. For such matrices one can choose quasiseparable generators that are complex numbers, so that more results can be deduced.

The first chapter focuses on characteristic polynomials and their derivatives, for which various recursion relations involving characteristic polynomials of the principal leading submatrices are established. These recursions permit to compute the coefficients of the characteristic polynomial by an $O(N^2)$ algorithm which generalizes the simpler case of a tridiagonal matrix. This special example, together with the case of upper Hessenberg unitary matrices, are fully treated. Also, an analog of the Sturm property is proved.

The second chapter deals with eigenvalues of geometric multiplicity one. An $O(N)$ algorithm for computing eigenvectors is devised. The theory is specified for Hermitian matrices. Other examples are discussed.

Next we consider eigenspaces of higher dimensions. Since the matrix $A - \lambda_0 I$ has the same upper and lower quasiseparable generators as the ones which are given for A itself, this amounts to studying in details the kernels of quasiseparable of order one matrices. This study is performed in the next chapter.

The last chapter studies partitions of eigenspaces. Under some assumptions we obtain the partition of an eigenspace in a direct sum of elementary subspaces. The main result consists in building the eigenspaces under some alternative assumptions, which for a Hermitian matrix cover all the possibilities.

The part contains a series of different examples.

Chapter 21

Quasiseparable of Order One Matrices. Characteristic Polynomials

Here we study in detail and in the three subsequent chapters the properties of eigenvalues and eigenvectors of scalar matrices with quasiseparable order one. For such matrices one can use quasiseparable generators with orders equal to one. This means that here quasiseparable generators are complex numbers.

§21.1 Recursive relations

In this section we derive recursion relations for the characteristic polynomials

$$\gamma_k(\lambda) = \det(A(1:k,1:k)-\lambda I),\ \theta_k(\lambda) = \det(A(k:N,k:N)-\lambda I)\ (k=1,\dots,N)$$

of the principal submatrices of a matrix A with quasiseparable order one.

Theorem 21.1. *Let A be a scalar matrix with quasiseparable generators $p(i)$ $(i = 2,\dots,N)$, $q(j)$ $(j = 1,\dots,N-1)$, $a(k)$ $(k = 2,\dots,N-1)$; $g(i)$ $(i = 1,\dots,N-1)$, $h(j)$ $(j = 2,\dots,N)$, $b(k)$ $(k = 2,\dots,N-1)$; $d(k)$ $(k = 1,\dots,N)$ with orders equal to one. Using these generators consider the following expressions linear in λ:*

$$c_k(\lambda) = (d(k)-\lambda)a(k)b(k) - q(k)p(k)b(k) - h(k)g(k)a(k), \quad k = 2,\dots,N-1. \tag{21.1}$$

Then the following recursion relations hold:

$$\gamma_1(\lambda) = d(1)-\lambda, \quad f_1(\lambda) = q(1)g(1), \tag{21.2}$$
$$\gamma_k(\lambda) = (d(k)-\lambda)\gamma_{k-1}(\lambda) - p(k)h(k)f_{k-1}(\lambda), \tag{21.3}$$
$$f_k(\lambda) = c_k(\lambda)f_{k-1}(\lambda) + q(k)g(k)\gamma_{k-1}(\lambda), \quad k = 2,\dots,N-1, \tag{21.4}$$
$$\gamma_N(\lambda) = (d(N)-\lambda)\gamma_{N-1}(\lambda) - p(N)h(N)f_{N-1}(\lambda) \tag{21.5}$$

and

$$\theta_N(\lambda) = d(N) - \lambda, \quad z_N(\lambda) = h(N)p(N), \tag{21.6}$$
$$\theta_k(\lambda) = (d(k) - \lambda)\theta_{k+1}(\lambda) - g(k)q(k)z_{k+1}(\lambda), \tag{21.7}$$
$$z_k(\lambda) = c_k(\lambda)z_{k+1}(\lambda) + \theta_{k+1}(\lambda)h(k)p(k), \quad k = N-1, \dots, 2, \tag{21.8}$$
$$\theta_1(\lambda) = (d(1) - \lambda)\theta_2(\lambda) - q(1)g(1)z_2(\lambda). \tag{21.9}$$

Here $f_k(\lambda), z_k(\lambda)$ are auxiliary polynomials.

Proof. $A - \lambda I$ is a matrix with quasiseparable order one with the same generators $p(i), q(j), a(k); g(i), h(j), b(k)$ as the matrix A and diagonal entries $d(k) - \lambda$. Applying the relations (19.2) one can derive easily the recursions (21.2)–(21.5). Similarly, (21.6)–(21.9) follow from (19.3). □

From the relations (21.2)–(21.5) one can easily derive the recursions for the derivatives of the corresponding polynomials.

Corollary 21.2. *Under the conditions of Theorem* 21.1*, the derivatives of the polynomials $\gamma_k(\lambda)$, $f_k(\lambda)$ satisfy the recursion relations*

$$\gamma_1'(\lambda) = -1, \quad f_1'(\lambda) = 0, \tag{21.10}$$
$$\gamma_k'(\lambda) = (d(k) - \lambda)\gamma_{k-1}'(\lambda) - \gamma_{k-1}(\lambda) - p(k)h(k)f_{k-1}'(\lambda), \tag{21.11}$$
$$f_k'(\lambda) = c_k(\lambda)f_{k-1}'(\lambda) - a(k)b(k)f_{k-1}(\lambda) + q(k)g(k)\gamma_{k-1}'(\lambda), \quad k = 2, \dots, N-1, \tag{21.12}$$
$$\gamma_N'(\lambda) = -\gamma_{N-1}(\lambda) - (d(N) - \lambda)\gamma_{N-1}'(\lambda) - p(N)h(N)f_{N-1}'(\lambda). \tag{21.13}$$

The relations (21.2)–(21.5) and (21.10)–(21.13) yield an $O(N)$ algorithm for the evaluation of the characteristic polynomials of the principal submatrices of a quasiseparable matrix and of their derivatives. This algorithm can be used for a fast implementation of the Newton iteration method (for the Newton method see, for instance, [35]).

Using the recursions (21.2)–(21.5) one can easily derive an $O(N^2)$ algorithm for the computation of the coefficients of the characteristic polynomial of the matrix A. This generalizes the corresponding algorithm for the simplest case of tridiagonal matrices.

In the sequel, to simplify the exposition, we use the auxiliary expressions linear in λ

$$d_k(\lambda) = d(k) - \lambda, \quad k = 1, \dots, N,$$
$$c_k(\lambda) = (d(k) - \lambda)a(k)b(k) - q(k)p(k)b(k) - h(k)g(k)a(k),$$
$$l_k(\lambda) = (d(k) - \lambda)a(k) - q(k)p(k),$$
$$\delta_k(\lambda) = (d(k) - \lambda)b(k) - h(k)g(k), \quad k = 2, \dots, N-1.$$

The expressions $l_k(\lambda), \delta_k(\lambda)$ play an essential role below. One can check easily that

$$d_k(\lambda)c_k(\lambda) = l_k(\lambda)\delta_k(\lambda) - q(k)p(k)h(k)g(k). \tag{21.14}$$

Under the assumption that some generators of the matrix A are zero or not zero, the recursion relations (21.2)–(21.5) may be given in a form more convenient for our next purposes.

Theorem 21.3. *Let A be a scalar matrix with quasiseparable generators $p(i)$ $(i = 2,\dots,N)$, $q(j)$ $(j = 1,\dots,N-1)$, $a(k)$ $(k = 2,\dots,N-1)$; $g(i)$ $(i = 1,\dots,N-1)$, $h(j)$ $(j = 2,\dots,N)$, $b(k)$ $(k = 2,\dots,N-1)$; $d(k)$ $(k = 1,\dots,N)$ with orders equal to one. Let for some $k \in \{3,\dots,N\}$ the inequality $p(k-1)h(k-1) \neq 0$ hold. Then the polynomials $\gamma_j(\lambda) = \det(A(1:j,1:j) - \lambda I)$, $j = k-2, k-1, k$ satisfy the three-term recursion relation*

$$\gamma_k(\lambda) = \psi_k(\lambda)\gamma_{k-1}(\lambda) - \varphi_k(\lambda)\gamma_{k-2}(\lambda), \tag{21.15}$$

where

$$\begin{aligned} \psi_k(\lambda) &= d(k) - \lambda + \frac{p(k)h(k)}{p(k-1)h(k-1)}c_{k-1}(\lambda), \\ \varphi_k(\lambda) &= \frac{p(k)h(k)}{p(k-1)h(k-1)}l_{k-1}(\lambda)\delta_{k-1}(\lambda). \end{aligned} \tag{21.16}$$

Proof. Substituting $k-1$ instead of k in (21.4) and (21.3) we obtain

$$f_{k-1}(\lambda) = c_{k-1}(\lambda)f_{k-2}(\lambda) + q(k-1)g(k-1)\gamma_{k-2}(\lambda), \tag{21.17}$$

$$\gamma_{k-1}(\lambda) = d_{k-1}(\lambda)\gamma_{k-2}(\lambda) - p(k-1)h(k-1)f_{k-2}(\lambda). \tag{21.18}$$

The relation (21.18) implies

$$f_{k-2}(\lambda) = \frac{1}{p(k-1)h(k-1)}[d_{k-1}(\lambda)\gamma_{k-2}(\lambda) - \gamma_{k-1}(\lambda)]. \tag{21.19}$$

Substituting (21.19) in (21.17) and taking into account (21.14), we obtain

$$\begin{aligned} &f_{k-1}(\lambda) \\ &= \frac{c_{k-1}(\lambda)}{p(k-1)h(k-1)}[d_{k-1}(\lambda)\gamma_{k-2}(\lambda) - \gamma_{k-1}(\lambda)] + q(k-1)g(k-1)\gamma_{k-2}(\lambda) \\ &= -\frac{c_{k-1}(\lambda)}{p(k-1)h(k-1)}\gamma_{k-1}(\lambda) \\ &\quad + \left[\frac{c_{k-1}(\lambda)}{p(k-1)h(k-1)}d_{k-1}(\lambda) + g(k-1)q(k-1)\right]\gamma_{k-2}(\lambda) \\ &= -\frac{c_{k-1}(\lambda)}{p(k-1)h(k-1)}\gamma_{k-1}(\lambda) + \frac{1}{p(k-1)h(k-1)}l_{k-1}(\lambda)\delta_{k-1}(\lambda)\gamma_{k-2}(\lambda). \end{aligned}$$

Substitution of the last expression in (21.3) yields (21.15). □

Now assume that the conditions $p(i)h(i) \neq 0$ is valid for all i. In this case we obtain the recursion relations for the characteristic polynomials $\gamma_k(\lambda)$ without using the auxiliary polynomials $f_k(\lambda)$.

Corollary 21.4. *Let A be a scalar matrix with quasiseparable generators $p(i)$ ($i = 2,\dots,N$), $q(j)$ ($j = 1,\dots,N-1$), $a(k)$ ($k = 2,\dots,N-1$); $g(i)$ ($i = 1,\dots,N-1$), $h(j)$ ($j = 2,\dots,N$), $b(k)$ ($k = 2,\dots,N-1$); $d(k)$ ($k = 1,\dots,N$) with orders equal to one. Let $p(i)h(i) \neq 0$ ($i = 2,\dots,N-1$). Then the polynomials $\gamma_k(\lambda) = \det(A(1:k,1:k) - \lambda I)$ ($k = 1,\dots,N$) satisfy the three-term recursion relations:*

$$\gamma_1(\lambda) = d(1) - \lambda, \quad \gamma_2(\lambda) = (d(2)-\lambda)(d(1)-\lambda) - p(2)q(1)g(1)h(2), \tag{21.20}$$

$$\gamma_k(\lambda) = \psi_k(\lambda)\gamma_{k-1}(\lambda) - \varphi_k(\lambda)\gamma_{k-2}(\lambda), \quad k = 3,\dots,N, \tag{21.21}$$

where

$$\psi_k(\lambda) = d(k) - \lambda + \frac{p(k)h(k)}{p(k-1)h(k-1)} c_{k-1}(\lambda),$$

$$\varphi_k(\lambda) = \frac{p(k)h(k)}{p(k-1)h(k-1)} l_{k-1}(\lambda)\delta_{k-1}(\lambda).$$

Proof. The relations (21.20) directly follow from (21.2) and from (21.3) with $k = 2$. For $k = 3,\dots,N$, the relations (21.21) follow from Theorem 21.3. □

Next we combine the cases $p(k)h(k) = 0$ and $p(k)h(k) \neq 0$. We use here the following auxiliary relations.

Lemma 21.5. *For any $2 \le s \le k \le N-1$ the polynomials $f_k(\lambda)$ defined in (21.2)–(21.4) satisfy the relations*

$$f_k(\lambda) = (c_{k+1,s-1}(\lambda))^{>} f_{s-1}(\lambda) + \sum_{j=s}^{k} (c_{k+1,j}(\lambda))^{>} q(j)g(j)\gamma_{j-1}(\lambda). \tag{21.22}$$

Proof. The proof proceeds by induction on k. For $k = s$, by (21.4),

$$\begin{aligned} f_k(\lambda) = f_s(\lambda) &= c_s(\lambda)f_{s-1}(\lambda) + q(s)g(s)\gamma_{s-1}(\lambda) \\ &= (c_{s+1,s-1}(\lambda))^{>} f_{s-1}(\lambda) + (c_{s+1,j}(\lambda))^{>} q(s)g(s)\gamma_{s-1}(\lambda) \end{aligned}$$

which implies (21.22). Assume that (21.22) is valid for some $k \ge s$. By (21.4) and the equality

$$c_{k+1}(\lambda)(c_{k+1,s-1}(\lambda))^{>} = (c_{k+2,s-1}(\lambda))^{>},$$

we have

$$\begin{aligned} f_{k+1}(\lambda) &= c_{k+1}(\lambda)f_k(\lambda) + q(k+1)g(k+1)(\lambda)\gamma_k(\lambda) \\ &= c_{k+1}(\lambda)\left[(c_{k+1,s-1}(\lambda))^{>} f_{s-1}(\lambda) + \sum\nolimits_{j=s}^{k} (c_{k+1,j}(\lambda))^{>} q(j)g(j)\gamma_{j-1}(\lambda)\right] \\ &\quad + q(k+1)g(k+1)\gamma_k(\lambda) \end{aligned}$$

$$
\begin{aligned}
&= (c_{k+2,s-1}(\lambda))^{>} f_{s-1}(\lambda) + \sum_{j=s}^{k} (c_{k+2,j}(\lambda))^{>} q(j) g(j) \gamma_{j-1}(\lambda) \\
&\quad + (c_{k+2,k+1}(\lambda))^{>} q(k+1) g(k+1) \gamma_k(\lambda) \\
&= (c_{k+2,s-1}(\lambda))^{>} f_{s-1}(\lambda) + \sum_{j=s}^{k+1} (c_{k+2,j}(\lambda))^{>} q(j) g(j) \gamma_{j-1}(\lambda),
\end{aligned}
$$

which completes the proof of the lemma. □

Now we derive a generalization of Theorem 21.3.

Theorem 21.6. *Let A be a scalar matrix with quasiseparable generators $p(i)$ ($i = 2,\dots,N$), $q(j)$ ($j = 1,\dots,N-1$), $a(k)$ ($k = 2,\dots,N-1$); $g(i)$ ($i = 1,\dots,N-1$), $h(j)$ ($j = 2,\dots,N$), $b(k)$ ($k = 2,\dots,N-1$); $d(k)$ ($k = 1,\dots,N$) with orders equal to one. Assume that for some integers k, s with $2 \le s < k-1 \le N-1$ the relations*

$$
p(s+1)h(s+1) = \cdots = p(k-1)h(k-1) = 0, \quad p(s)h(s) \neq 0
$$

hold.

Then

$$
\gamma_k(\lambda) = \psi_{k,s}(\lambda)\gamma_s(\lambda) - \varphi_{k,s}(\lambda)\gamma_{s-1}(\lambda), \tag{21.23}
$$

where $\gamma_j(\lambda) = \det(A(1:j,1:j) - \lambda I)$ and

$$
\begin{aligned}
\psi_{k,s}(\lambda) &= (d_{k+1,s}(\lambda))^{>} + \frac{p(k)h(k)}{p(s)h(s)} (c_{k,s-1}(\lambda))^{>} \\
&\qquad - p(k)h(k) \sum_{j=s+1}^{k-1} (c_{k,j}(\lambda))^{>} q(j) g(j) (d_{j,s}(\lambda))^{>}, \\
\varphi_{k,s}(\lambda) &= \frac{p(k)h(k)}{p(s)h(s)} (c_{k,s}(\lambda))^{>} l_s(\lambda) \delta_s(\lambda).
\end{aligned}
$$

Proof. By Lemma 21.5,

$$
\begin{aligned}
f_{k-1}(\lambda) &= (c_{k,s-1}(\lambda))^{>} f_{s-1}(\lambda) \\
&\quad + \sum_{j=s+1}^{k-1} (c_{k,j}(\lambda))^{>} q(j) g(j) \gamma_{j-1}(\lambda) + (c_{k,s}(\lambda))^{>} q(s) g(s) \gamma_{s-1}(\lambda).
\end{aligned} \tag{21.24}
$$

As a direct consequence of (21.3) we get that if for some $1 \le i < j \le N$, $p(i+1)h(i+1) = \cdots = p(j)h(j) = 0$, then

$$
\gamma_j(\lambda) = (d(j) - \lambda) \dots (d(i+1) - \lambda)\gamma_i(\lambda). \tag{21.25}
$$

Hence,

$$
\gamma_{j-1}(\lambda) = \gamma_s(\lambda)(d_{j,s}(\lambda))^{>}, \quad j = s+1,\dots,k. \tag{21.26}
$$

Next, by (21.3),

$$f_{s-1}(\lambda) = \frac{1}{p(s)h(s)}[(d(s)-\lambda)\gamma_{s-1}(\lambda) - \gamma_s(\lambda)]. \tag{21.27}$$

Substituting (21.27) and (21.26) in (21.24) we obtain

$$\begin{aligned} f_{k-1}(\lambda) = {} & \frac{(c_{k,s-1}(\lambda))^>}{p(s)h(s)}(d(s)-\lambda)\gamma_{s-1}(\lambda) - \frac{(c_{k,s-1}(\lambda))^>}{p(s)h(s)}\gamma_s(\lambda) \\ & + \sum_{j=s+1}^{k-1} (c_{k,j}(\lambda))^> q(j)g(j)(d_{j,s}(\lambda))^> \gamma_s(\lambda) + (c_{k,s}(\lambda))^> q(s)g(s)\gamma_{s-1}(\lambda). \end{aligned}$$

From here using the equalities

$$(c_{k,s-1}(\lambda))^> = (c_{k,s}(\lambda))^> c_s(\lambda), \quad \frac{c_s(\lambda)}{p(s)h(s)}(d(s)-\lambda) + q(s)g(s) = \frac{l_s(\lambda)\delta_s(\lambda)}{p(s)h(s)}$$

we obtain

$$\begin{aligned} f_{k-1}(\lambda) = {} & \left[-\frac{(c_{k,s-1}(\lambda))^>}{p(s)h(s)} + \sum_{j=s+1}^{k-1} (c_{k,j}(\lambda))^> q(j)g(j)(d_{j,s}(\lambda))^> \right] \gamma_s(\lambda) \\ & + \frac{(c_{k,s}(\lambda))^>}{p(s)h(s)} l_s(\lambda)\delta_s(\lambda)\gamma_{s-1}(\lambda). \end{aligned} \tag{21.28}$$

Now using (21.26) we get

$$(d(k)-\lambda)\gamma_{k-1}(\lambda) = (d(k)-\lambda)(d_{k,s}(\lambda))^> \gamma_s(\lambda) = (d_{k+1,s}(\lambda))^> \gamma_s(\lambda). \tag{21.29}$$

Substituting (21.28) and (21.29) in (21.3) we obtain (21.23). □

Example 21.7. Consider the matrix

$$A = \begin{pmatrix} 1 & 0 & 1 & 1 & 1 \\ 0 & 0 & 0 & 0 & 0 \\ 1 & 0 & 1 & 1 & 1 \\ 1 & 0 & 1 & 0 & 1 \\ 1 & 0 & 1 & 1 & 1 \end{pmatrix}$$

and take as quasiseparable generators the numbers

$$\begin{aligned} & p(2) = q(2) = g(2) = h(2) = d(2) = d(4) = 0, \\ & p(3) = p(4) = p(5) = q(1) = q(3) = q(4) = g(1) = g(3) = g(4) \\ & \qquad = h(3) = h(4) = h(5) = 1, \\ & d(1) = d(3) = d(5) = 1, \;\; a(2) = a(3) = a(4) = b(2) = b(3) = b(4) = 1. \end{aligned}$$

We will compute for the above representation of A the characteristic polynomials of its principal leading submatrices in three ways: with the definition, which is also given at the beginning of the section, with Theorem 21.1, and with Theorem 21.3.

With Theorem 21.1 one gets

$$c_2(\lambda) = -\lambda, \quad c_3(\lambda) = 1 - \lambda - 1 - 1 = -\lambda - 1, \quad c_4(\lambda) = -\lambda - 1 - 1 = -\lambda - 2,$$
$$\gamma_1(\lambda) = 1 - \lambda, \quad f_1(\lambda) = 1, \quad \gamma_2(\lambda) = -\lambda(1 - \lambda), \quad f_2(\lambda) = -\lambda \cdot 1,$$
$$\gamma_3(\lambda) = -(1 - \lambda)^2\lambda + \lambda = -\lambda^2(\lambda - 2),$$
$$f_3(\lambda) = (\lambda + 1)\lambda - \lambda(1 - \lambda) = 2\lambda^2,$$
$$\gamma_4(\lambda) = \lambda^2(1 - \lambda)^2 - \lambda^2 - 2\lambda^2 = \lambda^2(\lambda^2 - 2\lambda - 2),$$
$$f_4(\lambda) = -(\lambda + 2)2\lambda^2 - \lambda^2(\lambda - 2) = -\lambda^2(3\lambda + 2)$$

and

$$\gamma_5(\lambda) = (1 - \lambda)\lambda^2(\lambda^2 - 2\lambda - 2) + \lambda^2(3\lambda + 2) = \lambda^3(-\lambda^2 + 3\lambda + 3).$$

With Theorem 21.3 one can obtain $\gamma_4(\lambda)$:

$$\psi_4(\lambda) = -\lambda + c_3(\lambda) = -2\lambda - 1, \quad \varphi_4(\lambda) = l_3(\lambda)\delta_3(\lambda) = (1 - \lambda - 1)(1 - \lambda - 1) = \lambda^2,$$

therefore

$$\gamma_4(\lambda) = -(2\lambda + 1)\gamma_3(\lambda) - \lambda^2\gamma_2(\lambda) = (2\lambda + 1)\lambda^2(\lambda - 2) + \lambda^3(1 - \lambda) = \lambda^2(\lambda^2 - 2\lambda - 2)$$

and since the theorem is also true for $N = 5$ one can also obtain $\gamma_5(\lambda)$:

$$\psi_5(\lambda) = 1 - \lambda + c_4(\lambda) = -2\lambda - 1,$$
$$\varphi_5(\lambda) = l_4(\lambda)\delta_4(\lambda) = (-\lambda - 1)(-\lambda - 1) = \lambda^2 + 2\lambda + 1,$$

therefore

$$\begin{aligned}\gamma_5(\lambda) &= -(2\lambda + 1)\lambda^2(\lambda^2 - 2\lambda - 2) - (\lambda + 1)^2\lambda^2(2 - \lambda) \\ &= -\lambda^3(\lambda^2 - 3\lambda - 3).\end{aligned}$$

◊

§21.2 Tridiagonal and upper Hessenberg unitary matrices

Here we study two important particular cases.

Example 21.8. Consider an irreducible tridiagonal matrix A, i.e., a tridiagonal matrix with non-zero entries on the lower and upper subdiagonals. In this case we have $a(k) = b(k) = 0$, $A_{k+1,k} = p(k+1)q(k) \neq 0$, $A_{k,k+1} = g(k)h(k+1) \neq 0$ $(k = 1, \dots, N-1)$. Applying Corollary 21.4, we have

$$c_k(\lambda) = 0, \quad l_k(\lambda) = -p(k)q(k), \; \delta_k(\lambda) = -g(k)h(k),$$

and so

$$\psi_k(\lambda) = d(k) - \lambda,$$
$$\varphi_k(\lambda) = \frac{p(k)h(k)}{p(k-1)h(k-1)}p(k-1)q(k-1)g(k-1)h(k-1)$$
$$= p(k)q(k-1)g(k-1)h(k).$$

Hence, one can write down the recursions (21.20), (21.21) in the form

$$\gamma_0(\lambda) = 1, \quad \gamma_1(\lambda) = d(1) - \lambda;$$

$$\gamma_k(\lambda) = (d(k) - \lambda)\gamma_{k-1}(\lambda) - (p(k)q(k-1))(g(k-1)h(k))\gamma_{k-2}(\lambda), \quad k = 2, \dots, N.$$

Note that in the case

$$d(k) = \overline{d(k)},\ k = 1, \dots, N, \quad (p(k)q(k-1))(g(k-1)h(k)) > 0, \quad k = 2, \dots, N-1,$$

these recursions defines polynomials orthogonal with some weight on the real line (see, for instance, [1, p. 121]). ◊

Example 21.9. Consider the unitary upper Hessenberg matrix U defined in Section §7.4,

$$U = \begin{bmatrix} -\rho_1\rho_0^* & -\rho_2\mu_1\rho_0^* & -\rho_3\mu_2\mu_1\rho_0^* & \cdots & -\rho_{N-1}\mu_{N-2}\cdots\mu_1\rho_0^* & -\rho_N\mu_{N-1}\cdots\mu_1\rho_0^* \\ \mu_1 & -\rho_2\rho_1^* & -\rho_3\mu_2\rho_1^* & \cdots & -\rho_{N-1}\mu_{N-2}\cdots\mu_2\rho_1^* & -\rho_N\mu_{N-1}\cdots\mu_2\rho_1^* \\ 0 & \mu_2 & -\rho_3\rho_2^* & \cdots & -\rho_{N-1}\mu_{N-2}\cdots\mu_3\rho_2^* & -\rho_N\mu_{N-1}\cdots\mu_3\rho_2^* \\ \vdots & \ddots & \mu_3 & & \vdots & \vdots \\ \vdots & & \ddots & \ddots & -\rho_{N-1}\rho_{N-2}^* & -\rho_N\mu_{N-1}\rho_{N-2}^* \\ 0 & \cdots & \cdots & 0 & \mu_{N-1} & -\rho_N\rho_{N-1}^* \end{bmatrix}, \tag{21.30}$$

with $\mu_k > 0$, $|\rho_k|^2 + \mu_k^2 = 1$ $(k = 1, \dots, N-1)$, $\rho_0 = -1$, $|\rho_N| = 1$. The matrix U has quasiseparable generators $d(k) = -\rho_k\rho_{k-1}^*$, $a(k) = 0$, $b(k) = \mu_{k-1}$, $p(i) = 1$, $q(j) = \mu_j$, $g(i) = -\rho_{i-1}^*$, $h(j) = \mu_{j-1}\rho_j$ with orders equal to one. Next consider the recursions (21.2)–(21.4). We have

$$\gamma_1(\lambda) = -\rho_1\rho_0^* - \lambda, \quad f_1(\lambda) = -\mu_1\rho_0^*, \tag{21.31}$$
$$\gamma_k(\lambda) = (-\rho_k\rho_{k-1}^* - \lambda)\gamma_{k-1}(\lambda) - \mu_{k-1}\rho_k f_{k-1}(\lambda), \tag{21.32}$$
$$f_k(\lambda) = -\mu_{k-1}\mu_k f_{k-1}(\lambda) - \mu_k\rho_{k-1}^*\gamma_{k-1}(\lambda), \quad k = 2, \dots, N-1. \tag{21.33}$$

To compare with some known results we consider the polynomials $\widetilde{\gamma}_k(\lambda) = \det(\lambda I - A(1:k, 1:k))$ instead of $\gamma_k(\lambda) = \det(A(1:k,1:k) - \lambda I)$, i.e., $\widetilde{\gamma}_k(\lambda) = (-1)^k\gamma_k(\lambda)$ and set $\widetilde{f}_k(\lambda) = (-1)^{k+1}f_k(\lambda)$. From the recursions (21.31)–(21.33) we obtain

$$\widetilde{\gamma}_0(\lambda) = 1, \quad \widetilde{f}_0(\lambda) = 0, \tag{21.34}$$
$$\widetilde{\gamma}_k(\lambda) = (\lambda + \rho_k\rho_{k-1}^*)\widetilde{\gamma}_{k-1}(\lambda) - \mu_{k-1}\rho_k\widetilde{f}_{k-1}(\lambda), \tag{21.35}$$
$$\widetilde{f}_k(\lambda) = \mu_{k-1}\mu_k\widetilde{f}_{k-1}(\lambda) - \mu_k\rho_{k-1}^*\widetilde{\gamma}_{k-1}(\lambda), \quad k = 1, \dots, N-1 \tag{21.36}$$

(here we take $\mu_0 = 0$). Set

$$G_k(\lambda) = -\rho_k^* \widetilde{\gamma}_k(\lambda) + \mu_k \widetilde{f}_k(\lambda), \quad k = 0, \dots, N-1. \tag{21.37}$$

Let us show that the polynomials $\widetilde{\gamma}_k(\lambda), G_k(\lambda)$ satisfy the two-terms recursion relations

$$\begin{aligned} \begin{pmatrix} G_0(\lambda) \\ \widetilde{\gamma}_0(\lambda) \end{pmatrix} &= \begin{pmatrix} 1 \\ 1 \end{pmatrix}, \quad \begin{pmatrix} G_k(\lambda) \\ \widetilde{\gamma}_k(\lambda) \end{pmatrix} \\ &= \begin{pmatrix} 1 & -\rho_k^* \lambda \\ -\rho_k & \lambda \end{pmatrix} \begin{pmatrix} G_{k-1}(\lambda) \\ \widetilde{\gamma}_{k-1}(\lambda) \end{pmatrix}, \quad k = 1, \dots, N-1. \end{aligned} \tag{21.38}$$

These recursions define Szegö polynomials orthogonal with the corresponding weight on the unit circle (see, for instance, [1, p. 176]). From (21.35) and (21.37) we obtain

$$\widetilde{\gamma}_k(\lambda) = \lambda \widetilde{\gamma}_{k-1}(\lambda) - \rho_k G_{k-1}(\lambda), \quad k = 1, 2, \dots, N.$$

Next, using (21.37), (21.35), (21.36) and the equalities $\mu_k^2 + |\rho_k|^2 = 1$, we obtain

$$\begin{aligned} G_k(\lambda) = &- \rho_k^* \widetilde{\gamma}_k(\lambda) + \mu_k \widetilde{f}_k(\lambda) \\ = &- \rho_k^* (\lambda + \rho_k \rho_{k-1}^*) \widetilde{\gamma}_{k-1}(\lambda) + \mu_{k-1} \rho_k^* \rho_k \widetilde{f}_{k-1}(\lambda) \\ &+ \mu_k (\mu_{k-1} \mu_k \widetilde{f}_{k-1}(\lambda) - \mu_k \rho_{k-1}^* \widetilde{\gamma}_{k-1}(\lambda)) \\ = &- \rho_k^* \lambda \widetilde{\gamma}_{k-1}(\lambda) - |\rho_k|^2 \rho_{k-1}^* \widetilde{\gamma}_{k-1}(\lambda) + \mu_{k-1} |\rho_k|^2 \widetilde{f}_{k-1}(\lambda) \\ &+ \mu_k^2 \mu_{k-1} \widetilde{f}_{k-1}(\lambda) - \mu_k^2 \rho_{k-1}^* \widetilde{\gamma}_{k-1}(\lambda) \\ = &- \rho_k^* \lambda \widetilde{\gamma}_{k-1}(\lambda) - \rho_{k-1}^* \widetilde{\gamma}_{k-1}(\lambda) + \mu_{k-1} \widetilde{f}_{k-1}(\lambda) \\ = &- \rho_k^* \lambda \widetilde{\gamma}_{k-1}(\lambda) + G_{k-1}(\lambda), \quad k = 1, \dots, N-1 \end{aligned}$$

which completes the proof of (21.38). ◊

§21.3 The Sturm property

In this section we establish an analog of the well-known Sturm property for the characteristic polynomials of principal leading submatrices of a Hermitian matrix with quasiseparable order one. This property yields information on the location of the eigenvalues of matrices with quasiseparable order one and is a basis for the bisection method (for the bisection method see, for instance, [35, p. 301]). To obtain this result we use the following Gundelfinger theorem (see, for instance, [31, p. 298]).

Theorem 21.10 (Gundelfinger). *Let A be an $N \times N$ Hermitian scalar matrix with determinants of leading principal submatrices $\gamma_k = \det A(1 : k, 1 : k)$, $k = 1, \dots, N$. Set $\gamma_0 = 1$. Assume that $\gamma_N \neq 0$ and the sequence*

$$\gamma_N, \gamma_{N-1}, \dots, \gamma_1, \gamma_0 \tag{21.39}$$

contains no two successive zeros.

Then the number of negative eigenvalues of A (counting each eigenvalue in accordance with its multiplicity) is equal to the number of sign changes in the sequence (21.39), *if any sign is assigned to each (isolated) zero γ_i in* (21.39).

Thus we obtain the following result.

Theorem 21.11. *Let A be a Hermitian scalar matrix with quasiseparable generators $p(i)$ $(i = 2, \dots, N)$, $q(j)$ $(j = 1, \dots, N-1)$, $a(k)$ $(k = 2, \dots, N-1)$; $g(i)$ $(i = 1, \dots, N-1)$, $h(j)$ $(j = 2, \dots, N)$, $b(k)$ $(k = 2, \dots, N-1)$; $d(k)$ $(k = 1, \dots, N)$ with orders equal to one. Assume that $p(k)h(k) \neq 0$, $k = 2, \dots, N-2$. Let $\gamma_0(\lambda) \equiv 1$ and $\gamma_k(\lambda) = \det(A(1:k, 1:k) - \lambda I)$, $k = 1, \dots, N$, be the characteristic polynomials of the principal leading submatrices of A. Let $\nu(\lambda)$ be the number of sign changes in the sequence*

$$\gamma_N(\lambda), \gamma_{N-1}(\lambda), \dots, \gamma_1(\lambda), \gamma_0(\lambda) \tag{21.40}$$

and let (α, β) be an interval on the real axis such that $\gamma_N(\alpha) \neq 0$, $\gamma_N(\beta) \neq 0$. Then:

1) *$\nu(\beta) \geq \nu(\alpha)$;*
2) *the difference $\nu(\beta) - \nu(\alpha)$ equals the number of eigenvalues of the matrix A in the interval (α, β), counting each eigenvalue in accordance with its multiplicity.*

Proof. Let us show that if $\gamma_N(\lambda_0) \neq 0$ and $\gamma_i(\lambda_0) = 0$ for some i, $1 \leq i \leq N-1$, then $\gamma_{i+1}(\lambda_0) \neq 0$. If this is not the case then, by Theorem 21.3, we have also $\gamma_{i+2}(\lambda_0) = 0$. Next we may continue to apply Theorem 21.3 and obtain $\gamma_N(\lambda_0) = 0$, a contradiction.

Thus we may apply the Gundelfinger theorem to the matrix $A - \beta I$ and conclude that the number of negative eigenvalues of this matrix or, equivalently, the number of eigenvalues of the matrix A which are less than β, is equal to $\nu(\beta)$. Similarly, we obtain that the number of eigenvalues of the matrix A which are less than α is equal to $\nu(\alpha)$, which completes the proof of the theorem. □

§21.4 Comments

The recursion relations for principal leading submatrices of tridiagonal matrices and of unitary Hessenberg matrices are well known (see for instance [1]). Some results of this chapter were obtained by D. Fasino and L. Gemignani in [26] in the particular case of symmetric diagonal plus Green of order one matrices.

A more detailed study of characteristic polynomials of the principal leading submatrices of quasiseparable matrices and related methods for inversion of some Vandermonde-type matrices for such polynomials was done recently in the papers [7, 8, 9]. The extension of these methods to matrices with greater than one quasiseparable orders was performed by T. Bella, V. Olshevsky and P. Zhlobich in [13, 14].

Chapter 22

Eigenvalues with Geometric Multiplicity One

In this chapter we study conditions under which the eigenspaces of quasiseparable of order one matrices are one-dimensional. We also derive explicit formulas for the corresponding eigenvectors.

§22.1 The main theorem

We use a special partition of a matrix with given quasiseparable generators.

Lemma 22.1. *Let $A = \{A_{ij}\}_{i,j=1}^N$ be a matrix with lower quasiseparable generators $p(i)$ $(i = 2, \dots, N)$, $q(j)$ $(j = 1, \dots, N-1)$, $a(k)$ $(k = 2, \dots, N-1)$, upper quasiseparable generators $g(i)$ $(i = 1, \dots, N-1)$, $h(j)$ $(j = 2, \dots, N)$, $b(k)$ $(k = 2, \dots, N-1)$, and diagonal entries $d(k)$ $(k = 1, \dots, N)$. With the lower and upper quasiseparable generators define the matrices P_k, Q_k and G_k, H_k via the relations (5.1), (5.2) and (5.5), (5.6). Set $A_k = A(1:k, 1:k)$, $B_k = A(k:N, k:N)$.*

Then the matrix A admits the partitions

$$A = \begin{pmatrix} A_{k-1} & G_{k-1}h(k) & G_{k-1}b(k)H_{k+1} \\ p(k)Q_{k-1} & d(k) & g(k)H_{k+1} \\ P_{k+1}a(k)Q_{k-1} & P_{k+1}q(k) & B_{k+1} \end{pmatrix}. \tag{22.1}$$

Proof. Using Lemma 5.1 and Lemma 5.4 one obtains, for every k, $1 \le k \le N-1$, the representation

$$A = \begin{pmatrix} A_k & G_k H_{k+1} \\ P_{k+1}Q_k & B_{k+1} \end{pmatrix}. \tag{22.2}$$

Moreover using (5.11) and (5.14) one obtains the partitions of the principal leading submatrices $A_k = A(1:k, 1:k)$ in the form

$$A_k = \begin{pmatrix} A_{k-1} & G_{k-1}h(k) \\ p(k)Q_{k-1} & d(k) \end{pmatrix}, \quad k = 2, \dots, N. \tag{22.3}$$

Combining the relations (5.3), (5.7) and the equalities (22.2), (22.3) we obtain (22.1). □

Next we obtain conditions under which an eigenvalue of a quasiseparable of order one matrix has geometric multiplicity one and derive explicit formulas for the corresponding eigenvector.

Theorem 22.2. *Let A be a scalar matrix with quasiseparable order one and with quasiseparable generators $p(i)$ $(i = 2, \dots, N)$, $q(j)$ $(j = 1, \dots, N-1)$, $a(k)$ $(k = 2, \dots, N-1)$; $g(i)$ $(i = 1, \dots, N-1)$, $h(j)$ $(j = 2, \dots, N)$, $b(k)$ $(k = 2, \dots, N-1)$; $d(k)$ $(k = 1, \dots, N)$. Assume that $\lambda = \lambda_0$ is an eigenvalue of A and*

$$|d(1) - \lambda_0| + |q(1)| > 0, \quad |d(1) - \lambda_0| + |g(1)| > 0; \tag{22.4}$$

$$\begin{gathered} l_k(\lambda_0) = (d(k) - \lambda_0)a(k) - p(k)q(k) \neq 0, \\ \delta_k(\lambda_0) = (d(k) - \lambda_0)b(k) - g(k)h(k) \neq 0, \\ k = 2, \dots, N-1, \end{gathered} \tag{22.5}$$

$$|d(N) - \lambda_0| + |p(N)| > 0, \quad |d(N) - \lambda_0| + |h(N)| > 0. \tag{22.6}$$

Then the eigenvalue λ_0 has geometric multiplicity one. Moreover, an eigenvector v corresponding to λ_0 can be determined as follows.

1) *If $p(N)h(N) \neq 0$, then*

$$\gamma_{N-1}(\lambda_0) = \det(A(1:N-1, 1:N-1) - \lambda_0 I) \neq 0$$

and

$$w = \begin{pmatrix} A_{N-1}(\lambda_0)^{-1} x_N \\ 1 \end{pmatrix}, \tag{22.7}$$

where $A_{N-1}(\lambda_0) = A(1:N-1, 1:N-1) - \lambda_0 I$, $x_N = -A(1:N-1, N)$. Furthermore, the eigenvector may be also expressed in terms of the values of the polynomials $\gamma_k(\lambda)$, $f_k(\lambda)$ at the point λ_0 as follows:

$$\tilde{w} = \begin{pmatrix} -V_{N-1}(\lambda_0)h(N) \\ \gamma_{N-1}(\lambda_0) \end{pmatrix}, \tag{22.8}$$

where $V_{N-1}(\lambda_0) = \mathrm{col}(v_i(\lambda_0)\delta^{<}_{i,N}(\lambda_0))_{i=1}^{N-1}$,

$$v_1(\lambda_0) = g(1), \quad v_i(\lambda_0) = g(i)\gamma_{i-1}(\lambda_0) - p(i)f_{i-1}(\lambda_0)b(i), \ i = 2, \dots, N-1.$$

2) *If for some m, $2 \le m \le N-1$, we have $p(m)h(m) \neq 0$ and $p(m+1)h(m+1) = \dots = p(N)h(N) = 0$, then*

$$\begin{aligned} \gamma_{m-1}(\lambda_0) &= \det(A(1:m-1, 1:m-1) - \lambda_0 I) \neq 0, \\ \theta_{m+1}(\lambda_0) &= \det(A(m+1:N, m+1:N) - \lambda_0 I) \neq 0 \end{aligned}$$

and

$$v = \begin{pmatrix} A_{m-1}(\lambda_0)^{-1}x_m \\ 1 \\ B_{m+1}(\lambda_0)^{-1}y_m \end{pmatrix}, \tag{22.9}$$

where

$$A_{m-1}(\lambda_0) = A(1:m-1,1:m-1) - \lambda_0 I, \; x_m = -A(1:m-1,m),$$
$$B_{m+1}(\lambda_0) = A(m+1:N,m+1:N) - \lambda_0 I,$$
$$y_m = P_{m+1}(a(m)\frac{f_{m-1}(\lambda_0)}{\gamma_{m-1}(\lambda_0)}h(m) - q(m)), \quad P_{m+1} = \mathrm{col}(p(k)a^{>}_{km})_{k=m+1}^{N}.$$

3) *If* $p(2)h(2) = \cdots = p(N)h(N) = 0$, *then*

$$\theta_2(\lambda_0) = \det(A(2:N,2:N) - \lambda_0 I) \neq 0$$

and

$$v = \begin{pmatrix} 1 \\ B_2(\lambda_0)^{-1}y_2 \end{pmatrix}, \tag{22.10}$$

where $B_2(\lambda_0) = A(2:N,2:N) - \lambda_0 I, \; y_2 = -A(2:N,1)$.

Proof. We start proving Part 1). First we show that $\gamma_{N-1}(\lambda_0) \neq 0$. Assume that this is not true. We will show that in this case

$$\gamma_1(\lambda_0) = \gamma_2(\lambda_0) = \cdots = \gamma_{N-1}(\lambda_0) = \gamma_N(\lambda_0) = 0. \tag{22.11}$$

The proof proceeds by reverse induction.

We have $\gamma_N(\lambda_0) = \gamma_{N-1}(\lambda_0) = 0$, by assumption. Assume that for some k with $3 \leq k \leq N-1$ the relations

$$\gamma_{k-1}(\lambda_0) = \gamma_k(\lambda_0) = \cdots = \gamma_{N-1}(\lambda_0) = \gamma_N(\lambda_0) = 0$$

hold. We have the following cases.

(a) $p(k-1)h(k-1) = 0$. By virtue of (21.25),

$$\gamma_{k-1}(\lambda) = (d(k-1) - \lambda)\gamma_{k-2}(\lambda)$$

and since by virtue of the conditions (22.5) $d(k-1) - \lambda_0 \neq 0$, we obtain $\gamma_{k-2}(\lambda_0) = 0$.

(b) $p(k-1)h(k-1) \neq 0$. Let t be an index such that $t \leq k \leq N$ and $p(k)h(k) = \cdots = p(t-1)h(t-1) = 0$, $p(t)h(t) \neq 0$. Such a number t exists since $p(N)h(N) \neq 0$; it is possible that $t = k$. Applying Theorem 21.6, we obtain

$$\gamma_t(\lambda) = \psi_{t,k-1}(\lambda)\gamma_{k-1}(\lambda) - \varphi_{t,k-1}(\lambda)\gamma_{k-2}(\lambda), \tag{22.12}$$

where

$$\varphi_{t,k-1}(\lambda) = \frac{p(t)h(t)}{p(k-1)h(k-1)} c_{t-1}(\lambda) \cdots c_k(\lambda) l_{k-1}(\lambda) \delta_{k-1}(\lambda).$$

Here $p(t)h(t) \neq 0$ by assumption, $l_{k-1}(\lambda_0)\delta_{k-1}(\lambda_0) \neq 0$ by virtue of (22.5) and $c_j(\lambda_0) \neq 0$ $(j = k, \dots, t-1)$ by virtue of (22.5) and (21.14). Thus $\varphi_{t,k-1}(\lambda_0) \neq 0$ and since $\gamma_t(\lambda_0)=0$, $\gamma_{k-1}(\lambda_0)=0$, and from (22.12) we conclude that $\gamma_{k-2}(\lambda_0)=0$.

Thus the equalities (22.11) hold. Hence in particular follows that $\gamma_1(\lambda_0) = d(1)-\lambda_0 = 0$ and by virtue of the condition (22.4) and the second equality in (21.2) we have $f_1(\lambda_0) = q(1)g(1) \neq 0$. Let t be the minimal index such that $p(t)h(t) \neq 0$. Using (21.3) and (22.11) we obtain

$$\gamma_t(\lambda_0) = (d(t) - \lambda_0)\gamma_{t-1}(\lambda_0) - p(t)h(t)f_{t-1}(\lambda_0) = -p(t)h(t)f_{t-1}(\lambda_0). \quad (22.13)$$

By virtue of Lemma 21.5 and (22.11), we have

$$\begin{aligned} f_{t-1}(\lambda_0) &= (c_{t,1}(\lambda_0))^{>} f_1(\lambda_0) + \sum_{j=2}^{t-1} (c_{t,j}(\lambda_0))^{>} q(j)g(j)\gamma_{j-1}(\lambda_0) \\ &= c_{t-1}(\lambda_0) \cdots c_2(\lambda_0) f_1(\lambda_0). \end{aligned}$$

From the conditions (22.5) and the equalities (21.14) we obtain $c_j(\lambda_0) \neq 0$, $j = 2, \dots, t-1$, whence $f_{t-1}(\lambda_0) \neq 0$. But from (22.13) we conclude that $\gamma_t(\lambda_0) \neq 0$, which is a contradiction. Thus we have proved that $\gamma_{N-1}(\lambda_0) \neq 0$.

Next for the matrix $A - \lambda_0 I$ consider the partition

$$A - \lambda_0 I = \begin{pmatrix} A_{N-1}(\lambda_0) & -x_N \\ r_N & d(N) - \lambda_0 \end{pmatrix}, \quad (22.14)$$

where $A_{N-1}(\lambda_0) = A(1 : N-1, 1 : N-1) - \lambda_0 I$, $x_N = -A(1 : N-1, N)$, $r_N = A(N, 1 : N-1)$. Let w be an eigenvector corresponding to the eigenvalue λ_0. We represent this vector in the form $w = \begin{pmatrix} v' \\ \alpha \end{pmatrix}$, where v' is an $(N-1)$-dimensional vector and α is a scalar. Using (22.14) we obtain $v' = \alpha A_{N-1}^{-1}(\lambda_0)x_N$, whence $w = \alpha v_0$, where $v_0 = \begin{pmatrix} A_{N-1}^{-1}(\lambda_0)x_N \\ 1 \end{pmatrix}$. This implies that the eigenvalue λ_0 has the geometric multiplicity one. Taking $\alpha = 1$ we obtain (22.7). Next, by virtue of (5.14), $x_N = -A(1 : N-1, 1 : N-1) = -G_{N-1}h(N)$ and moreover using the formula

$$A_{N-1}(\lambda_0)^{-1} = \frac{1}{\gamma_{N-1}(\lambda_0)} (\operatorname{adj} A_{N-1}(\lambda_0))$$

we obtain

$$w = \frac{1}{\gamma_{N-1}(\lambda_0)} \begin{pmatrix} -V_{N-1}(\lambda_0)h(N) \\ \gamma_{N-1}(\lambda_0) \end{pmatrix},$$

where $V_{N-1}(\lambda_0) = (\operatorname{adj} A_{N-1}(\lambda_0))G_{N-1}$. Applying the first formula from (19.9), the relations (19.18) and the corresponding formulas from (19.2) to the matrix $A - \lambda_0 I$ and reducing the factor $\frac{1}{\gamma_{N-1}(\lambda_0)}$ we obtain the representation (22.8) for the eigenvector.

2) At the beginning let us show that $\theta_{m+1}(\lambda_0) \neq 0$. By (21.6) and the fact that $p(N)h(N) = 0$, we have $z_N(\lambda_0) = 0$. Then, by virtue of (21.8) we obtain $z_{N-1}(\lambda_0) = \cdots = z_{m+1}(\lambda_0) = 0$ and using (21.6)and (21.7) we obtain

$$\theta_{m+1}(\lambda_0) = (d(N) - \lambda_0) \cdots (d(m+1) - \lambda_0).$$

By virtue of conditions (22.5), (22.6) and equalities $p(m+1)h(m+1) = \cdots = p(N)h(N) = 0$ we conclude that $d(j) - \lambda_0 \neq 0,\ j = m+1, \dots, N$, and hence $\theta_{m+1}(\lambda_0) \neq 0$.

By (21.25) and the fact that λ_0 is an eigenvalue of the matrix A we have

$$(d(N) - \lambda_0) \cdots (d(m+1) - \lambda_0)\gamma_m(\lambda_0) = \gamma_N(\lambda_0) = 0. \tag{22.15}$$

The equalities $p(m+1)h(m+1) = \cdots = p(N)h(N) = 0$ and the conditions (22.5), (22.6) imply $d(j) - \lambda_0 \neq 0,\ j = m+1, \dots, N$ and therefore from (22.15) we conclude that $\gamma_m(\lambda_0) = 0$. Applying Part 1) to the matrix $A(1:m, 1:m)$ we conclude that $\gamma_{m-1}(\lambda_0) \neq 0$. Applying the partition (22.1) with $k = m$ to the matrix $A - \lambda_0 I$ we get

$$A - \lambda_0 I = \begin{pmatrix} A_{m-1}(\lambda_0) & G_{m-1}h(m) & G_{m-1}b(m)H_{m+1} \\ p(m)Q_{m-1} & d(m) - \lambda_0 & g(m)H_{m+1} \\ P_{m+1}a(m)Q_{m-1} & P_{m+1}q(m) & B_{m+1}(\lambda_0) \end{pmatrix}, \tag{22.16}$$

where $A_{m-1}(\lambda_0) = A(1:m-1, 1:m-1) - \lambda_0 I,\ B_{m+1}(\lambda_0) = A(m+1:N, m+1:N) - \lambda_0 I$ and the elements Q_m, G_m, P_m, H_m are defined by (5.1), (5.2), (5.5), (5.6). Moreover from (19.9), (19.10) we obtain

$$Q_{m-1}A_{m-1}^{-1}(\lambda_0)G_{m-1} = \frac{f_{m-1}(\lambda_0)}{\gamma_{m-1}(\lambda_0)}, \quad H_{m+1}B_{m+1}^{-1}(\lambda_0)P_{m+1} = \frac{z_{m+1}(\lambda_0)}{\theta_{m+1}(\lambda_0)}. \tag{22.17}$$

Let v be an eigenvector corresponding to λ_0. We represent this vector in the form $v = \begin{pmatrix} v' \\ \alpha \\ v'' \end{pmatrix}$, where v', v'' are $(m-1)$ and $(N-m)$-dimensional vectors and α is a scalar. From the equality $(A - \lambda_0 I)v = 0$, using the representation (22.16) we obtain

$$\begin{aligned} A_{m-1}(\lambda_0)v' + G_{m-1}(h(m)\alpha + b(m)H_{m+1}v'') &= 0, \\ P_{m+1}(a(m)Q_{m-1}v' + q(m)\alpha) + B_{m+1}(\lambda_0)v'' &= 0. \end{aligned} \tag{22.18}$$

Since the matrices $A_{m-1}(\lambda_0), B_{m+1}(\lambda_0)$ are invertible, we may rewrite (22.18) in the form

$$v' = -A_{m-1}^{-1}(\lambda_0)G_{m-1}(h(m)\alpha + b(m)H_{m+1}v''), \tag{22.19}$$

$$v'' = -B_{m+1}^{-1}(\lambda_0)P_{m+1}(a(m)Q_{m-1}v' + q(m)\alpha). \tag{22.20}$$

Multiplying (22.20) by H_{m+1} and using the second equality from (22.17) and the equality $z_{m+1}(\lambda_0) = 0$, we obtain $H_{m+1}v'' = 0$. Substituting this in (22.19) we obtain

$$v' = -A_{m-1}^{-1}(\lambda_0)G_{m-1}h(m)\alpha = A_{m-1}^{-1}(\lambda_0)x_m\alpha. \tag{22.21}$$

Substituting (22.21) in (22.20) we obtain

$$v'' = B_{m+1}^{-1}(\lambda_0)P_{m+1}(a(m)Q_{m-1}A_{m-1}^{-1}(\lambda_0)G_{m-1}h(m) - q(m))\alpha$$

and using the first equality from (22.17) we conclude that

$$v'' = B_{m+1}^{-1}(\lambda_0)P_{m+1}(a(m)\frac{f_{m-1}(\lambda_0)}{\gamma_{m-1}(\lambda_0)}h(m) - q(m))\alpha = B_{m+1}^{-1}(\lambda_0)y_m\alpha. \tag{22.22}$$

Thus from (22.21), (22.22) we conclude that $v = \alpha v_0$, where

$$v_0 = \begin{pmatrix} A_{m-1}^{-1}(\lambda_0)x_m \\ 1 \\ B_{m+1}^{-1}(\lambda_0)y_m \end{pmatrix}.$$

This implies that the eigenvalue λ_0 has geometric multiplicity one. Taking $\alpha = 1$ we obtain (22.9).

3) In the same way as in the proof of 2) we show that $\theta_2(\lambda_0) \neq 0$. Consider the partition of the matrix $A - \lambda_0 I$ in the form

$$A - \lambda_0 I = \begin{pmatrix} d(1) - \lambda_0 & r_2 \\ y_2 & B_2(\lambda_0) \end{pmatrix}, \tag{22.23}$$

where $B_2(\lambda_0) = A(2:N, 2:N) - \lambda_0 I$, $y_2 = -A(2:N, 1)$, $r_2 = A(1, 2:N)$. Let v be an eigenvector corresponding to λ_0. We represent v in the form $v = \begin{pmatrix} \alpha \\ v' \end{pmatrix}$, where v' is an $(N-1)$-dimensional vector and α is a scalar. From the equality $(A - \lambda_0 I)v = 0$ using the representation (22.23) we obtain $B_2(\lambda_0)v' - \alpha y_2 = 0$. This means $v' = B_2^{-1}(\lambda_0)\alpha y_2$, i.e., $v = \alpha v_0$, where $v_0 = \begin{pmatrix} 1 \\ B_2^{-1}(\lambda_0)y_2 \end{pmatrix}$. This implies that the eigenvalue λ_0 has geometric multiplicity one. Taking $\alpha = 1$ we obtain (22.10). □

Remark. Using Algorithm 19.6 and Algorithm 13.1 or Algorithm 20.11 one can compute the eigenvector given by the corresponding formula (22.7), (22.9) or (22.10) in $O(N)$ arithmetic operations. In the case when $p(N)h(N) \neq 0$ and $\gamma_k(\lambda_0) \neq 0$ $k = 1, \dots, N-1$, one can use also the algorithm presented in Section §18.3.

§22.2 Hermitian Matrices

Here we consider a specification of Theorem 22.2 for Hermitian matrices.

Theorem 22.3. *Let A be a Hermitian matrix with quasiseparable generators $p(i)$ $(i = 2, \dots, N)$, $q(j)$ $(j = 1, \dots, N-1)$, $a(k)$ $(k = 2, \dots, N-1)$; $g(i) = \overline{q(i)}$; $(i = 1, \dots, N-1)$, $h(j) = \overline{p(j)}$ $(j = 2, \dots, N)$, $b(k) = \overline{a(k)}$; $(k = 2, \dots, N-1)$; $d(k)$ $(k = 1, \dots, N)$ of orders equal to one.*

Assume that $\lambda = \lambda_0$ is an eigenvalue of the matrix A and

$$|d(1) - \lambda_0| + |q(1)| > 0;$$
$$l_k(\lambda_0) = (d(k) - \lambda_0)a(k) - p(k)q(k) \neq 0, \quad k = 2, \dots, N-1;$$
$$|d(N) - \lambda_0| + |p(N)| > 0.$$

Then the eigenvalue λ_0 has geometric multiplicity one. Moreover, the eigenvector v corresponding to λ_0 can be determined as follows.

1) *If $p(N) \neq 0$, then*

$$\gamma_{N-1}(\lambda_0) = \det(A(1:N-1, 1:N-1) - \lambda_0 I) \neq 0$$

and

$$v = \begin{pmatrix} A_{N-1}(\lambda_0)^{-1} x_N \\ 1 \end{pmatrix}, \tag{22.24}$$

where $A_{N-1}(\lambda_0) = A(1:N-1, 1:N-1) - \lambda_0 I$, $x_N = -A(1:N-1, N)$.

2) *If for some m, $2 \le m \le N-1$, we have $p(m) \neq 0$ and $p(m+1) = \cdots = p(N) = 0$, then*

$$\gamma_{m-1}(\lambda_0) = \det(A(1:m-1, 1:m-1) - \lambda_0 I) \neq 0$$

and

$$v = \begin{pmatrix} A_{m-1}(\lambda_0)^{-1} x_m \\ 1 \\ 0_{N-m} \end{pmatrix}, \tag{22.25}$$

where $A_{m-1}(\lambda_0) = A(1:m-1, 1:m-1) - \lambda_0 I$, $x_m = -A(1:m-1, m)$.

Proof. Part 1) of the theorem follows directly from Part 1) of Theorem 22.2.

Part 2) follows from Part 2) of Theorem 22.2 using the fact that in this case we have $P_{m+1} = \operatorname{col}(p(k)a^{>}_{km})_{k=m+1}^{N} = 0$ and therefore the vector y_m in the formula (22.9) equals zero. □

§22.3 Examples

We consider first some cases when the validity of the conditions of Theorem 22.2 is guaranteed for any eigenvalue.

Example 22.4. Let A be a Hermitian matrix with quasiseparable generators of orders one $p(i)$ $(i = 2, \dots, N)$, $q(j)$ $(j = 1, \dots, N-1)$, $a(k)$ $(k = 2, \dots, N-1)$; $\overline{q(i)}$ $(i = 1, \dots, N-1)$, $\overline{p(j)}$ $(j = 2, \dots, N)$, $a(k)$ $(k = 2, \dots, N-1)$; $d(k)$ $(k = 1, \dots, N)$ that satisfy the conditions

$$\overline{p(k)q(k)} \neq p(k)q(k),\ \overline{a(k)} = a(k), \quad k = 2, \dots, N-1.$$

In this case we have

$$\begin{aligned} l_k(\lambda) &= (d(k) - \lambda)a(k) - p(k)q(k),\ \delta_k(\lambda) \\ &= (d(k) - \lambda)a(k) - \overline{p(k)q(k)}, \quad k = 2, \dots, N-1. \end{aligned}$$

If λ is an eigenvalue of A, then the number $(\lambda - d(k))a(k)$ is real. The numbers $p(k)q(k)$, $\overline{p(k)q(k)}$ are not real and therefore the condition (22.5) holds for any eigenvalue of A. Hence, all the eigenvalues of A have geometric multiplicity one.

Assume that $q(1) \neq 0$, $p(N) \neq 0$. Then all the conditions of Theorem 22.2 hold and since for a Hermitian matrix the geometric multiplicity of any eigenvalue is equal to its algebraic multiplicity we conclude that all eigenvalues of the matrix A are simple. ◊

Example 22.5. Let A be an irreducible tridiagonal matrix, i.e., a tridiagonal matrix with non-zero entries on the lower and upper subdiagonals. In this case we have $a(k) = b(k) = 0$, $A_{k+1,k} = p(k+1)q(k) \neq 0$, $A_{k,k+1} = g(k)h(k+1) \neq 0$ $(k = 1, \dots, N-1)$ and therefore

$$l_k(\lambda) = -p(k)q(k) \neq 0,\ \delta_k(\lambda) = -g(k)h(k) \neq 0, \quad k = 2, \dots, N-1.$$

and hence the conditions of Theorem 22.2 hold for any λ.

Assume additionally that the conditions

$$\overline{d(i)} = d(i), \quad i = 1, \dots, N; \quad \frac{p(i+1)q(i)}{g(i)h(i+1)} > 0, \quad i = 1, \dots, N-1 \tag{22.26}$$

are satisfied. Let $D = \operatorname{diag}(\rho_i)_{i=1}^N$ be a nonsingular diagonal matrix with the entries

$$\rho_1 = 1, \quad \rho_k = \sqrt{\frac{p(k)q(k-1)}{g(k-1)h(k)}}\rho_{k-1},\ k = 2, \dots, N, \tag{22.27}$$

and let $Q = D^{-1}AD$. The matrix Q is a tridiagonal Hermitian matrix. Indeed the entries of Q have the form

$$Q_{ij} = \begin{cases} \rho_{j+1}^{-1}p(j+1)q(j)\rho_j, & i = j+1, j = 1, \dots, N-2, \\ d_i, & 1 \leq i = j \leq N, \\ \rho_i^{-1}g(i)h(i+1)\rho_{i+1}, & j = i+1, i = 1, \dots, N-1, \\ 0, & |i-j| > 1. \end{cases}$$

The relations (22.27) imply

$$p(i+1)q(i)\rho_i^2 = \overline{g(i)h(i+1)}\rho_{i+1}^2, \quad i = 1, \dots, N-1, \tag{22.28}$$

which means

$$\rho_{i+1}^{-1}p(i+1)q(i)\rho_i = \overline{\rho_i^{-1}g(i)h(i+1)\rho_{i+1}}, \quad i = 1, \dots, N-1.$$

Thus from here and Theorem 22.2 we obtain the known result that under the conditions (22.26) the tridiagonal matrix A is diagonalizable and its eigenvalues are real and simple. ◊

Example 22.6. Consider the invertible matrix A with quasiseparable generators of orders one satisfying the conditions

$$\begin{gathered} a(k) \neq 0, b(k) \neq 0, \quad d(k)a(k) - p(k)q(k) = 0, \\ d(k)b(k) - g(k)h(k) = 0, \quad k = 2, \dots, N-1. \end{gathered} \tag{22.29}$$

One can easily check that A is a Green matrix of the first order (see Section §6.3).

For such a matrix we have

$$l_k(\lambda) = -\lambda a(k), \; \delta_k(\lambda) = -\lambda b(k), \quad k = 2, \dots, N-1,$$

and since by assumption zero is not an eigenvalue of A, the condition (22.5) holds for any eigenvalue of the matrix. Suppose also that $p(k+1)q(k) \neq 0, \; g(k)h(k+1) \neq 0$ $(k = 1, \dots, N-1)$. Then all the conditions of Theorem 22.2 hold for any eigenvalue of the matrix A.

Now as in the previous example assume that $p(k+1)q(k) \neq 0, \; g(k)h(k+1) \neq 0, \; k = 1, \dots, N-1$ and the conditions (22.26) are satisfied. We define the nonsingular diagonal matrix $D = \operatorname{diag}\,(\rho_i)_{i=1}^N$ via the relations (22.27). Let us show that in this case, like in the previous example, $Q = D^{-1}AD$ is a Hermitian matrix. Indeed the entries of the matrix Q have the form

$$Q_{ij} = \begin{cases} \rho_i^{-1}p(i)a_{ij}^{>}q(j)\rho_j, & 1 \leq j < i \leq N, \\ d(i), & 1 \leq i = j \leq N, \\ \rho_i^{-1}g(i)b_{ij}^{<}h(j)\rho_j, & 1 \leq i < j \leq N. \end{cases} \tag{22.30}$$

We need to check that

$$p(i)a_{ij}^{>}q(j)\rho_j^2 = \overline{g(j)b_{ij}^{>}h(i)}\rho_i^2, \quad 1 \leq j < i \leq N. \tag{22.31}$$

From (22.28) it follows that

$$\frac{\rho_i^2}{\rho_j^2} = \frac{\rho_i^2}{\rho_{i-1}^2} \cdots \frac{\rho_{j+1}^2}{\rho_j^2} = \frac{p(i)q(i-1)}{\overline{g(i-1)h(i)}} \frac{p(i-1)q(i-2)}{\overline{g(i-2)h(i-1)}} \cdots \frac{p(j+1)q(j)}{\overline{g(j)h(j+1)}},$$

which means that

$$\frac{\rho_i^2}{\rho_j^2} = \frac{p(i)}{\overline{h(i)}} \frac{q(j)}{\overline{g(j)}} \frac{p_{ij}^> q_{ij}^>}{\overline{g_{ij}^> h_{ij}^>}}, \quad 1 \le j < i \le N. \tag{22.32}$$

By virtue of (22.29) and the assumption that the numbers $d(k)$ are real,

$$d(k) = \frac{p(k)q(k)}{a(k)} = \frac{g(k)h(k)}{b(k)} = \frac{\overline{g(k)h(k)}}{\overline{b(k)}}, \quad k = 2, \dots, N-1. \tag{22.33}$$

It follows that

$$\frac{p_{ij}^> q_{ij}^>}{\overline{g_{ij}^> h_{ij}^>}} = \frac{a_{ij}^>}{\overline{b_{ij}^>}}, \quad 1 \le j < i \le N. \tag{22.34}$$

Comparing (22.32) and (22.34) we obtain

$$\frac{\rho_i^2}{\rho_j^2} = \frac{p(i)q(j)}{\overline{g(j)h(i)}} \frac{a_{ij}^>}{\overline{b_{ij}^>}}, \quad 1 \le j < i \le N,$$

which is equivalent to (22.31). Thus we conclude that in the case under consideration the matrix A is similar to a Hermitian matrix and hence A is diagonalizable and its eigenvalues are real. Moreover, by Theorem 22.2, the eigenvalues of A are simple. ◊

Example 22.7. Consider the unitary upper Hessenberg matrix U defined in Section §7.4. U has quasiseparable generators $d(k) = -\rho_k \rho_{k-1}^*$, $a(k) = 0$, $b(k) = \mu_{k-1}$, $p(i) = 1$, $q(j) = \mu_j$, $g(i) = -\rho_{i-1}^*$, $h(j) = \mu_{j-1}\rho_j$, with orders equal to one. We have $p(k)q(k) = \mu_k$, $g(k)h(k) = -\rho_{k-1}^* \mu_{k-1}\rho_k$ and

$$\begin{aligned} l_k(\lambda) = -\mu_k, \quad \delta_k(\lambda) &= -(\lambda + \rho_k \rho_{k-1}^*)\mu_{k-1} + \rho_{k-1}^* \mu_{k-1}\rho_k \\ &= -\lambda \mu_{k-1}, \quad k = 2, \dots, N-1. \end{aligned}$$

From here, since the matrix U is invertible, we conclude that the condition (22.5) holds for any eigenvalue of this matrix. The other conditions of Theorem 22.2 also hold for any eigenvalue of U. Since for a unitary matrix the geometric multiplicity of any eigenvalue is equal to its algebraic multiplicity, Theorem 22.2 implies that all eigenvalues of the matrix U are simple. ◊

Example 22.8. Consider the Toeplitz matrix

$$A = \begin{pmatrix} d & a & a^2 & \dots & a^{n-1} \\ b & d & a & \dots & a^{n-2} \\ b^2 & b & d & \dots & a^{n-3} \\ \vdots & \vdots & \vdots & \ddots & \vdots \\ b^{n-1} & b^{n-2} & b^{n-3} & \dots & d \end{pmatrix},$$

with $a \neq 0$, $b \neq 0$. A is with quasiseparable order one and with the quasiseparable generators that have been computed in Example 5.14:

$$\begin{array}{llll} p(i) = b^{i-1} & (i = 2, \dots, N), & q(j) = b^{1-j} & (j = 1, \dots, N-1), \\ a(k) = 1 & (k = 2, \dots, N-1), & g(i) = a^{1-i} & (i = 1, \dots, N-1), \\ h(j) = a^{j-1} & (j = 2, \dots, N), & b(k) = 1 & (k = 2, \dots, N-1), \\ d(k) = d & (k = 1, \dots, N). & & \end{array}$$

We have

$$l_k(\lambda) = \delta_k(\lambda) = d - \lambda - 1, \quad k = 2, \dots, N-1.$$

Assume that $ab \neq 1$. Then (see Example 10.3.2) the matrix

$$A - (d-1)I = \begin{pmatrix} 1 & a & a^2 & \dots & a^{n-1} \\ b & 1 & a & \dots & a^{n-2} \\ b^2 & b & 1 & \dots & a^{n-3} \\ \vdots & \vdots & \vdots & \ddots & \vdots \\ b^{n-1} & b^{n-2} & b^{n-3} & \dots & 1 \end{pmatrix} \tag{22.35}$$

is invertible, i.e., the number $d-1$ is not an eigenvalue of A. Thus in this case the conditions of Theorem 22.2 are satisfied for any eigenvalue of the matrix A and hence all eigenvalues have geometric multiplicity one.

Assume now that the number d is real and the number $b/\overline{a}$ is real and positive. In this case

$$\frac{p(i+1)q(i)}{\overline{g(i)h(i+1)}} = \frac{b^i b^{1-i}}{\overline{a^{1-i}a^i}} = \frac{b}{\overline{a}} > 0, \quad i = 1, \dots, N-1,$$

and hence the matrix A satisfies the conditions (22.26). Applying the formulas (22.27) define the nonsingular diagonal matrix $D = \operatorname{diag}(\rho_i)_{i=1}^N$ with the entries

$$\rho_1 = 1, \quad \rho_k = \sqrt{\frac{b}{\overline{a}}}\rho_{k-1} = \left(\frac{b}{\overline{a}}\right)^{\frac{k-1}{2}}, \quad k = 2, \dots, N. \tag{22.36}$$

Moreover,

$$\frac{p(k)q(k)}{a(k)} = \frac{\overline{g(k)h(k)}}{\overline{b(k)}} = 1, \quad k = 2, \dots, N-1$$

and hence in the same way as in Examples 22.5 or 22.6 we obtain that $Q = D^{-1}AD$ is a Hermitian matrix. Furthermore, using the formulas (22.30) and (22.36) we conclude that Q is a matrix with quasiseparable order one with quasiseparable generators $p^Q(i) = (\overline{a}b)^{(i-1)/2}$ $(i = 2, \dots, N)$, $q^Q(j) = (\overline{a}b)^{(1-j)/2}$ $(j = 1, \dots, N-1)$, $a^Q(k) = 1$ $(k = 2, \dots, N-1)$; $g^Q(i) = (\overline{b}a)^{(1-i)/2}$ $(i = 1, \dots, N-1)$, $h^Q(j) = (\overline{b}a)^{(j-1)/2}$ $(j = 2, \dots, N)$, $b^Q(k) = 1$ $(k = 2, \dots, N-1)$; $d^Q(k) = d$ $(k =$

$1, \ldots, N$). This implies that Q is a Hermitian Toeplitz matrix. It follows that in the case under consideration the matrix A is diagonalizable and moreover from Theorem 22.2 we conclude that all eigenvalues of A are simple.

The arguments above imply, in particular, that for a real d and for $ab > 0$, $ab \neq 1$ the Toeplitz matrix A is diagonalizable and its eigenvalues are simple. Let us show that in the case $ab < 0$ the matrix A may have multiple eigenvalues. By Theorem 22.2, this implies that A is not diagonalizable. As an example we take the 3×3 matrix

$$A = \begin{pmatrix} d & a & a^2 \\ b & d & a \\ b^2 & b & d \end{pmatrix},$$

with $ab = -8$. Checking directly or applying the formula (21.21) we obtain the characteristic polynomial of this matrix:

$$\begin{aligned}\gamma_3(\lambda) &= (d-\lambda)^3 - (a^2b^2 + 2ab)(d-\lambda) + 2a^2b^2 \\ &= (d-\lambda)^3 - 48(d-\lambda) + 128 = -(\lambda - d + 4)^2(\lambda - d - 8).\end{aligned}$$

It is clear that this polynomial has a multiple eigenvalue.

Next we consider the case $ab = 1$, in which the relations (21.20), (21.21) have the form

$$\begin{aligned}&\gamma_1(\lambda) = d - \lambda, \quad \gamma_2(\lambda) = (d-\lambda)^2 - 1 = (d - \lambda + 1)(d - \lambda - 1), \\ &\gamma_k(\lambda) = 2(d - \lambda - 1)\gamma_{k-1}(\lambda) - (d - \lambda - 1)^2\gamma_{k-2}(\lambda), \quad k = 3, \ldots, N.\end{aligned}$$

From here one can easily derive by induction that

$$\gamma_k(\lambda) = (d - \lambda + k - 1)(d - \lambda - 1)^{k-1}, \quad k = 2, \ldots, N.$$

It follows that in this case the matrix A has the simple eigenvalue $\lambda = d + N - 1$ and the multiple eigenvalue $\lambda = d - 1$. For $\lambda = d + N - 1$ the eigenvector is determined by the relation (22.7). For $\lambda = d - 1$ we obtain directly from (22.35) the eigenvectors

$$u_1 = \begin{pmatrix} 1 \\ -b \\ 0 \\ \vdots \\ 0 \end{pmatrix}, \; u_2 = \begin{pmatrix} 0 \\ 1 \\ -b \\ \vdots \\ 0 \end{pmatrix}, \; \ldots, \; u_{N-1} = \begin{pmatrix} 0 \\ 0 \\ \vdots \\ 1 \\ -b \end{pmatrix}.$$

Hence, for $ab = 1$ the Toeplitz matrix A is diagonalizable. ◊

Example 22.9. Consider the matrix

$$A = \begin{pmatrix} 1 & 0 & 4 \\ 0 & 2 & 5 \\ 2 & -2 & 3 \end{pmatrix}.$$

A set of generators for A is

$$g(1)=4,\ g(2)=5,\ h(2)=0,\ h(3)=1,\quad d(1)=1,\ d(2)=2,\ d(3)=3,$$
$$p(2)=0,\ p(3)=1,\ q(1)=2,\ q(2)=-2,\ a(2)=b(2)=1.$$

Let us use for this matrix the formulas of Chapter 22 and Chapter 21, which are applicable to it.

By Theorem 21.1,

$$\begin{gathered}\gamma_1(\lambda)=1-\lambda,\quad \gamma_2(\lambda)=(1-\lambda)(2-\lambda),\\ \gamma_3(\lambda)=(1-\lambda)(2-\lambda)(3-\lambda)-8(2-\lambda)+10(1-\lambda),\\ \theta_1=\gamma_3,\ \ \theta_2=(2-\lambda)(3-\lambda)+10,\ \ \theta_3=3-\lambda.\end{gathered}$$

Also, $f_1(\lambda)=8$, $c_2(\lambda)=2-\lambda$, $f_2(\lambda)=8(2-\lambda)-10(1-\lambda)$, $z_3(\lambda)=1$, $z_2(\lambda)=2-\lambda$.

One can check that $\det A=0$, in other words, $\lambda_0=0$ is an eigenvalue of A. Using the Sturm property in Theorem 21.11 for $\alpha=-1,\beta=0.5$ we obtain that in the interval $(-1,0.5)$ the matrix A has exactly one eigenvalue of multiplicity one, since the number of sign changes for the sequence

$$\gamma_3(-1)=2\cdot3\cdot4-8\cdot3+10\cdot2=20,\ \gamma_2(-1)=2\cdot3=6,\ \gamma_1(-1)=2,\ \gamma_0(-1)=1$$

is $\nu(\alpha)=0$, while the number of sign changes for the sequence

$$\begin{gathered}\gamma_2\left(\frac{1}{2}\right)=\frac{1}{2}\cdot\frac{3}{2}\cdot\frac{5}{2}-8\cdot\frac{3}{2}+10\cdot\frac{1}{2}=\frac{15-96+40}{8}<0,\\ \gamma_2\left(\frac{1}{2}\right)=\frac{1}{2}\cdot\frac{3}{2}>0,\quad \gamma_1\left(\frac{1}{2}\right)=\frac{1}{2},\ \gamma_0\left(\frac{1}{2}\right)=1\end{gathered}$$

is $\nu(\beta)=1$.

The matrix A with the above generators satisfies the conditions of Theorem 22.2 for the eigenvalue $\lambda_0=0$, since

$$\begin{aligned}|d(1)-\lambda_0|+|q(1)|&=1+2>0, & |d(1)-\lambda_0|+|g(1)|&=1+4>0,\\ |d(3)-\lambda_0|+|p(3)|&=3+1>0, & |d(3)-\lambda_0|+|h(3)|&=3+1>0,\end{aligned}$$

$$\begin{gathered}l_2(0)=(d(2)-0)a(2)-p(2)q(2)=(2-0)\cdot1-0=2\neq0,\\ \delta_2(0)=(d(2)-0)b(2)-g(2)h(2)=(2-0)\cdot1-0=2\neq0.\end{gathered}$$

Therefore the eigenvalue has indeed multiplicity one, while $\gamma_2(\lambda_0)=2\neq0$ and then the inverse of the corresponding matrix is

$$A_2^{-1}(0)=\begin{pmatrix}1-0 & 0\\ 0 & 2-0\end{pmatrix}^{-1}=\begin{pmatrix}1 & 0\\ 0 & 0.5\end{pmatrix}.$$

By the formula (22.7) from Theorem 22.2, the eigenvector w is obtained as

$$w = \begin{pmatrix} -\begin{pmatrix} 1 & 0 \\ 0 & 0.5 \end{pmatrix}\begin{pmatrix} 4 \\ 5 \end{pmatrix} \\ 1 \end{pmatrix} = \begin{pmatrix} -4 \\ -2.5 \\ 1 \end{pmatrix}.$$

By the formula (22.8) from the same theorem, a multiple of w can also be obtained using

$$v_1(\lambda_0) = 4, \quad v_2(\lambda_0) = 5 \cdot 1 = 5, \quad \gamma_2(\lambda_0) = 2, \quad \delta_2(\lambda_0) = 2, \quad h(3) = 1$$

as $\begin{pmatrix} -4 \cdot 2 \cdot 1 \\ -5 \cdot 1 \\ 2 \end{pmatrix} = \begin{pmatrix} -8 \\ -5 \\ 2 \end{pmatrix} = 2w.$ ◊

Example 22.10. We will show with this example that the conditions of Theorem 22.2 are essential.

Consider the matrix $\begin{pmatrix} 1 & 4 & 0 \\ 2 & 2 & 0 \\ 0 & 0 & 3 \end{pmatrix}$ for which it is easy to see that $\lambda_0 = 3$ is an eigenvalue. Some of its generators could be taken as $p(2) = h(2) = 1$, $p(3) = h(3) = 0$, $d(3) = 3$, as in Case 2) of Theorem 22.2. However, here $|\lambda_0 - d(3)| + |h(3)| = 0$, so this matrix does not satisfy all the conditions of the theorem in Case 2), and indeed $\theta_3(\lambda_0) = 3 - \lambda_0 = 0$ and the matrix $B_3(3) = 0$ is not invertible.

Consider the matrix

$$\begin{pmatrix} 1 & 2 & 0 & 6 \\ 0 & 0 & 0 & 6 \\ 3 & 3 & 4 & 6 \\ 0 & 0 & 0 & 5 \end{pmatrix}$$

for which it is easy to see that $\lambda_0 = 0$ is an eigenvalue. Its generators can be taken as

$$\begin{aligned} &p(2) = 0, \quad p(3) = 3, \quad p(4) = 0, \quad q(1) = q(2) = q(3) = a(2) = a(3) = 1, \\ &h(2) = 2, \quad h(3) = 0, \quad h(4) = 6, \quad g(1) = g(2) = g(3) = b(2) = b(3) = 1, \end{aligned}$$

therefore $h(2)p(2) = h(3)p(3) = h(4)p(4) = 0$ and if the conditions of the theorem were satisfied, this is Case 3). However, $l_2(0) = 0$ and indeed it is not true that $\theta_2(\lambda_0) \neq 0$. ◊

Example 22.11. Let N be an odd integer, $N = 2r + 1$, and $a \neq 0$ and d be scalars, and consider the $N \times N$ matrix

$$A = \begin{pmatrix} d & a & a & \cdots & a & a \\ -a & d & a & \cdots & a & a \\ -a & -a & d & \cdots & a & a \\ \vdots & \vdots & \vdots & \ddots & \vdots & \vdots \\ -a & -a & -a & \cdots & d & a \\ -a & -a & -a & \cdots & -a & d \end{pmatrix}.$$

In Example 15.2 we found the inverse of such an $(N-1)\times(N-1)$ matrix with $d=0$. This inverse will be of use here.

For the matrix A one can use the quasiseparable generators

$$\begin{array}{ll} p(i)=a,\ i=2,\dots,N, & q(j)=-1,\ j=1,\dots,N-1, \\ a(k)=1,\ k=2,\dots,N-1, & g(j)=1,\quad j=1,\dots,N-1, \\ h(i)=a,\ i=2,\dots,N, & b(k)=1,\quad k=2,\dots,N-1, \\ d(k)=d,\ k=1,\dots,N. & \end{array}$$

One can check that $\lambda_0=d$ is an eigenvalue with the eigenvector

$$w=\begin{pmatrix} 1 & -1 & 1 & \cdots & -1 & 1 \end{pmatrix}^T,$$

i.e., the vector with components $w(k)=(-1)^{k+1}$. In particular, for $k=N$, which is odd, $w(N)=1$.

Let us show that using Theorem 22.2 one can find again precisely this eigenvector. Indeed,

$$\begin{gathered} |d(1)-\lambda_0|+|q(1)|=|d-d|+1>0, \quad |d(1)-\lambda_0|+|g(1)|=|d-d|+1>0, \\ |d(N)-\lambda_0|+|p(N)|=|d-d|+|a|>0, \ |d(N)-\lambda_0|+|h(N)|=|d-d|+|a|>0, \\ l_k(d)=(d(k)-d)a(k)-p(k)q(k)=(d-d)\cdot 1-a\neq 0, k=2,\dots,N-1 \\ \delta_k(d)=(d(k)-d)b(k)-g(k)h(k)=(d-d)\cdot 1-a\neq 0, k=2,\dots,N-1, \end{gathered}$$

so that the above representation fulfills the conditions of the theorem and therefore the eigenvalue $\lambda_0=d$ is simple and the matrix $A_{N-1}(d)$ is invertible.

We did see in Example 15.2 that

$$A_{N-1}(d)^{-1}=\begin{pmatrix} 0 & -\frac{1}{a} & \frac{1}{a} & -\frac{1}{a} & \cdots & \frac{1}{a} & -\frac{1}{a} \\ \frac{1}{a} & 0 & -\frac{1}{a} & \frac{1}{a} & \cdots & \frac{1}{a} & \frac{1}{a} \\ -\frac{1}{a} & \frac{1}{a} & 0 & -\frac{1}{a} & \cdots & \frac{1}{a} & -\frac{1}{a} \\ \frac{1}{a} & -\frac{1}{a} & \frac{1}{a} & 0 & \cdots & \frac{1}{a} & \frac{1}{a} \\ \vdots & \vdots & \vdots & \vdots & \ddots & \vdots & \vdots \\ -\frac{1}{a} & \frac{1}{a} & -\frac{1}{a} & \frac{1}{a} & \cdots & 0 & -\frac{1}{a} \\ \frac{1}{a} & -\frac{1}{a} & \frac{1}{a} & -\frac{1}{a} & \cdots & \frac{1}{a} & 0 \end{pmatrix}.$$

One can verify that the vector $x_N=-A(1:N-1,N)$ is the $(N-1)$-dimensional vector with all components equal to $-a$. Therefore $A_{N-1}(d)^{-1}x_N$ is the vector whose components are equal to -1 if the corresponding row of $A_{N-1}(d)^{-1}$ has more $\frac{1}{a}$ elements and to 1 if it has more $-\frac{1}{a}$ elements. It is easy to see that one obtains the first $N-1$ components of the eigenvector w.

If the representation falls in Theorem 22.2, case 1) then the theorem yields yet another eigenvector $\tilde{w}$, which has as its last component the determinant of the matrix $A_{N-1}(d)$, namely $\tilde{w}(N)=\gamma_{N-1}(d)$ instead of $w(N)=1$. Since for an eigenvalue of multiplicity one the eigenvector is unique up to multiplication by a scalar, it follows that

$$\tilde{w} = \gamma_{N-1}(d)w. \tag{22.37}$$

Compute now the factor $\gamma_{N-1}(d)$. By Theorem 21.1,

$$\gamma_1(d) = 0,\ f_1(d) = -1,\ \gamma_2(d) = -a^2 f_1(d) = a^2,\ f_2(d) = c_2(d)\cdot 1 - \gamma_1(d) = 0,$$

since $c_k(d) = a - a = 0$. One can prove by induction that

$$\gamma_{2m+1}(d) = 0,\ \gamma_{2m}(d) = a^{2m},\ \ f_{2m}(d) = 0,\ f_{2m+1}(d) = -a^{2m},\ \ m = 0, 1, \ldots, r.$$

In particular, $\gamma_{N-1} = a^{N-1}$, hence by (22.37) the components of the eigenvector $\tilde{w}$ are

$$\tilde{w}(k) = (-1)^{k+1}a^{N-1}, \quad 1 \le k \le N. \tag{22.38}$$

The second eigenvector given by Theorem 22.2 is precisely this $\tilde{w}$. Indeed, since $d(i) = d$, it follows that

$$\delta_i(d) = (d(i) - d)b(i) - h(i)g(i) = -a.$$

Hence

$$\delta^{<}_{i,N}(\lambda_0) = \delta_{i+1}(d)\cdot\delta_{i+2}\cdot\cdots\cdot\delta_{N-1} = (-1)^{N-i-1}a^{N-i-1}. \tag{22.39}$$

Also,

$$v_1(d) = 1, \quad v_i(d) = \gamma_{i-1}(d) - af_{i-1}(d),$$

and, by induction, $v_k(d) = a^{k-1}$. Now this formula together with formulas (22.38) and (22.39) show that the eigenvector $\tilde{w}$ is the one given by the theorem. $\Diamond$

Example 22.12. Consider the $(N+1)\times(N+1)$ matrix

$$A = \begin{pmatrix} d & \alpha_1 & \alpha_2 & \cdots & \alpha_{N-1} & \alpha_N \\ \alpha_1 & d & \alpha_2 & \cdots & \alpha_{N-1} & \alpha_N \\ \alpha_1 & \alpha_2 & d & \cdots & \alpha_{N-1} & \alpha_N \\ \vdots & \vdots & \vdots & \ddots & \vdots & \vdots \\ \alpha_1 & \alpha_2 & \alpha_3 & \cdots & d & \alpha_N \\ \alpha_1 & \alpha_2 & \alpha_3 & \cdots & \alpha_N & d \end{pmatrix},$$

with $\alpha_k \neq 0,\ k = 1, \ldots, N$ and $\alpha_k - \alpha_j \neq 0,\ k, j = 1, \ldots, N,\ k \neq j$.

If one adds all the rows to the last one, one obtains on the last row all the elements equal to $d + \alpha_1 + \alpha_2 + \cdots + \alpha_N$, therefore this is a common factor in all the $(N+1)!$ products which give $\det A$, so that $\det A$ is divisible by this factor. If one takes instead of d the number α_1, then the first row coincides with the second, if one takes instead of d the number α_2, then the second row coincides with the third, and so on. A determinant with two identical rows is equal to zero, and this means that $d-\alpha_1, d-\alpha_2, \dots$ divide $\det A$. Since in $\det A$ the highest power of d is $N+1$, it follows that

$$\det A = (d+\alpha_1+\alpha_2+\cdots+\alpha_N)(d-\alpha_1)(d-\alpha_2)\cdot\cdots\cdot(d-\alpha_N). \tag{22.40}$$

Therefore the characteristic polynomial of the matrix A is

$$(d-\lambda+\alpha_1+\alpha_2+\cdots+\alpha_N)(d-\lambda-\alpha_1)(d-\lambda-\alpha_2)\cdot\cdots\cdot(d-\lambda-\alpha_N)$$

and then the matrix A has $N+1$ eigenvalues of multiplicity one:

$$\lambda_0 = d-(\alpha_1+\alpha_2+\ldots+\alpha_N), \quad \lambda_k = d-\alpha_k, \; k=1,\ldots,N.$$

Consider the simple eigenvalue $\lambda_1 = d-\alpha_1$. It is easy to find an eigenvector w corresponding to λ_1 such that its last component $w(N+1)=1$, since if we denote by S the sum

$$S = \alpha_1+\alpha_2+\ldots+\alpha_N \tag{22.41}$$

and by $\mu = -\frac{S}{\alpha_1}$, then

$$\begin{pmatrix} \alpha_1 & \alpha_1 & \alpha_2 & \cdots & \alpha_{N-1} & \alpha_N \\ \alpha_1 & \alpha_1 & \alpha_2 & \cdots & \alpha_{N-1} & \alpha_N \\ \alpha_1 & \alpha_2 & \alpha_1 & \cdots & \alpha_{N-1} & \alpha_N \\ \vdots & \vdots & \vdots & \ddots & \vdots & \vdots \\ \alpha_1 & \alpha_2 & \alpha_3 & \cdots & \alpha_1 & \alpha_N \\ \alpha_1 & \alpha_2 & \alpha_3 & \cdots & \alpha_N & \alpha_1 \end{pmatrix} \begin{pmatrix} \mu \\ 1 \\ 1 \\ \vdots \\ 1 \\ 1 \end{pmatrix} = \begin{pmatrix} 0 \\ 0 \\ 0 \\ \vdots \\ 0 \\ 0 \end{pmatrix}.$$

If one tries to obtain this eigenvector with Theorem 22.2, one finds that the matrix $A_N(d-\alpha_1)$ is not invertible. Indeed, by formula (22.40) with $d=\alpha_1$ and $\alpha_1-\alpha_{N-1}$ as its Nth and last factor, one obtains that $\det A_N(d-\alpha_1)=0$. This means that one or more of the conditions of the theorem are not fulfilled.

Indeed, for the matrix A one can use the quasiseparable generators

$$\begin{aligned} &p(i)=1, && i=2,\ldots,N+1, && q(j)=\alpha_j,\; j=1,\ldots,N, \\ &a(k)=1, && k=2,\ldots,N, && g(j)=1, \quad j=1,\ldots,N, \\ &h(i)=\alpha_{i-1}, && i=2,\ldots,N+1, && b(k)=1, \quad k=2,\ldots,N, \\ &d(k)=d, && k=1,\ldots,N+1. \end{aligned}$$

Then $\delta_k(d-\alpha_1) = (d-d+\alpha_1)b(k)-g(k)h(k)$, therefore $\delta_2(d-\alpha_1)=\alpha_1-\alpha_1=0.\diamondsuit$

Example 22.13. Consider the 3×3 matrix

$$A = \begin{pmatrix} 3 & 3.6 & 4 \\ 5 & 6 & 7 \\ 0 & 0 & 8 \end{pmatrix},$$

which has the eigenvalue $\lambda_0 = 0$. For A consider the following quasiseparable generators:

$$p(2) = 1, p(3) = 0,\ q(1) = 5,\ q(2) = 0,\quad a(2) = 1, d(1) = 3,\ d(2) = 6,$$
$$d(3) = 8, g(1) = \frac{4}{7}, g(2) = 1, h(2) = 6.3, h(3) = 7, b(2) = 1.$$

It is easy to see that the matrix A satisfies the case 2) of Theorem 22.2 with $m = 2$. Indeed, $p(2)h(2) \neq 0$, while $p(3)h(3) = 0$. Also,

$$|d(1) - \lambda_0| + |q(1)| = |3 - 0| + 5 > 0, \quad |d(1) - \lambda_0| + |g(1)| = |3 - 0| + \frac{4}{7} > 0,$$
$$|d(3) - \lambda_0| + |p(3)| = |8 - 0| + 0 > 0, \quad |d(3) - \lambda_0| + |h(3)| = |8 - 0| + 7 > 0,$$
$$l_k(0) = (d(k) - 0)a(k) - p(k)q(k) = (6 - 0) \cdot 1 - 1 \cdot 0 \neq 0,$$
$$\delta_k(0) = (d(k) - 0)b(k) - g(k)h(k) = (6 - 0) \cdot 1 - 1 \cdot 6.3 \neq 0.$$

Then $\gamma_1(\lambda_0) = 3 \neq 0$ and $\theta_3(\lambda_0) = 8 \neq 0$, as Theorem 22.2 predicts.

In order to find an eigenvector notice that $x_2 = -A(1:1, 2) = -3.6$ and $P_3 = 0$, therefore $y_3 = 0$ and the eigenvector given by the case 2) of Theorem 22.2 is $v = \begin{pmatrix} -1.2 \\ 1 \\ 0 \end{pmatrix}$. Indeed, a direct check gives $Av = 0$. ◊

Example 22.14. Consider the 3×3 matrix

$$A = \begin{pmatrix} 3 & 0 & 4 \\ 5 & 6 & 7 \\ 0 & 0 & 8 \end{pmatrix},$$

which has the eigenvalue $\lambda_0 = 3$. For A consider the following quasiseparable generators:

$$p(2) = 1, p(3) = 0, q(1) = 5, q(2) = 0, a(2) = 1, d(1) = 3,\ d(2) = 6,$$
$$d(3) = 8, g(1) = 4, g(2) = 7, h(2) = 0, h(3) = 1, b(2) = 1.$$

It is easy to see that the matrix A satisfies the Case 3) of Theorem 22.2. Indeed $p(2)h(2) = 0$ and $p(3)h(3) = 0$. Also,

$$|d(1) - \lambda_0| + |q(1)| = |3 - 3| + 5 > 0, \quad |d(1) - \lambda_0| + |g(1)| = |3 - 3| + 4 > 0,$$
$$|d(3) - \lambda_0| + |p(3)| = |8 - 3| + 0 > 0, \quad |d(3) - \lambda_0| + |h(3)| = |8 - 3| + 1 > 0,$$
$$l_k(3) = (d(k) - 3)a(k) - p(k)q(k) = (6 - 3) \cdot 1 - 1 \cdot 0 \neq 0,$$
$$\delta_k(3) = (d(k) - 3)b(k) - g(k)h(k) = (6 - 3) \cdot 1 - 7 \cdot 0 \neq 0.$$

Then $\theta_2(3) = \det B_2(3) = \det \begin{pmatrix} 3 & 7 \\ 0 & 5 \end{pmatrix} = 15 \neq 0$, as Theorem 22.2 predicts.

In order to find an eigenvector notice that $y_2 = -A(2:3,1) = \begin{pmatrix} -5 \\ 0 \end{pmatrix}$ and $B_2(3)^{-1} = \begin{pmatrix} \frac{1}{3} & -\frac{7}{15} \\ 0 & \frac{1}{5} \end{pmatrix}$ applied to y_2 gives $\begin{pmatrix} -\frac{5}{3} \\ 0 \end{pmatrix}$, so that the eigenvector given by the case 3) of Theorem 22.2 is $v = \begin{pmatrix} 1 \\ -\frac{5}{3} \\ 0 \end{pmatrix}$. Indeed, a direct check gives $Av = 0$. ◊

§22.4 Comments

The results in this chapter first appeared in [24].

Chapter 23

Kernels of Quasiseparable of Order One Matrices

Next in this part we study the eigenspaces of quasiseparable of order one matrices. Notice that for a matrix A the matrix $A - \lambda_0 I$ has the same lower and upper quasiseparable generators. This is the reason we study the properties of kernels of the considered matrices.

§23.1 The partition theorem

We derive some auxiliary relations for principal submatrices of a matrix with given quasiseparable generators and some submatrices defined via the generators.

Lemma 23.1. *Let $A = \{A_{ij}\}_{i,j=1}^{N}$ be a matrix with lower quasiseparable generators $p(i)$ $(i = 2, \dots, N)$, $q(j)$ $(j = 1, \dots, N-1)$, $a(k)$ $(k = 2, \dots, N-1)$, upper quasiseparable generators $g(i)$ $(i = 1, \dots, N-1)$, $h(j)$ $(j = 2, \dots, N)$, $b(k)$ $(k = 2, \dots, N-1)$, and diagonal entries $d(k)$ $(k = 1, \dots, N)$. Set $h(1) = b(1) = p(1) = a(1) = q(N) = g(N) = a(N) = b(N) = 0$. Set*

$$A_{ij} = A(i:j, i:j), \quad 1 \le i, j \le N,$$

and using lower and upper quasiseparable generators define the row vector

$$H_{ij} = \mathrm{row}(b^{<}_{i-1,t}h(t))_{t=i}^{j}, \quad Q_{ij} = \mathrm{row}(a^{>}_{j+1,t}q(t))_{t=i}^{j}, \quad 1 \le i \le j \le N, \tag{23.1}$$

and the column vector

$$P_{ij} = \mathrm{col}(p(t)a^{>}_{t,i-1})_{t=i}^{j}, \quad G_{ij} = \mathrm{col}(g(t)b^{<}_{t,j+1})_{t=i}^{j}, \quad 1 \le i \le j \le N. \tag{23.2}$$

Let $1 \le k \le m < n \le N$.

Then the matrix $A_{kn} = A(k:n, k:n)$ admits the partition

$$A_{kn} = \begin{pmatrix} A_{km} & G_{km}H_{m+1,n} \\ P_{m+1,n}Q_{km} & A_{m+1,n} \end{pmatrix}, \tag{23.3}$$

and the matrices H_{kn} and Q_{kn} admit the partitions

$$H_{kn} = \begin{pmatrix} H_{km} & b^<_{k-1,m+1} H_{m+1,n} \end{pmatrix}, \tag{23.4}$$

$$Q_{kn} = \begin{pmatrix} a^>_{n+1,m} Q_{km} & Q_{m+1,n} \end{pmatrix}. \tag{23.5}$$

Proof. Using Lemmas 5.1 and 5.4 we get

$$A = \begin{pmatrix} A_{1m} & G_m H_{m+1} \\ P_{m+1} Q_m & A_{m+1,N} \end{pmatrix},$$

where

$$P_m = \operatorname{col}(p(t) a^>_{t,m-1})_{t=m}^N, \quad Q_m = \operatorname{row}(a^>_{m+1,t} q(t))_{t=1}^m,$$
$$G_m = \operatorname{col}(g(t) b^<_{t,m+1})_{t=1}^m, \quad H_m = \operatorname{row}(b^<_{m-1,t} h(t))_{t=m}^N.$$

This yields

$$A_{kn} = \begin{pmatrix} A_{1m}(k:m, k:m) & G_m(k:m,:) H_{m+1}(:, 1:n-m) \\ P_{m+1}(1:n-m,:) Q_m(:, k:m) & A_{m+1,N}(1:n-m, 1:n-m) \end{pmatrix}. \tag{23.6}$$

One can easily check that

$$\begin{aligned} A_{1m}(k:m, k:m) &= A_{km}, & A_{m+1,N}(1:n-m, 1:n-m) &= A_{m+1,n}, \\ G_m(k:m,:) &= G_{km}, & Q_m(:, k:m) &= Q_{km}, \\ P_{m+1}(1:n-m,:) &= P_{m+1,n}, & H_{m+1}(:, 1:n-m) &= H_{m+1,n}. \end{aligned} \tag{23.7}$$

Inserting the expressions (23.7) in (23.6) we obtain (23.3).

Notice that

$$H_{kn} = \operatorname{row}(b^<_{k-1,t} h(t))_{t=k}^n = \begin{pmatrix} \operatorname{row}(b^<_{k-1,t} h(t))_{t=k}^m & \operatorname{row}(b^<_{k-1,t} h(t))_{t=m+1}^n \end{pmatrix}.$$

For $t = m+1, \dots, n$ we use the identity $b^<_{k-1,t} = b^<_{k-1,m+1} b^<_{mt}$ and obtain

$$H_{kn} = \begin{pmatrix} \operatorname{row}(b^<_{k-1,t} h(t))_{t=k}^m & b^<_{k-1,m+1} \operatorname{row}(b^<_{m,t} h(t))_{t=m+1}^n \end{pmatrix},$$

which implies (23.4).

Notice that

$$Q_{kn} = \operatorname{row}(a^>_{n+1,t} q(t))_{t=k}^n = \begin{pmatrix} \operatorname{row}(a^>_{n+1,t} q(t))_{t=k}^m & \operatorname{row}(a^>_{n+1,t} q(t))_{t=m+1}^n \end{pmatrix}.$$

For $t = k, \dots, m$ we use the identity $a^>_{n+1,t} = a^>_{n+1,m} a^>_{m+1,t}$ and obtain

$$Q_{kn} = \begin{pmatrix} a^>_{n+1,m} \operatorname{row}(a^>_{m+1,t} q(t))_{t=k}^m & \operatorname{row}(a^>_{n+1,t} q(t))_{t=m+1}^n \end{pmatrix},$$

which implies (23.5). □

Next let us show that under some conditions one can reduce the computation of the kernel of a matrix with quasiseparable order one to the computation of kernels for matrices which also have quasiseparable order one, but are of smaller dimensions.

Theorem 23.2. *Let A be a matrix with quasiseparable generators $p(i)$ $(i = 2, \dots, N)$, $q(j)$ $(j = 1, \dots, N-1)$, $a(k)$ $(k = 2, \dots, N-1)$; $g(i)$ $(i = 1, \dots, N-1)$, $h(j)$ $(j = 2, \dots, N)$, $b(k)$ $(k = 2, \dots, N-1)$; $d(k)$ $(k = 1, \dots, N)$ with orders equal to one. Set $h(1) = b(1) = p(1) = a(1) = q(N) = g(N) = a(N) = b(N) = 0$. For indices $1 \le i \le j \le N$ we define the row vectors (23.1) and the submatrices $A_{ij} = A(i:j, i:j)$ and $W_{ij} = \begin{pmatrix} H_{ij} \\ A_{ij} \\ Q_{ij} \end{pmatrix}$. Set*

$$U_{ij} = \operatorname{Ker} W_{ij}.$$

Assume that for some index m with $1 < m < N$ the relations

$$d(m)a(m) - p(m)q(m) = 0, \quad d(m)b(m) - g(m)h(m) = 0 \tag{23.8}$$

and

$$|p(m)| + |d(m)| + |g(m)| > 0, \quad |q(m)| + |d(m)| + |h(m)| > 0 \tag{23.9}$$

hold.

Then for k, n with $1 \le k < m < n \le N$

$$U_{kn} = M_{km} \oplus M_{mn}, \tag{23.10}$$

where

$$M_{km} = \left\{ \begin{pmatrix} u_1 \\ 0_{n-m} \end{pmatrix} : u_1 \in U_{km} \right\}, \quad M_{mn} = \left\{ \begin{pmatrix} 0_{m-k} \\ u_2 \end{pmatrix} : u_2 \in U_{mn} \right\}.$$

Proof. Step 1. Partition of matrices.

In this step we derive different partitions of the matrix

$$W_{kn} = \begin{pmatrix} H_{kn} \\ A_{kn} \\ Q_{kn} \end{pmatrix} \tag{23.11}$$

and some partitions of the matrices

$$W_{km} = \begin{pmatrix} H_{km} \\ A_{km} \\ Q_{km} \end{pmatrix}, \quad W_{mn} = \begin{pmatrix} H_{mn} \\ A_{mn} \\ Q_{mn} \end{pmatrix}. \tag{23.12}$$

They are used in the proof of the theorem. The list of these partitions is presented here.

Substituting (23.4), (23.5) and (23.3) in (23.11) we get

$$W_{kn} = \begin{pmatrix} H_{km} & b^{<}_{k-1,m+1}H_{m+1,n} \\ A_{km} & G_{km}H_{m+1,n} \\ P_{m+1,n}Q_{km} & A_{m+1,n} \\ a^{>}_{n+1,m}Q_{km} & Q_{m+1,n} \end{pmatrix}. \tag{23.13}$$

Changing m by $m-1$ we obtain

$$W_{kn} = \begin{pmatrix} H_{k,m-1} & b^{<}_{k-1,m}H_{mn} \\ A_{k,m-1} & G_{k,m-1}H_{mn} \\ P_{mn}Q_{k,m-1} & A_{mn} \\ a^{>}_{n+1,m-1}Q_{k,m-1} & Q_{mn} \end{pmatrix}. \tag{23.14}$$

Using (23.4) with $k = m$ and the equalities $H_{mm} = h(m)$, $b^{<}_{m-1,m+1} = b(m)$, we get

$$H_{mn} = \begin{pmatrix} h(m) & b(m)H_{m+1,n} \end{pmatrix}. \tag{23.15}$$

Using (23.5) with $k = m$ and the equality $Q_{mm} = q(m)$, we get

$$Q_{mn} = \begin{pmatrix} a^{>}_{n+1,m}q(m) & Q_{m+1,n} \end{pmatrix}. \tag{23.16}$$

Using (23.3) with $k = m$ and the equalities $A_{mm} = d(m)$, $Q_{mm} = q(m)$, $G_{mm} = g(m)$, we get

$$A_{mn} = \begin{pmatrix} d(m) & g(m)H_{m+1,n} \\ P_{m+1,n}q(m) & A_{m+1,n} \end{pmatrix}. \tag{23.17}$$

Substituting (23.15), (23.16) and (23.17) in the second formula in (23.12) we obtain

$$W_{mn} = \begin{pmatrix} h(m) & b(m)H_{m+1,n} \\ d(m) & g(m)H_{m+1,n} \\ P_{m+1,n}q(m) & A_{m+1,n} \\ a^{>}_{n+1,m}q(m) & Q_{m+1,n} \end{pmatrix}. \tag{23.18}$$

Changing in (23.4) and (23.5) m to $m-1$ we obtain the relations

$$H_{kn} = \begin{pmatrix} H_{k,m-1} & b^{<}_{k-1,m}H_{mn} \end{pmatrix}, \tag{23.19}$$

and

$$Q_{kn} = \begin{pmatrix} a^{>}_{n+1,m-1}Q_{k,m-1} & Q_{mn} \end{pmatrix}. \tag{23.20}$$

Using (23.19) with $n = m$ and the equality $H_{mm} = h(m)$, we get

$$H_{km} = \begin{pmatrix} H_{k,m-1} & b^{<}_{k-1,m}h(m) \end{pmatrix}. \tag{23.21}$$

Using (23.20) with $n = m$ and the equalities $Q_{mm} = q(m)$, $a^{>}_{m+1,m-1} = a(m)$, we get

$$Q_{km} = \begin{pmatrix} a(m)Q_{k,m-1} & q(m) \end{pmatrix}. \tag{23.22}$$

Changing in (23.3) m to $m-1$ we get

$$A_{km} = \begin{pmatrix} A_{k,m-1} & G_{k,m-1}H_{mn} \\ P_{mn}Q_{k,m-1} & A_{mn} \end{pmatrix}.$$

Taking in this equality $n = m$ and using $A_{mm} = d(m)$, $P_{mm} = p(m)$, $H_{mm} = h(m)$, we obtain

$$A_{km} = \begin{pmatrix} A_{k,m-1} & G_{k,m-1}h(m) \\ p(m)Q_{k,m-1} & d(m) \end{pmatrix}. \tag{23.23}$$

Substituting (23.21), (23.22) and (23.23) in the first formula in (23.12) we obtain

$$W_{km} = \begin{pmatrix} H_{k,m-1} & b^{<}_{k-1,m}h(m) \\ A_{k,m-1} & G_{k,m-1}h(m) \\ p(m)Q_{k,m-1} & d(m) \\ a(m)Q_{k,m-1} & q(m) \end{pmatrix}. \tag{23.24}$$

Inserting (23.15) in (23.19) we get

$$H_{kn} = \begin{pmatrix} H_{k,m-1} & b^{<}_{k-1,m}h(m) & b^{<}_{k-1,m}b(m)H_{m+1,n} \end{pmatrix}. \tag{23.25}$$

Inserting (23.22) in (23.5) we get

$$Q_{kn} = \begin{pmatrix} a^{>}_{n+1,m}a(m)Q_{k,m-1} & a^{>}_{n+1,m}q(m) & Q_{m+1,n} \end{pmatrix}. \tag{23.26}$$

Substituting (23.23) in (23.3) and using (23.22) and $G_{km} = \begin{pmatrix} G_{k,m-1}b(m) \\ g(m) \end{pmatrix}$, we get

$$A_{kn} = \begin{pmatrix} A_{k,m-1} & G_{k,m-1}h(m) & G_{k,m-1}h(m)H_{m+1,n} \\ p(m)Q_{k,m-1} & d(m) & g(m)H_{m+1,n} \\ P_{m+1,n}a(m)Q_{k,m-1} & P_{m+1,n}q(m) & A_{m+1,n} \end{pmatrix}. \tag{23.27}$$

Substituting (23.25), (23.26) and (23.27) in (23.13) we obtain the partition

$$W_{kn} = \begin{pmatrix} H_{k,m-1} & b^{<}_{k-1,m}h(m) & b^{<}_{k-1,m}b(m)H_{m+1,n} \\ A_{k,m-1} & G_{k,m-1}h(m) & G_{k,m-1}b(m)H_{m+1,n} \\ p(m)Q_{k,m-1} & d(m) & g(m)H_{m+1,n} \\ P_{m+1,n}a(m)Q_{k,m-1} & P_{m+1,n}q(m) & A_{m+1,n} \\ a^{>}_{n+1,m}a(m)Q_{k,m-1} & a^{>}_{n+1,m}q(m) & Q_{m+1,n} \end{pmatrix}. \tag{23.28}$$

Step 2. $M_{km}, M_{mn} \subseteq U_{kn}$. We show that M_{km}, M_{mn} are subspaces of U_{kn}. Let

$$W_{km}u_1 = \begin{pmatrix} H_{km} \\ A_{km} \\ Q_{km} \end{pmatrix} u_1 = 0.$$

Using (23.13) we get $W_{kn}\begin{pmatrix} u_1 \\ 0_{n-m} \end{pmatrix} = 0$. Hence, $M_{km} \subseteq U_{kn}$. Let

$$W_{mn}u_2 = \begin{pmatrix} H_{mn} \\ A_{mn} \\ Q_{mn} \end{pmatrix} u_2 = 0.$$

Using (23.14) we get $W_{kn}\begin{pmatrix} 0_{m-k} \\ u_2 \end{pmatrix} = 0$. Hence, $M_{mn} \subseteq U_{kn}$.

Step 3. $M_{km} \cap M_{mn} = \{0\}$. Obviously, $M_{km} \cap M_{mn} = \begin{pmatrix} 0_{m-k} \\ \theta \\ 0_{n-m} \end{pmatrix}$, where θ is a number and

$$\begin{pmatrix} 0_{m-k} \\ \theta \end{pmatrix} \in U_{km}, \quad \begin{pmatrix} \theta \\ 0_{n-m} \end{pmatrix} \in U_{mn}. \tag{23.29}$$

Assume that $\theta \neq 0$. Using (23.18) and the second relation in (23.29), we get $d(m) = h(m) = 0$. Using (23.24) and the first relation in (23.29), we get $q(m) = 0$. Thus we get $d(m) = q(m) = h(m) = 0$ contradicting the second inequality in (23.9). It follows that $M_{km} \cap M_{mn} = \{0\}$.

Step 4. $U_{kn} = M_{km} + M_{mn}$. We show that if a vector $x = \begin{pmatrix} x_1 \\ \theta \\ x_2 \end{pmatrix}$, where x_1, x_2 are vectors of the sizes $m-k, n-m$, respectively, and θ is a number, belongs to U_{kn}, then there exist numbers θ_1, θ_2 such that $\theta_1 + \theta_2 = \theta$ and

$$\begin{pmatrix} x_1 \\ \theta_1 \end{pmatrix} \in U_{km}, \quad \begin{pmatrix} \theta_2 \\ x_2 \end{pmatrix} \in U_{mn}. \tag{23.30}$$

We denote

$$\hat{A}_{k,m-1} = \begin{pmatrix} H_{k,m-1} \\ A_{k,m-1} \end{pmatrix}, \quad \hat{G}_{k,m-1} = \begin{pmatrix} b^{<}_{k-1,m} \\ G_{k,m-1} \end{pmatrix}, \tag{23.31}$$

$$\hat{P}_{m+1,n} = \begin{pmatrix} P_{m+1,n} \\ a^{>}_{n+1,m} \end{pmatrix}, \quad \hat{A}_{m+1,n} = \begin{pmatrix} A_{m+1,n} \\ Q_{m+1,n} \end{pmatrix} \tag{23.32}$$

and using (23.28) we obtain

$$\begin{cases} \hat{A}_{k,m-1}x_1 + \hat{G}_{k,m-1}h(m)\theta + \hat{G}_{k,m-1}b(m)H_{m+1,n}x_2 = 0, \\ p(m)Q_{k,m-1}x_1 + d(m)\theta + g(m)H_{m+1,n}x_2 = 0, \\ \hat{P}_{m+1,n}a(m)Q_{k,m-1}x_1 + \hat{P}_{m+1,n}q(m)\theta + \hat{A}_{m+1,n}x_2 = 0. \end{cases} \tag{23.33}$$

From the first inequality in (23.9) we obtain that at least one of the conditions $|p(m)| + |d(m)| > 0$ or $|g(m)| + |d(m)| > 0$ holds.

Case 1. $|p(m)| + |d(m)| > 0$. Then the first equality from (23.8) implies

$$\det \begin{pmatrix} p(m) & d(m) \\ a(m) & q(m) \end{pmatrix} = 0 \tag{23.34}$$

and hence there exists a number α such that

$$\begin{pmatrix} a(m) & q(m) \end{pmatrix} = \alpha \begin{pmatrix} p(m) & d(m) \end{pmatrix}. \tag{23.35}$$

Multiplying the second equation in (23.33) by $\hat{P}_{m+1,n}\alpha$ and subtracting the result from the third equation and using (23.35), we obtain the equivalent system

$$\begin{cases} \hat{A}_{k,m-1}x_1 + \hat{G}_{k,m-1}h(m)\theta + \hat{G}_{k,m-1}b(m)H_{m+1,n}x_2 = 0, \\ p(m)Q_{k,m-1}x_1 + d(m)\theta + g(m)H_{m+1,n}x_2 = 0, \\ -\hat{P}_{m+1,n}\alpha g(m)H_{m+1,n}x_2 + \hat{A}_{m+1,n}x_2 = 0. \end{cases} \tag{23.36}$$

Notice that $|p(m)| + |d(m)| > 0$ implies $|h(m)| + |d(m)| > 0$, otherwise we get $d(m) = 0$, hence by virtue of (23.35) $q(m) = 0$ and therefore $|q(m)| + |d(m)| + |h(m)| = 0$, contradicting the second inequality from (23.9). The second equality in (23.8) implies

$$\det \begin{pmatrix} h(m) & b(m) \\ d(m) & g(m) \end{pmatrix} = 0 \tag{23.37}$$

and hence there exists a number β such that

$$\begin{pmatrix} b(m) \\ g(m) \end{pmatrix} = \beta \begin{pmatrix} h(m) \\ d(m) \end{pmatrix}. \tag{23.38}$$

Set $\theta_2 = -\beta H_{m+1,n}x_2$, $\theta_1 = \theta - \theta_2$. We need to show that $\begin{pmatrix} \theta_2 \\ x_2 \end{pmatrix} \in U_{mn}$. We use the partition (23.18) of the matrix W_{mn}. Using (23.38) and then the definition of θ_2 we get

$$\begin{pmatrix} h(m) & b(m)H_{m+1,n} \\ d(m) & g(m)H_{m+1,n} \end{pmatrix} \begin{pmatrix} \theta_2 \\ x_2 \end{pmatrix} = \begin{pmatrix} h(m) \\ d(m) \end{pmatrix} (\theta_2 + \beta H_{m+1,n}x_2) = 0. \tag{23.39}$$

Using the definition of θ_2, the relations (23.35), (23.38) and the third equation from (23.36) we obtain

$$\begin{aligned} \hat{P}_{m+1,n}q(m)\theta_2 + \hat{A}_{m+1,n}x_2 &= -\hat{P}_{m+1}\alpha d(m)\beta H_{m+1,n}x_2 + \hat{A}_{m+1,n}x_2 \\ &= -\hat{P}_{m+1,n}\alpha g(m)H_{m+1,n}x_2 + \hat{A}_{m+1,n}x_2 \\ &= 0. \end{aligned} \tag{23.40}$$

From the equalities (23.39) and (23.40) it follows that

$$\begin{pmatrix} h(m) & b(m)H_{m+1,n} \\ d(m) & g(m)H_{m+1,n} \\ \hat{P}_{m+1,n}q(m) & \hat{A}_{m+1,n} \end{pmatrix} \begin{pmatrix} \theta_2 \\ x_2 \end{pmatrix} = 0. \tag{23.41}$$

From here using (23.32) and (23.18) we obtain the second relation from (23.30).

Let us prove the first relation in (23.30). Using the first and the second equations in (23.36) we get

$$\hat{A}_{k,m-1}x_1 + \hat{G}_{k,m-1}h(m)\theta_1 + \hat{G}_{k,m-1}h(m)\theta_2 + \hat{G}_{k,m-1}b(m)H_{m+1,n}x_2 = 0$$

and

$$p(m)Q_{k,m-1}x_1 + d(m)\theta_1 + d(m)\theta_2 + g(m)H_{m+1,n}x_2 = 0,$$

i.e.,

$$\begin{pmatrix} \hat{A}_{k,m-1} & \hat{G}_{k,m-1}h(m) \\ p(m)Q_{k,m-1} & d(m) \end{pmatrix} \begin{pmatrix} x_1 \\ \theta_1 \end{pmatrix} + \begin{pmatrix} \hat{G}_{k,m-1} & 0 \\ 0 & I \end{pmatrix} \begin{pmatrix} h(m) & b(m)H_{m+1,n} \\ d(m) & g(m)H_{m+1,n} \end{pmatrix} \begin{pmatrix} \theta_2 \\ x_2 \end{pmatrix} = 0.$$

From here using (23.39) we get

$$\begin{pmatrix} \hat{A}_{k,m-1} & \hat{G}_{k,m-1}h(m) \\ p(m)Q_{k,m-1} & d(m) \end{pmatrix} \begin{pmatrix} x_1 \\ \theta_1 \end{pmatrix} = 0. \tag{23.42}$$

This relation implies

$$\begin{pmatrix} p(m)Q_{k,m-1} & d(m) \end{pmatrix} \begin{pmatrix} x_1 \\ \theta_1 \end{pmatrix} = 0$$

and therefore using (23.35) we get

$$\begin{pmatrix} a(m)Q_{k,m-1} & q(m) \end{pmatrix} \begin{pmatrix} x_1 \\ \theta_1 \end{pmatrix} = 0. \tag{23.43}$$

From the equalities (23.42) and (23.43) it follows that

$$\begin{pmatrix} \hat{A}_{k,m-1} & \hat{G}_{k,m-1}h(m) \\ p(m)Q_{k,m-1} & d(m) \\ a(m)Q_{k,m-1} & q(m) \end{pmatrix} \begin{pmatrix} x_1 \\ \theta \end{pmatrix} = 0. \tag{23.44}$$

Hence using (23.31) and (23.24) we obtain $\begin{pmatrix} x_1 \\ \theta_1 \end{pmatrix} \in U_{km}$.

Case 2. $|g(m)| + |d(m)| > 0$. We rewrite the system (23.33) in the form

$$\begin{cases} \hat{A}_{m+1,n}x_2 + \hat{P}_{m+1,n}q(m)\theta + \hat{P}_{m+1,n}a(m)Q_{k,m-1}x_1 = 0, \\ g(m)H_{m+1,n}x_2 + d(m)\theta + p(m)Q_{k,m-1}x_1 = 0, \\ \hat{G}_{k,m-1}b(m)H_{m+1,n}x_2 + \hat{G}_{k,m-1}h(m)\theta + \hat{A}_{k,m-1}x_1 = 0. \end{cases}$$

Applying the same arguments as in Case 1) we obtain (23.30). □

Remark. The statement of Theorem 23.2 fails to be true if the condition (23.9) does not hold. Indeed, consider the matrix

$$A = \begin{pmatrix} 2 & 1 & 1 & 1 \\ 1 & 1 & 1 & 1 \\ 0 & 0 & 0 & 0 \\ 1 & 2 & 1 & 2 \end{pmatrix}.$$

This is a matrix with quasiseparable order one with fquasiseparable generators $p(2) = 1,\ p(3) = 0,\ p(4) = 1,\ q(1) = 1,\ q(2) = 2,\ q(3) = 1,\ a(2) = 1,\ a(3) = 1;\ g(1) = 1,\ g(2) = 1,\ g(3) = 0,\ h(2) = 1,\ h(3) = 1,\ h(4) = 1,\ b(2) = 1,\ b(3) = 1;\ d(1) = 2,\ d(2) = 1,\ d(3) = 0,\ d(4) = 1$. We have $d(3) - p(3)q(3) = 0,\ d(3) - g(3)h(3) = 0$, i.e., (23.8) holds. However, we have

$$|p(3)| + |d(3)| + |g(3)| = 0,$$

i.e., the condition (23.9) is not valid.

Next we get

$$\operatorname{Ker} W_{13} = \operatorname{Ker} \begin{pmatrix} 2 & 1 \\ 1 & 1 \\ 1 & 2 \end{pmatrix} = \{0\}, \quad \operatorname{Ker} W_{34} = \operatorname{Ker} \begin{pmatrix} 1 & 1 \\ 0 & 0 \\ 1 & 2 \end{pmatrix} = \{0\},$$

but $\operatorname{Ker} W_{14} = \operatorname{Ker} A \neq 0$ and therefore the partition (23.10) does not hold.

Next we use Theorem 23.2 to get a decomposition of the whole kernel of a quasiseparable of order one matrix.

Theorem 23.3. *Let A be a matrix with quasiseparable generators $p(i)$ $(i = 2, \dots, N)$, $q(j)$ $(j = 1, \dots, N-1)$, $a(k)$ $(k = 2, \dots, N-1)$; $g(i)$ $(i = 1, \dots, N-1)$, $h(j)$ $(j = 2, \dots, N)$, $b(k)$ $(k = 2, \dots, N-1)$; $d(k)$ $(k = 1, \dots, N)$ with orders equal to one. Suppose that for some integer t, $1 \le t \le N-2$, there exists a set of indices $j_1, \dots, j_t \in \{2, \dots, N-1\}$, $j_1 < j_2 < \dots < j_t$ such that for $m = j_1, j_2, \dots, j_t$ the equalities*

$$d(m)a(m) - p(m)q(m) = 0, \quad d(m)b(m) - g(m)h(m) = 0 \tag{23.45}$$

and the inequalities

$$|p(m)| + |d(m)| + |g(m)| > 0, \quad |q(m)| + |d(m)| + |h(m)| > 0 \tag{23.46}$$

hold.

Set $j_0 = 1$, $j_{t+1} = N$ and define

$$\begin{aligned} A^{(s)} &= A(j_{s-1} : j_s, j_{s-1} : j_s), && s = 1, \dots, t+1, \\ Q^{(s)} &= \operatorname{row}(a^{>}_{j_s+1,i} q(i))_{i=j_{s-1}}^{j_s}, && s = 1, \dots, t, \\ H^{(s)} &= \operatorname{row}(b^{<}_{j_{s-1}-1,i} h(i))_{i=j_{s-1}}^{j_s}, && s = 2, \dots, t+1 \end{aligned}$$

and next

$$W^{(1)} = \begin{pmatrix} A^{(1)} \\ Q^{(1)} \end{pmatrix}, \quad W^{(s)} = \begin{pmatrix} H^{(s)} \\ A^{(s)} \\ Q^{(s)} \end{pmatrix}, s = 2, \dots, t, \quad W^{(t+1)} = \begin{pmatrix} H^{(t+1)} \\ A^{(t+1)} \end{pmatrix},$$

$$U^{(s)} = \operatorname{Ker} W^{(s)}, \quad s = 1, \dots, t.$$

Then the subspace Ker A *admits the decomposition*

$$\operatorname{Ker} A = M_1 \oplus M_2 \oplus \cdots \oplus M_{t+1}, \tag{23.47}$$

where

$$M_s = \left\{ \begin{pmatrix} 0_{j_{s-1}-1} \\ u_s \\ 0_{N-j_s} \end{pmatrix} : u_s \in \operatorname{Ker} W^{(s)} \right\}, \quad s = 1, \dots, t+1. \tag{23.48}$$

Proof. Using Theorem 23.2 with $k = 1$, $n = N$ and $m = j_1$, we get

$$U_{1N} = M_{1,j_1} \oplus M_{j_1,N},$$

where $U_{1N} = \operatorname{Ker} A$,

$$M_{1,j_1} = \left\{ \begin{pmatrix} u_1 \\ 0_{N-j_1} \end{pmatrix} : u_1 \in \operatorname{Ker} \begin{pmatrix} A_{1,j_1} \\ Q_{1,j_1} \end{pmatrix} \right\},$$

i.e., $M_{1,j_1} = M_1$ and

$$M_{j_1,N} = \left\{ \begin{pmatrix} 0_{j_1-1} \\ \tilde{u}_2 \end{pmatrix} : \tilde{u}_2 \in U_{j_1,N} \right\}, \tag{23.49}$$

where $U_{j_1,N} = \operatorname{Ker} \begin{pmatrix} H_{j_1,N} \\ A_{j_1,N} \end{pmatrix}$. We denote $\tilde{M}_2 = M_{j_1,N}$ and obtain the decomposition

$$\operatorname{Ker} A = M_1 \oplus \tilde{M}_2. \tag{23.50}$$

Next we apply the same procedure to the matrix $\tilde{M}_2$. Using Theorem 23.2 with $k = j_1$, $n = N$ and $m = j_2$ we get

$$U_{j_1,N} = M_{j_1,j_2} \oplus M_{j_2,N},$$

where

$$M_{j_1,j_2} = \left\{ \begin{pmatrix} u_2 \\ 0_{N-j_2} \end{pmatrix} : u_2 \in U_{j_1,j_2} \right\}, \tag{23.51}$$

with

$$U_{j_1,j_2} = \operatorname{Ker} \begin{pmatrix} H_{j_1,j_2} \\ A_{j_1,j_2} \\ Q_{j_1,j_2} \end{pmatrix} = \operatorname{Ker} W^{(2)}$$

and

$$M_{j_2,N} = \left\{ \begin{pmatrix} 0_{j_2-j_1} \\ \tilde{u}_3 \end{pmatrix} : \tilde{u}_3 \in U_{j_2,N} \right\},$$

with $U_{j_2,N} = \operatorname{Ker} \begin{pmatrix} H_{j_2,N} \\ A_{j_2,N} \end{pmatrix}$. Using (23.49) and (23.48) we have

$$\tilde{M}_2 = \left\{ \begin{pmatrix} 0_{j_1-1} \\ \tilde{u}_2 \end{pmatrix} : u_2 \in U_{j_1,N} \right\}, \quad M_2 = \left\{ \begin{pmatrix} 0_{j_1-1} \\ \hat{u}_2 \end{pmatrix} : \hat{u}_2 \in M_{j_1,j_2} \right\}$$

and setting

$$\tilde{M}_3 = \left\{ \begin{pmatrix} 0_{j_2-1} \\ \tilde{u}_3 \end{pmatrix} : \tilde{u}_3 \in U_{j_2,N} \right\}$$

we obtain

$$\tilde{M}_2 = M_2 \oplus \tilde{M}_3. \tag{23.52}$$

Inserting (23.52) in (23.50) we obtain

$$M = M_1 \oplus M_2 \oplus \tilde{M}_3.$$

Next we apply the same procedure to the matrix $\tilde{M}_3$ and so on. Finally we obtain the partition (23.47). □

§23.2 The nontrivial kernels

Here we derive conditions in which the subspaces $\operatorname{Ker} W_{km}$ used in the previous section are non-trivial.

We use the following partitions of the matrices W_{km}.

Lemma 23.4. *Let A be a matrix with quasiseparable generators $p(i)$ $(i = 2, \dots, N)$, $q(j)$ $(j = 1, \dots, N-1)$, $a(k)$ $(k = 2, \dots, N-1)$; $g(i)$ $(i = 1, \dots, N-1)$, $h(j)$ $(j = 2, \dots, N)$, $b(k)$ $(k = 2, \dots, N-1)$; $d(k)$ $(k = 1, \dots, N)$. Set $h(1) = b(1) = p(1) = a(1) = q(N) = g(N) = a(N) = b(N) = 0$. Consider the matrices*

$$W_{km} = \begin{pmatrix} H_{km} \\ A_{km} \\ Q_{km} \end{pmatrix}, \quad 1 \le k < m \le N, \tag{23.53}$$

where

$$H_{km} = \operatorname{row}(b^{<}_{k-1,i} h(i))_{i=k}^{m}, \quad Q_{km} = \operatorname{row}(a^{>}_{m+1,i} q(i))_{i=k}^{m}, \tag{23.54}$$

and $A_{km} = A(k:m, k:m)$, and set also

$$P_{ij} = \operatorname{col}(p(t) a^{>}_{t,i-1})_{t=i}^{j}, \quad G_{ij} = \operatorname{col}(g(t) b^{<}_{t,j+1})_{t=i}^{j}, \quad 1 \le i \le j \le N. \tag{23.55}$$

Then the partitions

$$W_{km} = \begin{pmatrix} h(k) & b(k)H_{k+1,m-1} & b(k)b^<_{km}h(m) \\ d(k) & g(k)H_{k+1,m-1} & g(k)b^<_{km}h(m) \\ P_{k+1,m-1}q(k) & A_{k+1,m-1} & G_{k+1,m-1}h(m) \\ p(m)a^>_{mk}q(k) & p(m)Q_{k+1,m-1} & d(m) \\ a(m)a^>_{mk}q(k) & a(m)Q_{k+1,m-1} & q(m) \end{pmatrix}, \quad 1 \le k < m \le N \tag{23.56}$$

hold.

Proof. Using the first equality from (23.54) we get

$$H_{km} = \begin{pmatrix} b^<_{k-1,k}h(k) & \mathrm{row}(b^<_{k-1,i}h(i))_{i=k+1}^{m-1} & b^<_{k-1,m}h(m) \end{pmatrix}.$$

Therefore, since $b^<_{k-1,k} = 1$, $b^<_{k-1,i} = b(k)b^<_{ki}$, $b^<_{k-1,m} = b(k)b^<_{km}$ and

$$\mathrm{row}(b^<_{k-1,i}h(i))_{i=k+1}^{m-1} = b(k)\mathrm{row}(b^<_{ki}h(i))_{i=k+1}^{m-1} = b(k)H_{k+1,m-1},$$

we obtain

$$H_{km} = \begin{pmatrix} h(k) & b(k)H_{k+1,m-1} & b(k)b^<_{km}h(m) \end{pmatrix}. \tag{23.57}$$

Using the second equality from (23.54) we get

$$Q_{km} = \begin{pmatrix} a^>_{m+1,k}q(k) & \mathrm{row}(a^>_{m+1,i}q(i))_{i=k+1}^{m-1} & a^>_{m+1,m}q(m) \end{pmatrix}.$$

Therefore, since $a^>_{m+1,m} = 1$, $a^>_{m+1,i} = a(m)a^>_{mi}$, $a^>_{m+1,k} = a(m)a^>_{mk}$ and

$$\mathrm{row}(a^>_{m+1,i}q(i))_{i=k+1}^{m-1} = a(m)\mathrm{row}(a^>_{mi}q(i))_{i=k+1}^{m-1} = a(m)Q_{k+1,m-1},$$

we obtain

$$Q_{km} = \begin{pmatrix} a(m)a^>_{mk}q(k) & a(m)Q_{k+1,m-1} & q(m) \end{pmatrix}. \tag{23.58}$$

To represent the matrix A_{km} we use the partition (23.23):

$$A_{km} = \begin{pmatrix} A_{k,m-1} & G_{k,m-1}h(m) \\ p(m)Q_{k,m-1} & d(m) \end{pmatrix}. \tag{23.59}$$

Changing in (23.17) m to k and n to $m-1$ we get

$$A_{k,m-1} = \begin{pmatrix} d(k) & g(k)H_{k+1,m-1} \\ P_{k+1,m-1}q(k) & A_{k+1,m-1} \end{pmatrix}. \tag{23.60}$$

Using the second equality from (23.54) we have

$$Q_{k,m-1} = \mathrm{row}(a^>_{mi}q(i))_{i=k}^{m-1} = \begin{pmatrix} a^>_{mk}q(k) & \mathrm{row}(a^>_{mi}q(i))_{i=k+1}^{m-1} \end{pmatrix},$$

i.e.,

$$Q_{k,m-1} = \begin{pmatrix} a^>_{mk}q(k) & Q_{k+1,m-1} \end{pmatrix}. \tag{23.61}$$

Next, using the second equality from (23.55) we have

$$G_{k,m-1} = \operatorname{col}(g(i)b^{<}_{im})_{i=k}^{m-1} = \begin{pmatrix} g(k)b^{<}_{km} \\ \operatorname{col}(g(i)b^{<}_{im})_{i=k+1}^{m-1} \end{pmatrix},$$

i.e.,

$$G_{k,m-1} = \begin{pmatrix} g(k)b^{<}_{km} \\ G_{k+1,m-1} \end{pmatrix}. \tag{23.62}$$

Finally, inserting (23.60), (23.61) and (23.62) in (23.59) we obtain

$$A_{km} = \begin{pmatrix} d(k) & g(k)H_{k+1,m-1} & g(k)b^{<}_{km}h(m) \\ P_{k+1,m-1}q(k) & A_{k+1,m-1} & G_{k+1,m-1}h(m) \\ p(m)a^{>}_{mk}q(k) & p(m)Q_{k+1,m-1} & d(m) \end{pmatrix}. \tag{23.63}$$

Inserting (23.57), (23.58) and (23.63) in (23.53) we obtain the partition (23.56). □

Now we present the main result of this section.

Theorem 23.5. *Let A be a matrix with quasiseparable generators $p(i)$ $(i = 2,\dots,N)$, $q(j)$ $(j = 1,\dots,N-1)$, $a(k)$ $(k = 2,\dots,N-1)$; $g(i)$ $(i = 1,\dots,N-1)$, $h(j)$ $(j = 2,\dots,N)$, $b(k)$ $(k = 2,\dots,N-1)$; $d(k)$ $(k = 1,\dots,N)$ with orders equal to one. Set $h(1) = b(1) = p(1) = a(1) = q(N) = g(N) = a(N) = b(N) = 0$.*

Let for some indices k, m, $1 \le k < m \le N$, the equalities

$$d(k)b(k) - g(k)h(k) = 0, \tag{23.64}$$

$$d(m)a(m) - p(m)q(m) = 0 \tag{23.65}$$

hold. Consider the matrix W_{km} defined in (23.53).

1) *Let the inequalities*

$$|d(k)| + |g(k)| > 0, \tag{23.66}$$

$$|p(m)| + |d(m)| > 0, \tag{23.67}$$

hold.

Then the subspace $\operatorname{Ker} W_{km}$ *is nontrivial if and only if* $\det A_{km} = 0$. *A vector $u \ne 0$ belongs to* $\operatorname{Ker} W_{km}$ *if and only if u belongs to the kernel of the matrix A_{km}.*

2) *Let the equalities*

$$p(m) = 0, \quad d(m) = 0, \quad h(m) = 0 \tag{23.68}$$

and the relations (23.66) and $q(m) \ne 0$ hold.

Then the subspace $\operatorname{Ker} W_{km}$ *is nontrivial if and only if* $\det A_{k,m-1} = 0$. *A vector $u \ne 0$ belongs to* $\operatorname{Ker} W_{km}$ *if and only if $u = \begin{pmatrix} x \\ \theta \end{pmatrix}$, where x belongs to the kernel of the matrix $A_{k,m-1}$ and $\theta = -\frac{a(m)}{q(m)} Q_{k,m-1} x$.*

3) *Let the equalities*

$$d(k) = 0, \quad g(k) = 0, \quad q(k) = 0 \tag{23.69}$$

and the relations (23.67) *and* $h(k) \neq 0$ *hold.*

Then the subspace $\operatorname{Ker} W_{km}$ *is nontrivial if and only if* $\det A_{k+1,m} = 0$. *A vector* $u \neq 0$ *belongs to* $\operatorname{Ker} W_{km}$ *if and only if* $u = \begin{pmatrix} \theta \\ x \end{pmatrix}$, *where* x *belongs to the kernel of the matrix* $A_{k+1,m}$ *and* $\theta = -\frac{b(k)}{h(k)} H_{k+1,m} x$.

4) *Let the equalities*

$$p(m) = 0, \quad d(m) = 0, \quad h(m) = 0, \tag{23.70}$$

$$d(k) = 0, \quad g(k) = 0, \quad q(k) = 0 \tag{23.71}$$

and the relations $q(m) \neq 0$, $h(k) \neq 0$ *hold.*

Then the subspace $\operatorname{Ker} W_{km}$ *is nontrivial if and only if* $\det A_{k+1,m-1} = 0$. *A vector* $u \neq 0$ *belongs to* $\operatorname{Ker} W_{km}$ *if and only if* $u = \begin{pmatrix} \theta_1 \\ x \\ \theta_2 \end{pmatrix}$, *where* x *belongs to the kernel of the matrix* $A_{k+1,m-1}$ *and* $\theta_1 = -\frac{b(k)}{h(k)} H_{k+1,m-1} x$, $\theta_2 = -\frac{a(m)}{q(m)} Q_{k+1,m-1} x$.

Proof. At first notice that (23.65) implies

$$\det \begin{pmatrix} p(m) & d(m) \\ a(m) & q(m) \end{pmatrix} = 0$$

and moreover, if (23.67) holds, then there exists a number α such that

$$\begin{pmatrix} a(m) & q(m) \end{pmatrix} = \alpha \begin{pmatrix} p(m) & d(m) \end{pmatrix}. \tag{23.72}$$

Similarly, (23.64) implies

$$\det \begin{pmatrix} h(k) & b(k) \\ d(k) & g(k) \end{pmatrix} = 0$$

and moreover, if (23.66), holds then there exists a number β such that

$$\begin{pmatrix} h(k) & b(k) \end{pmatrix} = \beta \begin{pmatrix} d(k) & g(k) \end{pmatrix}. \tag{23.73}$$

Case 1). The relation (23.72) implies

$$\begin{pmatrix} a(m)a^{>}_{mk}q(k) & a(m)Q_{k+1,m-1} & q(m) \end{pmatrix} = \alpha \begin{pmatrix} p(m)a^{>}_{mk}q(k) & p(m)Q_{k+1,m-1} & d(m) \end{pmatrix}. \tag{23.74}$$

and (23.73) implies

$$\begin{pmatrix} h(k) & b(k)H_{k+1,m-1} & b(k)b^{<}_{km}h(m) \end{pmatrix} = \beta \begin{pmatrix} d(k) & g(k)H_{k+1,m-1} & g(k)b^{<}_{km}h(m) \end{pmatrix}. \tag{23.75}$$

Thus combining the equalities (23.74), (23.75) and the partitions (23.56) and (23.63) we conclude that

$$\operatorname{Ker} W_{km} = \operatorname{Ker} A_{km}.$$

It follows that $\operatorname{Ker} W_{km}$ is nontrivial if and only if $\det A_{km} = 0$; also, a vector $u \neq 0$ belongs to $\operatorname{Ker} W_{km}$ if and only if u belongs to $\operatorname{Ker} A_{km}$.

Case 2). Inserting (23.68) in the formula (23.56) we get

$$W_{km} = \begin{pmatrix} h(k) & b(k)H_{k+1,m-1} & 0 \\ d(k) & g(k)H_{k+1,m-1} & 0 \\ P_{k+1,m-1}q(k) & A_{k+1,m-1} & 0 \\ 0 & 0 & 0 \\ a(m)a^{>}_{mk}q(k) & a(m)Q_{k+1,m-1} & q(m) \end{pmatrix}$$

which implies

$$\operatorname{Ker} W_{km} = \operatorname{Ker} \begin{pmatrix} h(k) & b(k)H_{k+1,m-1} & 0 \\ d(k) & g(k)H_{k+1,m-1} & 0 \\ P_{k+1,m-1}q(k) & A_{k+1,m-1} & 0 \\ a(m)a^{>}_{mk}q(k) & a(m)Q_{k+1,m-1} & q(m) \end{pmatrix}. \tag{23.76}$$

The equality (23.73) implies

$$\begin{pmatrix} h(k) & b(k)H_{k+1,m-1} \end{pmatrix} = \beta \begin{pmatrix} d(k) & g(k)H_{k+1,m-1} \end{pmatrix},$$

and therefore using (23.76) we obtain

$$\operatorname{Ker} W_{km} = \operatorname{Ker} \begin{pmatrix} d(k) & g(k)H_{k+1,m-1} & 0 \\ P_{k+1,m-1}q(k) & A_{k+1,m-1} & 0 \\ a(m)a^{>}_{mk}q(k) & a(m)Q_{k+1,m-1} & q(m) \end{pmatrix}.$$

Using (23.60) and (23.61) we obtain

$$\operatorname{Ker} W_{km} = \operatorname{Ker} \begin{pmatrix} A_{k,m-1} & 0 \\ a(m)Q_{k,m-1} & q(m) \end{pmatrix}.$$

From here using $q(m) \neq 0$ we conclude that the subspace $\operatorname{Ker} W_{km}$ is nontrivial if and only if $\det A_{k,m-1} = 0$, and a vector $u \neq 0$ belongs to $\operatorname{Ker} W_{km}$ if and only if $u = \begin{pmatrix} x \\ \theta \end{pmatrix}$, where x belongs to the kernel of the matrix $A_{k,m-1}$ and $\theta = -\frac{a(m)}{q(m)} Q_{k,m-1} x$.

Case 3). Inserting (23.69) in the formula (23.56) we get

$$W_{km} = \begin{pmatrix} h(k) & b(k)H_{k+1,m-1} & b(k)b^{<}_{km}h(m) \\ 0 & 0 & 0 \\ 0 & A_{k+1,m-1} & G_{k+1,m-1}h(m) \\ 0 & p(m)Q_{k+1,m-1} & d(m) \\ 0 & a(m)Q_{k+1,m-1} & q(m) \end{pmatrix}$$

which implies

$$\operatorname{Ker} W_{km} = \operatorname{Ker} \begin{pmatrix} h(k) & b(k)H_{k+1,m-1} & b(k)b^{<}_{km}h(m) \\ 0 & A_{k+1,m-1} & G_{k+1,m-1}h(m) \\ 0 & p(m)Q_{k+1,m-1} & d(m) \\ 0 & a(m)Q_{k+1,m-1} & q(m) \end{pmatrix}. \tag{23.77}$$

Further, (23.72) implies

$$\begin{pmatrix} a(m)Q_{k+1,m-1} & q(m) \end{pmatrix} = \alpha \begin{pmatrix} p(m)Q_{k+1,m-1} & d(m) \end{pmatrix},$$

and therefore using (23.77) we obtain

$$\operatorname{Ker} W_{km} = \operatorname{Ker} \begin{pmatrix} h(k) & b(k)H_{k+1,m-1} & b(k)b^{<}_{km}h(m) \\ 0 & A_{k+1,m-1} & G_{k+1,m-1}h(m) \\ 0 & p(m)Q_{k+1,m-1} & d(m) \end{pmatrix}. \tag{23.78}$$

Changing k to $k+1$ in (23.23) and (23.21) we get

$$\begin{pmatrix} H_{k+1,m-1} & b^{<}_{km}h(m) \end{pmatrix} = H_{k+1,m}, \quad \begin{pmatrix} A_{k,m-1} & G_{k,m-1}h(m) \\ p(m)Q_{k,m-1} & d(m) \end{pmatrix} = A_{km}.$$

Inserting this in (23.78) we get

$$\operatorname{Ker} W_{km} = \operatorname{Ker} \begin{pmatrix} h(k) & b(k)H_{k+1,m} \\ 0 & A_{k+1,m} \end{pmatrix}.$$

Hence, since $h(k) \neq 0$, we conclude that the subspace $\operatorname{Ker} W_{km}$ is nontrivial if and only if $\det A_{k+1,m} = 0$, and a vector $u \neq 0$ belongs to $\operatorname{Ker} W_{km}$ if and only if $u = \begin{pmatrix} \theta \\ x \end{pmatrix}$, where x belongs to the kernel of the matrix $A_{k+1,m}$ and $\theta = -\frac{b(k)}{h(k)} H_{k+1,m} x$.

Case 4). Inserting (23.70) and (23.71) in the formula (23.56) we get

$$W_{km} = \begin{pmatrix} h(k) & b(k)H_{k+1,m-1} & 0 \\ 0 & 0 & 0 \\ 0 & A_{k+1,m-1} & 0 \\ 0 & 0 & 0 \\ 0 & a(m)Q_{k+1,m-1} & q(m) \end{pmatrix}.$$

Consequently,

$$\operatorname{Ker} W_{km} = \operatorname{Ker} \begin{pmatrix} h(k) & b(k)H_{k+1,m-1} & 0 \\ 0 & A_{k+1,m-1} & 0 \\ 0 & a(m)Q_{k+1,m-1} & q(m) \end{pmatrix}.$$

Hence, since $h(k) \neq 0$, $q(m) \neq 0$ we conclude that $\operatorname{Ker} W_{km}$ is nontrivial if and only if $\det A_{k+1,m-1} = 0$. A vector $u \neq 0$ belongs to $\operatorname{Ker} W_{km}$ if and only if $u = \begin{pmatrix} \theta_1 \\ x \\ \theta_2 \end{pmatrix}$, where x belongs to the kernel of the matrix $A_{k+1,m-1}$ and $\theta_1 = -\frac{b(k)}{h(k)} H_{k+1,m-1}x$, $\theta_2 = -\frac{a(m)}{q(m)} Q_{k+1,m-1}x$. □

In the case of a Hermitian matrix any eigenvalue λ_0 is real and hence we have $l_m(\lambda_0) = \overline{\delta_m(\lambda_0)}$. This implies that the conditions (22.5) on one hand and (23.8) on the other hand cover all the possibilities.

§23.3 Comments

This chapter has a preliminary character. Such a presentation appears here for the first time.

Chapter 24

Multiple Eigenvalues

Here we apply results of previous chapter to get partitions of eigenspaces of quasiseparable of order one matrices.

§24.1 The algorithm

Based on Theorem 23.5 we derive an algorithm to determine the basis of an eigenspace of a quasiseparable of order one matrix.

Algorithm 24.1. Let A be a matrix with quasiseparable generators $p(i)$ $(i = 2,\dots,N)$, $q(j)$ $(j = 1,\dots,N-1)$, $a(k)$ $(k = 2,\dots,N-1)$; $g(i)$ $(i = 1,\dots,N-1)$, $h(j)$ $(j = 2,\dots,N)$, $b(k)$ $(k = 2,\dots,N-1)$; $d(k)$ $(k = 1,\dots,N)$ with orders equal to one. Set $h(1) = b(1) = p(1) = a(1) = q(N) = g(N) = a(N) = b(N) = 0$. Let λ_0 be an eigenvalue of A and suppose that for some integer t, $1 \le t \le N-2$, there is a set of indices $j_1,\dots,j_t \in \{2,\dots,N-1\}$, $j_1 < j_2 < \cdots < j_t$ such that for $i = j_1, j_2, \dots, j_t$,

$$l_i(\lambda_0) = (d(i) - \lambda_0)a(i) - p(i)q(i) = 0, \tag{24.1}$$

$$\delta_i(\lambda_0) = (d(i) - \lambda_0)b(i) - g(i)h(i) = 0, \tag{24.2}$$

and

$$|p(i)| + |d(i) - \lambda_0| + |g(i)| > 0, \tag{24.3}$$

$$|q(i)| + |d(i) - \lambda_0| + |h(i)| > 0, \tag{24.4}$$

and for the other values of $i \in \{2,\dots,N-1\}$,

$$l_i(\lambda_0) \neq 0, \quad \delta_i(\lambda_0) \neq 0. \tag{24.5}$$

Assume also that

$$|d(1) - \lambda_0| + |q(1)| > 0, \qquad |d(1) - \lambda_0| + |g(1)| > 0, \tag{24.6}$$

$$|d(N) - \lambda_0| + |p(N)| > 0, \quad |d(N) - \lambda_0| + |h(N)| > 0. \tag{24.7}$$

Then the linearly independent eigenvectors u_i of the matrix A corresponding to λ_0 are obtained via the following algorithm.

Set $j_0 = 1,\ j_{t+1} = N,\ n = 0$. For $s = 1, 2, \dots, t+1$ perform the following operations.

1. To simplify the presentation set $k = j_{s-1},\ m = j_s$. Set $\rho_s = 0$.

2.1. Case 1. If

$$|d(k) - \lambda_0| + |g(k)| > 0, \quad |p(m)| + |d(m) - \lambda_0| > 0, \tag{24.8}$$

determine the matrix $A_{km} = A(k:m, k:m)$. Using quasiseparable generators of A_{km}, compute with the formulas (21.2)–(21.5) the value $\det(A_{km} - \lambda_0 I)$.

If $\det(A_{km} - \lambda_0 I) = 0$, compute the eigenvector $\tilde{u}_s$ of the matrix A_{km} corresponding to the eigenvalue λ_0 using the corresponding formula from Theorem 22.2. Set $\rho_s = 1$.

2.2. Case 2. If

$$|d(k) - \lambda_0| + |g(k)| > 0, \quad |p(m)| + |d(m) - \lambda_0| = 0, \tag{24.9}$$

determine the matrix $A_{k,m-1} = A(k : m-1, k : m-1)$ and the row vector $Q_{k,m-1} = \mathrm{row}(a^{>}_{mi} q(i))_{i=k}^{m-1}$. Using quasiseparable generators of $A_{k,m-1}$, compute with the formulas (21.2)–(21.5) the value $\det(A_{k,m-1} - \lambda_0 I)$.

If $\det(A_{k,m-1} - \lambda_0 I) = 0$, compute the eigenvector x of the matrix $A_{k,m-1}$ corresponding to the eigenvalue λ_0 using the corresponding formula from Theorem 22.2. Compute

$$\theta = -\frac{a(m)}{q(m)} Q_{k,m-1} x.$$

Set $\tilde{u}_s = \begin{pmatrix} x \\ \theta \end{pmatrix}$. Set $\rho_s = 1$.

2.3. Case 3. If

$$|d(k) - \lambda_0| + |g(k)| = 0, \quad |p(m)| + |d(m) - \lambda_0| > 0, \tag{24.10}$$

determine the matrix $A_{k+1,m} = A(k+1 : m, k+1 : m)$ and the row vector $H_{k+1,m} = \mathrm{row}(b^{<}_{ki} h(i))_{i=k+1}^{m}$. Using quasiseparable generators of $A_{k+1,m}$, compute with the formulas (21.2)–(21.5) the value $\det(A_{k+1,m} - \lambda_0 I)$.

If $\det(A_{k+1,m} - \lambda_0 I) = 0$, compute the eigenvector x of the matrix $A_{k+1,m}$ corresponding to the eigenvalue λ_0 using the corresponding formula from Theorem 22.2. Compute

$$\theta = -\frac{b(k)}{h(k)} H_{k+1,m} x.$$

Set $\tilde{u}_s = \begin{pmatrix} \theta \\ x \end{pmatrix}$. Set $\rho_s = 1$.

2.4. Case 4. If

$$|d(k) - \lambda_0| + |g(k)| = 0, \quad |p(m)| + |d(m) - \lambda_0| = 0, \tag{24.11}$$

determine the matrix $A_{k+1,m-1} = A(k+1:m-1, k+1:m-1)$ and the row vectors $Q_{k+1,m-1} = \mathrm{row}(a^{>}_{mi}q(i))_{i=k+1}^{m-1}$, $H_{k+1,m-1} = \mathrm{row}(b^{<}_{ki}h(i))_{i=k+1}^{m-1}$. Using quasiseparable generators of $A_{k+1,m-1}$, compute with the formulas (21.2)–(21.5) the value $\det(A_{k+1,m-1} - \lambda_0 I)$.

If $\det(A_{k+1,m-1} - \lambda_0 I) = 0$, compute the eigenvector x of the matrix $A_{k+1,m-1}$ corresponding to the eigenvalue λ_0 using the corresponding formula from Theorem 22.2. Compute

$$\theta_1 = -\frac{b(k)}{h(k)} H_{k+1,m} x, \quad \theta_2 = -\frac{a(m)}{q(m)} Q_{k,m-1} x.$$

Set $\tilde{u}_s = \begin{pmatrix} \theta_1 \\ x \\ \theta_2 \end{pmatrix}$. Set $\rho_s = 1$.

3. If $\rho_s = 1$, set $n := n+1$ and determine the eigenvector u_n of the matrix A via

$$u_n = \begin{pmatrix} 0_{k-1} \\ \tilde{u}_s \\ 0_{N-m} \end{pmatrix}. \tag{24.12}$$

To justify the algorithm notice that by Theorem 23.3 to be applied to the matrix $A - \lambda_0 I$ the eigenspace $\mathrm{Ker}(A - \lambda_0 I)$ admits the decomposition

$$\mathrm{Ker}(A - \lambda_0 I) = M_1(\lambda_0) \oplus M_2(\lambda_0) \oplus \cdots \oplus M_{t+1}(\lambda_0),$$

where

$$M_s(\lambda_0) = \left\{ \begin{pmatrix} 0_{j_{s-1}-1} \\ u_s \\ 0_{N-j_s} \end{pmatrix} : u_s \in \mathrm{Ker}\, W^{(s)}(\lambda_0) \right\}, \quad s = 1, \dots, t+1,$$

with

$$W^{(1)}(\lambda_0) = \begin{pmatrix} A^{(1)}(\lambda_0) \\ Q^{(1)} \end{pmatrix},$$

$$W^{(s)}(\lambda_0) = \begin{pmatrix} H^{(s)} \\ A^{(s)}(\lambda_0) \\ Q^{(s)} \end{pmatrix}, \quad s = 2, \dots, t,$$

$$W^{(t+1)}(\lambda_0) = \begin{pmatrix} H^{(t+1)} \\ A^{(t+1)}(\lambda_0) \end{pmatrix}.$$

Here

$$\begin{aligned} A^{(s)} &= A(j_{s-1}:j_s, j_{s-1}:j_s),\ A^{(s)}(\lambda_0) = A^{(s)} - \lambda_0 I, && s = 1, \dots, t+1, \\ Q^{(s)} &= \mathrm{row}(a^{>}_{j_s+1,i} q(i))_{t=j_{s-1}}^{j_s}, && s = 1, \dots, t, \\ H^{(s)} &= \mathrm{row}(b^{<}_{j_{s-1}-1,i} h(i))_{t=j_{s-1}}^{j_s}, && s = 2, \dots, t+1. \end{aligned}$$

We show that the dimension of each subspace $\operatorname{Ker} W^{(s)}(\lambda_0)$ is not greater than one. In the case where this subspace is nontrivial we compute a nonzero vector $u_s \in \operatorname{Ker} W^{(s)}(\lambda_0)$ and then determine the corresponding eigenvector of the matrix A via (24.12).

We start setting $n = 0$, here n is the index of the eigenvector we determine.

For each $s = 1, \dots, t+1$ we proceed as follows. On Step 1 of the algorithm, we set $k = j_{s-1}$, $m = j_s$ to simplify the upcoming formulas. We set $\rho_s = 0$; here the value ρ_s indicates that the subspace $\operatorname{Ker} W^{(s)}(\lambda_0)$ is nontrivial. If this is the case, we set $\rho_s = 1$ in the corresponding cases 2.1–2.4.

For each $s = 1, \dots, t+1$ we have four possibilities.

Assume that the conditions (24.8) hold. Then we follow Case 2.1 in the algorithm. By Case 1 of Theorem 23.5, the subspace $\operatorname{Ker} W_{km}(\lambda_0)$ is nontrivial if and only if $\det(A_{km} - \lambda_0 I) = 0$. If this is the case a vector $\tilde{u}_s$ belongs to $\operatorname{Ker} W^{(s)}(\lambda_0)$ if and only if $\tilde{u}_s$ is an eigenvector of the matrix A_{km} corresponding to the eigenvalue λ_0. We show that the matrix A_{km} and the eigenvalue λ_0 satisfy the conditions of Theorem 22.2. More precisely we need to check that

$$|d(k) - \lambda_0| + |q(k)| > 0, \qquad |d(k) - \lambda_0| + |g(k)| > 0, \tag{24.13}$$

$$|d(m) - \lambda_0| + |p(m)| > 0, \quad |d(m) - \lambda_0| + |h(m)| > 0 \tag{24.14}$$

and

$$l_i(\lambda_0) \neq 0, \quad \delta_i(\lambda_0) \neq 0, \quad i = k+1, \dots, m-1. \tag{24.15}$$

If $k = 1$, the condition (24.13) coincides with (24.6). If $2 \leq k \leq N-1$, then from $|d(k) - \lambda_0| + |g(k)| > 0$ it follows that $|q(k)| + |d(k) - \lambda_0| > 0$, otherwise we have $d(k) - \lambda_0 = 0$, $g(k) \neq 0$ and using (24.4) with $i = k$ we get $h(k) \neq 0$, contradicting the equality (24.2) with $i = k$. If $m = N$, the condition (24.14) coincides with (24.7). If $2 \leq m \leq N-1$, then from $|d(m) - \lambda_0| + |p(m)| > 0$ it follows that $|h(m)| + |d(m) - \lambda_0| > 0$, otherwise we have $d(m) - \lambda_0 = 0$, $p(m) \neq 0$ and using (24.4) with $i = m$ we get $q(m) \neq 0$, contradicting the equality (24.1) with $i = m$. The relations (24.15) follow directly from (24.5). Thus the matrix A_{km} and the eigenvalue λ_0 satisfy all the conditions of Theorem 22.2. Hence, the geometric multiplicity of λ_0 is not greater than one and the eigenvector $\tilde{u}_s$ of A_{km} may be determined using the corresponding formula from Theorem 22.2.

Assume that the conditions (24.9) hold. Then we follow Case 2.2 in the algorithm. Using (24.3) with $i = m$ we get $g(m) \neq 0$ and moreover, using (24.2) with $i = m$, we get $h(m) = 0$. From here using (24.4) with $i = m$ we obtain $q(m) \neq 0$. By Case 2 of Theorem 23.5, the subspace $\operatorname{Ker} W_{km}(\lambda_0)$ is nontrivial if and only if $\det(A_{k,m-1} - \lambda_0 I) = 0$ and a vector $\tilde{u}_s \neq 0$ belongs to $\operatorname{Ker} W_{km}(\lambda_0)$ if and only if $\tilde{u}_s = \begin{pmatrix} x \\ \theta \end{pmatrix}$, where x is an eigenvector of $A_{k,m-1}$ corresponding to the eigenvalue λ_0 and $\theta = -\frac{a(m)}{q(m)} Q_{k,m-1} x$. We show that the matrix $A_{k,m-1}$ and the eigenvalue λ_0 satisfy the conditions of Theorem 22.2. More precisely, we need to check that

$$|d(k)-\lambda_0|+|q(k)|>0, \qquad |d(k)-\lambda_0|+|g(k)|>0, \quad (24.16)$$
$$|d(m-1)-\lambda_0|+|p(m-1)|>0, \quad |d(m-1)-\lambda_0|+|h(m-1)|>0, \quad (24.17)$$

and

$$l_i(\lambda_0)\neq 0, \quad \delta_i(\lambda_0)\neq 0, \quad i=k+1,\dots,m-2. \quad (24.18)$$

If $k=1$, the condition (24.16) coincides with (24.6). If $2\le k\le N-1$, then from $|d(k)-\lambda_0|+|g(k)|>0$ it follows that $|q(k)|+|d(k)-\lambda_0|>0$, otherwise we have $d(k)-\lambda_0=0$, $g(k)\neq 0$ and using (24.4) with $i=k$ we get $h(k)\neq 0$, contradicting the equality (24.2) with $i=k$. Next, using (24.5) with $i=m-1$, we get

$$(d(m-1)-\lambda_0)a(m-1)-p(m-1)q(m-1)\neq 0,$$
$$(d(m-1)-\lambda_0)b(m-1)-g(m-1)h(m-1)\neq 0,$$

which implies (24.17). The relations (24.18) follow directly from (24.5). Thus the matrix $A_{k,m-1}$ and the eigenvalue λ_0 satisfy all the conditions of Theorem 22.2. Hence the geometric multiplicity of λ_0 is not greater than one and the eigenvector x of the matrix $A_{k,m-1}$ may be determined using the corresponding formula from Theorem 22.2.

Assume that the conditions (24.10) hold. Then we follow Case 2.3 in the algorithm. Using (24.3) with $i=k$ we get $p(k)\neq 0$, and moreover using (24.1) with $i=k$, we get $q(k)=0$. From here using (24.4) with $i=k$ we obtain $h(k)\neq 0$. By Case 3) of Theorem 23.5, the subspace $\operatorname{Ker} W_{km}(\lambda_0)$ is nontrivial if and only if $\det(A_{k+1,m}-\lambda_0 I)=0$ and a vector $\tilde{u}_s\neq 0$ belongs to $\operatorname{Ker} W_{km}(\lambda_0)$ if and only if $\tilde{u}_s=\begin{pmatrix}\theta\\ x\end{pmatrix}$, where x is an eigenvector of the matrix $A_{k+1,m}$ corresponding to the eigenvalue λ_0 and $\theta=-\frac{b(k)}{h(k)}H_{k+1,m}x$. We show that the matrix $A_{k+1,m}$ and the eigenvalue λ_0 satisfy the conditions of Theorem 22.2. More precisely, we need to check that

$$|d(k+1)-\lambda_0|+|q(k+1)|>0, \quad |d(k+1)-\lambda_0|+|g(k+1)|>0, \quad (24.19)$$
$$|d(m)-\lambda_0|+|p(m)|>0, \qquad |d(m)-\lambda_0|+|h(m)|>0, \quad (24.20)$$

and

$$l_i(\lambda_0)\neq 0, \quad \delta_i(\lambda_0)\neq 0, \quad i=k+2,\dots,m-1. \quad (24.21)$$

If $m=N$, the condition (24.20) coincides with (24.7). If $2\le m\le N-1$, then from $|d(m)-\lambda_0|+|p(m)|>0$ it follows that $|h(m)|+|d(m)-\lambda_0|>0$, otherwise we have $d(m)-\lambda_0=0$, $p(m)\neq 0$ and using (24.4) with $i=m$ we get $q(m)\neq 0$, contradicting the equality (24.1) with $i=m$. Next using (24.5) with $i=k+1$ we get

$$(d(k+1)-\lambda_0)a(k+1)-p(k+1)q(k+1)\neq 0,$$
$$(d(k+1)-\lambda_0)b(k+1)-g(k+1)h(k+1)\neq 0,$$

which implies (24.19). The relations (24.21) follow directly from (24.5). Thus the matrix $A_{k+1,m}$ and the eigenvalue λ_0 satisfy all the conditions of Theorem 22.2. Hence the geometric multiplicity of λ_0 is not greater than one and the eigenvector x of the matrix $A_{k,m-1}$ may be determined using the corresponding formula from Theorem 22.2.

Assume that the conditions (24.11) hold. Then we follow Case 2.4 in the algorithm. Using (24.3) with $i = k$ we get $p(k) \neq 0$, and moreover using (24.1) with $i = k$, we get $q(k) = 0$. From here using (24.4) with $i = k$ we obtain $h(k) \neq 0$. Next, using (24.3) with $i = m$ we get $g(m) \neq 0$, and moreover using (24.2) with $i = m$ we get $h(m) = 0$. From here using (24.4) with $i = m$ we obtain $q(m) \neq 0$. By Case 4) of Theorem 23.5, the subspace $\operatorname{Ker} W_{km}(\lambda_0)$ is nontrivial if and only if $\det(A_{k+1,m-1} - \lambda_0 I) = 0$ and a vector $\tilde{u}_s \neq 0$ belongs to $\operatorname{Ker} W_{km}(\lambda_0)$ if and only if $u = \begin{pmatrix} \theta_1 \\ x \\ \theta_2 \end{pmatrix}$, where x is an eigenvector of the matrix $A_{k+1,m-1}$ corresponding to the eigenvalue λ_0 and $\theta_1 = -\frac{b(k)}{h(k)} H_{k+1,m-1}x$, $\theta_2 = -\frac{a(m)}{q(m)} Q_{k+1,m-1}x$. We show that the matrix $A_{k+1,m-1}$ and the eigenvalue λ_0 satisfy the conditions of Theorem 22.2. More precisely, we need to check that

$$|d(k+1) - \lambda_0| + |q(k+1)| > 0, \qquad |d(k+1) - \lambda_0| + |g(k+1)| > 0, \tag{24.22}$$

$$|d(m-1) - \lambda_0| + |p(m-1)| > 0, \quad |d(m-1) - \lambda_0| + |h(m-1)| > 0, \tag{24.23}$$

and

$$l_i(\lambda_0) \neq 0, \quad \delta_i(\lambda_0) \neq 0, \quad i = k+2, \dots, m-2. \tag{24.24}$$

If $k = 1$, the condition (24.16) coincides with (24.6). If $2 \leq k \leq N-1$, then from $|d(k) - \lambda_0| + |g(k)| > 0$ it follows that $|q(k)| + |d(k) - \lambda_0| > 0$, otherwise we have $d(k) - \lambda_0 = 0$, $g(k) \neq 0$. Using (24.4) with $i = k$ we get $h(k) \neq 0$, contradicting the equality (24.2) with $i = k$. Using (24.5) with $i = m-1$ and $i = k+1$ we get

$$(d(m-1) - \lambda_0)a(m-1) - p(m-1)q(m-1) \neq 0,$$
$$(d(m-1) - \lambda_0)b(m-1) - g(m-1)h(m-1) \neq 0$$

and

$$(d(k+1) - \lambda_0)a(k+1) - p(k+1)q(k+1) \neq 0,$$
$$(d(k+1) - \lambda_0)b(k+1) - g(k+1)h(k+1) \neq 0,$$

which implies (24.22) and (24.23). The relations (24.24) follow directly from (24.5). Thus the matrix $A_{k+1,m-1}$ and the eigenvalue λ_0 satisfy all the conditions of Theorem 22.2. Hence the geometric multiplicity of λ_0 is not greater than one and the eigenvector x of the matrix $A_{k+1,m-1}$ may be determined using the corresponding formula from Theorem 22.2.

On Step 3 we check if $\rho_s = 1$, and if this is the case we determine the eigenvector of the matrix A via (24.12).

Corollary 24.2. *Under the conditions of Algorithm* 24.1 *the geometric multiplicity of the eigenvalue* λ_0 *is less than or equal to* $t+1$.

Example 24.3. Consider the $N \times N$ matrix

$$A = \begin{pmatrix} \mu & h & h & h & \cdots & h \\ p & \alpha & \beta & \beta & \cdots & \beta \\ p & \beta & \alpha & \beta & \cdots & \beta \\ p & \beta & \beta & \alpha & \cdots & \beta \\ \vdots & \vdots & \vdots & \vdots & \ddots & \vdots \\ p & \beta & \beta & \beta & \cdots & \alpha \end{pmatrix},$$

with $\mu \neq 0, \beta \neq 0, h \neq 0, p \neq 0, \alpha \neq \beta, \mu \neq \alpha - \beta, (\mu - \alpha + \beta)\beta - ph \neq 0$.

For the matrix A one can use the quasiseparable generators

$$p(i) = p,\ i = 2, \dots, N,\ \ q(1) = 1,\ \ q(j) = \frac{\beta}{p},\ j = 2, \dots, N-1,$$

$$a(k) = 1,\ k = 2, \dots, N-1,\ \ g(1) = 1,\ \ g(j) = \frac{\beta}{h},\ j = 2, \dots, N-1,$$

$$h(i) = h,\ i = 2, \dots, N,\ \ b(k) = 1,\ k = 2, \dots, N-1,$$

$$d(1) = \mu,\ d(k) = \alpha,\ k = 2, \dots, N.$$

In order to compute $\det A$, subtract $\frac{\beta}{h}$ times the first row from each of the remaining rows, thus obtaining zeroes instead of β. Factor the fraction $\frac{ph-\mu\beta}{h(\alpha-\beta)}$ from the first column and subtract all the remaining columns from the first. The determinant of A will be

$$(\alpha - \beta)^{N-2}(\mu\alpha + (N-2)\mu\beta - (N-1)ph).$$

Therefore, the determinant which gives the characteristic polynomial of this matrix is

$$\gamma_N(\lambda) = (\alpha - \beta - \lambda)^{N-2}((\mu - \lambda)(\alpha - \lambda) + (N-2)(\mu - \lambda)\beta - (N-1)ph). \quad (24.25)$$

One can obtain the same formula by induction using Corollary 21.4. The formula (24.25) implies that $\lambda_0 = \alpha - \beta$ is an eigenvalue of multiplicity $N-2$ of the matrix A. In the sequel $N-2$ independent eigenvectors corresponding to this eigenvalue will be found using Algorithm 24.1.

Algorithm 24.1 can be applied since all its condition are fulfilled. Indeed, for any $i = 2, \dots, N-1$ we have

$$l_i(\lambda_0) = (d(i) - \lambda_0)a(i) - p(i)q(i) = (\alpha - (\alpha - \beta)) \cdot 1 - p \cdot \frac{\beta}{p} = 0$$

and

$$\delta_i(\lambda_0) = (d(i) - \lambda_0)b(i) - g(i)h(i) = (\alpha - (\alpha - \beta)) \cdot 1 - \frac{\beta}{h} \cdot h = 0,$$

which implies (24.1), (24.2) with $i = 2, \dots, N-1$. Moreover, we get

$$|p(i)| + |d(i) - \lambda_0| + |g(i)| = |p| + |\beta| + \left|\frac{\beta}{h}\right| > 0,$$

$$|q(i)| + |d(i) - \lambda_0| + |h(i)| = \left|\frac{\beta}{p}\right| + |\beta| + |h| > 0, \quad i = 2, \dots, N-1,$$

which means that the inequalities (24.3), (24.4) hold. Also, the inequalities (24.6) become

$$|d(1) - \lambda_0| + |q(1)| = |\mu - \alpha + \beta| + \left|\frac{\beta}{p}\right| > 0,$$

$$|d(1) - \lambda_0| + |g(1)| = |\mu - \alpha + \beta| + \left|\frac{\beta}{h}\right| > 0,$$

while the inequalities (24.7) are satisfied, since

$$|d(N) - \lambda_0| + |p(N)| = |\beta| + |p| > 0, \quad |d(N) - \lambda_0| + |h(N)| = |\beta| + |h| > 0.$$

Therefore, taking $t = N-2$ and $j_s = s+1$, $s = 0, 1, \dots, t$, $j_{N-1} = N$, we conclude that the conditions under which Algorithm 24.1 finds eigenvectors are fulfilled.

Set $n = 0$, $k = j_0 = 1$, $m = j_1 = 2$. Set $s = 1$ and $\rho_s = 0$. Then

$$|d(k) - \lambda_0| + |g(k)| = |\mu - \alpha + \beta| + \left|\frac{\beta}{h}\right| > 0, \quad |p(m)| + |d(m) - \lambda_0| = |p| + |\beta| > 0$$

so that Case 1 of the algorithm can be applied. Set $k = 1, m = 2$. Determine the matrix

$$A_{km} = A(k:m, k:m) = A(1:2, 1:2) = \begin{pmatrix} \mu & h \\ p & \alpha \end{pmatrix}$$

and compute $\det(A_{km} - \lambda_0 I)$. For A_{12} one has

$$\det\begin{pmatrix} \mu - (\alpha - \beta) & h \\ p & \alpha - (\alpha - \beta) \end{pmatrix} = (\mu - \alpha + \beta)\beta - ph \neq 0.$$

Since $\det(A_{12} - (\alpha - \beta)I) = \det(A_{km} - \lambda_0 I) \neq 0$, there exists no eigenvector $\tilde{u}_s = \tilde{u}_1$ of the matrix A_{12} corresponding to the eigenvalue $\alpha - \beta$.

For any $s = 2, \dots, N-1$, set $k = j_{s-1} = s$, $m = j_s = s+1$. Set $\rho_s = 0$. Then

$$|d(k) - \lambda_0| + |g(k)| = |\alpha - \alpha + \beta| + \left|\frac{\beta}{h}\right| > 0, \quad |p(m)| + |d(m) - \lambda_0| = |p| + |\beta| > 0,$$

so that Case 1 can be applied. Determine the matrix

$$A_{km} = A(k:m, k:m) = A(s:s+1, s:s+1) = \begin{pmatrix} \alpha & \beta \\ \beta & \alpha \end{pmatrix}$$

and compute the value $\det(A_{km} - \lambda_0 I)$. For $A_{s,s+1}$ one has

$$\det\begin{pmatrix} \alpha-(\alpha-\beta) & \beta \\ \beta & \alpha-(\alpha-\beta) \end{pmatrix} = \det\begin{pmatrix} \beta & \beta \\ \beta & \beta \end{pmatrix} = 0.$$

Since $\det(A_{s,s+1} - (\alpha-\beta)I) = \det(A_{km} - \lambda_0 I) = 0$, compute the eigenvector $\tilde{u}_s$ of the matrix A_{km} corresponding to the eigenvalue $\alpha-\beta$ using the corresponding formula from Theorem 22.2. For this example it is again easy to see that the eigenvector is $\tilde{u}_s = \begin{pmatrix} -1 \\ 1 \end{pmatrix}$. Set $\rho_s = 1$ and $n = s$. A new eigenvector of the matrix A is $u_s = \begin{pmatrix} 0_{s-1} \\ -1 \\ 1 \\ 0_{N-s-1} \end{pmatrix}$.

It follows that $N-2$ eigenvectors for the matrix A corresponding to the eigenvalue $\lambda_0 = \alpha - \beta$ are

$$\begin{pmatrix} 0 \\ -1 \\ 1 \\ 0_{N-3} \end{pmatrix}, \begin{pmatrix} 0 \\ 0 \\ -1 \\ 1 \\ 0_{N-4} \end{pmatrix}, \dots, \begin{pmatrix} 0_{s-1} \\ -1 \\ 1 \\ 0_{N-s-1} \end{pmatrix}, \dots, \begin{pmatrix} 0_{N-3} \\ -1 \\ 1 \\ 0 \end{pmatrix}, \begin{pmatrix} 0_{N-2} \\ -1 \\ 1 \end{pmatrix},$$

which are linearly independent. ◊

§24.2 Hermitian and similar matrices

§24.2.1 The algorithm for Hermitian matrices

Here based on Algorithm 24.1 we derive an algorithm to compute the set of all eigenvectors of an arbitrary Hermitian matrix with quasiseparable order one corresponding to any eigenvalue λ_0, as follows.

Algorithm 24.4. Let A be a Hermitian matrix with quasiseparable generators $p(i)$ $(i = 2, \dots, N)$, $q(j)$ $(j = 1, \dots, N-1)$, $a(k)$ $(k = 2, \dots, N-1)$; $g(i) = \overline{q(i)}$; $(i = 1, \dots, N-1)$, $h(j) = \overline{p(j)}$ $(j = 2, \dots, N)$, $b(k) = \overline{a(k)}$; $(k = 2, \dots, N-1)$; $d(k)$ $(k = 1, \dots, N)$ with orders equal to one. Set $h(1) = b(1) = p(1) = a(1) = q(N) = g(N) = a(N) = b(N) = 0$. Let λ_0 be a an eigenvalue of the matrix A and suppose that for some integer t, $1 \le t \le N-2$, there is a set of indices $j_1, \dots, j_t \in \{2, \dots, N-1\}$, $j_1 < j_2 < \dots < j_t$ such that for $i = j_1, j_2, \dots, j_t$ the equalities

$$l_i(\lambda_0) = (d(i) - \lambda_0)a(i) - p(i)q(i) = 0$$

hold and for the other values of $i \in \{2, \dots, N-1\}$ one has $l_i(\lambda_0) \neq 0$.

Then the linearly independent eigenvectors u_i of the matrix A corresponding to λ_0 are obtained via the following algorithm.

1. If $d(1) - \lambda_0 = q(1) = 0$, take e_1 as an eigenvector of A and consider the matrix $A(2:N, 2:N)$ instead of A.
2. If $d(N) - \lambda_0 = p(N) = 0$, take e_N as an eigenvector of A and consider the matrix $A(1:N-1, 1:N-1)$ instead of A.
3. For $s = 1, \dots, t$ set $i = j_s$. If $p(i) = d(i) - \lambda_0 = q(i)$ take e_i as an eigenvector of A and consider the matrix

$$\begin{pmatrix} A(1:i-1, 1:i-1) & A(1:i-1, i+1:N) \\ A(i+1:N, 1:i-1) & A(i+1:N, i+1:N) \end{pmatrix}$$

   (obtained from A by deleting the ith column and the ith row) instead of the matrix A.

   Here e_j means the jth column vector of the standard basis in $\mathbb{C}^N$.
4. Apply Algorithm 24.1 to the obtained matrix.

To justify the algorithm, assume at first that the inequalities

$$|p(i)| + |d(i) - \lambda_0| + |q(i)| > 0, \quad i = j_1, j_2, \dots, j_t \tag{24.26}$$

and

$$|d(1) - \lambda_0| + |q(1)| > 0, \quad |d(N) - \lambda_0| + |p(N)| > 0 \tag{24.27}$$

hold. In this case using the equalities

$$\begin{aligned} g(i) &= \overline{q(i)}, \quad i = 1, \dots, N-1, \\ h(j) &= \overline{p(j)}, \quad j = 2, \dots, N, \\ b(k) &= \overline{a(k)}, \quad k = 2, \dots, N-1 \end{aligned}$$

and the fact that A is a Hermitian matrix we conclude that all the conditions of Algorithm 24.1 hold.

Now if for some $i \in \{2, \dots, N-1\}$ the condition (24.26) is not valid, i.e., $p(i) = d(i) - \lambda_0 = q(i) = 0$, then the ith row and the ith column of the matrix $A - \lambda_0 I$ are zeros. Similarly, if the first condition in (24.27) is not valid, i.e., $d(1) - \lambda_0 = q(1) = 0$, then the first row and the first column of $A - \lambda_0 I$ are zeros and if the second condition in (24.27) is not valid, i.e., $d(N) - \lambda_0 = p(N) = 0$, then the last row and the last column of $A - \lambda_0 I$ are zeros. It is clear that after deleting from the matrix A all columns and rows which correspond to zero rows and columns in the matrix $A - \lambda_0 I$, the obtained matrix satisfies all the conditions (24.26), (24.27). Thus, we established Algorithm 24.4.

Example 24.5. As an example of a Hermitian matrix with quasiseparable order one with a multiple eigenvalue consider

$$A = \begin{pmatrix} \alpha & 1 & 1 & a & a & a \\ 1 & 2 & 1 & a & a & a \\ 1 & 1 & 1 & a & a & a \\ a & a & a & 1 & 1 & 1 \\ a & a & a & 1 & 2 & 1 \\ a & a & a & 1 & 1 & \beta \end{pmatrix}.$$

Here $\alpha, \beta, a \neq 0$ are parameters. This is a matrix with quasiseparable order one with quasiseparable generators $p(2) = p(3) = 1,\ p(4) = p(5) = p(6) = a;\ q(1) = q(2) = q(3) = 1,\ q(4) = q(5) = a^{-1};\ a(k) = 1,\ k = 2, \dots, 5;\ g(1) = g(2) = g(3) = 1,\ g(4) = g(5) = a^{-1};\ h(2) = h(3) = 1,\ h(4) = h(5) = h(6) = a,\ b(k) = 1,\ k = 2, \dots, 5;\ d(1) = \alpha,\ d(2) = 2,\ d(3) = d(4) = 1,\ d(5) = 2,\ d(6) = \beta$. Take $\lambda_0 = 0$. It follows that $l_3(\lambda_0) = \delta_3(\lambda_0) = l_4(\lambda_0) = \delta_4(\lambda_0) = 0$. By the Algorithm 24.4, $\lambda_0 = 0$ is an eigenvalue of the matrix A of multiplicity three if and only if

$$\det \begin{pmatrix} \alpha & 1 & 1 \\ 1 & 2 & 1 \\ 1 & 1 & 1 \end{pmatrix} = \det \begin{pmatrix} 1 & a \\ a & 1 \end{pmatrix} = \det \begin{pmatrix} 1 & 1 & 1 \\ 1 & 2 & 1 \\ 1 & 1 & \beta \end{pmatrix} = 0,$$

i.e., if $\alpha = 1,\ \beta = 1,\ a = \pm 1$. In the case $\alpha = 1,\ \beta = 1,\ a \neq \pm 1$ the multiplicity of $\lambda_0 = 0$ equals two. The other cases when the multiplicity of $\lambda_0 = 0$ equals two and also the cases when this eigenvalue is simple can be treated in the same way.

The subspace $\operatorname{Ker} W^{(1)}(\lambda_0)$ is defined as

$$\operatorname{Ker} W^{(1)}(\lambda_0) = \operatorname{Ker} \begin{pmatrix} \alpha & 1 & 1 \\ 1 & 2 & 1 \\ 1 & 1 & 1 \\ 1 & 1 & 1 \end{pmatrix} = \operatorname{Ker} \begin{pmatrix} \alpha & 1 & 1 \\ 1 & 2 & 1 \\ 1 & 1 & 1 \end{pmatrix},$$

so $\operatorname{Ker} W^{(1)}(\lambda_0)$ is not trivial if and only if $\alpha = 1$, and then

$$\tilde{u}_1 = \begin{pmatrix} -1 \\ 0 \\ 1 \end{pmatrix} \in \operatorname{Ker} W^{(1)}(\lambda_0)$$

is the eigenvector given by Theorem 22.2. The corresponding eigenvector for the matrix A is

$$u_1 = \begin{pmatrix} -1 & 0 & 1 & 0 & 0 & 0 \end{pmatrix}^T.$$

The subspace $\operatorname{Ker} W^{(2)}(\lambda_0)$ is defined as

$$\operatorname{Ker} W^{(2)}(\lambda_0) = \operatorname{Ker} \begin{pmatrix} 1 & a \\ 1 & a \\ a & 1 \\ 1 & a^{-1} \end{pmatrix} = \operatorname{Ker} \begin{pmatrix} 1 & a \\ a & 1 \end{pmatrix}$$

and it is not trivial if and only if $a = \pm 1$, and in this case

$$\tilde{u}_2 = \begin{pmatrix} \mp 1 \\ 1 \end{pmatrix} \in \operatorname{Ker} W^{(2)}(\lambda_0)$$

is the eigenvector given by Theorem 22.2. The corresponding eigenvector for the matrix A is

$$u_2 = \begin{pmatrix} 0 & 0 & \mp 1 & 1 & 0 & 0 \end{pmatrix}^T.$$

The subspace $\operatorname{Ker} W^{(3)}(\lambda_0)$ is defined as

$$\operatorname{Ker} W^{(3)}(\lambda_0) = \operatorname{Ker} \begin{pmatrix} a & a & a \\ 1 & 1 & 1 \\ 1 & 2 & 1 \\ 1 & 1 & \beta \end{pmatrix} = \operatorname{Ker} \begin{pmatrix} 1 & 1 & 1 \\ 1 & 2 & 1 \\ 1 & 1 & \beta \end{pmatrix},$$

so $\operatorname{Ker} W^{(3)}(\lambda_0)$ is not trivial if and only if $\beta = 1$ and then

$$\tilde{u}_3 = \begin{pmatrix} -1 \\ 0 \\ 1 \end{pmatrix} \in \operatorname{Ker} W^{(1)}(\lambda_0)$$

is the eigenvector given by Theorem 22.2. The corresponding eigenvector for the matrix A is

$$u_3 = \begin{pmatrix} 0 & 0 & 0 & -1 & 0 & 1 \end{pmatrix}^T.$$

In order to exemplify other cases of Algorithm 24.4, consider also the matrix

$$B = \begin{pmatrix} \alpha & 1 & 0 & a & a & a \\ 1 & 2 & 0 & a & a & a \\ 0 & 0 & 0 & a & a & a \\ a & a & a & 1 & 1 & 1 \\ a & a & a & 1 & 2 & 1 \\ a & a & a & 1 & 1 & \beta \end{pmatrix},$$

and take as quasiseparable generators the same generators as for the matrix A, except $p(3) = 0,\ d(3) = 0,\ h(3) = 0$.

In this case the subspace $\operatorname{Ker} W^{(1)}(\lambda_0)$ is defined as

$$\operatorname{Ker} W^{(1)}(\lambda_0) = \operatorname{Ker} \begin{pmatrix} \alpha & 1 & 0 \\ 1 & 2 & 0 \\ 0 & 0 & 0 \\ 1 & 1 & 1 \end{pmatrix} = \operatorname{Ker} \begin{pmatrix} \alpha & 1 & 0 \\ 1 & 2 & 0 \\ 1 & 1 & 1 \end{pmatrix},$$

so $\operatorname{Ker} W^{(1)}(\lambda_0)$ is not trivial if and only if $\alpha = \frac{1}{2}$, and then by Case 2.2 it follows that $x = \begin{pmatrix} -2 \\ 1 \end{pmatrix}$,

$$\theta = -\frac{a(3)}{q(3)} \begin{pmatrix} q(1) & q(2) \end{pmatrix} x = -\begin{pmatrix} 1 & 1 \end{pmatrix} \begin{pmatrix} -2 \\ 1 \end{pmatrix} = 1,$$

and therefore

$$\tilde{u}_1 = \begin{pmatrix} -2 \\ 1 \\ 1 \end{pmatrix} \in \operatorname{Ker} W^{(1)}(\lambda_0).$$

The corresponding eigenvector for the matrix B is

$$u_1 = \begin{pmatrix} -2 & 1 & 1 & 0 & 0 & 0 \end{pmatrix}^T.$$

The subspace $\operatorname{Ker} W^{(2)}(\lambda_0)$ is defined as

$$\operatorname{Ker} W^{(2)}(\lambda_0) = \operatorname{Ker} \begin{pmatrix} 0 & a \\ 0 & a \\ a & 1 \\ 1 & a^{-1} \end{pmatrix} = \operatorname{Ker} \begin{pmatrix} 0 & a \\ a & 1 \end{pmatrix}$$

and is trivial, since $a \neq 0$.

The subspace $\operatorname{Ker} W^{(3)}(\lambda_0)$ is the same as for the matrix A.

Finally, consider the matrix

$$C = \begin{pmatrix} \alpha & 1 & 1 & 0 & a & a \\ 1 & 2 & 1 & 0 & a & a \\ 1 & 1 & 1 & 0 & a & a \\ 0 & 0 & 0 & 0 & 1 & 1 \\ a & a & a & 1 & 2 & 1 \\ a & a & a & 1 & 1 & \beta \end{pmatrix}.$$

and take as quasiseparable generators the same generators as for the matrix A, except $p(4) = 0,\ d(4) = 0,\ h(4) = 0$.

In this case, the subspace $\operatorname{Ker} W^{(1)}(\lambda_0)$ is the same as for the matrix A.

The subspace $\operatorname{Ker} W^{(2)}(\lambda_0)$ is defined as

$$\operatorname{Ker} W^{(2)}(\lambda_0) = \operatorname{Ker} \begin{pmatrix} 1 & 0 \\ 1 & 0 \\ 0 & 0 \\ 1 & a^{-1} \end{pmatrix} = \operatorname{Ker} \begin{pmatrix} 1 & 0 \\ 1 & a^{-1} \end{pmatrix},$$

and it is clearly trivial.

The subspace $\operatorname{Ker} W^{(3)}(\lambda_0)$ is defined as

$$\operatorname{Ker} W^{(3)}(\lambda_0) = \operatorname{Ker} \begin{pmatrix} 0 & a & a \\ 0 & 1 & 1 \\ 1 & 2 & 1 \\ 1 & 1 & \beta \end{pmatrix} = \operatorname{Ker} \begin{pmatrix} 0 & 1 & 1 \\ 1 & 2 & 1 \\ 1 & 1 & \beta \end{pmatrix}$$

therefore the subspace $\operatorname{Ker} W^{(3)}(\lambda_0)$ is not trivial if and only if $\beta = 0$, and then

$$\tilde{u}_3 = \begin{pmatrix} 1 \\ -1 \\ 1 \end{pmatrix} \in \operatorname{Ker} W^{(1)}(\lambda_0).$$

The vector

$$u_3 = \begin{pmatrix} 0 & 0 & 0 & 1 & -1 & 1 \end{pmatrix}^T$$

is the corresponding eigenvector for the matrix C. ◊

§24.2.2 Additional properties

The following theorem completes the results of Theorem 22.2 and Algorithm 24.1 under some additional assumptions.

Theorem 24.6. *Let A be a matrix with quasiseparable generators $p(i)$ $(i = 2, \dots, N)$, $q(j)$ $(j = 1, \dots, N-1)$, $a(k)$ $(k = 2, \dots, N-1)$; $g(i)$ $(i = 1, \dots, N-1)$, $h(j)$ $(j = 2, \dots, N)$, $b(k)$ $(k = 2, \dots, N-1)$; $d(k)$ $(k = 1, \dots, N)$ with orders equal to one and assume that the generators of A satisfy the following conditions:*

$$\begin{gathered} \overline{d(i)} = d(i), \quad i = 1, \dots, N, \quad p(i+1)q(i) \neq 0, \; g(i)h(i+1) \neq 0, \\ \frac{p(i+1)q(i)}{\overline{g(i)h(i+1)}} > 0, \quad i = 1, \dots, N-1, \end{gathered} \tag{24.28}$$

$$\begin{gathered} a(k) \neq 0, \; b(k) \neq 0, \quad \frac{p(k)q(k)}{a(k)} = \frac{g(k)h(k)}{b(k)} = \frac{\overline{g(k)h(k)}}{\overline{b(k)}}, \\ k = 2, \dots, N-1. \end{gathered} \tag{24.29}$$

Let λ_0 be an eigenvalue of the matrix A.

If

$$l_k(\lambda_0) = (d(k) - \lambda_0)a(k) - p(k)q(k) \neq 0, \quad k = 2, \dots, N-1, \tag{24.30}$$

then the eigenvalue λ_0 is simple and the corresponding eigenvector is given by the formula (22.7).

If for some integer t, $1 \leq t \leq N-2$, there is a set of indices $j_1, \dots, j_t \in \{2, \dots, N-1\}$, $j_1 < j_2 < \dots < j_t$ such that for $m = j_1, j_2, \dots, j_t$,

$$l_m(\lambda_0) = (d(m) - \lambda_0)a(m) - p(m)q(m) = 0 \tag{24.31}$$

and for the other values of $m \in \{2, \dots, N-1\}$

$$l_m(\lambda_0) \neq 0, \tag{24.32}$$

then the multiplicity of λ_0 is less than or equal to $t+1$ and the corresponding eigenvectors may be computed by Algorithm 24.1.

Proof. In the same way as in Example 22.6 from Section §22.3, we show that the matrix A is similar to a Hermitian matrix. Hence, the matrix A is diagonalizable. Moreover, the condition (24.29) implies that for any λ,

$$\delta_k(\lambda) = (d(k) - \lambda)b(k) - g(k)h(k) = b(k)\left(d(k) - \lambda - \frac{g(k)h(k)}{b(k)}\right)$$
$$= b(k)\left(d(k) - \lambda - \frac{p(k)q(k)}{a(k)}\right) = \frac{b(k)}{a(k)}l_k(\lambda),\ k = 2, \dots, N-1.$$

It follows that for any λ and for any $k \in \{2, \dots, N-1\}$, $l_k(\lambda) = 0$ if and only if $\delta_k(\lambda) = 0$. Therefore, if the condition (24.30) holds, then taking also into account that by (24.28) we have $q(1) \neq 0,\ g(1) \neq 0,\ p(N) \neq 0,\ h(N) \neq 0$, we conclude that all the conditions of Theorem 22.2 are satisfied. Hence, by Theorem 22.2 and by virtue of the fact that the matrix A is diagonalizable, we conclude that the eigenvalue λ_0 is simple. Moreover, since $p(N)h(N) \neq 0$, the corresponding eigenvector is given by the formula (22.7).

In the case when the conditions (24.31), (24.32) are valid we have

$$l_m(\lambda_0) = \delta_m(\lambda_0) = 0$$

for $m = j_1, j_2, \dots, j_k,\ j_1, \dots, j_k \in \{2, \dots, N-1\}$ and

$$l_m(\lambda_0) \neq 0, \quad \delta_m(\lambda_0) \neq 0.$$

for the other values of $m \in \{2, \dots, N-1\}$. Taking into account that by virtue of condition (24.28) we have $q(k) \neq 0,\ g(k) \neq 0\ (k = 1, \dots, N-1),\ p(k) \neq 0,\ h(k) \neq 0\ (k = 2, \dots, N)$, we conclude that all the conditions of Algorithm 24.1 are satisfied. Hence, using also Corollary 24.2 and the fact that the matrix A is diagonalizable, we conclude that the multiplicity of λ_0 is less than or equal to $t+1$ and the corresponding eigenvectors may be computed by the Algorithm 24.1. □

§24.3 An intermediate case

Next we consider an intermediate case when at least one of the conditions (22.5) of Theorem 22.2 does not hold, but the other one may be satisfied or not, i.e., for some $i \in \{2, \dots, N-1\}$, $l_i(\lambda_0)\delta_i(\lambda_0) = 0$. We assume for simplicity that $d(i) - \lambda_0 \neq 0$, which implies the conditions (24.3) and (24.4). We obtain here an analog of Theorem 23.2 for the matrix $A - \lambda_0 I$. However, we do not suggest here an algorithm for the computation of the eigenvectors, as it was done in Algorithm 24.1.

At first we show that in the case under consideration the value of the characteristic polynomial of a matrix at the point λ_0 may be expressed via the product of the values at this point of the characteristic polynomials of the corresponding principal submatrices.

Lemma 24.7. *Let A be a matrix with quasiseparable generators $p(i)$ $(i = 2, \dots, N)$, $q(j)$ $(j = 1, \dots, N-1)$, $a(k)$ $(k = 2, \dots, N-1)$; $g(i)$ $(i = 1, \dots, N-1)$, $h(j)$ $(j = 2, \dots, N)$, $b(k)$ $(k = 2, \dots, N-1)$; $d(k)$ $(k = 1, \dots, N)$ with orders equal to one, and let λ_0 be a number such that for some $m \in \{2, \dots, N-1\}$ we have $d(m) - \lambda_0 \neq 0$ and $l_m(\lambda_0)\delta_m(\lambda_0) = 0$. Then*

$$\det(A - \lambda_0 I) = \frac{1}{d(m) - \lambda_0}\gamma_m(\lambda_0)\theta_m(\lambda_0), \tag{24.33}$$

where $\gamma_m(\lambda_0) = \det(A(1:m, 1:m) - \lambda_0 I)$ and $\theta_m(\lambda_0) = \det(A(m:N, m:N) - \lambda_0 I)$.

Proof. Assume that $l_m(\lambda_0) = 0$, i.e.,

$$\det\begin{pmatrix} p(m) & d(m) - \lambda_0 \\ a(m) & q(m) \end{pmatrix} = 0.$$

Since $d(m) - \lambda_0 \neq 0$, there exists a number α such that

$$\begin{pmatrix} a(m) & q(m) \end{pmatrix} = \alpha \begin{pmatrix} p(m) & d(m) - \lambda_0 \end{pmatrix}. \tag{24.34}$$

By virtue of the partition (22.16) we have

$$\det(A - \lambda_0 I) = \det\begin{pmatrix} A_{m-1}(\lambda_0) & G_{m-1}h(m) & G_{m-1}b(m)H_{m+1} \\ p(m)Q_{m-1} & d(m) - \lambda_0 & g(m)H_{m+1} \\ P_{m+1}a(m)Q_{m-1} & P_{m+1}q(m) & B_{m+1}(\lambda_0) \end{pmatrix},$$

where $A_{m-1}(\lambda_0) = A(1:m-1, 1:m-1) - \lambda_0 I$, $B_{m+1}(\lambda_0) = A(m+1:N, m+1:N) - \lambda_0 I$. Multiplying the second row of this determinant by $P_{m+1}\alpha$ and subtracting the result from the third row we obtain

$$\det(A - \lambda_0 I) = \det\begin{pmatrix} A_{m-1}(\lambda_0) & G_{m-1}h(m) & G_{m-1}b(m)H_{m+1} \\ p(m)Q_{m-1} & d(m) - \lambda_0 & g(m)H_{m+1} \\ 0 & 0 & \widetilde{B}_{m+1}(\lambda_0) \end{pmatrix},$$

where $\widetilde{B}_{m+1}(\lambda_0) = B_{m+1}(\lambda_0) - P_{m+1}\alpha g(m)H_{m+1}$, which implies

$$\det(A - \lambda_0 I) = \det A_m(\lambda_0) \cdot \det \widetilde{B}_{m+1}(\lambda_0) = \gamma_m(\lambda_0) \cdot \det \widetilde{B}_{m+1}(\lambda_0). \tag{24.35}$$

Next consider the determinant

$$\theta_m(\lambda_0) = \det\begin{pmatrix} d(m) - \lambda_0 & g(m)H_{m+1} \\ P_{m+1}q(m) & B_{m+1}(\lambda_0) \end{pmatrix}.$$

Multiplying the first row of this determinant by $P_{m+1}\alpha$ and subtracting the result from the second row, we obtain

$$\theta_m(\lambda_0) = \det\begin{pmatrix} d(m) - \lambda_0 & g(m)H_{m+1} \\ 0 & \widetilde{B}_{m+1}(\lambda_0) \end{pmatrix} = (d(m) - \lambda_0) \cdot \det \widetilde{B}_{m+1}(\lambda_0). \tag{24.36}$$

Comparing (24.35) and (24.36), we obtain (24.33).

In the same way one can proceed in the case $\delta_m(\lambda_0) = 0$. □

Next we provide an analog of Theorem 23.2 for the eigenvectors of a matrix satisfying for some $m \in \{2, \dots, N-1\}$ the condition $l_m(\lambda_0)\delta_m(\lambda_0) = 0$.

Theorem 24.8. *Let A be a matrix with quasiseparable generators $p(i)$ $(i = 2, \dots, N)$, $q(j)$ $(j = 1, \dots, N-1)$, $a(k)$ $(k = 2, \dots, N-1)$; $g(i)$ $(i = 1, \dots, N-1)$, $h(j)$ $(j = 2, \dots, N)$, $b(k)$ $(k = 2, \dots, N-1)$; $d(k)$ $(k = 1, \dots, N)$ with orders equal to one and let λ_0 be a number such that for some $m \in \{2, \dots, N-1\}$ we have $d(m) - \lambda_0 \neq 0$ and*

$$l_m(\lambda_0)\delta_m(\lambda_0) = 0, \tag{24.37}$$

where

$$l_m(\lambda) = (d(m) - \lambda_0)a(m) - p(m)q(m), \ \delta_m(\lambda) = (d(m) - \lambda_0)b(m) - g(m)h(m).$$

Then $\lambda = \lambda_0$ is an eigenvalue of A if and only if

$$\gamma_m(\lambda_0)\theta_m(\lambda_0) = 0, \tag{24.38}$$

where $\gamma_m(\lambda_0) = \det A_m(\lambda_0)$, $\theta_m(\lambda_0) = \det B_m(\lambda_0)$ and $A_m(\lambda_0) = A(1 : m, 1 : m) - \lambda_0 I$, $B_m(\lambda_0) = A(m : N, m : N) - \lambda_0 I$. Moreover if $\lambda = \lambda_0$ is an eigenvalue of A and $l_m(\lambda_0) = 0$, then the vector $x = \begin{pmatrix} x_1 \\ \theta \\ x_2 \end{pmatrix}$, where x_1, x_2 are vectors of the sizes $m-1, N-m$, respectively and θ is a number, is an eigenvector of A corresponding to λ_0 if and only if there exist numbers θ_1, θ_2 such that $\theta_1 + \theta_2 = \theta$ and

$$B_m(\lambda_0)\begin{pmatrix} \theta_2 \\ x_2 \end{pmatrix} = 0, \quad A_m(\lambda_0)\begin{pmatrix} x_1 \\ \theta_1 \end{pmatrix} = -\begin{pmatrix} G_{m-1} \\ 0 \end{pmatrix}\frac{\delta_m(\lambda_0)}{d(m) - \lambda_0}(H_{m+1}x_2) \tag{24.39}$$

and at least one of the vectors $\begin{pmatrix} x_1 \\ \theta_1 \end{pmatrix}, \begin{pmatrix} \theta_2 \\ x_2 \end{pmatrix}$ is not zero.

Furthermore, if $\lambda = \lambda_0$ is an eigenvalue of A and $\delta_m(\lambda_0) = 0$, then the vector x is an eigenvector of A corresponding to λ_0 if and only if there exist numbers θ_1, θ_2 such that $\theta_1 + \theta_2 = \theta$ and

$$A_m(\lambda_0)\begin{pmatrix} x_1 \\ \theta_1 \end{pmatrix} = 0, \quad B_m(\lambda_0)\begin{pmatrix} \theta_2 \\ x_2 \end{pmatrix} = -\begin{pmatrix} 0 \\ P_{m+1} \end{pmatrix}\frac{l_m(\lambda_0)}{d(m) - \lambda_0}(Q_{m-1}x_1), \tag{24.40}$$

and at least one of the vectors $\begin{pmatrix} x_1 \\ \theta_1 \end{pmatrix}, \begin{pmatrix} \theta_2 \\ x_2 \end{pmatrix}$ is not zero.

Here the vectors P_k, Q_k, G_k, H_k are defined via (5.1), (5.2), (5.5), (5.6).

Unfortunately we have no simple criteria or an algorithm at our disposal for solving (24.39), (24.40) with singular matrices and non-zero right-hand sides.

Proof. From Lemma 24.7 we conclude that $\lambda = \lambda_0$ is an eigenvalue of the matrix A if and only if (24.38) holds.

Assume that λ_0 is an eigenvalue of A. Let $l_m(\lambda_0) = 0$ and let the vectors $\begin{pmatrix} x_1 \\ \theta_1 \end{pmatrix}$, $\begin{pmatrix} \theta_2 \\ x_2 \end{pmatrix}$ satisfy the condition (24.39). Using the representations

$$A_m(\lambda_0) = \begin{pmatrix} A_{m-1}(\lambda_0) & G_{m-1}h(m) \\ p(m)Q_{m-1} & d(m) - \lambda_0 \end{pmatrix}, \quad B_m(\lambda_0) = \begin{pmatrix} d(m) - \lambda_0 & g(m)H_{m+1} \\ P_{m+1}q(m) & B_{m+1}(\lambda_0) \end{pmatrix}$$

we obtain

$$A_{m-1}(\lambda_0)x_1 + G_{m-1}h(m)\theta_1 + G_{m-1}\frac{\delta_m(\lambda_0)}{d(m) - \lambda_0}H_{m+1}x_2 = 0, \tag{24.41}$$

$$p(m)Q_{m-1}x_1 + (d(m) - \lambda_0)\theta_1 = 0, \tag{24.42}$$

$$(d(m) - \lambda_0)\theta_2 + g(m)H_{m+1}x_2 = 0, \tag{24.43}$$

$$P_{m+1}q(m)\theta_2 + B_{m+1}(\lambda_0)x_2 = 0. \tag{24.44}$$

From (24.43) we get $\theta_2 = -\frac{1}{d(m)-\lambda_0}g(m)H_{m+1}x_2$ and then substituting $\theta_1 = \theta - \theta_2$ in (24.41) we obtain

$$A_{m-1}(\lambda_0)x_1 + G_{m-1}h(m)(\theta_1+\theta_2) + G_{m-1}\left(\frac{g(m)h(m)}{d(m) - \lambda_0} + \frac{\delta_m(\lambda_0)}{d(m) - \lambda_0}\right)H_{m+1}x_2 = 0.$$

From here using the equality $\delta_m(\lambda_0) = b(m)(d(m) - \lambda_0) - g(m)h(m)$ and setting $\theta = \theta_1 + \theta_2$ we obtain

$$A_{m-1}(\lambda_0)x_1 + G_{m-1}h(m)\theta + G_{m-1}b(m)H_{m+1}x_2 = 0. \tag{24.45}$$

Further, adding (24.42), (24.43) it follows that

$$p(m)Q_{m-1}x_1 + (d(m) - \lambda_0)\theta + g(m)H_{m+1}x_2 = 0. \tag{24.46}$$

Next, using the relations (24.42) and (24.34) we obtain

$$a(m)Q_{m-1}x_1 + q(m)\theta_1 = 0,$$

whence

$$P_{m+1}a(m)Q_{m-1}x_1 + P_{m+1}q(m)\theta_1 = 0.$$

The last equality together with (24.44) yields

$$P_{m+1}a(m)Q_{m-1}x_1 + P_{m+1}q(m)\theta + B_{m+1}(\lambda_0)x_2 = 0. \tag{24.47}$$

From (24.45)–(24.47) using the partition (22.16) we conclude that the vector $x = \begin{pmatrix} x_1 \\ \theta \\ x_2 \end{pmatrix}$ belongs to $\mathrm{Ker}(A - \lambda_0 I)$.

Now assume that $l_m(\lambda_0) = 0$ and the vector $x = \begin{pmatrix} x_1 \\ \theta \\ x_2 \end{pmatrix}$, where x_1, x_2 are vectors of the sizes $m-1, N-m$, respectively, and θ is a number, is an eigenvector of the matrix A corresponding to the eigenvalue λ_0. Applying the formula (23.36) to the matrix $A - \lambda_0 I$ we get

$$\begin{cases} A_{m-1}(\lambda_0)x_1 + G_{m-1}h(m)\theta + G_{m-1}b(m)H_{m+1}x_2 = 0, \\ p(m)Q_{m-1}x_1 + (d(m) - \lambda_0)\theta + g(m)H_{m+1}x_2 = 0, \\ -P_{m+1}\alpha g(m)H_{m+1}x_2 + B_{m+1}(\lambda_0)x_2 = 0. \end{cases} \tag{24.48}$$

Moreover, the number α in the third equation of this system is defined via the relation (24.34), which implies $\alpha = q(m)/(d(m) - \lambda_0)$. Set

$$\theta_2 = -(g(m)H_{m+1}x_2)/(d(m) - \lambda_0). \tag{24.49}$$

Using the third equation in (24.48) and the equality $q(m)\theta_2 = -\alpha g(m)H_{m+1}x_2$ we obtain

$$B_m(\lambda_0)\begin{pmatrix} \theta_2 \\ x_2 \end{pmatrix} = \begin{pmatrix} d(m) - \lambda_0 & g(m)H_{m+1} \\ P_{m+1}q(m) & B_{m+1}(\lambda_0) \end{pmatrix}\begin{pmatrix} \theta_2 \\ x_2 \end{pmatrix} = 0.$$

Furthermore, setting $\theta_1 = \theta - \theta_2$ and using the first and the second equation from (24.48) and the equality (24.49), we get

$$\begin{cases} A_{m-1}(\lambda_0)x_1 + G_{m-1}h(m)\theta_1 + G_{m-1}\left(-\dfrac{h(m)g(m)}{d(m) - \lambda_0} + b(m)\right)H_{m+1}x_2 = 0, \\ p(m)Q_{m-1}x_1 + (d(m) - \lambda_0)\theta_1 = 0. \end{cases}$$

Hence,

$$\begin{pmatrix} A_{m-1}(\lambda_0) & G_{m-1}h(m) \\ p(m)Q_{m-1} & d(m) - \lambda_0 \end{pmatrix}\begin{pmatrix} x_1 \\ \theta_1 \end{pmatrix} = -\begin{pmatrix} G_{m-1}\frac{\delta_m(\lambda_0)}{d(m)-\lambda_0}(H_{m+1}x_2) \\ 0 \end{pmatrix},$$

which completes the proof of (24.39).

In the case $\delta_m(\lambda_0) = 0$ one can proceed in the same way. □

Example 24.9. Consider the $(N+1) \times (N+1)$ matrix A from Example 22.12.

Consider the eigenvalue $\lambda_0 = d - \alpha_1$. In Example 22.12 it is shown that an eigenvector corresponding to this eigenvalue is

$$\begin{pmatrix} -\frac{S}{\alpha_1} & 1 & 1 & \dots & 1 \end{pmatrix}^T, \tag{24.50}$$

with $S = \alpha_1 + \alpha_2 + \dots + \alpha_N$ given by (22.41).

In Example 22.12 it is proved that the matrix A does not satisfy for this eigenvalue the conditions of Theorem 22.2 and therefore the eigenvector (24.50) cannot be obtained using this theorem. In what follows it is shown that Theorem 24.8 gives the desired result.

Indeed, for $m = 2$ it follows that $d(m) - \lambda_0 = d - (d - \alpha_1) = \alpha_1 \neq 0$ and

$$\delta_m(\lambda_0) = \delta_2(d - \alpha_1) = (d(2) - \lambda_0)b(2) - g(2)h(2) = \alpha_1 - \alpha_1 = 0,$$

while

$$l_m(\lambda_0) = l_2(d - \alpha_1) = (d(2) - \lambda_0)a(2) - p(2)q(2) = \alpha_1 - \alpha_2 \neq 0.$$

Moreover,

$$A_m(\lambda_0) = A(1:m, 1:m) - \lambda_0 I = A_2(d - \alpha_1) = \begin{pmatrix} \alpha_1 & \alpha_1 \\ \alpha_1 & \alpha_1 \end{pmatrix}; \tag{24.51}$$

therefore, $\gamma_m(\lambda_0) = \det A_m(\lambda_0) = 0$ and (24.38) is satisfied.

The theorem then assures that if $\lambda_0 = d - \alpha_1$ is an eigenvalue of A and $\delta_2(d-\alpha_1) = 0$, then $x = \begin{pmatrix} x_1 \\ \theta \\ x_2 \end{pmatrix}$ is an eigenvector of the matrix A corresponding to this eigenvalue if and only if there exist numbers θ_1, θ_2 such that $\theta_1 + \theta_2 = \theta$,

$$A_2(d-\alpha_1)\begin{pmatrix} x_1 \\ \theta_1 \end{pmatrix} = 0, \quad B_2(d-\alpha_1)\begin{pmatrix} \theta_2 \\ x_2 \end{pmatrix} = -\begin{pmatrix} 0 \\ P_3 \end{pmatrix} \frac{l_2(d - \alpha_1)}{d(2) - (d - \alpha_1)}(Q_1 x_1),$$

and at least one of the vectors $\begin{pmatrix} x_1 \\ \theta_1 \end{pmatrix}$, $\begin{pmatrix} \theta_2 \\ x_2 \end{pmatrix}$ is not zero.

In order to obtain from Theorem 24.8 the eigenvector (24.50) one must solve the first equation above, for instance take $x_1 = -\frac{S}{\alpha_1}$, and then from (24.51) it follows that $\theta_1 = \frac{S}{\alpha_1}$.

According to Theorem 24.8, the right-hand side of the second equation above must be equal to

$$-\begin{pmatrix} 0 \\ P_3 \end{pmatrix} \frac{l_2(d - \alpha_1)}{d(2) - (d - \alpha_1)}(Q_1 x_1).$$

A computation gives $\frac{l_2(d-\alpha_1)}{d(2)-(d-\alpha_1)} = \frac{\alpha_1 - \alpha_2}{\alpha_1}$ and, by (5.1) and (5.2),

$$P_3 = \begin{pmatrix} 1 & 1 & \dots & 1 \end{pmatrix}^T, \quad Q_1 = \alpha_1.$$

Hence, the right-hand side with $x_1 = -\frac{S}{\alpha_1}$ is

$$\begin{pmatrix} 0 \\ \frac{\alpha_1 - \alpha_2}{\alpha_1} S \\ \frac{\alpha_1 - \alpha_2}{\alpha_1} S \\ \vdots \\ \frac{\alpha_1 - \alpha_2}{\alpha_1} S \end{pmatrix}. \tag{24.52}$$

Applying the matrix

$$B_2(d-\alpha_1) = \begin{pmatrix} \alpha_1 & \alpha_2 & \alpha_3 & \cdots & \alpha_{N-1} & \alpha_N \\ \alpha_2 & \alpha_1 & \alpha_3 & \cdots & \alpha_{N-1} & \alpha_N \\ \alpha_2 & \alpha_3 & \alpha_1 & \cdots & \alpha_{N-1} & \alpha_N \\ \vdots & \vdots & \vdots & \ddots & \vdots & \vdots \\ \alpha_2 & \alpha_3 & \alpha_4 & \cdots & \alpha_1 & \alpha_N \\ \alpha_2 & \alpha_3 & \alpha_4 & \cdots & \alpha_N & \alpha_1 \end{pmatrix}$$

to the vector $\begin{pmatrix} \theta_2 \\ x_2 \end{pmatrix}$ gives (24.52) if $\theta_2 = -\frac{\alpha_2+\alpha_3+\cdots+\alpha_N}{\alpha_1}$ and x_2 has all its $N-1$ components equal to 1.

Indeed, the result is

$$\begin{pmatrix} 0 \\ -\frac{\alpha_2(\alpha_2+\cdots+\alpha_N)}{\alpha_1} + \alpha_1 + \alpha_3 + \alpha_4 + \cdots + \alpha_{N-1} + \alpha_N \\ -\frac{\alpha_2(\alpha_2+\cdots+\alpha_N)}{\alpha_1} + \alpha_3 + \alpha_1 + \alpha_4 + \cdots + \alpha_{N-1} + \alpha_N \\ \vdots \\ -\frac{\alpha_2(\alpha_2+\cdots+\alpha_N)}{\alpha_1} + \alpha_3 + \alpha_4 + \alpha_5 + \cdots + \alpha_N + \alpha_1 \end{pmatrix}$$

$$= \begin{pmatrix} 0 \\ -\frac{\alpha_2}{\alpha_1}(S-\alpha_1) + S - \alpha_2 \\ -\frac{\alpha_2}{\alpha_1}(S-\alpha_1) + S - \alpha_2 \\ \vdots \\ -\frac{\alpha_2}{\alpha_1}(S-\alpha_1) + S - \alpha_2 \end{pmatrix}.$$

Therefore, all the components of the result except the first one are equal to

$$\frac{-\alpha_2 S + \alpha_1\alpha_2 + \alpha_1 S - \alpha_1\alpha_2}{\alpha_1} = \frac{\alpha_1 - \alpha_2}{\alpha_1} S,$$

and so (24.52) holds.

Since $\theta = \theta_1 + \theta_2 = \frac{\alpha_1}{\alpha_1} = 1$, it follows that the eigenvector of A given by Theorem 24.8 is the eigenvector (24.50). ◊

§24.4 Comments

The presentation of results here follows [24].

Part VI

Divide and Conquer Method for the Eigenproblem

Introduction to Part VI

The method for the eigenvalue decomposition we study here consists of two steps, the divide step and the conquer step. Accordingly, the first two chapters of the present part are dedicated each to one of these two steps. In the third chapter we apply and use quasiseparable generators for the divide and conquer method for Hermitian matrices, especially order one quasiseparable Hermitian matrices. This method is well known for tridiagonal matrices, which are an example of quasiseparable matrices of order one. The last section is devoted to the divide and conquer method for unitary Hessenberg matrices.

The divide step consists in splitting a single problem into two smaller independent problems with size roughly half the size of the original problem. This is done recursively, until the obtained problems are of a convenient size, which is small enough so that they can be solved by standard techniques. We perform the divide step for any square matrix by using its quasiseparable representation and we obtain the quasiseparable generators of the resulting two smaller matrices. The order of the quasiseparable generators does not increase in the process.

In the case when the initial matrix belongs to one of certain important classes of matrices, like Hermitian, tridiagonal matrices, companion matrices, or unitary Hessenberg matrices, we also show that the two smaller matrices also belong to the same class.

After the division step of the algorithm is completed and the eigenstructure of the smallest matrices has been found, we perform the conquer step, in which the division tree is climbed back and we obtain the eigenstructure of a larger matrix A from knowledge of the eigenstructure of two smaller matrices B and C. To do this we need to compute the eigenstructure of a small sized matrix function with size equal to the order of perturbation. While performing the conquer step we impose increasingly restrictive conditions in order to obtain more results. Complete algorithms for computing the eigenvalue decomposition are obtained for quasiseparable of order one Hermitian matrices and for unitary upper Hessenberg matrices.

Chapter 25

Divide Step

The divide step consists in splitting a single problem into two smaller independent problems with half the size of the original problem. This is done recursively, until the obtained problems are of a convenient size which is small enough so that they can be solved by standard techniques.

Consider a large matrix A which is in quasiseparable representation. We write A in the form

$$A = \begin{pmatrix} B & 0 \\ 0 & C \end{pmatrix} + GH$$

with smaller sized matrices B and C and a perturbation matrix GH of small rank that depends on the order of quasiseparability. The matrices B and C have in turn at most the same order of quasiseparability and can therefore be divided further in the same way, until one obtains small enough matrices for which the eigenproblem can be solved conveniently. The inheritance of the structure by the smaller matrices B, C holds also for important subclasses, such as diagonal plus semiseparable, tridiagonal and unitary Hessenberg matrices, and for some particular cases.

We will apply the divide step to an initial matrix with quasiseparable structure. In order to specify the divide step of a divide and conquer algorithm we must show that, once completed, the resulting divided and thus smaller matrices have the same quasiseparable order as the initial matrix, and develop an algorithm to compute quasiseparable generators of these smaller matrices. Preserving this quasiseparable structure and order permits us to apply recursively the divide step to the obtained smaller matrices.

§25.1 The divide step via quasiseparable generators

We can assume that for an $N \times N$ matrix the orders of the lower and of the upper quasiseparable generators are the same, $r_k^L = r_k^U$ $(k = 1, \dots, N-1)$, since otherwise one can pad the smaller ones with zeroes. It follows that we can ask this as a condition for the following theorem, without loss of generality.

Theorem 25.1. *Let m, N be two positive integers such that $m < N$ and $A = \{A_{ij}\}_{i,j=1}^{N}$ be a block matrix with entries of sizes $m_i \times m_j$, with lower quasiseparable generators $p(i)$ $(i = 2, \dots, N)$, $q(j)$ $(j = 1, \dots, N-1)$, $a(k)$ $(k = 2, \dots, N-1)$ of orders r_k $(k = 1, \dots, N-1)$, upper quasiseparable generators $g(i)$ $(i = 1, \dots, N-1)$, $h(j)$ $(j = 2, \dots, N)$, $b(k)$ $(k = 2, \dots, N-1)$ of the same orders r_k $(k = 1, \dots, N-1)$, and diagonal entries $d(k)$ $(k = 1, \dots, N)$.*

Then the matrix A is a perturbation of rank at most r_m of a 2×2 block diagonal matrix

$$\begin{pmatrix} B & 0 \\ 0 & C \end{pmatrix} \tag{25.1}$$

with submatrices B of size $m \times m$ and C of size $(N-m) \times (N-m)$ which have quasiseparable generators of orders r_k, $k = 1, \dots, m-1$ and of orders r_k, $k = m+1, \dots, N-1$, respectively.

In fact, using quasiseparable generators define the matrices P_{m+1}, Q_m, G_m, H_{m+1} via the relations

$$Q_m = \mathrm{row}(a^{>}_{m+1,k} q(k))_{k=1}^{m}, \quad G_m = \mathrm{col}(g(k) b^{<}_{k,m+1})_{k=1}^{m}, \tag{25.2}$$

and

$$P_{m+1} = \mathrm{col}(p(k) a^{>}_{km})_{k=m+1}^{N}, \; H_{m+1} = \mathrm{row}(b^{<}_{mk} h(k))_{k=m+1}^{N}, \tag{25.3}$$

and taking some invertible $r_m \times r_m$ matrices α, β one can represent the matrix A in the form

$$A = \begin{pmatrix} B & 0 \\ 0 & C \end{pmatrix} + V_1 V_2, \tag{25.4}$$

where

$$V_1 = \begin{pmatrix} G_m \beta \\ P_{m+1} \alpha \end{pmatrix}, \quad V_2 = \begin{pmatrix} \alpha^{-1} Q_m & \beta^{-1} H_{m+1} \end{pmatrix}, \tag{25.5}$$

and

$$\begin{aligned} B = B_m &= A(1:m, 1:m) - (G_m \beta)(\alpha^{-1} Q_m), \\ C = C_m &= A(m+1:N, m+1:N) - (P_{m+1}\alpha)(\beta^{-1} H_{m+1}). \end{aligned} \tag{25.6}$$

Moreover, the matrix B has quasiseparable generators

$$p_B(i) = p(i) - g(i) b^{<}_{i,m+1} \beta \alpha^{-1} a^{>}_{m+1,i-1}, \qquad i = 2, \dots, m, \tag{25.7}$$

$$q_B(j) = q(j),\ j = 1, \dots, m-1,\ a_B(k) = a(k),\ k = 2, \dots, m-1;$$

$$g_B(i) = g(i), \qquad i = 1, \dots, m-1,$$

$$h_B(j) = h(j) - b^{<}_{j-1,m+1} \beta \alpha^{-1} a^{>}_{m+1,j} q(j), \qquad j = 2, \dots, m, \tag{25.8}$$

$$b_B(k) = b(k) \qquad k = 2, \dots, m-1;$$

$$d_B(k) = d(k) - g(k) b^{<}_{k,m+1} \beta \alpha^{-1} a^{>}_{m+1,k} q(k), \qquad k = 1, \dots, m \tag{25.9}$$

of orders r_k, $k=1,\dots,m-1$, *and the matrix* C *has the quasiseparable generators*

$$\begin{aligned}
p_C(i-m) &= p(i), && i=m+2,\dots,N, && (25.10)\\
q_C(j-m) &= q(j)-a^{>}_{j+1,m}\alpha\beta^{-1}b^{<}_{mj}h(j), && j=m+1,\dots,N-1, && (25.11)\\
a_C(k-m) &= a(k), && k=m+2,\dots,N-1; && (25.12)\\
g_C(i-m) &= g(i)-p(i)a^{>}_{im}\alpha\beta^{-1}b^{<}_{m,i+1}, && i=m+1,\dots,N-1, && (25.13)\\
h_C(j-m) &= h(j), && j=m+2,\dots,N, && (25.14)\\
b_C(k-m) &= b(k), && k=m+2,\dots,N-1, && (25.15)\\
d_C(k-m) &= d(k)-p(k)a^{>}_{km}\alpha\beta^{-1}b^{<}_{mk}h(k), && k=m+1,\dots,N, && (25.16)
\end{aligned}$$

of orders r_k, $k=m+1,\dots,N-1$.

Proof. If α is an invertible $r_m\times r_m$ matrix, then using (5.9) we get

$$A(m+1:N,1:m)=P_{m+1}\alpha\alpha^{-1}Q_m,\quad m=1,\dots,N-1. \tag{25.17}$$

If β is an invertible $r_m\times r_m$ matrix, then using (5.12) we get

$$A(1:m,m+1:N)=G_m\beta\beta^{-1}H_{m+1},\quad m=1,\dots,N. \tag{25.18}$$

It follows that the matrix A may be partitioned as

$$A=\begin{pmatrix} A(1:m,1:m) & G_m\beta\beta^{-1}H_{m+1}\\ P_{m+1}\alpha\alpha^{-1}Q_m & A(m+1:N,m+1:N)\end{pmatrix}. \tag{25.19}$$

Using (25.19), one can represent A in the form

$$A=\begin{pmatrix} B & 0\\ 0 & C\end{pmatrix}+\begin{pmatrix} G_m\beta\\ P_{m+1}\alpha\end{pmatrix}\begin{pmatrix}\alpha^{-1}Q_m & \beta^{-1}H_{m+1}\end{pmatrix}, \tag{25.20}$$

where B and C are given by (25.6). Thus we have represented the matrix A as a sum of a block diagonal 2×2 matrix and a matrix of rank at most r_m.

It remains to show that the matrix B has quasiseparable generators of orders r_k, $k=1,\dots,m-1$ and the matrix C has quasiseparable generators of orders r_k, $k=m+1,\dots,N-1$, and obtain the formulas for these generators. We will proceed first for the matrix B. We have

$$B(i,j)=A(i,j)-G_m(i,1:r_m)\beta\alpha^{-1}Q_m(1:r_m,j),\quad 1\le i,j\le m. \tag{25.21}$$

For $1\le j<i\le m$ we have

$$B(i,j)=p(i)a^{>}_{ij}q(j)-g(i)b^{<}_{i,m+1}\beta\alpha^{-1}a^{>}_{m+1,j}q(j).$$

Using the equality (4.7) we get

$$a^{>}_{m+1,j}=a^{>}_{m+1,i-1}a^{>}_{ij},$$

and therefore

$$B(i,j) = (p(i) - g(i)b^{<}_{i,m+1}\beta\alpha^{-1}a^{>}_{m+1,i-1})a^{>}_{ij}q(j), \quad 1 \leq j < i \leq m.$$

Thus the matrix B has lower quasiseparable generators

$$p(i) - g(i)b^{<}_{i,m+1}\beta\alpha^{-1}a^{>}_{m+1,i-1} \quad (i = 2,\dots,m),$$
$$q(j) \ \ (j = 1,\dots,m-1), \quad a(k) \ \ (k = 2,\dots,m-1)$$

of orders r_k, $k = 1,\dots,m-1$.

Similarly, for $1 \leq j < i \leq m$ we have

$$B(i,j) = g(i)b^{<}_{ij}h(j) - g(i)b^{<}_{i,m+1}\beta\alpha^{-1}a^{>}_{m+1,j}q(j).$$

Using the formula (4.8) we get

$$b^{<}_{i,m+1} = b^{<}_{ij}b^{<}_{j-1,m+1},$$

and therefore

$$B(i,j) = g(i)b^{<}_{ij}(h(j) - b^{<}_{j-1,m+1}\beta\alpha^{-1}a^{>}_{m+1,j}q(j)), \quad 1 \leq i < j \leq m.$$

Thus the matrix B has upper quasiseparable generators $g(i)$ $(i = 1,\dots,m-1)$, $h(j) - b^{<}_{j-1,m+1}\beta\alpha^{-1}a^{>}_{m+1,j}q(j)$ $(j = 2,\dots,m)$, $b(k)$ $(k = 2,\dots,m-1)$ of orders r_k, $k = 1,\dots,m-1$.

Next, using (25.21) with $i = j = k$ we obtain the formula (25.9) for the diagonal entries of the matrix B.

In the same way we get that the matrix C has quasiseparable generators of orders r_k, $k = m+1,\dots,N-1$, and we obtain the formulas for these generators. We have

$$C(i-m,j-m) = A(i,j) - P_{m+1}(i,1:r_m)\alpha\beta^{-1}H_{m+1}(1:r_m,j), \quad m+1 \leq i,j \leq N. \tag{25.22}$$

For $m+1 \leq j < i \leq N$ we have

$$C(i-m,j-m) = p(i)a^{>}_{ij}q(j) - p(i)a^{>}_{im}\alpha\beta^{-1}b^{<}_{mj}h(j).$$

Using the equality (4.7) we get

$$a^{>}_{im} = a^{>}_{ij}a^{>}_{j+1,m},$$

and therefore

$$C(i-m,j-m) = p(i)a^{>}_{ij}(q(j) - a^{>}_{j+1,m}\alpha\beta^{-1}b^{<}_{mj}h(j)), \quad m+1 \leq j < i \leq N.$$

Thus the matrix C has lower quasiseparable generators of orders r_k, $k = m+1,\dots,N-1$, defined by (25.10)–(25.12).

Similarly, for $m+1 \le j < i \le N$ we have

$$C(i-m, j-m) = g(i)b^{<}_{ij}h(j) - p(i)a^{>}_{im}\alpha\beta^{-1}b^{<}_{mj}h(j).$$

Using the formula (4.8) we get

$$b^{<}_{mj} = b^{<}_{m,i+1}b^{<}_{ij},$$

and therefore

$$C(i-m, j-m) = (g(i) - p(i)a^{>}_{im}\alpha\beta^{-1}b^{<}_{m,i+1})b^{<}_{ij}h(j), \quad m+1 \le i < j \le N.$$

Thus the matrix C has upper quasiseparable generators of orders r_k, $k = m+1, \dots, N-1$, defined by (25.13)–(25.15).

Next, using (25.22) with $i = j = k$ we obtain the formula (25.16) for the diagonal entries of the matrix C. □

The following algorithm obtains in an efficient manner quasiseparable generators of suitable orders for the matrices B and C.

Algorithm 25.2. Algorithm for suitable quasiseparable generators for the divided matrices and for building the perturbation matrices V_1, V_2

Let $A = \{A_{ij}\}_{i,j=1}^{N}$ be a block matrix with entries of sizes $m_i \times m_j$, with lower quasiseparable generators $p(i)$ $(i = 2, \dots, N)$, $q(j)$ $(j = 1, \dots, N-1)$, $a(k)$ $(k = 2, \dots, N-1)$ of orders r_k $(k = 1, \dots, N-1)$, upper quasiseparable generators $g(i)$ $(i = 1, \dots, N-1)$, $h(j)$ $(j = 2, \dots, N)$, $b(k)$ $(k = 2, \dots, N-1)$ of the same orders r_k $(k = 1, \dots, N-1)$, and diagonal entries $d(k)$ $(k = 1, \dots, N)$. Let the matrices B and C be given by (25.6).

Then a set of quasiseparable generators $p_B(i)$ $(i = 2, \dots, m)$, $q_B(j)$ $(j = 1, \dots, m-1)$, $a_B(k)$ $(k = 2, \dots, m-1)$; $g_B(i)$ $(i = 1, \dots, m-1)$, $h_B(j)$ $(j = 2, \dots, m)$, $b_B(k)$ $(k = 2, \dots, m-1)$; $d_B(k)$ $(k = 1, \dots, m)$ of the matrix B, and a set of quasiseparable generators $p_C(i)$ $(i = 2, \dots, N-m)$, $q_C(j)$ $(j = 1, \dots, N-m-1)$, $a_C(k)$ $(k = 2, \dots, N-m-1)$; $g_C(i)$ $(i = 1, \dots, N-m-1)$, $h_C(j)$ $(j = 2, \dots, N-m)$, $b_C(k)$ $(k = 2, \dots, N-m-1)$; $d_C(k)$ $(k = 1, \dots, N-m)$ of the matrix C which have the same orders as the generators of the matrix A, as well as perturbation matrices V_1, V_2 are obtained with the following algorithm.

1. Find the quasiseparable generators of B and the entries $V_1(i)$, $V_2(i)$, $i = 1, \dots, m$.

1.1. Compute

$$u_m = \alpha^{-1}a(m), \quad v_m = b(m)\beta, \tag{25.23}$$

$$z_m = g(m)\beta, \quad w_m = \alpha^{-1}q(m), \tag{25.24}$$

$$\begin{aligned} d_B(m) &= d(m) - z_m w_m, \quad p_B(m) = p(m) - z_m u_m, \\ h_B(m) &= h(m) - v_m w_m, \end{aligned} \tag{25.25}$$

and set

$$V_2(m) = w_m, \quad V_1(m) = z_m. \tag{25.26}$$

1.2. For $k = m-1, \dots, 2$ compute

$$u_k = u_{k+1}a(k), \quad v_k = b(k)v_{k+1}, \tag{25.27}$$

$$z_k = g(k)v_{k+1}, \quad w_k = u_{k+1}q(k), \tag{25.28}$$

$$\begin{aligned} d_B(k) &= d(k) - z_k w_k, \quad p_B(k) = p(k) - z_k u_k, \\ h_B(k) &= h(k) - v_k w_k, \end{aligned} \tag{25.29}$$

and set

$$V_2(k) = w_k, \quad V_1(k) = z_k \tag{25.30}$$

$$q_B(k) = q(k), \quad a_B(k) = a(k), \quad b_B(k) = b(k), \quad g_B(k) = g(k). \tag{25.31}$$

1.3. Compute

$$z_1 = g(1)v_2, \qquad w_1 = u_2 q(1), \tag{25.32}$$

$$d_B(1) = d(1) - z_1 w_1, \tag{25.33}$$

and set

$$V_1(1) = z_1, \qquad V_2(1) = w_1, \tag{25.34}$$

$$q_B(1) = q(1), \quad g_B(1) = g(1). \tag{25.35}$$

2. Find the quasiseparable generators of C and the entries $V_1(i)$, $V_2(i)$, $i = m+1, \dots, N$.

2.1. Compute

$$t = m+1, \quad u_t = a(t)\alpha, \quad v_t = \beta^{-1}b(t), \tag{25.36}$$

$$z_t = p(t)\alpha, \quad w_t = \beta^{-1}h(t), \tag{25.37}$$

$$d_C(1) = d(t) - z_t w_t, \quad q_C(1) = q(t) - u_t w_t, \quad g_C(1) = g(t) - w_t v_t, \tag{25.38}$$

and set

$$V_1(t) = z_t, \quad V_2(t) = w_t. \tag{25.39}$$

2.2. For $k = m+2, \dots, N-1$ compute

$$s = k - m, \quad u_k = a(k)u_{k-1}, \quad v_k = v_{k-1}b(k), \tag{25.40}$$

$$z_k = p(k)u_{k-1}, \quad w_k = v_{k-1}h(k), \tag{25.41}$$

$$q_C(s) = q(k) - u_k w_k, \quad d_C(s) = d(k) - z_k w_k, \quad g_C(s) = g(k) - z_k v_k, \tag{25.42}$$

and set

$$V_1(k) = z_k, \quad V_2(k) = w_k, \tag{25.43}$$

$$a_C(s) = a(k), \quad p_C(s) = p(k), \quad b_C(s) = b(k), \quad h_C(s) = h(k). \tag{25.44}$$

2.3. Compute

$$z_N = p(N)u_{N-1}, \qquad w_N = v_{N-1}h(N), \tag{25.45}$$
$$s = N - m, \qquad d_C(s) = d(N) - z_N w_N, \tag{25.46}$$

and set

$$V_1(N) = z_N, \qquad V_2(N) = w_N, \tag{25.47}$$
$$p_C(s) = p(N), \qquad h_C(s) = h(N). \tag{25.48}$$

Proof. To obtain a fast algorithm based on the formulas from Theorem 25.1, we introduce the auxiliary variables

$$u_k = \alpha^{-1}a^{>}_{m+1,k-1}, \; v_k = b^{<}_{k-1,m+1}\beta, \quad k = m, \dots, 2, \tag{25.49}$$
$$u_k = a^{>}_{k+1,m}\alpha, \qquad v_k = \beta^{-1}b^{<}_{m,k+1}, \quad k = m+1, \dots, N-1. \tag{25.50}$$

Using (4.7), (4.8) one can easily check that these auxiliary variables satisfy the recurrence relations (25.23), (25.27) and (25.36), (25.40).

Moreover, inserting (25.49) in (25.7), (25.8), (25.9) we obtain the relations (25.25), (25.29), (25.33) to determine generators of the matrix B with the auxiliary variables z_k, w_k defined in (25.24), (25.28), (25.32). Furthermore, using the formulas (25.5) and (25.2) we obtain the formulas (25.26), (25.30), (25.34) for the matrices $V_1(1:m,:) = G_m\beta$, $V_2(:,1:m) = \alpha^{-1}Q_m$.

Similarly, inserting (25.50) in (25.11), (25.13), (25.16) we obtain the relations (25.38), (25.42), (25.46) to determine generators of the matrix C with the auxiliary variables z_k, w_k defined in (25.37), (25.41), (25.45). Furthermore, using the formulas (25.5) and (25.3) we obtain the formulas (25.39), (25.43), (25.47) for the matrices $V_1(m+1:N,:) = P_{m+1}\alpha$, $V_2(:,m+1:N) = \beta^{-1}H_{m+1}$. □

In order to compute the complexity of the Algorithm 25.2 we will constrain ourselves to a detailed analysis of Steps 1.2 and 2.2, which are performed not just once, but $m-2$ and respectively $N-m-2$ times, thus about $N/2$ times each. We will make use of the fact that the multiplication of a $k \times n$ matrix by an $n \times s$ matrix requires kns arithmetical multiplications and $k(n-1)s$ arithmetical additions, thus a total of $k(2n-1)s$ arithmetical operations.

In Step 1.2, formulas (25.27)–(25.29) perform seven matrix multiplications and three matrix subtractions, which require

$$r_m(2r_k-1)m_k + r_m(2r_k-1)r_{k-1} + m_k(2r_k-1)r_m + m_k(2r_m-1)m_k$$
$$+ m_k(2r_m-1)r_{k-1} + r_{k-1}(2r_k-1)r_m + r_{k-1}(2r_m-1)m_k$$

and respectively $m_k^2 + m_k r_{k-1} + r_{k-1}m_k$ arithmetical operations.

If we denote $r = \max_{k=1,\dots,N-1} r_k$ and $n = \max_{k=1,\dots,N} m_k$, then Step 1.2 requires less than

$$(2r^3 + 4r^2 n + rn^2 - r^2 - rn)N$$

arithmetical operations, while Steps 1.1 and 1.3 require $4r^2n + 2rn^2$ and respectively $2r^3 + 4r^2n + n^2 - 2rn - r^2$ arithmetical operations only.

In Step 2.2, formulas (25.40)–(25.42) perform a number subtraction and seven matrix multiplications and three matrix subtractions, which require

$$m_k(2r_k - 1)r_m + r_m(2r_{k-1} - 1)m_k + r_k(2r_{k-1} - 1)r_m + r_k(2r_m - 1)m_k \\ + m_k(2r_m - 1)m_k + r_m(2r_{k-1} - 1)r_k + m_k(2r_m - 1)r_k$$

and respectively $r_km_k + m_k^2 + m_kr_k$ arithmetical operations.

Then Step 2.2 requires less than

$$(r^3 + 3r^2n + rn^2 - rn)N$$

arithmetical operations, while Steps 2.1 and 2.3 require $4r^2n + 2rn^2$ and respectively $2r^3 + 4r^2n + n^2 - 2rn - r^2 + 1$ arithmetical operations only.

Therefore, the overall complexity of Algorithm 25.2 is less than

$$(3r^3+7r^2n+2rn^2-r^2-2rn+1)N+4r^3+16r^2n+4rn^2-2r^2-4rn+2n^2+1. \quad (25.51)$$

We apply Theorem 25.1 or Corollary 27.1 below for m which is roughly speaking half of N, for instance $m = [\frac{N}{2}]$. Then we may apply the same procedure to the matrices B, C and so on. On each step we obtain the partition of the corresponding matrix in the form (25.20) with matrices with given quasiseparable generators. The number of the needed steps until we receive small, say 2×2, matrices that we are prepared to deal with in the conquer stage of the algorithm is $\log_2 N$. The cost of this step is $O(N)$ and the overall cost of all the steps of this algorithm is $O(N \log_2 N)$ operations.

§25.2 Diagonal plus semiseparable matrices

In the case of a diagonal plus semiseparable matrix A it follows from the next corollary to Theorem 25.1 that the divided smaller matrices B and C are also diagonal plus semiseparable of at most the same order.

Corollary 25.3. *Let m, N be two positive integers such that $m < N$ and $A = \{A_{ij}\}_{i,j=1}^N$ be a block matrix with entries of sizes $m_i \times m_j$ with lower semiseparable generators $p(i)$ $(i = 2, \dots, N)$, $q(j)$ $(j = 1, \dots, N-1)$ of order r, upper semiseparable generators $g(i)$ $(i = 1, \dots, N-1)$, $h(j)$ $(j = 2, \dots, N)$ of the same order r, and diagonal entries $d(k)$ $(k = 1, \dots, N)$.*

Then the matrix A is a perturbation of rank at most r of a 2×2 block diagonal matrix

$$\begin{pmatrix} B & 0 \\ 0 & C \end{pmatrix}, \quad (25.52)$$

with submatrices B of size $m \times m$ and C of size $(N-m) \times (N-m)$ which have semiseparable generators of order r.

In fact, if one takes two invertible $r \times r$ matrices α, β, one can represent A in the form

$$A = \begin{pmatrix} B & 0 \\ 0 & C \end{pmatrix} + V_1 V_2, \tag{25.53}$$

where

$$\begin{aligned} V_1 &= \begin{pmatrix} g(1)\beta & g(2)\beta & \dots & g(m)\beta & p(m+1)\alpha & \dots & p(N)\alpha \end{pmatrix}^T, \\ V_2 &= \begin{pmatrix} \alpha^{-1}q(1) & \alpha^{-1}q(2) & \dots & \alpha^{-1}q(m) & \beta^{-1}h(m+1) & \dots & \beta^{-1}h(N) \end{pmatrix}, \end{aligned} \tag{25.54}$$

and

$$\begin{aligned} B = B_m &= A(1:m, 1:m) - (g(i)q(j))_{i,j=1}^{m}, \\ C = C_m &= A(m+1:N, m+1:N) - (p(i+m)h(j+m))_{i,j=1}^{N-m}. \end{aligned} \tag{25.55}$$

Moreover, the matrix B has the semiseparable generators

$$\begin{aligned} p_B(i) &= p(i) - g(i)\beta\alpha^{-1}, & i &= 2, \dots, m, \\ q_B(j) &= q(j), & j &= 1, \dots, m-1, \\ g_B(i) &= g(i), & i &= 1, \dots, m-1, \\ h_B(j) &= h(j) - \beta\alpha^{-1}q(j), & j &= 2, \dots, m, \\ d_B(k) &= d(k) - g(k)\beta\alpha^{-1}q(k), & k &= 1, \dots, m \end{aligned}$$

of order r, and the matrix C has the semiseparable generators

$$\begin{aligned} p_C(i-m) &= p(i), & i &= m+2, \dots, N, \\ q_C(j-m) &= q(j) - \alpha\beta^{-1}h(j), & j &= m+1, \dots, N-1, \\ g_C(i-m) &= g(i) - p(i)\alpha\beta^{-1}, & i &= m+1, \dots, N-1, \\ h_C(j-m) &= h(j), & j &= m+2, \dots, N, \\ d_C(k-m) &= d(k) - p(k)\alpha\beta^{-1}h(k), & k &= m+1, \dots, N \end{aligned}$$

of the same order r.

Proof. One can regard a diagonal plus semiseparable matrix as a quasiseparable one, which has $a(k)$, $k = 1, 2, \dots, N$, and $b(k)$, $k = 1, 2, \dots, N$, equal to the identity matrix I_r. □

§25.3 Block tridiagonal matrices

Here we obtain the well-known result for tridiagonal matrices as a corollary of the main Theorem 25.1.

Corollary 25.4. *Let m, N be two positive integers such that $m < N$ and let $A = \{A_{ij}\}_{i,j=1}^{N}$ be a block tridiagonal matrix*

$$\begin{pmatrix} \gamma_1 & \beta_1 & 0 & \cdots & 0 & 0 & 0 \\ \alpha_1 & \gamma_2 & \beta_2 & \cdots & 0 & 0 & 0 \\ 0 & \alpha_2 & \gamma_3 & \cdots & 0 & 0 & 0 \\ \vdots & \vdots & \vdots & \ddots & \vdots & \vdots & \vdots \\ 0 & 0 & 0 & \cdots & \alpha_{N-2} & \gamma_{N-1} & \beta_{N-1} \\ 0 & 0 & 0 & \cdots & 0 & \alpha_{N-1} & \gamma_N, \end{pmatrix}, \tag{25.56}$$

where γ_k are $n_k \times n_k$ matrices, $k = 1, \dots, N$ and α_i, β_i are $n_{i+1} \times n_i$ and $n_i \times n_{i+1}$ matrices, respectively, $i = 1, \dots, N-1$. The matrix A has the block quasiseparable generators

$$\begin{aligned} p(i) &= \alpha_{i-1},\ i = 2, \dots, N, & q(j) &= I_{n_j}, & j &= 1, \dots, N-1, \\ & & a(k) &= 0_{n_{k+1} \times n_k}, & k &= 2, \dots, N-1, \\ g(j) &= I_{n_j},\ \ j = 1, \dots, N-1, & h(i) &= \beta_{i-1}, & i &= 2, \dots, N, \\ & & b(k) &= 0_{n_k \times n_{k+1}}, & k &= 2, \dots, N-1, \end{aligned} \tag{25.57}$$

and the diagonal entries $d(k) = \gamma_k$, $k = 1, \dots, N$.

Then the matrix A is a perturbation of block rank one of a 2×2 block diagonal matrix

$$\begin{pmatrix} B & 0 \\ 0 & C \end{pmatrix} \tag{25.58}$$

with block tridiagonal submatrices B of block size $m \times m$ and C of block size $(N-m) \times (N-m)$ which preserve the quasiseparable generators of block order one of A and differ from it only on the diagonal entries. Namely, taking some $n_m \times n_m$ size invertible matrices α, β, one can represent the matrix A in the form

$$A = \begin{pmatrix} B & 0 \\ 0 & C \end{pmatrix} + V_1 V_2, \tag{25.59}$$

where

$$V_1 = \begin{pmatrix} 0_{\chi_m \times n_m} \\ \beta \\ \alpha_m \alpha \\ 0_{\eta_m \times n_m} \end{pmatrix}, \quad V_2 = \begin{pmatrix} 0_{n_m \times \chi_m} & \alpha^{-1} & \beta^{-1}\beta_m & 0_{n_m \times \eta_m} \end{pmatrix}, \tag{25.60}$$

with $\chi_m = \sum_{i=1}^{m-1} n_i$, $\eta_m = \sum_{i=m+1}^{N} n_i$, and

$$B = \begin{pmatrix} \gamma_1 & \beta_1 & 0 & \cdots & 0 & 0 & 0 \\ \alpha_1 & \gamma_2 & \beta_2 & \cdots & 0 & 0 & 0 \\ 0 & \alpha_2 & \gamma_3 & \cdots & 0 & 0 & 0 \\ \vdots & \vdots & \vdots & \ddots & \vdots & \vdots & \vdots \\ 0 & 0 & 0 & \cdots & \alpha_{m-2} & \gamma_{m-1} & \beta_{m-1} \\ 0 & 0 & 0 & \cdots & 0 & \alpha_{m-1} & \gamma_m - \beta\alpha^{-1} \end{pmatrix}, \tag{25.61}$$

$$C = \begin{pmatrix} \gamma_{m+1} - \alpha_m \alpha \beta^{-1} \beta_m & \beta_{m+1} & 0 & \cdots & 0 & 0 & 0 \\ \alpha_{m+1} & \gamma_{m+2} & \beta_{m+2} & \cdots & 0 & 0 & 0 \\ 0 & \alpha_{m+2} & \gamma_{m+3} & \cdots & 0 & 0 & 0 \\ \vdots & \vdots & \vdots & \ddots & \vdots & \vdots & \vdots \\ 0 & 0 & 0 & \cdots & \alpha_{N-2} & \gamma_{N-1} & \beta_{N-1} \\ 0 & 0 & 0 & \cdots & 0 & \alpha_{N-1} & \gamma_N \end{pmatrix}, \tag{25.62}$$

i.e., the matrices B and C have lower and upper quasiseparable generators

$$\begin{aligned} p_B(i) &= \alpha_{i-1}, & i &= 2, \dots, m, \\ q_B(j) &= I_{n_j}, & j &= 1, \dots, m-1, \\ a_B(k) &= 0_{n_{k+1} \times n_k}, & k &= 2, \dots, m-1, \\ g_B(j) &= I_{n_j}, & j &= 1, \dots, m-1, \\ h_B(i) &= \beta_{i-1}, & i &= 2, \dots, m, \\ b_B(k) &= 0_{n_k \times n_{k+1}}, & k &= 2, \dots, m-1; \\ p_C(i-m) &= \alpha_{i-1}, & i &= m+2, \dots, N, \\ q_C(j-m) &= I_{n_j}, & j &= m+1, \dots, N-1, \\ a_C(k-m) &= 0_{n_{k+1} \times n_k}, & k &= m+2, \dots, N-1, \\ g_C(j-m) &= I_{n_j}, & j &= m+1, \dots, N-1, \\ h_C(i-m) &= \beta_{i-1}, & i &= m+2, \dots, N, \\ b_C(k-m) &= 0_{n_k \times n_{k+1}}, & k &= m+2, \dots, N-1 \end{aligned} \tag{25.63}$$

and diagonal entries

$$\begin{aligned} d_B(k) &= \gamma_k,\ k = 1, \dots, m-1, & d_B(m) &= \gamma_m - \beta \alpha^{-1}, \\ d_C(1) &= \gamma_{m+1} - \alpha_m \alpha \beta^{-1} \beta_m, & d_C(k-m) &= \gamma_k,\ k = m+2, \dots, N. \end{aligned} \tag{25.64}$$

Proof. Inserting zero matrices $a(k), b(k)$ in (25.2), (25.3) and in (25.7)–(25.16), we obtain the relations

$$\begin{aligned} P_{m+1} &= \mathrm{col}(p(k) a^{>}_{km})_{k=m+1}^{N} &&= \begin{pmatrix} \alpha_m \\ 0_{\eta_m \times n_m} \end{pmatrix}, \\ Q_m &= \mathrm{row}(a^{>}_{m+1,k} q(k))_{k=1}^{m} &&= \begin{pmatrix} 0_{n_m \times \chi_m} & I_{n_m} \end{pmatrix}, \\ G_m &= \mathrm{col}(g(k) b^{<}_{k,m+1})_{k=1}^{m} &&= \begin{pmatrix} 0_{\chi_m \times n_m} \\ I_{n_m} \end{pmatrix}, \end{aligned}$$

and

$$H_{m+1} = \mathrm{row}(b^{<}_{mk} h(k))_{k=m+1}^{N} = \begin{pmatrix} \beta_m & 0_{n_m \times \eta_m} \end{pmatrix},$$

and the formulas (25.63), (25.64) to determine quasiseparable generators of the matrices B and C. The formulas (25.63), (25.64) imply that the desired B and C satisfy (25.61) and (25.62). Thus B differs from the submatrix $A(1:m, 1:m)$ only on the entry $d(m)$, while C differs from the submatrix $A(m+1:N, m+1:N)$ only on the entry $d(m+1)$. It follows that the new matrices are also tridiagonal and that they preserve lower and upper quasiseparable generators and most of the diagonal of entries of A given in (25.57). Moreover, the perturbation matrices given in (25.5) are of block rank one and defined via (25.60). □

§25.4 Unitary upper Hessenberg matrices

Another important class of matrices invariant under the divide step is that of unitary upper Hessenberg matrices, defined in Section §7.4. Recall that such a matrix has the form

$$A = \begin{bmatrix} -\rho_1\rho_0^* & -\rho_2\mu_1\rho_0^* & -\rho_3\mu_2\mu_1\rho_0^* & \cdots & -\rho_{N-1}\mu_{N-2}\cdots\mu_1\rho_0^* & -\rho_N\mu_{N-1}\cdots\mu_1\rho_0^* \\ \mu_1 & -\rho_2\rho_1^* & -\rho_3\mu_2\rho_1^* & \cdots & -\rho_{N-1}\mu_{N-2}\cdots\mu_2\rho_1^* & -\rho_N\mu_{N-1}\cdots\mu_2\rho_1^* \\ 0 & \mu_2 & -\rho_3\rho_2^* & \cdots & -\rho_{N-1}\mu_{N-2}\cdots\mu_3\rho_2^* & -\rho_N\mu_{N-1}\cdots\mu_3\rho_2^* \\ 0 & 0 & \mu_3 & \cdots & -\rho_{N-1}\mu_{N-2}\cdots\mu_4\rho_3^* & -\rho_N\mu_{N-1}\cdots\mu_4\rho_3^* \\ \vdots & \vdots & \vdots & \ddots & \vdots & \vdots \\ 0 & 0 & 0 & \cdots & -\rho_{N-1}\rho_{N-2}^* & -\rho_N\mu_{N-1}\rho_{N-2}^* \\ 0 & 0 & 0 & \cdots & \mu_{N-1} & -\rho_N\rho_{N-1}^* \end{bmatrix}, \tag{25.65}$$

where

$$\mu_k \geq 0,\ |\rho_k|^2 + \mu_k^2 = 1,\ k = 1, \dots, N-1,\ \rho_0 = -1,\ |\rho_N| = 1. \tag{25.66}$$

For the matrix A in (25.65) we will use lower quasiseparable generators of order one:

$$p(i) = \mu_{i-1},\ i = 2, \dots, N,\ \ q(j) = 1,\ j = 1, \dots, N-1,\ \ a(k) = 0,\ k = 2, \dots, N-1, \tag{25.67}$$

upper quasiseparable generators of the same orders

$$\begin{aligned} &g(i) = -\rho_{i-1}^*,\ i = 1, \dots, N-1, \quad h(j) = \mu_{j-1}\rho_j,\ j = 2, \dots, N, \\ &b(k) = \mu_{k-1},\ \ k = 2, \dots, N-1, \end{aligned} \tag{25.68}$$

and diagonal entries

$$d(k) = -\rho_k\rho_{k-1}^*,\ k = 1, \dots, N. \tag{25.69}$$

The complex numbers $\rho_0, \rho_1, \dots, \rho_N$ are called *the Schur parameters* of the unitary upper Hessenberg matrix, while $\mu_1, \mu_2, \dots, \mu_{N-1}$ are called *the complementary parameters* of the matrix. The Schur parameters completely determine the matrix A and the first Schur parameter is always $\rho_0 = -1$, so that we can denote

$$A = A(\rho_1, \rho_2, \dots, \rho_N). \tag{25.70}$$

It is clear that if one of the subdiagonal entries (one of the complementary parameters) is null, then the unitary upper Hessenberg matrix A can be easily split there into two unitary upper Hessenberg matrices. That is why we can suppose that the matrix A is *unreduced*, i.e., no subdiagonal entry is zero.

The following result is a corollary to Theorem 25.1.

Theorem 25.5. *Let m, N be two positive integers such that $m < N$ and let A be an $N \times N$ unitary unreduced upper Hessenberg matrix with the Schur parameters $\rho_0, \rho_1, \ldots, \rho_N$ and complementary parameters $\mu_1, \mu_2, \ldots, \mu_{N-1}$.*

Then the matrix A is a perturbation of rank at most one of a 2×2 block diagonal matrix

$$\begin{pmatrix} B & 0 \\ 0 & C \end{pmatrix} \tag{25.71}$$

with unitary upper Hessenberg submatrices B of size $m \times m$ and C of size $(N - m) \times (N - m)$.

In fact, if $\mu_m \neq 0$ one can represent the matrix A in the form

$$A = \begin{pmatrix} B & 0 \\ 0 & C \end{pmatrix} + V_1 V_2, \tag{25.72}$$

with

$$V_1 = \begin{pmatrix} G_0\beta & \mu_m\alpha & 0_{N-m-1} \end{pmatrix}^T, \quad V_2 = \begin{pmatrix} 0_{m-1} & \alpha^{-1} & \beta^{-1}\mu_m H_0 \end{pmatrix}, \tag{25.73}$$

where

$$G_0 = \begin{pmatrix} -\mu_{m-1} \cdot \cdots \cdot \mu_3\mu_2\mu_1\rho_0^* \\ -\mu_{m-1} \cdot \cdots \cdot \mu_3\mu_2\rho_1^* \\ -\mu_{m-1} \cdot \cdots \cdot \mu_3\rho_2^* \\ \vdots \\ -\mu_{m-1}\rho_{m-2}^* \\ -\rho_{m-1}^* \end{pmatrix}, \tag{25.74}$$

$$H_0 = \begin{pmatrix} \rho_{m+1} & \rho_{m+2}\mu_{m+1} & \ldots & \rho_{N-1}\mu_{N-2}\cdots\mu_{m+2}\mu_{m+1} & \rho_N\mu_{N-1}\cdots\mu_{m+2}\mu_{m+1} \end{pmatrix} \tag{25.75}$$

and $\alpha, \beta \neq 0$ that satisfy

$$\beta\alpha^{-1} = \rho_m + \frac{\rho_m}{|\rho_m|}, \tag{25.76}$$

if $0 < \mu_m < 1$, and

$$|\alpha| = |\beta| \neq 0, \tag{25.77}$$

if $\mu_m = 1$. Here B, C are defined with the Schur parameters as in (25.70) via

$$B = B(\rho_1, \rho_2, \ldots, \rho_{m-1}, \tau), \tag{25.78}$$

$$C = C(\rho_{m+1}\tau^*, \rho_{m+2}\tau^*, \ldots, \rho_N\tau^*), \tag{25.79}$$

where

$$\tau = \begin{cases} -\frac{\rho_m}{|\rho_m|}, & \text{if} \quad 0 < \mu_m < 1 \\ -\frac{\beta}{\alpha}, & \text{if} \quad \mu_m = 1, \end{cases} \tag{25.80}$$

Thus in any of the cases one gets

$$B = \begin{bmatrix} -\rho_1\rho_0^* & -\rho_2\mu_1\rho_0^* & -\rho_3\mu_2\mu_1\rho_0^* & \cdots & -\rho_{m-1}\mu_{m-2}\cdots\mu_1\rho_0^* & -\tau\mu_{m-1}\cdots\mu_1\rho_0^* \\ \mu_1 & -\rho_2\rho_1^* & -\rho_3\mu_2\rho_1^* & \cdots & -\rho_{m-1}\mu_{m-2}\cdots\mu_2\rho_1^* & -\tau\mu_{m-1}\cdots\mu_2\rho_1^* \\ 0 & \mu_2 & -\rho_3\rho_2^* & \cdots & -\rho_{m-1}\mu_{m-2}\cdots\mu_3\rho_2^* & -\tau\mu_{m-1}\cdots\mu_3\rho_2^* \\ 0 & 0 & \mu_3 & \cdots & -\rho_{m-1}\mu_{m-2}\cdots\mu_4\rho_3^* & -\tau\mu_{m-1}\cdots\mu_4\rho_3^* \\ \vdots & \vdots & \vdots & \ddots & \vdots & \vdots \\ 0 & 0 & 0 & \cdots & -\rho_{m-1}\rho_{m-2}^* & -\tau\mu_{m-1}\rho_{m-2}^* \\ 0 & 0 & 0 & \cdots & \mu_{m-1} & -\tau\rho_{m-1}^* \end{bmatrix}, \tag{25.81}$$

$$C = \begin{bmatrix} \rho_{m+1}\tau^* & \rho_{m+2}\mu_{m+1}\tau^* & \rho_{m+3}\mu_{m+2}\mu_{m+1}\tau^* & \cdots & \rho_N\mu_{N-1}\cdots\mu_{m+1}\tau^* \\ \mu_{m+1} & -\rho_{m+2}\rho_{m+1}^* & -\rho_{m+3}\mu_{m+2}\rho_{m+1}^* & \cdots & -\rho_N\mu_{N-1}\cdots\mu_{m+2}\rho_{m+1}^* \\ 0 & \mu_{m+2} & -\rho_{m+3}\rho_{m+2}^* & \cdots & -\rho_N\mu_{N-1}\cdots\mu_{m+3}\rho_{m+2}^* \\ 0 & 0 & \mu_{m+3} & \cdots & -\rho_N\mu_{N-1}\cdots\mu_{m+4}\rho_{m+3}^* \\ \vdots & \vdots & \vdots & \ddots & \vdots \\ 0 & 0 & 0 & \cdots & -\rho_N\mu_{N-1}\rho_{N-2}^* \\ 0 & 0 & 0 & \cdots & -\rho_N\rho_{N-1}^* \end{bmatrix}. \tag{25.82}$$

Proof. We use Theorem 25.1 with quasiseparable generators (25.67), (25.68), (25.69). Inserting these generators in (25.2), (25.3), (25.5) we obtain the formulas (25.73)–(25.75) for the perturbation vectors V_1, V_2. Applying the formulas (25.7)–(25.16), we obtain quasiseparable generators of the matrices B, C by the recipes

$$\begin{aligned} p_B(i) &= p(i) = \mu_{i-1}, \ i = 2, \dots, m, \\ q_B(j) &= q(j) = 1, \quad j = 1, \dots, m-1, \\ a_B(k) &= a(k) = 0, \quad k = 2, \dots, m-1, \end{aligned} \tag{25.83}$$

$$\begin{aligned} g_B(i) &= g(i) = -\rho_{i-1}^*, & i &= 1, \dots, m-1, \\ h_B(j) &= h(j) = \mu_{j-1}\rho_j, & j &= 2, \dots, m-1, \\ h_B(m) &= h(m) - b(m)\beta\alpha^{-1}q(m) = \mu_{m-1}(\rho_m - \beta\alpha^{-1}), \\ b_B(k) &= b(k) = \mu_{k-1}, & k &= 2, \dots, m-1, \end{aligned} \tag{25.84}$$

$$\begin{aligned} d_B(k) &= d(k) = -\rho_k\rho_{k-1}^*, \ k = 1, \dots, m-1, \\ d_B(m) &= d(m) - g(m)\beta\alpha^{-1}q(m) = -\rho_{m-1}^*(\rho_m - \beta\alpha^{-1}), \end{aligned} \tag{25.85}$$

and

$$\begin{aligned} p_C(i-m) &= p(i) = \mu_{i-1}, \ i = m+2, \dots, N, \\ q_C(j-m) &= q(j) = 1, \quad j = m+1, \dots, N-1, \\ a_C(k-m) &= 0, \quad k = m+2, \dots, N-1, \end{aligned} \tag{25.86}$$

$$\begin{aligned} g_C(1) &= g(m+1) - p(m+1)\alpha\beta^{-1}b(m+1) = -(\rho_m^* + \alpha\beta^{-1}\mu_m^2), \\ g_C(i-m) &= g(i) = -\rho_{i-1}^*, & i &= m+2, \dots, N-1, \\ h_C(j-m) &= h(j) = \mu_{j-1}\rho_j, & j &= m+2, \dots, N, \\ b_C(k-m) &= b(k)\mu_k, & k &= m+2, \dots, N-1, \end{aligned} \tag{25.87}$$

$$\begin{aligned} d_C(1) &= d(m+1) - p(m+1)\alpha\beta^{-1}h(m+1) \\ &= -\rho_{m+1}(\rho_m^* + \mu_m^2\alpha\beta^{-1}), \\ d_C(k-m) &= d(k) = -\rho_k\rho_{k-1}^*, \; k = m+2, \dots, N. \end{aligned} \tag{25.88}$$

Set

$$\tau = \rho_m - \beta\alpha^{-1}, \quad \eta = -\rho_m^* - \alpha\beta^{-1}\mu_m^2. \tag{25.89}$$

For $0 < \mu_m < 1$, (25.76) implies

$$|\tau| = -\frac{\rho_m}{|\rho_m|}, \quad \eta = \tau^*. \tag{25.90}$$

In particular, the last equality takes place since

$$-\eta = \rho_m^* + (1-|\rho_m|^2)\frac{|\rho_m|}{\rho_m(1+|\rho_m|)} = \rho_m^* + \frac{(1-|\rho_m|)|\rho_m|}{\rho_m} = \frac{|\rho_m|}{\rho_m} = \frac{\rho_m^*}{|\rho_m|} = -\tau^*.$$

For $\mu_m = 1$ we have $\rho_m = 0$, and hence, using (25.77), we get

$$\tau = -\frac{\beta}{\alpha}, \quad \eta = -\frac{\alpha}{\beta} = \tau^*. \tag{25.91}$$

Thus, combining (25.83)–(25.88) and (25.89), (25.90), (25.91) we obtain (25.81), (25.82). Moreover using $|\tau| = 1$ we conclude that B, C are unitary matrices and that the relations (25.78) and (25.79) hold. □

§25.5 Divide step for some classes of matrices

Here we consider some special classes of matrices which are invariant under divide step

§25.5.1 Diagonal plus semiconstant and constant matrices

Consider the $N \times N$ matrix

$$A = \begin{pmatrix} d_1 & a & a & \cdots & a & a \\ b & d_2 & a & \cdots & a & a \\ b & b & d_3 & \cdots & a & a \\ \vdots & \vdots & \vdots & \ddots & \vdots & \vdots \\ b & b & b & \cdots & d_{N-1} & a \\ b & b & b & \cdots & a & d_N \rightarrow \end{pmatrix}, \tag{25.92}$$

where the entries are complex numbers.

We denote this matrix by $A = A(a, b, d_1, d_2, \ldots, d_N)$ and call it a *diagonal plus semiconstant matrix*. In the particular case where $a = b$, the matrix is called *diagonal plus constant*.

We will use for A the following semiseparable generators of order one:

$$\begin{aligned} p(i) &= b,\ i = 2, \ldots, N, & q(j) &= 1,\ j = 1, \ldots, N-1, \\ g(j) &= 1,\ j = 1, \ldots, N-1, & h(i) &= a,\ i = 2, \ldots, N, \end{aligned} \tag{25.93}$$

and the diagonal entries $d(k) = d_k,\ k = 1, \ldots, N$.

Applying Corollary 25.3 we obtain that the diagonal plus semiconstant matrix A admits the representation (25.4) with diagonal plus semiconstant submatrices $B = B(a-1, b-1, d_1 - 1, d_2 - 1, \ldots, d_m - 1)$ and $C = C(a - ab, b - ab, d_{m+1} - ab, d_{m+2} - ab, \ldots, d_N - ab)$ and perturbation vectors

$$\begin{aligned} V_1 &= \begin{pmatrix} 1 & 1 & \ldots & 1 & a & a & \ldots & a \end{pmatrix}^T \\ V_2 &= \begin{pmatrix} 1 & 1 & \ldots & 1 & b & b & \ldots & b \end{pmatrix}. \end{aligned}$$

We will consider now a complex number $a \neq 0, a \neq 1$ and the symmetric diagonal plus constant $N \times N$ matrix

$$A = A(a, a, d_1, d_2, \ldots, d_N) = \begin{pmatrix} d_1 & a & a & \cdots & a & a \\ a & d_2 & a & \cdots & a & a \\ a & a & d_3 & \cdots & a & a \\ \vdots & \vdots & \vdots & \ddots & \vdots & \vdots \\ a & a & a & \cdots & d_{N-1} & a \\ a & a & a & \cdots & a & d_N \end{pmatrix}, \tag{25.94}$$

but instead of the symmetric semiseparable generators which would have resulted from (25.93) we will use the following non-symmetric generators:

$$\begin{aligned} p(i) &= a,\ i = 2, \ldots, N, & q(j) &= 1,\ j = 1, \ldots, N-1, \\ g(j) &= a,\ j = 1, \ldots, N-1, & h(i) &= 1,\ i = 2, \ldots, N \end{aligned} \tag{25.95}$$

and the diagonal entries $d(k) = d_k,\ k = 1, \ldots, N$. These generators are not Hermitian since $g(j) = a \neq 1 = (q(j))^*,\ j = 1, \ldots, N-1$ and also $h(i) = 1 \neq a = (p(i))^*,\ i = 2, \ldots, N$.

We will compute the matrices B and C obtained by dividing A. These matrices will be different from the matrices B and C computed by making use of semiseparable representations (25.93). Using Corollary 25.3, it follows that the desired B and C satisfy

$$B = B_m = \text{diag}\ (d_1 - a, d_2 - a, \ldots, d_m - a)$$

and

$$C = \text{diag}\ (d_{m+1} - a, d_{m+2} - a, \ldots, d_N - a),$$

and the perturbation vectors are

$$V_1 = \begin{pmatrix} a & \dots & a \end{pmatrix}^T, \quad V_2 = \begin{pmatrix} 1 & \dots & 1 \end{pmatrix}.$$

We know therefore the spectral data of B and C and we will combine again the results for the two matrices in order to build the spectral data of A.

§25.5.2 Companion matrices

We will now show that a companion matrix A' is divided into two companion matrices B' and C'.

For the monic scalar polynomial $P(x) = x^N + \alpha_{N-1}x^{N-1} + \dots + \alpha_1 x + \alpha_0$ with complex coefficients α_k, $k = 0, \dots, N-1$, the companion $N \times N$ matrix associated to it is

$$A' = \begin{pmatrix} 0 & 0 & \dots & 0 & -\alpha_0 \\ 1 & 0 & \dots & 0 & -\alpha_1 \\ \vdots & \ddots & \ddots & & \vdots \\ 0 & 0 & \dots & 0 & -\alpha_{N-2} \\ 0 & 0 & \dots & 1 & -\alpha_{N-1} \end{pmatrix}. \tag{25.96}$$

We will use the quasiseparable generators

$$\begin{aligned} &p(i) = 1,\ i = 2, \dots, N, && q(j) = 1, \quad j = 1, \dots, N-1, \\ &a(k) = 0,\ k = 2, \dots, N-1, && g(j) = \alpha_{j-1},\ j = 1, \dots, N-1, \\ &h(i) = 0,\ i = 2, \dots, N-1, && h(N) = -1, \\ &b(k) = 1,\ k = 2, \dots, N-1, \end{aligned}$$

and the diagonal entries $d(k) = 0,\ k = 1, \dots, N-1,\ d(N) = -\alpha_{N-1}$.

Corollary 25.6. *A companion matrix A' is a perturbation of rank one of a 2×2 block diagonal matrix*

$$\begin{pmatrix} B' & 0 \\ 0 & C' \end{pmatrix},$$

where the submatrices B' and C' are the companion matrices of the polynomials

$$b(x) = x^m + \alpha_{m-1}x^{m-1} + \dots + \alpha_1 x + \alpha_0$$

and respectively

$$c(x) = x^{N-m} + \alpha_{N-1}x^{N-m-1} + \dots + \alpha_{m+2}x^2 + \alpha_{m+1}x + (\alpha_m - 1),$$

and the perturbation vectors are

$$\begin{aligned} V_1 &= \begin{pmatrix} \alpha_0 & \alpha_1 & \dots & \alpha_{m-1} & 1 & 0 & \dots & 0 \end{pmatrix}^T, \\ V_2 &= \begin{pmatrix} 0_{1\times(m-1)} & 1 & 0_{1\times(N-m-1)} & -1 \end{pmatrix}. \end{aligned} \tag{25.97}$$

Proof. It follows by (25.2), (25.3) that

$$P_{m+1} = \operatorname{col}(p(k)a^{>}_{km})_{k=m+1}^{N} = \begin{pmatrix} 1 \\ 0_{(N-m-1)\times 1} \end{pmatrix},$$
$$Q_m = \operatorname{row}(a^{>}_{m+1,k}q(k))_{k=1}^{m} = \begin{pmatrix} 0_{1\times(m-1)} & 1 \end{pmatrix},$$
$$G_m = \operatorname{col}(g(k)b^{<}_{k,m+1})_{k=1}^{m} = \begin{pmatrix} \alpha_0 & \alpha_1 & \dots & \alpha_{m-1} \end{pmatrix}^T,$$

and

$$H_{m+1} = \operatorname{row}(b^{<}_{mk}h(k))_{k=m+1}^{N} = \begin{pmatrix} 0_{1\times(N-m-1)} & -1 \end{pmatrix}.$$

Using (25.6) (with $\alpha = \beta = 0$) it follows that the desired B' satisfies

$$B' = B'_m = A'(1:m, 1:m) - G_m Q_m$$
$$= \begin{pmatrix} 0 & 0 & \dots & 0 & 0 \\ 1 & 0 & \dots & 0 & 0 \\ \vdots & \ddots & \ddots & & \vdots \\ 0 & 0 & \dots & 0 & 0 \\ 0 & 0 & \dots & 1 & 0 \end{pmatrix} - \begin{pmatrix} 0 & 0 & \dots & 0 & \alpha_0 \\ 0 & 0 & \dots & 0 & \alpha_1 \\ \vdots & \ddots & \ddots & & \vdots \\ 0 & 0 & \dots & 0 & \alpha_{m-2} \\ 0 & 0 & \dots & 0 & \alpha_{m-1} \end{pmatrix}, \tag{25.98}$$

which is the companion matrix of the polynomial

$$b(x) = x^m + \alpha_{m-1}x^{m-1} + \dots + \alpha_1 x + \alpha_0, \tag{25.99}$$

and that the desired C' satisfies

$$C' = C'_m = A'(m+1:N, m+1:N) - P_{m+1}H_{m+1}$$
$$= \begin{pmatrix} 0 & 0 & \dots & 0 & -\alpha_m \\ 1 & 0 & \dots & 0 & -\alpha_{m+1} \\ \vdots & \ddots & \ddots & & \vdots \\ 0 & 0 & \dots & 0 & -\alpha_{N-2} \\ 0 & 0 & \dots & 1 & -\alpha_{N-1} \end{pmatrix} - \begin{pmatrix} 0 & 0 & \dots & 0 & -1 \\ 0 & 0 & \dots & 0 & 0 \\ \vdots & \ddots & \ddots & & \vdots \\ 0 & 0 & \dots & 0 & 0 \\ 0 & 0 & \dots & 0 & 0 \end{pmatrix}, \tag{25.100}$$

which is the companion matrix of the polynomial

$$c(x) = x^{N-m} + \alpha_{N-1}x^{N-m-1} + \dots + \alpha_{m+2}x^2 + \alpha_{m+1}x + (\alpha_m - 1). \tag{25.101}$$

Moreover the perturbation defined in (25.5) is given by the following two vectors (of rank one):

$$V_1 = \begin{pmatrix} G_m \\ P_{m+1} \end{pmatrix} = \begin{pmatrix} \alpha_0 & \alpha_1 & \dots & \alpha_{m-1} & 1 & 0 & \dots & 0 \end{pmatrix}^T,$$
$$V_2 = \begin{pmatrix} Q_m & H_{m+1} \end{pmatrix} = \begin{pmatrix} 0_{1\times(m-1)} & 1 & 0_{1\times(N-m-1)} & -1 \end{pmatrix}. \qquad \square$$

§25.6 Comments

The algorithm for the divide step for quasiseparable matrices of arbitrary order was presented in [9]. The idea of using of the parameters α, β for symmetric tridiagonal matrices and taking $\alpha = \beta = \pm 1$ is well known for diagonal matrices, see, for instance, the monograph by G.L. Golub and C.F. Van Loan [29]. The more general treatment in Theorem 25.1 as well as application to unitary Hessenberg matrices in (25.5) appear here for the first time.

Chapter 26

Conquer Step and Eigenproblem for Rational Matrix Functions

In the conquer step, the solutions of the smaller problems into which a larger sized problem has been decomposed (or split) are successfully combined two by two to solutions of the next larger problem.

We want to find the solution to the eigenproblem for a larger matrix A when one knows the eigenvalues and the eigenvectors of the two smaller matrices B and C which resulted in the divide step from A due to its representation as

$$A = \begin{pmatrix} B & 0 \\ 0 & C \end{pmatrix} + V_1 V_2$$

and we also know the small rank perturbation matrix V_1V_2. We constrain ourselves to the cases where the matrix A as well as the matrices B and C are diagonalazible. The matrices with Jordan forms were studied in the paper [25].

We show that if A on one hand, and B and C on the other hand, have different eigenvalues, then in order to solve the eigenproblem for A one must solve the eigenproblem for a small sized rational matrix function. Its size is the rank of V_1V_2 and the matrix function is in fact a scalar function in the case of an order one quasiseparable matrix A.

§26.1 The link between the eigenproblem for A and an eigenproblem for a rational matrix function

Suppose that for the smaller divided matrices B and C of sizes $m \times m$ and respectively $(N-m) \times (N-m)$ we already have their spectral data, i.e., we have $m \times m$ and $(N-m) \times (N-m)$ invertible matrices P_B and respectively P_C, such that $P_B^{-1} B P_B = J_B$ and $P_C^{-1} C P_C = J_C$, where the matrices J_B and J_C are diagonal. We must compute the spectral data of the twice larger matrix A which satisfies (25.4) with the known $N \times r_m$ and respectively $r_m \times N$ matrices V_1, V_2.

Denote

$$U = \begin{pmatrix} P_B & 0 \\ 0 & P_C \end{pmatrix}.$$

Then U is invertible and

$$U^{-1}AU = U^{-1}\left(\begin{pmatrix} B & 0 \\ 0 & C \end{pmatrix} + V_1V_2\right)U = J + z_1z_2,$$

where

$$J = \begin{pmatrix} J_B & 0 \\ 0 & J_C \end{pmatrix},$$

while

$$z_1 = U^{-1}V_1, \quad z_2 = V_2U \tag{26.1}$$

are small-rank $N \times r_m$ and respectively $r_m \times N$ matrices.

We must now find an invertible V which brings the matrix

$$K = J + z_1z_2 \tag{26.2}$$

to its diagonal form, i.e., such that

$$V^{-1}(J + z_1z_2)V = J_A,$$

where J_A is a diagonal matrix. We then set $P = UV$ to obtain

$$P^{-1}AP = J_A.$$

Thus, we have to study the eigensystem of the matrix K defined in (26.2).

Consider the $r_m \times r_m$ matrix function

$$F(\lambda) = I_{r_m} - z_2(\lambda I_N - J)^{-1}z_1. \tag{26.3}$$

The complex number λ_0 is called a *zero* (or an *eigenvalue*) of the rational matrix function $F(\lambda)$ if $\det F(\lambda_0) = 0$, and $\phi \neq 0$ is called an *eigenvector* of $F(\lambda)$ corresponding to λ_0 if $F(\lambda_0)\phi = 0$.

We will show that the eigenproblem for the potentially large $N \times N$ matrix K can be reduced to the eigenproblem for a small sized $r_m \times r_m$ matrix function $F(\lambda)$. Once we found zeroes for $\det F(\lambda)$, eigenvectors for the small sized matrix which is obtained when we substitute a zero value in $F(\lambda)$ is all that we need.

Theorem 26.1. *Suppose that J is an $N \times N$ square matrix, z_1 is an $N \times r_m$ and z_2 is an $r_m \times N$ matrix and that the matrices J and $K = J + z_1z_2$ have no common eigenvalues.*

Then λ_0 is an eigenvalue of the $N \times N$ matrix K if and only if λ_0 is a zero of the function

$$F(\lambda) = I_{r_m} - z_2(\lambda I_N - J)^{-1}z_1;$$

Moreover, if ϕ_0 is a corresponding eigenvector of $F(\lambda)$, then

$$y_0 = (\lambda_0 I_N - J)^{-1} z_1 \phi_0 \tag{26.4}$$

is an eigenvector of K corresponding its eigenvalue λ_0.

Proof. Let λ_0 be a zero of $F(\lambda)$ and let $\phi_0 \neq 0$ be an eigenvector corresponding to λ_0. Then

$$\begin{aligned}(\lambda_0 I_N - K)(\lambda_0 I_N - J)^{-1} z_1 \phi_0 &= ((\lambda_0 I_N - J) - z_1 z_2)(\lambda_0 I_N - J)^{-1} z_1 \phi_0 \\ &= (I - z_1 z_2 (\lambda_0 I_N - J)^{-1}) z_1 \phi_0 = z_1 F(\lambda_0)\phi_0 = 0.\end{aligned}$$

In order to prove that $(\lambda_0 I_N - J)^{-1} z_1 \phi_0$ is an eigenvector of K and λ_0 is one of its eigenvalues, it remains only to prove that $(\lambda_0 I_N - J)^{-1} z_1 \phi_0 \neq 0$. Indeed, since $\phi_0 \neq 0$, it follows that

$$z_2 (\lambda_0 I_N - J)^{-1} z_1 \phi_0 = -F(\lambda_0)\phi_0 + I_{rm}\phi_0 = 0 + \phi_0 \neq 0,$$

and therefore

$$(\lambda_0 I_N - J)^{-1} z_1 \phi_0 \neq 0.$$

So we have proved that λ_0 is an eigenvalue of K and $(\lambda_0 I_N - J)^{-1} z_1 \phi_0$ is one of its eigenvectors.

Conversely, let now λ_0 be an eigenvalue of K. Since K and J have no common eigenvalues, it follows that $\lambda_0 I_N - J$ is invertible.

Let x_0 be an eigenvector of K corresponding to its eigenvalue λ_0. Then $(K - \lambda_0 I_N)x_0 = 0$, so that

$$(J - \lambda_0 I_N)x_0 = -z_1 z_2 x_0, \tag{26.5}$$

and it follows that

$$F(\lambda_0) z_2 x_0 = I_{rm} z_2 x_0 - z_2 (\lambda_0 I_N - J)^{-1} z_1 z_2 x_0$$

which by (26.5) is equal to

$$z_2 x_0 + z_2 (\lambda_0 I_N - J)^{-1} (J - \lambda_0 I_N) x_0 = 0.$$

In order to show that λ_0 is an eigenvalue of $F(\lambda)$ and that $z_2 x_0$ as an eigenvector, it is sufficient to prove that this vector is not zero. Indeed, since λ_0 is not an eigenvalue of J and $x_0 \neq 0$, we have $(\lambda_0 I_N - J)x_0 \neq 0$ and, by (26.5), $z_1 z_2 x_0 \neq 0$, whence $z_2 x_0 \neq 0$. □

If the matrices J and K in the above theorem have common eigenvalues, then all the other eigenvalues of K still correspond to eigenvalues of $F(\lambda)$, while the eigenvalues of K which were not found by solving the eigenproblem for $F(\lambda)$ are readily found among the eigenvalues of J.

§26.2 Order one quasiseparable scalar matrices

If A is an order one quasiseparable matrix, or at least $r_m = 1$, then the perturbations V_1 and V_2 from (25.5) are vectors and then z_1, z_2 defined in (26.1) are vectors too and the rational function $F(\lambda)$ from (26.3) is a scalar function. If A is a scalar order one quasiseparable matrix. then its generators are complex numbers. Assume also that J has no equal diagonal entries.

In this case

$$F(\lambda) = 1 + \frac{(z_2)_1(z_1)_1}{d_1 - \lambda} + \frac{(z_2)_2(z_1)_2}{d_2 - \lambda} + \cdots + \frac{(z_2)_N(z_1)_N}{d_N - \lambda}, \tag{26.6}$$

where $(z_1)_i, (z_2)$, $i = 1, \ldots, N$ are the components of the vectors z_1, z_2, and

$$d_N > d_{N-1} > d_{N-2} > \cdots > d_2 > d_1$$

are the distinct diagonal entries of the diagonal matrix J.

The next lemma gives a sufficient condition under which Theorem 26.1 holds true.

Lemma 26.2. *Let J be a diagonal $N \times N$ complex matrix and z_1, z_2 be vectors with N complex components each. Suppose that J has no equal diagonal entries and that z_1, z_2 have no zero components. Then J and the matrix $K = J + z_1 z_2$ given by* (26.2) *have no common eigenvalues.*

Proof. Suppose, on the contrary, that J and K have a common eigenvalue λ and that v is an eigenvector of K corresponding to this eigenvalue. Then

$$Kv = (J + z_1 z_2)v = \lambda v \tag{26.7}$$

and $v \neq 0$. Since J is a diagonal matrix, $J = \operatorname{diag}(d_1, d_2, \ldots, d_N)$, it follows that its eigenvalue λ is one of the entries d_i, $1 \leq i \leq N$. Then, if e_i is the corresponding vector in the standard basis of $\mathbb{C}^N$, we have that

$$e_i^*(J - d_i I)w = 0 \tag{26.8}$$

for any vector w. By (26.7),

$$0 = Kv - \lambda v = (J - \lambda I)v + z_1(z_2 v) = (J - d_i I)v + z_1(z_2 v), \tag{26.9}$$

and by (26.8)

$$0 = e_i^*((J - d_i I)v + z_1(z_2 v)) = 0 + e_i^* z_1(z_2 v) = (z_1)_i z_2 v, \tag{26.10}$$

where $(z_1)_i$ is the component i of the vector z_1, which cannot be zero, by the assumptions of the lemma. It follows from (26.10) that

$$z_2 v = 0. \tag{26.11}$$

But then (26.9) shows that $(J - d_i I)v = 0$, therefore v is also an eigenvector of J for the same eigenvalue λ. Hence, since eigenvalues of J are distinct, we get $v = \alpha e_i$ for a complex scalar $\alpha \neq 0$. Therefore $z_2 v = \alpha(z_2)_i$, where $(z_2)_i$ is the component i of the vector z_2, which must be non-zero. Therefore $z_2 v \neq 0$ which contradicts (26.11). □

Example 26.3. Let us illustrate the use of Lemma 26.2 for complex diagonal plus constant matrices and two distinct perturbation vectors $z_1 \neq z_2$.

Let A be the matrix defined in (25.94), namely

$$A = \begin{pmatrix} d_1 & a & a & \cdots & a & a \\ a & d_2 & a & \cdots & a & a \\ a & a & d_3 & \cdots & a & a \\ \vdots & \vdots & \vdots & \ddots & \vdots & \vdots \\ a & a & a & \cdots & d_{N-1} & a \\ a & a & a & \cdots & a & d_N \end{pmatrix},$$

where $a \neq 1$, $a \neq 0$ and $d_1, d_2, \dots, d_N$ are (distinct) complex numbers. We will use this matrix with the quasiseparable generators from (25.95), namely

$$\begin{array}{lll} p(i) = a,\, i = 2, \dots, N, & q(j) = 1, j = 1, \dots, N-1, & a(k) = 1, k = 2, \dots, N-1, \\ g(j) = a, j = 1, \dots, N-1, & h(i) = 1, i = 2, \dots, N, & b(k) = 1, k = 2, \dots, N-1 \end{array}$$

and the diagonal entries $d(k) = d_k,\ k = 1, \dots, N$. In Subsection §25.5.1 we already computed its divided matrices B and C and we obtained two diagonal matrices

$$B = \text{diag}\ (d_1 - a, d_2 - a, \dots, d_m - a)$$

and

$$C = \text{diag}\ (d_{m+1} - a, d_{m+2} - a, \dots, d_m - a).$$

There is therefore no need to divide the children B and C of the matrix A again. We also obtained the perturbation vectors

$$V_1 = \begin{pmatrix} a & \dots & a \end{pmatrix}^T, \quad V_2 = \begin{pmatrix} 1 & \dots & 1 \end{pmatrix}.$$

Note also that B and C are symmetric diagonal matrices which have trivial Schur decompositions

$$U_B^* B U_B = D_B, \quad U_C^* B U_C = D_C,$$

with $U_B = I_m$, $U_C = I_{N-m}$, $D_B = B$, $D_C = C$.

In the conquer step we suppose that the spectral data, i.e., the Schur decompositions, of the two matrices B and C are known and we can build the Schur decomposition of A, where

$$A = \begin{pmatrix} B & 0 \\ 0 & C \end{pmatrix} + V_1 V_2 \tag{26.12}$$

has the eigenvalues $\lambda_1, \lambda_2, \ldots, \lambda_N$ which we must find, together with their corresponding eigenvectors.

Denote now

$$U = \begin{pmatrix} U_B & 0 \\ 0 & U_C \end{pmatrix}.$$

Then $U = I_N$ and is of course unitary, and

$$U^*AU = U^*\left(\begin{pmatrix} B & 0 \\ 0 & C \end{pmatrix} + V_1V_2\right)U = D + z_1z_2,$$

where

$$D = \begin{pmatrix} D_B & 0 \\ 0 & D_C \end{pmatrix} \tag{26.13}$$

is a diagonal matrix, while

$$z_1 = U^*V_1, z_2 = V_2U.$$

In our case $z_1 = V_1, z_2 = V_2$ and they have only non-zero components, since $a \neq 0$. Then, by Lemma 26.2, $K = D + z_1z_2$ and D have no common eigenvalues. We can now use Theorem 26.1 and we find a complete correspondence between the eigenvalues and the eigenvectors of K and those of the (now) scalar rational function

$$F(\lambda) = 1 - z_2(\lambda I - D)^{-1}z_1. \tag{26.14}$$

Exploring the properties of this function we find an invertible matrix V which brings the matrix $K = D + z_1z_2$ to its diagonal similar form, i.e.,

$$V^{-1}(D + z_1z_2)V = D_A,$$

where D_A is the diagonalization of the original matrix A. We then set $P = UV$ to obtain

$$P^{-1}AP = D_A = \operatorname{diag}(\lambda_1, \lambda_2, \ldots, \lambda_N),$$

as needed.

We know from Theorem 26.1 that $F(\lambda)$ and K have the same eigenvalues and if λ_0 is an eigenvalue of $F(\lambda)$ and ϕ_0 is an eigenvector of $F(\lambda)$ corresponding to λ_0, which in our case means the scalar $\phi_0 = 1$, then an eigenvector of K is given by (26.4), which with our current notation is

$$y = (\lambda_0 - D)^{-1}z_1. \tag{26.15}$$

In fact,

$$F(\lambda) = 1 + a\left(\frac{1}{d_1 - a - \lambda} + \frac{1}{d_2 - a - \lambda} + \cdots + \frac{1}{d_N - a - \lambda}\right). \tag{26.16}$$

In order to find approximative intervals for the roots, suppose in the following that the entries of the matrix A are real. The derivative of $F(\lambda)$ has constant sign between the poles d_i, $i = 1, \dots, N$, (the sign is sign$(-a)$) since

$$F'(\lambda) = a\left(-\frac{1}{(d_1 - a - \lambda)^2} - \frac{1}{(d_2 - a - \lambda)^2} - \dots - \frac{1}{(d_N - a - \lambda)^2}\right) \tag{26.17}$$

so that $F(\lambda)$ is monotone in the intervals between its poles. Moreover, $F(\lambda)$ takes all the real values between each two poles, including the value zero. It follows that $F(\lambda)$ has exactly N roots λ_i, $i = 1, \dots, N$, and they satisfy (after we order the diagonal elements of D in decreasing order,

$$\lambda_N > d_N - a > \lambda_{N-1} > d_{N-1} - a > \dots > \lambda_1 > d_1 - a. \tag{26.18}$$

Moreover, from (26.15) we have that the eigenvectors corresponding to the eigenvalues λ_i, $i = 1, \dots, N$, are

$$\begin{aligned} v_1 = -\begin{pmatrix} \frac{a}{d_1-a-\lambda_1} \\ \frac{a}{d_2-a-\lambda_1} \\ \dots \\ \frac{a}{d_N-a-\lambda_1} \end{pmatrix}, \quad v_2 = -\begin{pmatrix} \frac{a}{d_1-a-\lambda_2} \\ \frac{a}{d_2-a-\lambda_2} \\ \dots \\ \frac{a}{d_N-a-\lambda_2} \end{pmatrix}, \\ \dots \qquad , v_N = -\begin{pmatrix} \frac{a}{d_1-a-\lambda_N} \\ \frac{a}{d_2-a-\lambda_N} \\ \dots \\ \frac{a}{d_N-a-\lambda_N} \end{pmatrix}. \end{aligned} \tag{26.19}$$

§26.3 Comments

The presentation of the conquer step given here for quasiseparable of order one matrices is similar to one for the case of tridiagonal matrices. A more detailed study of conquer step for quasiseparable of any order matrices and related eigenvalue problems for rational matrix functions are presented in [25]. A description of the conquer step which is based on the using of arrowhead matrices instead of diagonal ones may be found in the monograph by R. Vandebril, M. Van Barel and N. Mastronardi [36].

Chapter 27

Complete Algorithm for Hermitian Matrices

In this chapter we study in detail the divide and conquer method for Hermitian matrices with quasiseparable representations. For Hermitian quasiseparable of order one matrices we obtain a complete algorithm to compute the eigenvalue decomposition of a matrix.

§27.1 The divide step for Hermitian matrices

Let A be a block Hermitian matrix with given lower quasiseparable generators $p(i)$ $(i = 2, \dots, N)$, $q(j)$ $(j = 1, \dots, N-1)$, $a(k)$ $(k = 2, \dots, N-1)$. The relations

$$\begin{gathered} g(j) = (q(j))^*,\ j = 1, \dots, N-1, \quad h(i) = (p(i))^*,\ i = 2, \dots, N \\ b(k) = (a(k))^*,\ k = 2, \dots, N-1 \end{gathered} \tag{27.1}$$

yield upper quasiseparable generators of A with the same orders as for the lower ones. It is clear that if the matrix A is Hermitian, then this property is also preserved for the matrices B and C in (25.1). In this case, only the computation of the lower quasiseparable generators is needed, so that the complexity is less than in the general one.

Inserting (27.1) in Theorem 25.1 we obtain the following result.

Corollary 27.1. *Let m, N be two positive integers such that $m < N$ and $A = \{A_{ij}\}_{i,j=1}^N$ be a block Hermitian matrix with entries of sizes $m_i \times m_j$ with lower quasiseparable generators $p(i)$ $(i = 2, \dots, N)$, $q(j)$ $(j = 1, \dots, N-1)$, $a(k)$ $(k = 2, \dots, N-1)$ of orders r_k $(k = 1, \dots, N-1)$ and diagonal entries $d(k)$ $(k = 1, \dots, N)$.*

Then the matrix A is a perturbation of rank at most r_m of a 2×2 block diagonal matrix

$$\begin{pmatrix} B & 0 \\ 0 & C \end{pmatrix} \tag{27.2}$$

with Hermitian submatrices B of size $m \times m$ and C of size $(N-m) \times (N-m)$ which have lower quasiseparable generators of orders r_k, $k = 1, \dots, m-1$ and of orders r_k, $k = m+1, \dots, N-1$ respectively.

In fact, using quasiseparable generators define the matrices Q_m, P_{m+1} via the relations

$$Q_m = \mathrm{row}(a^{>}_{m+1,k} q(k))_{k=1}^{m}, \quad P_{m+1} = \mathrm{col}(p(k) a^{>}_{km})_{k=m+1}^{N} \tag{27.3}$$

and taking some invertible $r_m \times r_m$ matrix α one can represent the matrix A in the form

$$A = \begin{pmatrix} B & 0 \\ 0 & C \end{pmatrix} + V_1 V_1^*, \tag{27.4}$$

where

$$V_1 = \begin{pmatrix} Q_m^* (\alpha^*)^{-1} \\ P_{m+1} \alpha \end{pmatrix} \tag{27.5}$$

and

$$\begin{aligned} B = B_m &= A(1:m, 1:m) - Q_m^* (\alpha\alpha^*)^{-1} Q_m, \\ C = C_m &= A(m+1:N, m+1:N) - P_{m+1} \alpha\alpha^* P_{m+1}^*. \end{aligned} \tag{27.6}$$

Moreover, the matrix B has lower quasiseparable generators

$$\begin{aligned} &p_B(i) = p(i) - (q(i))^* (a^{>}_{m+1,i})^* (\alpha\alpha^*)^{-1} a^{>}_{m+1,i-1}, \quad i = 2, \dots, m \\ &q_B(j) = q(j), \quad j = 1, \dots, m-1, \quad a_B(k) = a(k), \quad k = 2, \dots, m-1 \end{aligned}$$

of orders r_k, $k = 1, \dots, m-1$, and diagonal entries

$$d_B(k) = d(k) - (q(k))^* (a^{>}_{m+1,k})^* (\alpha\alpha^*)^{-1} a^{>}_{m+1,k} q(k), \quad k = 1, \dots, m,$$

and the matrix C has lower quasiseparable generators

$$\begin{aligned} &p_C(i-m) = p(i), && i = m+2, \dots, N), \\ &q_C(j-m) = q(j) - a^{>}_{j+1,m} \alpha\alpha^* (a^{>}_{j,m})^* (p(j))^*, && j = m+1, \dots, N-1, \\ &a_C(k-m) = a(k), && k = m+2, \dots, N-1 \end{aligned}$$

of orders r_k, $k = m+1, \dots, N-1$, and diagonal entries

$$d_C(k-m) = d(k) - p(k) a^{>}_{km} \alpha\alpha^* (a^{>}_{km})^* (p(k))^*, \; k = m+1, \dots, N.$$

Inserting (27.1) in Algorithm 25.2 we obtain a divide method for Hermitian matrices via quasiseparable generators.

Algorithm 27.2. Algorithm for suitable quasiseparable generators for the divided matrices and for the perturbation matrices in the Hermitian case

Let $A = \{A_{ij}\}_{i,j=1}^{N}$ be a block matrix with entries of sizes $m_i \times m_j$, with lower quasiseparable generators $p(i)$ $(i = 2, \dots, N)$, $q(j)$ $(j = 1, \dots, N-1)$, $a(k)$ $(k =$

$2, \dots, N-1)$ of orders r_k $(k = 1, \dots, N-1)$, and diagonal entries $d(k)$ $(k = 1, \dots, N)$. Let the matrices B and C be given by (27.6).

Then sets of lower quasiseparable generators and diagonal entries $p_B(i)$ $(i = 2, \dots, m)$, $q_B(j)$ $(j = 1, \dots, m-1)$, $a_B(k)$ $(k = 2, \dots, m-1)$; $d_B(k)$ $(k = 1, \dots, m)$ for the matrix B, and $p_C(i)$ $(i = 2, \dots, N-m)$, $q_C(j)$ $(j = 1, \dots, N-m-1)$, $a_C(k)$ $(k = 2, \dots, N-m-1)$; $d_C(k)$ $(k = 1, \dots, N-m)$ for the matrix C which have the same orders as the generators of the matrix A, as well as the perturbation matrix V_1 are obtained with the following algorithm.

1. Find lower quasiseparable generators and diagonal entries of B and the entries $V_1(i)$ $i = 1, \dots, m$.

   1.1
$$\begin{aligned} u_m &= \alpha^{-1} a(m), & w_m &= \alpha^{-1} q(m), \\ d_B(m) &= d(m) - w_m^* w_m, & p_B(m) &= p(m) - w_m^* u_m, \\ V_1(m) &= w_m^*. \end{aligned} \tag{27.7}$$

   1.2. For $k = m-1, \dots, 2$ perform the following. Compute
$$\begin{aligned} w_k &= u_{k+1} q(k), & u_k &= u_{k+1} a(k), \\ d_B(k) &= d(k) - w_k^* w_k, & p_B(k) &= p(k) - w_k^* u_k, \end{aligned} \tag{27.8}$$
   and set
$$q_B(k) = q(k), \quad a_B(k) = a(k), \quad V_1(k) = w_k^*. \tag{27.9}$$

   1.3. Compute
$$w_1 = u_2 q(1), \quad d_B(1) = d(1) - w_1^* w_1, \tag{27.10}$$
   and set
$$V_1(1) = w_1^*, \quad q_B(1) = q(1), \quad g_B(1) = g(1). \tag{27.11}$$

2. Find lower quasiseparable generators and diagonal entries of C and the entries $V_1(i)$, $i = m+1, \dots, N$.

   2.1. Compute
$$t = m+1, \qquad u_t = a(t)\alpha, \qquad z_t = p(t)\alpha, \tag{27.12}$$
$$d_C(1) = d(t) - z_t z_t^*, \quad q_C(1) = q(t) - u_t z_t^*, \tag{27.13}$$
   and set
$$V_1(t) = z_t. \tag{27.14}$$

   2.2. For $k = m+2, \dots, N-1$ compute
$$\begin{aligned} s &= k - m, & u_k &= a(k) u_{k-1}, \\ z_k &= p(k) u_{k-1}, & q_C(s) &= q(k) - u_k z_k^*, \end{aligned} \tag{27.15}$$
$$d_C(s) = d(k) - z_k z_k^* \tag{27.16}$$

and set

$$V_1(k) = z_k, \quad a_C(s) = a(k), \quad p_C(s) = p(k). \tag{27.17}$$

2.3. Compute

$$z_N = p(N)u_{N-1}, \quad s = N - m, \quad d_C(s) = d(N) - z_N z_N^* \tag{27.18}$$

and set

$$V_1(N) = z_N^*, \quad p_C(s) = p(N). \tag{27.19}$$

In order to compute the complexity of Algorithm 27.2 we will analyse in detail only Steps 1.2 and 2.2, which are performed not just once, but $m-2$ and respectively $N-m-2$ times, thus about $N/2$ times each.

In Step 1.2, formula (27.8) performs a conjugation of an $r_m \times m_k$ matrix and four matrix multiplications and two matrix subtractions, which require

$$r_m(2r_k - 1)m_k + r_m(2r_k - 1)r_{k-1} + m_k(2r_m - 1)m_k + m_k(2r_m - 1)r_{k-1}$$

and respectively $m_k^2 + m_k r_{k-1}$ arithmetical operations. If we denote

$$r = \max_{k=1,\dots,N-1} r_k \quad \text{and} \quad n = \max_{k=1,\dots,N} m_k,$$

then Step 1.2 asks for less than $(r^3 + 2r^2n + rn^2)N$ arithmetical operations, while Steps 1.1 and 1.3 require $4r^2n + 2rn^2 + rn$ and respectively $2rn^2$ arithmetical operations only.

In Step 2.2, formula (27.15) performs a number subtraction, a conjugation of an $r_m \times m_k$ matrix and mainly three matrix multiplications and a matrix subtraction, which require

$$m_k(2r_k - 1)r_m + r_k(2r_{k-1} - 1)r_m + r_k(2r_m - 1)m_k$$

and respectively $r_k m_k$ arithmetical operations. Formula (27.16) performs a matrix multiplication and a matrix subtraction, which require $m_k(2r_m - 1)m_k$ and respectively m_k^2 arithmetical operations. Then Step 2.2 requires less than $(r^3 + 2r^2n + rn^2 - rn)N$ arithmetical operations, while Steps 2.1 and 2.3 require $4r^2n + 2rn^2 + rn$ and respectively $2r^2n + 1$ arithmetical operations only.

Therefore the overall complexity of Algorithm 27.2 is less than

$$(2r^3 + 4r^2n + 2rn^2 - rn + 1)N + 10r^2n + 5rn^2 + rn + 1,$$

which is about half of the complexity computed in (25.51).

It follows from Corollaries 27.1 and 25.4 that in the case of a matrix A which is both tridiagonal and Hermitian the obtained matrices B and C also belong to the same class.

§27.2 Conquer step for Hermitian quasiseparable of order one matrices

If the initial matrix A is Hermitian and its quasiseparable generators satisfy (27.1), then by Corollary 27.1 the smaller matrices B and C are Hermitian too. In this case the results proved in [29] and references therein for the special case of a tridiagonal symmetric matrix A can be generalized for the larger class quasiseparable of order one Hermitian matrices.

Suppose that for the divided Hermitian matrices B and C of sizes $m \times m$ and respectively $(N-m) \times (N-m)$ we already have their Schur decompositions, i.e., we have $m \times m$ and $(N-m) \times (N-m)$ unitary matrices Q_B and respectively Q_C such that $Q_B^* B Q_B = D_B$ and $Q_C^* C Q_C = D_C$, where D_B and D_C are diagonal matrices. We must compute the spectral data of the twice larger matrix A which satisfies (27.4) with the known column vector V_1.

If we denote

$$U = \begin{pmatrix} Q_B & 0 \\ 0 & Q_C \end{pmatrix},$$

then U is unitary and

$$U^* A U = U^* \left(\begin{pmatrix} B & 0 \\ 0 & C \end{pmatrix} + V_1 V_1^* \right) U = D + z_1 z_1^*,$$

where

$$D = \begin{pmatrix} D_B & 0 \\ 0 & D_C \end{pmatrix}$$

and $z_1 = U^* V_1$ is a column vector which will be now denoted simply by z.

We must now find a unitary V which brings the matrix K from (26.2), which now becomes $K = D + zz^*$, to its diagonal form, i.e., such that

$$V^*(D + zz^*)V = D_A,$$

where D_A is the diagonal matrix in the Schur decomposition of the original matrix A. We then set $P = UV$ to obtain

$$P^* A P = D_A.$$

The conditions of Lemma 26.2 are fulfilled if all the diagonal entries of D are distinct and the vector z has no zero components.

In this case, the rational scalar function $F(\lambda)$ in (26.6) becomes

$$F(\lambda) = 1 + \frac{|z_1|^2}{d_1 - \lambda} + \frac{|z_2|^2}{d_2 - \lambda} + \cdots + \frac{|z_N|^2}{d_N - \lambda}, \tag{27.20}$$

where z_i are the components of the vector z and

$$d_N > d_{N-1} > d_{N-2} > \cdots > d_2 > d_1$$

are the distinct diagonal entries of the diagonal matrix D. Since A is a Hermitian matrix and $z \neq 0$, the derivative of $F(\lambda)$ is negative between the poles d_i, $i = 1, \ldots, N$:

$$F'(\lambda) = -\frac{|z_1|^2}{(d_1-\lambda)^2} - \frac{|z_2|^2}{(d_2-\lambda)^2} - \cdots - \frac{|z_N|^2}{(d_N-\lambda)^2}, \tag{27.21}$$

so that $F(\lambda)$ is monotone between its poles. Moreover, $F(\lambda)$ takes all the real values between each two poles, including the value zero. It follows that $F(\lambda)$ has exactly N roots λ_i, $i = 1, \ldots, N$, and they satisfy

$$\lambda_N > d_N > \lambda_{N-1} > d_{N-1} > \lambda_{N-2} > \cdots > d_2 > \lambda_1 > d_1. \tag{27.22}$$

Moreover, from (26.4) we have that the eigenvectors corresponding to the eigenvalues λ_i, $i = 1, \ldots, N$, are

$$v_1 = \begin{pmatrix} \frac{z_1}{d_1-\lambda_1} \\ \frac{z_2}{d_2-\lambda_1} \\ \cdots \\ \frac{z_N}{d_N-\lambda_1} \end{pmatrix}, \quad v_2 = \begin{pmatrix} \frac{z_1}{d_1-\lambda_2} \\ \frac{z_2}{d_2-\lambda_2} \\ \cdots \\ \frac{z_N}{d_N-\lambda_2} \end{pmatrix}, \quad \ldots \quad, v_N = \begin{pmatrix} \frac{z_1}{d_1-\lambda_N} \\ \frac{z_2}{d_2-\lambda_N} \\ \cdots \\ \frac{z_N}{d_N-\lambda_N} \end{pmatrix}, \tag{27.23}$$

which must be normalized to obtain the desired unitary matrix V.

§27.3 The rational function approximation method and the convexifying method for finding zeroes

§27.3.1 Methods for finding zeroes

In order to find the zeroes of the function $F(\lambda)$ in (27.20) we will now summarize two known methods: the local approximation in the region of a root by simple rational functions whose zeroes are easy to compute, and the use of convexifying transformations which precede the search for a root. This methods, which are due to Bunch, Nielsen and Sorensen [12] and respectively to Melman [32], have been especially conceived for rational functions of this type.

In both methods, for finding the ith root of

$$F(\lambda) = 1 + \frac{|z_1|^2}{d_1-\lambda} + \frac{|z_2|^2}{d_2-\lambda} + \cdots + \frac{|z_N|^2}{d_N-\lambda},$$

where $i = 1, 2, \ldots, N-1$, a linear change of variables

$$\mu = d_i - \lambda \tag{27.24}$$

is performed first. (Note that the case $i = N$ needs a different treatment, as (27.22) suggests.) This change of variables has numerical advantages for the accurate determination of the updated eigenvectors. After (27.24) the problem becomes to find the zero μ_i of the function

$$F_i(\mu) = 1 + \frac{|z_1|^2}{\delta_1 - \mu} + \frac{|z_2|^2}{\delta_2 - \mu} + \cdots + \frac{|z_N|^2}{\delta_N - \mu}, \tag{27.25}$$

where $\delta_j = d_j - d_i$, $j = 1, 2, \ldots, N$ and the root we look for must lie in the interval

$$0 < \mu_i < \min\left(\delta_{i+1},\ 1 - \sum_{j=1}^{i-1} \mu_j\right), \tag{27.26}$$

as it is stated before formula (3.2) in [12].

The method in [12], which is also recommended in [29], is as follows. Denote

$$\Psi(t) = \sum_{j=1}^{i} \frac{|z_j|^2}{\delta_j - t}, \quad \Phi(t) = \sum_{j=i+1}^{N} \frac{|z_j|^2}{\delta_j - t}. \tag{27.27}$$

Then (27.25) becomes

$$-\Psi(\mu_i) = \Phi(\mu_i) + 1 \tag{27.28}$$

and both sides are convex, but the left side is decreasing and the right side is increasing on (27.26). In order to find the root, suppose that at a certain stage we already have the approximation t_k between 0 and μ_i. The problem is to find a $t_{k+1} \in (t_k, \mu_i)$, i.e., a better approximation. To this end, the two functions in (27.27) are approximated by interpolating simpler rational functions $\frac{p}{q-t}$ and $r + \frac{s}{\delta_{i+1}-t}$ such that

$$\frac{p}{q - t_k} = \Psi(t_k), \quad r + \frac{s}{\delta_{i+1} - t_k} = \Phi(t_k),$$
$$\frac{p}{(q - t_k)^2} = \Psi'(t_k), \quad \frac{s}{(\delta_{i+1} - t_k)^2} = \Phi'(t_k).$$

It is easy to compute p, q, r, s and then solve the quadratic equation

$$-\frac{p}{q - t_{k+1}} = 1 + r + \frac{s}{\delta_{i+1} - t_{k+1}},$$

which is an approximation of (27.28). In fact,

$$t_{k+1} = t_k + \frac{2b}{a + \sqrt{a^2 - 4b}}, \tag{27.29}$$

where

$$a = \frac{\Delta(1 + \Phi_k) + \Psi_k^2/\Psi_k'}{c} + \frac{\Psi_k}{\Psi_k'}, \qquad b = \frac{\Delta w \Psi_k}{\Psi_k' c},$$
$$c = 1 + \Phi_k - \Delta\Phi_k', \qquad w = 1 + \Phi_k + \Psi_k, \qquad \Delta = \delta_{i+1} - t_k,$$
$$\Phi_k = \Phi(t_k),\ \Psi_k = \Psi(t_k), \qquad \Phi_k' = \Phi'(t_k), \qquad \Psi_k' = \Psi'(t_k).$$

The reasons for arranging the calculations in this way are: w must be computed anyway for a convergence test, cancellation is minimized, and t_{k+1} has an unambiguous sign. In [12] it is proved that starting with any $0 < t_0 < \mu_i$, the sequence obtained recursively by (27.29) converges increasingly to μ_i quadratically (namely $|t_{k+1} - \mu_i| = O(|t_k - \mu_i|^2)$).

Finally, the case when $i = N$ and we look for the last root is treated. In this case equation (27.28) becomes $-\Psi(t) = 1$ and accordingly the iterations for obtaining t_k are simpler:

$$t_{k+1} = t_k + \frac{(1 + \Psi_k)\Psi_k}{\Psi_k'}.$$

We will now describe a method proposed by Melman in [32]. This method is faster, which is important since it has to be used numerous times. It first performs a further transformation of variables, besides (27.24), and then the function becomes one for which both the Newton method and the secant method converge from any suitably chosen initial points and they do it faster. The secant method needs two initial points t_1 and t_2, and then

$$t_{k+1} = t_k - \frac{t_k - t_{k-1}}{f(t_k) - f(t_{k-1}}f(t_k), \quad k = 2, 3, \dots.$$

More specifically, a class of transformations of variables is considered which change the function into a convex one. These transformations must be twice continuously differentiable and also proper, i.e., they are one-to-one and their range (possibly including ∞) is sufficient to cover the values of the original variable. Such a transformation is, for instance, $w(\gamma) = \gamma^p$ for $0 < p \leq 1$.

It is shown that if $w''(\gamma) \leq 0$ for all γ such that $w(\gamma) > \frac{1}{\delta_{i+1}}$, then the function $F_i(\mu)$ from (27.25) becomes a convex function $F_i(\frac{1}{w(\gamma)})$. It is also mentioned in Theorem 2.1 there that if $F(x)$ is convex and decreasing (respectively increasing) on a closed interval $[a, b]$ and $F(a)F(b) < 0$, then Newton's method converges monotonically to the unique solution x^* of $F(x) = 0$ from any initial point in $[a, x^*]$ (respectively $[x^*, b]$). Moreover, denote the unique solution of $F_i(\frac{1}{w(\gamma)}) = 0$ by γ^* and suppose that $w'(\gamma)$ has a constant sign for each γ such that $w(\gamma) > \frac{1}{\delta_{i+1}} > 0$. Then (in Theorem 2.2. of [32] it is proved that) Newton's method applied to the function $F_i(\frac{1}{w(\gamma)})$ in this interval converges monotonically from any point

$$\zeta_0 \in \left(w^{-1}\left(\frac{1}{\delta_{i+1}}\right), \gamma^*\right] \tag{27.30}$$

or from the point $\zeta_0 \in [\gamma^*, w^{-1}(\frac{1}{\delta_{i+1}}))$, depending on whether w is increasing or decreasing.

Suppose that w is increasing and that we start from a point ζ_0 as in (27.30). Denote

$$R_i(\gamma) = 1 + \sum_{i \neq j=1}^{N} \frac{|z_j|^2}{\delta_j} - |z_i|^2 w(\gamma) + \sum_{j=1, j\neq i, i+1}^{N} \frac{(\frac{|z_j|}{\delta_j})^2}{w(\gamma) - \frac{1}{\delta_j}}, \tag{27.31}$$

which is the rest to remain from $F_i(\frac{1}{w(\gamma)})$ after its dominant most troublesome part

$$D_i(\gamma) = \frac{(\frac{|z_{i+1}|}{\delta_{i+1}})^2}{w(\gamma) - \frac{1}{\delta_{i+1}}} \tag{27.32}$$

is discarded. Then a sequence ζ_k with $k = 0, 1, 2, \ldots$ which converges to the root γ^* (faster than the Newton method which starts with the same ζ_0) satisfies

$$R_i(\zeta_k) + R_i'(\zeta_k)(\zeta_{k+1} - \zeta_k) + D_i(\zeta_{k+1}),$$

where $R_i(\gamma), D_i(\gamma)$ have been defined in (27.31), (27.32) and ζ_{k+1} stays in the interval defined in (27.30).

For $i = N$, the final part of Section 3 of [32] states that the function $F_N(\frac{1}{w(\gamma)})$ is almost the same but, since $\mu = \frac{1}{w(\gamma)}$, it is defined on $[w^{-1}(0), \infty)$ and $w^{-1}(0)$ can be a starting point (in which the function equals 1). Its dominant part is

$$D_N(\gamma) = \frac{(\frac{|z_1|}{\delta_1})^2}{w(\gamma) - \frac{1}{\delta_1}}$$

instead of the formula in (27.32).

§27.3.2 Repeated diagonal entries and zero components for z

Here we show that one can relax the conditions of Lemma 26.2 that z has only non-zero components and that the diagonal elements of D are all distinct.

Assume that the vector z has a zero component $z_i = 0$. Then one can see easily that the diagonal entry d_i of the diagonal matrix D is an eigenvalue of the matrix $D + zz^*$ with the corresponding vector e_i, which is an ith column vector from the standard basis.

Assume that the matrix D has equal diagonal entries $d_i = d_j$, $i < j$. Take the complex Givens rotation matrix G_{ij} such that $G_{ij}^* \begin{pmatrix} z_i \\ z_j \end{pmatrix} = \begin{pmatrix} w_i \\ 0 \end{pmatrix}$. We obviously have

$$G_{ij}^* \begin{pmatrix} d_i & 0 \\ 0 & d_j \end{pmatrix} G_{ij} = \begin{pmatrix} d_i & 0 \\ 0 & d_j \end{pmatrix}.$$

Finally, using permutation matrices one can move all zero components to the bottom of the vector z and sort the different diagonal entries of D in increasing order. Thus we obtain the following algorithm.

Lemma 27.3. *Let $D = \mathrm{diag}(\mu_1, \mu_2, \ldots, \mu_N)$ be a diagonal real matrix and let z be a vector with N components. Then there exists an integer $n \le N$ and a unitary matrix Q_1 such that $D_1 = Q_1^* D Q_1 = \mathrm{diag}(\mu_1, \mu_2, \ldots, \mu_N)$ and $w = Q_1^* z$, with*

$$\mu_1 < \mu_2 < \cdots < \mu_n,$$

$w_i \neq 0,\ i = 1, \ldots, n,$ and $w_i = 0,\ i = n+1, \ldots, N$.

The desired unitary matrix Q_1, the diagonal matrix D_1 and the vector w are obtained as follows.

Set $Q_1 := I_N$.

1. *For $i = 1, 2, \ldots, N-1$.*
   For $j = i+1, i+2, \ldots, N$.

   1.1. *If $\mu_i = \mu_j$, find the complex Givens rotation matrix*

$$G_{ij} = \begin{pmatrix} c_{11}^{(ij)} & c_{12}^{(ij)} \\ c_{21}^{(ij)} & c_{22}^{(ij)} \end{pmatrix}$$

   and the number w_i such that $G_{ij}^* \begin{pmatrix} z_i \\ z_j \end{pmatrix} = \begin{pmatrix} w_i \\ 0 \end{pmatrix}$.

   1.2. *Determine the unitary $N \times N$ matrix*

$$J_{ij} = \begin{pmatrix} 1 & 0 & 0 & \cdots & 0 & \cdots & 0 & 0 & 0 \\ 0 & \ddots & \vdots & & \vdots & & \vdots & & 0 \\ \vdots & & c_{11}^{(ij)} & \cdots & 0 & \cdots & c_{12}^{(ij)} & & \vdots \\ \vdots & & \vdots & \ddots & \vdots & & \vdots & & \vdots \\ 0 & \cdots & 0 & \cdots & 1 & \cdots & 0 & \cdots & 0 \\ \vdots & & \vdots & & \vdots & \ddots & \vdots & & \vdots \\ \vdots & & c_{21}^{(ij)} & \cdots & 0 & \cdots & c_{22}^{(ij)} & & \vdots \\ 0 & & \vdots & & \vdots & & \vdots & \ddots & 0 \\ 0 & 0 & 0 & \cdots & 0 & \cdots & 0 & 0 & 1 \end{pmatrix} \tag{27.33}$$

   and set $Q_1 := Q_1 J_{ij}$.

2. *Set $k = 0$.*
   For $i = 1, 2, \ldots, N$.
   If $w_i = 0$, swap between w_i, w_{N-k} and between μ_i, μ_{N-k} and set $Q_1 = Q_1 P$, where P is the corresponding permutation matrix. Set $k := k+1$.
   At the end set $n := N - k$.
3. *For $i = 1, 2, \ldots, n-1$.*
   For $j = i+1, i+2, \ldots, n$.
   Set $k := TRUE$.
   while $k = TRUE$ perform the following.

Set : $k = FALSE$. *If* $\mu_i > \mu_j$, *swap* μ_i, μ_j *and* w_i, w_j *and set* $Q_1 = Q_1 P$, *where* P *is the corresponding permutation matrix. Set* $k{:=}TRUE$.

When applying Lemma 27.3 in practice, we need to determine up to machine precision when two diagonal entries of D are distinct and when an entry of z is not zero. To this end, suppose that *tol* is a small multiple of the machine precision, for instance $tol = u(\|D\|_2 + \|z\|_2)$. By Lemma 27.3, we can determine a unitary matrix Q_1 and an integer $1 \le n \le N$ such that $Q_1^* D Q_1 = \mathrm{diag}(\mu_1, \dots, \mu_N)$ and a vector $w = Q_1^* z$ such that $\mu_{i+1} - \mu_i \ge tol$ for $i = 1, \dots, n-1$, $|w_i| \ge tol$ for $1 \le i \le n$, and $|w_i| < tol$ otherwise.

§27.3.3 Diagonalizing $D + zz^*$

After Lemma 27.3 it follows that the first n entries of the diagonal matrix D are in strictly increasing order and that the first n entries of z are not zero. We can therefore apply Theorem 26.1 to an $n \times n$ problem, so that we must first determine the n distinct zeroes of the rational function $F(\lambda)$ in (27.20), but with only n poles, namely

$$F(\lambda) = 1 + \frac{|z_1|^2}{\mu_1 - \lambda} + \frac{|z_2|^2}{\mu_2 - \lambda} + \cdots + \frac{|z_n|^2}{\mu_n - \lambda}.$$

If λ_i, $i = 1, \dots, n$, are these zeroes, then we complete them by

$$\lambda_i = \mu_i, \quad i = n+1, n+2, \dots, N. \tag{27.34}$$

We then build ith column of an $n \times n$ matrix $\tilde{Q}_2$ by normalizing v_i, $i = 1, \dots, n$ from (27.23), namely

$$v_i = \frac{(D - \lambda_i)^{-1} z}{\|(D - \lambda_i)^{-1} z\|}, \quad i = 1, 2, \dots, n.$$

Then we set $Q_2 = \mathrm{diag}(\tilde{Q}_2, I_{N-n})$.

Finally we compute $V = Q_1 Q_2$. The unitary matrix Q_1 is as in the preceding Lemma 27.3.

Thus we obtain the following algorithm:

Algorithm 27.4. Let $D = \mathrm{diag}(d_1, d_2, \dots, d_N)$ be a diagonal real matrix and let $z = (z_i)_{i=1}^N$ be a column vector. Then the unitary matrix V and the real diagonal matrix Λ such that

$$D + zz^* = V \Lambda V^*$$

are obtained by the following algorithm.

1. Using Lemma 27.3, determine an integer $n \le N$ and a unitary matrix Q_1 such that $D_1 = Q_1^* D Q_1 = \mathrm{diag}(\mu_1, \mu_2, \dots, \mu_N)$ and $w = Q_1^* z$, with

$$\mu_1 < \mu_2 < \cdots < \mu_n,$$

$w_i \ne 0$, $i = 1, \dots, n$, and $w_i = 0$, $i = n+1, \dots, N$.

2.1. Set $\hat{D} := \mathrm{diag}(\mu_1, \mu_2, \dots, \mu_n)$, $\hat{w} = (z_i)_{i=1}^n$ and using one of the iteration methods in Section §27.3, compute the n eigenvalues $\lambda_1, \lambda_2, \dots, \lambda_n$ of the matrix $\hat{D} + \hat{w}(\hat{w})^*$ with n instead of N.

2.2. Find n eigenvectors $v_1, v_2, \dots, v_n$ with formula (27.23).

2.3. Compute the normalized eigenvectors $v_1^{(0)}, v_2^{(0)}, \dots, v_n^{(0)}$ by dividing v_1, v_2, …, v_n, respectively by

$$\sqrt{-F'(\lambda_i)} = \sqrt{\frac{|z_1|^2}{(\mu_1 - \lambda_i)^2} + \frac{|z_2|^2}{(\mu_2 - \lambda_i)^2} + \dots + \frac{|z_n|^2}{(\mu_n - \lambda_i)^2}}, \quad (27.35)$$
$$i = 1, 2, \dots, n.$$

3. Set

$$\hat{Q}_2 := \begin{bmatrix} v_1^{(0)} & v_2^{(0)} & \dots & v_n^{(0)} \end{bmatrix}, \quad \Lambda := \mathrm{diag}(\lambda_1, \dots, \lambda_n, \mu_{n+1}, \dots, \mu_N);$$

then set $Q_2 := \mathrm{diag}(\tilde{Q}_2, I_{N-n})$ and compute $V = Q_1 Q_2$.

§27.4 The complete algorithm

Now we are ready to present the complete divide and conquer algorithm to compute the eigendecomposition of a Hermitian matrix with quasiseparable of order one representation.

Algorithm 27.5. Let $A = \{A_{ij}\}_{i,j=1}^N$ be an $N \times N$ Hermitian matrix where $N = 2^M$, with lower quasiseparable generators $p(i)$ $(i = 2, \dots, N)$, $q(j)$ $(j = 1, \dots, N-1)$, $a(k)$ $(k = 2, \dots, N-1)$ of order one and diagonal entries $d(k)$ $(k = 1, \dots, N)$. Then the N eigenvalues $\lambda_N, \lambda_{N-1}, \dots, \lambda_1$ of A and a unitary matrix P such that

$$P^* A P = \mathrm{diag}\,(\lambda_N, \lambda_{N-1}, \dots, \lambda_1)$$

are obtained by the following algorithm.

1. To perform the divide step:
Set

$$p^{(0,1)}(i) := p(i), \ \ i = 2, \dots, N, \qquad q^{(0,1)}(j) := q(j), \ j = 1, \dots, N-1,$$
$$a^{(0,1)}(k) := a(k), \ k = 2, \dots, N-1, \quad d^{(0,1)}(k) := d(k), \ k = 1, \dots, N.$$

For $n = 1, \dots, M$ perform the following:
Set $s := 2^{n-1}$, $m := 2^{M-n}$. For $j = 1, 2, \dots, s$ using lower quasiseparable generators

$$p^{(n-1,j)}(i), \ i = 2, \dots, 2m, \quad q^{(n-1,j)}(i), \ i = 1, \dots, 2m-1,$$
$$a^{(n-1,j)}(i), \ i = 2, \dots, 2m-1,$$

and diagonal entries $d^{(n-1,j)}(i)$, $i = 1, \dots, 2m$, of the matrix $A^{(n-1,j)}$ compute via Algorithm 27.2 lower quasiseparable generators

$$p^{(n,2j-1)}(i), p^{(n,2j)}(i),\ i = 2, \dots, m,\ q^{(n,2j-1)}(i), q^{(n,2j)}(i),\ i = 1, \dots, m-1,$$
$$a^{(n,2j-1)}(i), a^{(n,2j)}(i),\ i = 2, \dots, m-1,$$

and diagonal entries $d^{(n,2j-1)}(i)$, $d^{(n,2j)}(i)$, $i = 1, \dots, m$, of the matrices $A^{(n,2j-1)}$, $A^{(n,2j)}$ and the vectors $y^{(n-1,j)}$ such that

$$A^{(n-1,j)} = \begin{pmatrix} A^{(n,2j-1)} & 0 \\ 0 & A^{(n,2j)} \end{pmatrix} + y^{(n-1,j)}(y^{(n-1,j)})^*.$$

2. For perform the conquer step:
   Set
   $$\Lambda^{(0,t)} := d^{(M,t)}(1), \quad P^{(0,t)} := 1, \quad t = 1, \dots, N.$$
   For $n = 1, \dots, M$ perform the following. Set $m = 2^{M-n}$.
   For $j = 1, \dots, m$ perform the following.
   2.1. Compute
   $$z^{(n,j)} = \begin{pmatrix} (P^{(n-1,2j-1)})^* & 0 \\ 0 & (P^{(n-1,2j)})^* \end{pmatrix} y^{(M-n,j)}$$
   and set $D^{(n,j)} := \Lambda^{(n-1,2j-1)} \oplus \Lambda^{(n-1,2j)}$.
   2.2. Using Algorithm 27.4 determine the eigendecomposition
   $$D^{(n,j)} + z^{(n,j)}(z^{(n,j)})^* = V^{(n,j)}\Lambda^{(n,j)}(V^{(n,j)})^*,$$
   with a unitary matrix $V^{(n,j)}$ and a real diagonal matrix $\Lambda^{(n,j)}$.
   2.3. Compute
   $$P^{(n,j)} = \begin{pmatrix} P^{(n-1,2j-1)} & 0 \\ 0 & P^{(n-1,2j)} \end{pmatrix} V^{(n,j)}.$$
3. Set $P =: P^{(M,1)}$, $\Lambda := \Lambda^{(M,1)}$.

§27.5 Comments

The divide and conquer method for symmetric tridiagonal matrices was studied by J.R. Bunch, C.P. Nielsen and D.C. Sorensen in [12], by J. Cuppen in [15], by M. Gu and S. Eisenstat in [27]. For symmetric diagonal and banded plus semiseparable of order one matrices the divide and conquer method was developed by N. Mastronardi, E. Van Camp and M. Van Barel in [34], and by S. Chandrasekaran and M. Gu in [16]. The results for Hermitian quasiseparable of any order matrices were obtained in [9]; the complete algorithm appears here for the first time.

Chapter 28

Complete Algorithm for Unitary Hessenberg Matrices

We study the divide and conquer method for computing the eigenstructure of unitary upper Hessenberg matrices using their quasiseparable representations. In doing so we use essentially the fact that in Theorem 25.1 we proved the results using $r_m \times r_m$ matrices α and β which may be different from the identity. This generalization allows us to completely solve the problem for unitary upper Hessenberg matrices and to obtain as a particular case the results known in the literature. Since we consider scalar matrices which are order one quasiseparable, their quasiseparable generators are complex numbers, and so are α and β. In the conquer step we show that to reconstruct the eigenstructure of A from those of B and C amounts to the study of the zeroes of a scalar function which is a particular case of the rational matrix function that we had to deal with in Chapter 26.

§28.1 The divide step for unitary Hessenberg matrices

Based on Theorem 25.5 we obtain the following algorithm.

Algorithm 28.1. Algorithm for suitable Schur and complementary parameters for divided upper Hessenberg unitary matrices and auxiliary parameters

Let $A = \{A_{ij}\}_{i,j=1}^N$ be an unreduced upper Hessenberg unitary matrix with the Schur parameters $\rho_0 = -1, \rho_1, \dots, \rho_N$ and complementary parameters μ_1, $\mu_2, \dots,$ μ_{N-1}. Let B and C be the matrices from Theorem 25.5.

Then a set of Schur parameters $\rho_0^B, \rho_1^B, \dots, \rho_m^B$ and complementary parameters $\mu_1^B, \mu_2^B, \dots, \mu_{m-1}^B$ of the matrix B and a set of Schur parameters ρ_0^C, ρ_1^C, $\dots$, ρ_{N-m}^C and complementary parameters $\mu_1^C, \mu_2^C, \dots, \mu_{N-m-1}^C$ of the matrix C as well as the auxiliary parameters α, β in Theorem 25.5 are obtained with the following algorithm.

0. If $0 < \mu_m < 1$ set $\tau = -\frac{\rho_m}{|\rho_m|}$, while if $\mu_m = 1$ set $\tau = -1$.

Compute the parameters

$$\alpha = \sqrt{\frac{2}{1+|\rho_m|}}, \quad \beta = -\tau\sqrt{2(1+|\rho_m|)}.$$

1. Find the Schur parameters and complementary parameters of B.
   Set
   $$\rho_0^B = -1, \quad \rho_i^B = \rho_i, \; i=1,\dots,m-1, \quad \rho_m^B = \tau,$$
   $$\mu_i^B = \mu_i, \; i=1,\dots,m-1.$$
2. Find the Schur parameters and complementary parameters of C.
   Set
   $$\rho_0^C = -1, \quad \rho_i^C = \rho_{i+m}\tau^*, \; i=1,\dots,N-m,$$
   $$\mu_i^C = \mu_{i+m}, \quad i=1,\dots,N-m-1.$$
3. Keep the value of μ_n and the value of $|\rho_m|$ to be used later in the conquer step.

The special choice of the parameters α, β will be explained below.

Algorithm 28.1 uses less than $10(N-2)+22$ arithmetic operations. A few of the operations used in the algorithm are multiplications, but most of them are assignments. So the computational complexity is linear, less than $10N$.

§28.2 Conquer step for unitary Hessenberg matrices

We suppose that the matrix A is unreduced. One of the consequences is that all the eigenvalues have geometric multiplicity one. This is because the $N-1$ row vectors which have each a non-zero subdiagonal entry in a position where all the next rows have a zero entry are independent, so that the rank of $A-\lambda I$ is at least $N-1$.

Suppose that for the divided matrices B and C of sizes $m\times m$ and respectively $(N-m)\times(N-m)$ we already have their Schur decompositions, i.e., we have $m\times m$ and $(N-m)\times(N-m)$ unitary matrices Q_B and respectively Q_C so that $Q_B^*BQ_B = D_B$ and $Q_C^*CQ_C = D_C$ where D_B and D_C are diagonal matrices. We must compute the spectral data of the twice larger matrix A which satisfies (25.72) with the known column vector V_1 and row vector V_2 given by (25.73)–(25.75).

If we denote

$$U = \begin{pmatrix} Q_B & 0 \\ 0 & Q_C \end{pmatrix}, \tag{28.1}$$

then U is unitary and

$$U^*AU = U^*\left(\begin{pmatrix} B & 0 \\ 0 & C \end{pmatrix} + V_1V_2\right)U = D + z_1z_2,$$

where

$$D = \begin{pmatrix} D_B & 0 \\ 0 & D_C \end{pmatrix},$$

while

$$z_1 = U^* V_1, \quad z_2 = V_2 U.$$

Since

$$V_1 = \begin{pmatrix} G_0\beta & \mu_m\alpha & 0 \end{pmatrix}^T, \quad V_2 = \begin{pmatrix} 0 & \alpha^{-1} & \beta^{-1}\mu_m H_0 \end{pmatrix},$$

it follows that

$$\begin{aligned} U^*AU &= D + z_1 z_2 \qquad (28.2) \\ &= \begin{pmatrix} D_B & 0 \\ 0 & D_C \end{pmatrix} + \begin{pmatrix} Q_B^* G_0 \beta \\ Q_C^* \begin{pmatrix} \mu_m \alpha \\ 0 \end{pmatrix} \end{pmatrix} \left(\begin{pmatrix} 0 & \alpha^{-1} \end{pmatrix} Q_B \quad \beta^{-1}\mu_m H_0 Q_C \right). \end{aligned}$$

We must now find a unitary V which brings the matrix K from (26.2), which now becomes $K = D + z_1 z_2$, to its diagonal form, i.e., such that

$$V^*(D + z_1 z_2)V = D_A,$$

where D_A is the diagonal matrix in the Schur decomposition of the original matrix A. We then set $P = UV$ to obtain

$$P^*AP = D_A.$$

§28.2.1 Representation of the matrix $K = D + z_1 z_2$

Since the matrix A is unitary and so are the matrices B and C, their eigenvalues all lie on the unit circle, so that the diagonal entries of the diagonal matrices D, D_B, D_C are complex numbers of modulus 1. In this case (28.2) further implies that

$$K = D\left(I_N + \begin{pmatrix} D_B^* Q_B^* G_0 \beta \\ D_C^* Q_C^* \begin{pmatrix} \mu_m \alpha \\ 0 \end{pmatrix} \end{pmatrix} \left(\begin{pmatrix} 0 & \alpha^{-1} \end{pmatrix} Q_B \quad \beta^{-1}\mu_m H_0 Q_C \right) \right). \qquad (28.3)$$

But using (25.81) and (25.74) we get

$$D_B^* Q_B^* G_0 = Q_B^*(Q_B D_B^* Q_B^*) G_0 = Q_B^* B^* G_0 = Q_B^* \begin{pmatrix} 0 \\ \tau^* \end{pmatrix},$$

where τ is given in (25.80). Also using (25.82) and (25.75) we obtain

$$H_0 Q_C = H_0(Q_C D_C^* Q_C^*) Q_C D_C^* = H_0 C^* Q_C D_C = \begin{pmatrix} \tau & 0 \end{pmatrix} Q_C D_C.$$

Therefore, (28.3) becomes

$$K = D\left(I_N + \begin{pmatrix} Q_B^*\begin{pmatrix} 0 \\ \tau^*\beta \end{pmatrix} \\ D_C^* Q_C^* \begin{pmatrix} \mu_m\alpha \\ 0 \end{pmatrix} \end{pmatrix} \left(\begin{pmatrix} 0 & \alpha^{-1} \end{pmatrix} Q_B \quad \begin{pmatrix} \beta^{-1}\mu_m\tau & 0 \end{pmatrix} Q_C D_C \right)\right). \tag{28.4}$$

Here α, β are arbitrary nonzero complex numbers satisfying (25.76) or (25.77). We will impose on α, β some additional conditions to get the matrix K in a form suitable for an effective iteration eigenvalue algorithm.

Corollary 28.2. *Let A be a unitary upper Hessenberg matrix with $\mu_m \neq 0$. Then one can choose $\alpha \in \mathbb{R}$ and β in Theorem 25.5 such that K from (26.2) satisfies*

$$K = D(I_N - 2zz^*) \tag{28.5}$$

with a vector z such that

$$z^*z = 1. \tag{28.6}$$

To this end one can take

$$z = \left(\begin{pmatrix} 0_{1\times(m-1)} & \sqrt{\frac{1+|\rho_m|}{2}} \end{pmatrix} Q_B \quad \begin{pmatrix} -\frac{\mu_m}{\sqrt{2(1+|\rho_m|)}} & 0_{1\times(m-1)} \end{pmatrix} Q_C D_C \right)^*, \tag{28.7}$$

where $\mu_m, |\rho_m|$ are the values which have been saved in Step 3 of Algorithm 28.1.

Proof. Denote

$$z = \left(\begin{pmatrix} 0 & \alpha^{-1} \end{pmatrix} Q_B \quad \begin{pmatrix} \beta^{-1}\mu_m\tau & 0 \end{pmatrix} Q_C D_C \right)^*. \tag{28.8}$$

We require that this z fulfills all the conditions of the corollary, one of which is

$$\begin{pmatrix} Q_B^*\begin{pmatrix} 0 \\ \tau^*\beta \end{pmatrix} \\ D_C^* Q_C^* \begin{pmatrix} \alpha\mu_m \\ 0 \end{pmatrix} \end{pmatrix} = -2z = -2\begin{pmatrix} Q_B^*\begin{pmatrix} 0 \\ (\alpha^*)^{-1} \end{pmatrix} \\ D_C^* Q_C^* \begin{pmatrix} \tau^*\mu_m(\beta^*)^{-1} \\ 0 \end{pmatrix} \end{pmatrix}, \tag{28.9}$$

which means

$$\tau^*\alpha^*\beta = -2. \tag{28.10}$$

The condition $z^*z = 1$ implies

$$\frac{1}{|\alpha|^2} + \frac{\mu_m^2}{|\beta|^2} = 1, \tag{28.11}$$

and (25.76), (25.77), (25.80) mean that

$$\frac{\beta}{\alpha} = -\tau(1 + |\rho_m|). \tag{28.12}$$

Now one can easily check that the parameters α, β given by

$$\alpha = \sqrt{\frac{2}{1+|\rho_m|}}, \quad \beta = -\tau\sqrt{2(1+|\rho_m|)}. \tag{28.13}$$

satisfy the conditions (28.10)–(28.12). Inserting (28.13) in (28.8) we obtain (28.7). □

§28.2.2 Associated scalar rational function and its zeroes

Compare the matrix K from (26.2) in the general case with the same matrix K in the unitary upper Hessenberg case given by (28.5). Note that we can take $z_1 = Dz, z_2 = -2z^*$ and then the rational function from Theorem 26.1 becomes

$$F(\lambda) = 1 - 2z^*(D - \lambda I_N)^{-1}Dz.$$

By Theorem 26.1, all the eigenvalues of K, counted with multiplicities, are the zeroes of $F(\lambda)$ except possibly common eigenvalues of D and K. According to (28.6) it follows that if d_j, $j = 1, 2, \dots, N$ are the diagonal entries of D, then

$$F(\lambda) = \sum_{j=1}^{N} |z_j|^2 \frac{\lambda + d_j}{\lambda - d_j}, \tag{28.14}$$

where z_j are the components of the vector z.

Recall that the diagonal entries of D are eigenvalues of the unitary matrices B and C and so they lie on the unit circle. The eigenvalues we look for lie on the unit circle too. We can therefore take $d_j = e^{i\theta_j}$, $j = 1, 2, \dots, N$, with $-\pi < \theta_j \le \pi$ and $\lambda = e^{i\psi}$. Then $F(\lambda)$ from (28.14) becomes

$$F(\lambda) = -i\sum_{j=1}^{N} |z_j|^2 \cot\frac{\psi - \theta_j}{2} := -i\Phi(\psi), \tag{28.15}$$

whose derivative is

$$\Phi'(\psi) = -\frac{1}{2}\sum_{j=1}^{N} |z_j|^2 \frac{1}{\sin^2\frac{\psi-\theta_j}{2}}. \tag{28.16}$$

Suppose that the conditions of Lemma 26.2 are fulfilled. Then D has N distinct components and z_1, z_2 have no zero components. It follows from Lemma 26.2 and from (28.15) and (28.16) that $\Phi(\psi)$ has N distinct poles at θ_j, $j = 1, 2, \dots, N$, and between two consecutive poles it is monotonically decreasing. Therefore, its N roots $\psi_1, \psi_2, \dots, \psi_N$ interlace with the poles

$$\psi_N > \theta_N > \psi_{N-1} > \theta_{N-1} > \cdots > \theta_2 > \psi_1 > \theta_1. \tag{28.17}$$

§28.2.3 An iterative root finder for the function $\Phi(\psi)$ from (28.15)

To solve the secular equation one can use an iterative root finder suggested by W.B. Gragg and L. Reichel [30].

Note first that since $\sin \frac{\psi-\theta_j}{2} \leq 1$ in (28.16) and since the norm of z is 1, the derivative $\Phi'(\psi)$ has an absolute value of at least $\frac{1}{2}$, therefore the strictly interlacing zeroes of $\Phi(\psi)$ can be determined accurately.

Suppose that $-\pi < \theta_1 < \theta_2 < \cdots < \theta_N \leq \pi$ and suppose that an initial interpolant $\psi^{(0)}$ of a root satisfies $\theta_N - 2\pi < \psi^{(0)} < \theta_1$. Since in the interval between two poles the derivative has a constant sign, it follows that there exists there exactly one root ψ_1. Assume for the moment that $\Phi(\psi^{(0)}) < 0$ and introduce

$$\hat{\Phi}(\psi) = \rho + \varrho \cot\left(\frac{\theta_1 - \psi}{2}\right),$$

where the coefficients ρ, ϱ are determined by oscillatory interpolation, i.e.,

$$\hat{\Phi}(\psi^{(0)}) = \Phi(\psi^{(0)}), \quad \hat{\Phi}'(\psi^{(0)}) = \Phi'(\psi^{(0)}),$$

which yields

$$\rho = \Phi(\psi^{(0)}) - \Phi'(\psi^{(0)}) \sin(\theta_1 - \psi^{(0)}), \quad \varrho = 2\Phi'(\psi^{(0)}) \sin^2\left(\frac{\theta_1 - \psi^{(0)}}{2}\right).$$

Then the zero $\psi^{(1)}$ of $\hat{\Phi}(\psi)$ is the next approximation to the true zero ψ_1 of $\Phi(\psi)$.

§28.2.4 No zero components for z

Since $z_1 = Dz, z_2 = -2z$ and all the entries of D are non-zero (they are on the unit circle), the fact that z_1 and z_2 have no zero components, as Lemma 26.2 asks, is equivalent with the fact that z has no zero components. Since z is given by (28.7) and since D_C^* has no zero entries, this happens if and only if the last row of the unitary matrix Q_B which brings the unitary matrix B to its diagonal form has no zero entries, and neither has the first row of the unitary matrix Q_C which brings the unitary matrix C to its diagonal form. As we will see, all the eigenvectors of A which are built with the present algorithm have only non-zero components, so that if B and C are themselves a result of the present algorithm in its earlier stages we can assume that our z now has only non-zero components. Moreover, the following Proposition 28.5 shows that if B and C are unreduced and all their subdiagonal entries are less than 1, then z has no zero entries. But first we will prove two lemmas.

Lemma 28.3. *If B is an upper Hessenberg unreduced matrix, then the last component of all its eigenvectors is not zero.*

Proof. We will show that if the last component of an eigenvector x corresponding to an eigenvalue λ is zero, then all its entries are zero, a contradiction. To this end suppose that $2 \leq k \leq N$ and that the last components $x_k, x_{k+1}, \ldots, x_N$ of the eigenvector x are zero. We claim that $x_{k-1} = 0$, too. Indeed, λx_k is zero and on the other hand this number is given by row k of the matrix B as

$$0 \cdot x_1 + 0 \cdot x_2 + \cdots + 0 \cdot x_{k-2} + \sigma x_{k-1} + B_{k,k} x_k + B_{k,k+1} x_{k+1} + \cdots + B_{k,N} x_N = \lambda x_k = 0,$$

where $\sigma \neq 0$ is the subdiagonal entry on row k of B. It follows that $\sigma x_{k-1} = 0$, hence $x_{k-1} = 0$. □

Lemma 28.4. *If C is a unitary upper Hessenberg unreduced matrix with no subdiagonal entry equal to* 1*, then the first component of all its eigenvectors is not zero.*

Proof. We will use for the unitary upper Hessenberg matrix C the representation (25.65). We will show that if the first component of an eigenvector x corresponding to an eigenvalue λ is zero, then all its components are zero, a contradiction. To this end suppose that $1 \leq k < N$ and the first k components $x_1, x_2, \ldots, x_k$ of the eigenvector x are zero. We will prove that $x_{k+1} = 0$ too.

Denote

$$\begin{aligned} S_k = {} & \rho_{k+1} \mu_k \rho_{k-1}^* x_{k+1} + \rho_{k+2} \mu_{k+1} \mu_k \rho_{k-1}^* x_{k+2} \\ & + \cdots + \rho_N \mu_{N-1} \cdots \mu_{k+1} \mu_k \rho_{k-1}^* x_N. \end{aligned}$$

Denote also $F_k = \mu_{k-1} x_{k-1}$, $k = 2, 3, \ldots, N-1$, and $F_1 = 0$. We have that $\lambda x_k = 0$ and on the other hand this number is obtained from the row k of C as

$$F_k - \rho_k \rho_{k-1}^* x_k - S_k = 0,$$

and since the first k components of x are zero, it follows that $S_k = 0$. But then

$$\begin{aligned} \lambda x_{k+1} = {} & -\rho_{k+1} \rho_k^* x_{k+1} - \rho_{k+2} \mu_{k+1} \rho_k^* x_{k+2} - \rho_{k+3} \mu_{k+2} \mu_{k+1} \rho_k^* x_{k+3} \\ & - \cdots - \rho_N \mu_{N-1} \cdots \mu_{k+1} \rho_k^* x_N = -\frac{\rho_k^*}{\mu_k \rho_{k-1}^*} S_k = 0, \end{aligned}$$

while $|\lambda| = 1$, so that $x_{k+1} = 0$ too. We were able to divide by μ_k since the matrix C is unreduced and by ρ_{k-1}^* since $|\rho_{k-1}^*|^2 + \mu_{k-1}^2 = 1$ and μ_{k-1} is less than 1. □

Proposition 28.5. *If A is a unitary upper Hessenberg unreduced matrix with no subdiagonal entries equal to* 1*, then the vector z given by* (28.7) *has only non-zero components.*

Proof. It follows that both B and C are unitary upper Hessenberg unreduced matrices with no subdiagonal entries equal to 1 and then, by Lemmas 28.3 and 28.4, all their eigenvectors start and end with non-zero components. Therefore, if we build the unitary matrices Q_B and Q_C which bring B and respectively C to their diagonal form, we have that the last row of Q_B and the first row of Q_C have only non-zero entries, since the columns of these matrices are eigenvectors of B and respectively C. We already saw that in this case the vector z given by (28.7) has only non-zero components. □

In the case when C has at least one complementary parameter equal to 1, it is possible for z to have zero components.

§28.2.5 Repeated diagonal entries and zero components for z

We treat the case when the conditions of Lemma 26.2 are not valid in a similar way as for Hermitian matrices in Subsection §27.3.2.

Lemma 28.6. *Let $D = \operatorname{diag}\left(e^{i\theta_1}, e^{i\theta_2}, \dots, e^{i\theta_N}\right)$ be a real diagonal matrix and let z be a vector with N components. Then there exist an integer $n \leq N$ and a unitary matrix Q_1 such that $D_1 = Q_1^* D Q_1 = \operatorname{diag}\left(e^{i\theta_1}, e^{i\theta_2}, \dots, e^{i\theta_N}\right)$ and $w = Q_1^* z$ with*

$$\theta_1 < \theta_2 < \cdots < \theta_n,$$

$w_i \neq 0,\ i = 1, \dots, n$ and $w_i = 0,\ i = n+1, \dots, N$.

The desired unitary matrix Q_1, the diagonal matrix D_1, and the vector w are obtained as follows.

Set $Q_1 := I_N$.

1. *For $i = 1, 2, \dots, N-1$,*
   For $j = i+1, i+2, \dots, N$,
   1.1. *If $\theta_i = \theta_j$ find the complex Givens rotation matrix*

   $$G_{ij} = \begin{pmatrix} c_{11}^{(ij)} & c_{12}^{(ij)} \\ c_{21}^{(ij)} & c_{22}^{(ij)} \end{pmatrix}$$

   and the number w_i such that $G_{ij}^* \begin{pmatrix} z_i \\ z_j \end{pmatrix} = \begin{pmatrix} w_i \\ 0 \end{pmatrix}$.

   1.2. *Determine the unitary $N \times N$ matrix J_{ij} via (27.33) and set $Q_1 := Q_1 J_{ij}$.*
2. *Set $k = 0$.*
   For $i = 1, 2, \dots, N$.
   If $w_i = 0$ swap between w_i, w_{N-k} and between θ_i, θ_{N-k} and set $Q_1 = Q_1 P$, where P is the corresponding matrix of permutation. Set $k := k+1$.
   At the end set $n := N - k$.
3. *For $i = 1, 2, \dots, n-1$,*
   For $j = i+1, i+2, \dots, n$,
   Set $k := TRUE$.
   while $k = TRUE$ perform the following.
   Set : $k = FALSE$. If $\theta_i > \theta_j$ swap θ_i, θ_j and w_i, w_j and set $Q_1 = Q_1 P$, where P is the corresponding matrix of permutation. Set $k := TRUE$.

§28.2.6 The eigenvectors of A

By Theorem 26.1 (with $z_1 = Dz$) the kth component, $k = 1, 2, \ldots, n$, of the eigenvectors

$$v_j,\ j = 1, 2, \ldots, n$$

of A corresponding to its eigenvalues $\lambda_j = e^{i\psi_j}$, $j = 1, 2, \ldots, n$, respectively, are

$$\begin{aligned}(v_j)_k &= ((\lambda_j I_n - D)^{-1} z)_k = ((e^{i\psi_j} I_n - D)^{-1} Dz)_k \\ &= \left(\tfrac{e^{i\theta_k}}{e^{i\psi_j} - e^{i\theta_k}} z_k \right), \quad j = 1, 2, \ldots, n.\end{aligned} \tag{28.18}$$

This vectors are already orthogonal to one another since they correspond to different eigenvalues. All their components are non-zero, which will assure in the next, superior, step of the conquer procedure that z has no zero components.

Since

$$\begin{aligned}e^{i\psi_j} - e^{i\theta_k} &= \left(e^{i\frac{\psi_j - \theta_k}{2}} - e^{i\frac{\theta_k - \psi_j}{2}} \right) e^{i\frac{\psi_j + \theta_k}{2}} \\ &= 2i \sin \frac{\psi_j - \theta_k}{2} e^{i\frac{\psi_j + \theta_k}{2}} = 2 \sin \frac{\psi_j - \theta_k}{2} e^{i\frac{\psi_j + \theta_k + \pi}{2}}\end{aligned}$$

we obtain, according to (28.18), the eigenvector

$$v_j = \left(\frac{e^{(\theta_1 - \psi_j - \pi)\frac{i}{2}}}{2 \sin \frac{\psi_j - \theta_1}{2}} z_1 \quad \frac{e^{(\theta_2 - \psi_j - \pi)\frac{i}{2}}}{2 \sin \frac{\psi_j - \theta_2}{2}} z_2 \quad \cdots \quad \frac{e^{(\theta_n - \psi_j - \pi)\frac{i}{2}}}{2 \sin \frac{\psi_j - \theta_n}{2}} z_n \right)^T. \tag{28.19}$$

As a consequence of (28.19) the square of the norm of v_j is

$$\sum_{k=1}^{n} |(v_j)_k|^2 = \sum_{k=1}^{n} \frac{|z_k|^2}{4 \sin^2 \frac{\psi_j - \theta_n}{2}} = -\frac{1}{2} \Phi'(\psi_j), \tag{28.20}$$

which means that in order to normalize the eigenvector v_j one has to compute $\Phi'(\psi_j)$.

§28.2.7 Direct computation of the coefficients of the rational scalar function

As we saw in the computation of $\Phi(\psi)$ in (28.15), in the computation (28.16) of its derivative and in (28.20), when normalizing the eigenvectors, we use only the modulus of the components of the vector z rather than the values of these components. This permits to simplify the computations. If we are interested only in the eigenvalues of A and not in the eigenvectors, then we don't need to compute more than the modulus.

Proposition 28.7. *If Q_B and Q_C are the given unitary diagonalizing matrices for B and C, respectively, then*

$$\Phi(\psi) = \sum_{j=1}^{n} \zeta_j \cot \frac{\psi - \theta_j}{2} \tag{28.21}$$

and the derivative is

$$\Phi'(\psi) = -\frac{1}{2} \sum_{j=1}^{n} \zeta_j \frac{1}{\sin^2 \frac{\psi - \theta_j}{2}}, \tag{28.22}$$

where

$$\zeta_j = \begin{cases} \frac{1+|\rho_m|}{2} |(Q_B)_{mj}|^2, & \text{if} \quad 1 \leq j \leq m \\ \frac{1-|\rho_m|}{2} |(Q_C)_{1,j-m}|^2, & \text{if} \quad m < j \leq n \end{cases} \tag{28.23}$$

are positive numbers. The numbers $|\rho_m|$ are those which have been saved in Step 3 *of Algorithm* 28.1.

Proof. Consider the vector z given by (28.7) or (28.9) and denote $\zeta_j = |z_j|^2$, $j = 1, \dots, n$. As we are interested for the time being only in the modulus of the components of z and the modulus of all the diagonal entries of D_C^* is equal to the modulus of the diagonal entries of D_C, which in turn is equal to one since D_C contains the eigenvalues of the unitary matrix C, it follows that we can neglect the matrix D_C^* from the computation (28.9). Moreover, the modulus of the entries Q_B^* is the same if we do not perform conjugation but only transpose, and since Q_B^* multiplies a vector which has all its first $m-1$ components equal to zero, it follows that only the last component, which is $\frac{1+|\rho_m|}{2}$, counts, and from Q_B only the entries of its last row, which proves the result for $1 \leq j \leq m$. For the same reason, from Q_C only its first row counts and the coefficient of its entries is

$$\frac{\mu_m^2}{2(1+|\rho_m|)} = \frac{1-|\rho_m|^2}{2(1+|\rho_m|)} = \frac{1-|\rho_m|}{2}. \qquad \square$$

§28.2.8 Diagonalizing the matrix $D(I_N - zz^*)$

Thus we obtain the following algorithm to diagonalize the matrix $D(I_N - zz^*)$.

Algorithm 28.8. Let $D = \operatorname{diag}\left(e^{i\theta_1}, e^{i\theta_2}, \dots, e^{i\theta_N}\right)$ be a diagonal unitary matrix and let $z = (z_i)_{i=1}^N$ be a column vector. Then the unitary matrix V and the unitary diagonal matrix Λ such that

$$D(I_N - zz^*) = V\Lambda V^*$$

are obtained by the following algorithm.

1. Using Lemma 28.6, determine an integer $n \leq N$ and a unitary matrix Q_1 such that $D_1 = Q_1^* D Q_1 = \operatorname{diag}\left(e^{i\theta_1}, e^{i\theta_2}, \dots, e^{i\theta_N}\right)$ and $w = Q_1^* z$, with

$$\theta_1 < \theta_2 < \cdots < \theta_n,$$

$w_i \neq 0$, $i = 1, \dots, n$, and $w_i = 0$, $i = n+1, \dots, N$.

2. Compute the positive numbers $\zeta_j = |z_j|^2$, $j = 1, \dots, n$. By an iteration method compute the roots ψ_j, $j = 1, \dots, n$, of the complex function given in (28.21), namely

$$\Phi(\psi) = \sum_{j=1}^{n} \zeta_j \cot \frac{\psi - \theta_j}{2} \tag{28.24}$$

whose derivative is negative between to consecutive θ_j, $j = 1, \dots, n$, as it is given by (28.22) to be

$$\Phi'(\psi) = -\frac{1}{2} \sum_{j=1}^{n} \zeta_j \frac{1}{\sin^2 \frac{\psi - \theta_j}{2}}. \tag{28.25}$$

3. Compute the n corresponding eigenvectors v_j, $j = 1, \dots, n$, by using (28.19), namely

$$v_j = \begin{pmatrix} \frac{e^{(\theta_1 - \psi_j - \pi)\frac{i}{2}}}{2 \sin \frac{\psi_j - \theta_1}{2}} z_1 & \frac{e^{(\theta_2 - \psi_j - \pi)\frac{i}{2}}}{2 \sin \frac{\psi_j - \theta_2}{2}} z_2 & \cdots & \frac{e^{(\theta_n - \psi_j - \pi)\frac{i}{2}}}{2 \sin \frac{\psi_j - \theta_n}{2}} z_n \end{pmatrix}^T, \tag{28.26}$$

and normalize them using (28.25) above and the fact that

$$||v_j||^2 = -\frac{1}{2} \Phi'(\psi_j), \quad j = 1, \dots, n. \tag{28.27}$$

Denote the normalized vectors by $v_j^{(0)}$, $j = 1, \dots, n$.

4. Set

$$\hat{Q}_2 := \begin{bmatrix} v_1^{(0)} & v_2^{(0)} & \dots & v_n^{(0)} \end{bmatrix}, \quad \Lambda := \operatorname{diag}(\lambda_1, \dots, \lambda_n, \mu_{n+1}, \dots, \mu_N),$$

then set $Q_2 := \operatorname{diag}(\tilde{Q}_2, I_{N-n})$ and compute $V = Q_1 Q_2$.

§28.3 The complete algorithm

We can now present the complete divide and conquer algorithm for computing eigendecomposition of an upper Hessenberg unitary scalar matrix.

Algorithm 28.9. Let $A = \{A_{ij}\}_{i,j=1}^N$ be an unreduced unitary upper Hessenberg $N \times N$ matrix, where $N = 2^M$, with the Schur parameters $\rho_0 = -1, \rho_1, \dots, \rho_N$ and the complementary parameters $\mu_1, \mu_2, \dots, \mu_{N-1}$. Then the N eigenvalues $\lambda_N, \lambda_{N-1}, \dots, \lambda_1$ of A and a unitary matrix P such that

$$P^* A P = \operatorname{diag}(\lambda_N, \lambda_{N-1}, \dots, \lambda_1)$$

are obtained by the following algorithm.

1. For performing the divide step.
   Set
$$\rho^{(0,1)}(i) := \rho_i,\ i = 1,\dots,N, \quad \mu^{(0,1)}(j) := \mu_j,\ j = 1,\dots,N-1.$$
   For $p = 1,\dots,M$, perform the following.
   Set $s := 2^{p-1}, m := 2^{M-p}$. For $j = 1,2,\dots,s$, using the Schur parameters and the complementary parameters
$$\rho^{(p-1,j)}(i),\ i = 1,\dots,2m, \quad \mu^{(p-1,j)}(i),\ i = 1,\dots,2m-1,$$
   of the matrix $A^{(p-1,j)}$, compute via Algorithm 28.1 the Schur parameters and the complementary parameters
$$\rho^{(p,2j-1)}(i), \rho^{(p,2j)}(i),\ i = 1,\dots,m, \quad \mu^{(p,2j-1)}(i), \mu^{(p,2j)}(i),\ i = 1,\dots,m-1,$$
   of the matrices $A^{(p,2j-1)}, A^{(p,2j)}$. Set
$$\sigma^{(p-1,j)} = \mu^{(p-1,j)}(m), \quad c^{(p-1,j)} = \rho^{(p-1,j)}(m).$$
   For each application of Algorithm 28.1, the matrices $A^{(p,2j-1)}, A^{(p,2j)}$ are the ones denoted by B and C in the cited algorithm.
2. For performing the conquer step.
   Set
$$\Lambda^{(0,t)} := \rho^{(0,t)}(1), \quad Q^{(0,t)}(1) := 1, \quad t = 1,\dots,N.$$
   For $p = 1,2,\dots,M$ perform the following. Set $m = 2^{M-p},\ l = 2^{p-1}$.
   For $j = 1,\dots,m$, perform the following. Set $\sigma = \sigma^{(p-1,j)},\ \rho = |c^{(p-1,j)}|$.
   2.1. Compute $z^{(p,j)}$ as
$$\left(\begin{pmatrix} 0_{1\times(l-1)} & \sqrt{\frac{1+\rho}{2}} \end{pmatrix} Q^{(p-1,2j-1)} \quad \begin{pmatrix} -\frac{\sigma}{\sqrt{2(1+\rho)}} & 0_{1\times(l-1)} \end{pmatrix} Q^{(p-1,2j)}\Lambda^{(p-1,2j)}\right)^* \tag{28.28}$$
   and set $D^{(p,j)} := \Lambda^{(p-1,2j-1)} \oplus \Lambda^{(p-1,2j)}$.
   2.2. Using Algorithm 28.8, determine the eigendecomposition
$$D^{(p,j)} + z^{(p,j)}(z^{(p,j)})^* = V^{(p,j)}\Lambda^{(p,j)}(V^{(p,j)})^*$$
   with a unitary matrix $V^{(p,j)}$ and a diagonal matrix with unit entries $\Lambda^{(p,j)}$.
   2.3. Compute
$$Q^{(p,j)} = \begin{pmatrix} Q^{(p-1,2j-1)} & 0 \\ 0 & Q^{(p-1,2j)} \end{pmatrix} V^{(p,j)}.$$
3. Set $P =: Q^{(M,1)},\ D := D^{(M,1)}$.

§28.4 Example

Example 28.10. Consider the 6×6 unitary upper Hessenberg matrix

$$A = \begin{pmatrix} 0 & 0 & 0 & 0 & 0 & i \\ 1 & 0 & 0 & 0 & 0 & 0 \\ 0 & 1 & 0 & 0 & 0 & 0 \\ 0 & 0 & 1 & 0 & 0 & 0 \\ 0 & 0 & 0 & 1 & 0 & 0 \\ 0 & 0 & 0 & 0 & 1 & 0. \end{pmatrix}. \tag{28.29}$$

This matrix is also the companion matrix of the polynomial $\lambda^6 - i$, therefore its eigenvalues are the order-6 roots of the complex number $i = \cos\frac{\pi}{2} + i\sin\frac{\pi}{2}$, which are known to be

$$\lambda_j = \cos\frac{\frac{\pi}{2} + 2j\pi}{6} + i\sin\frac{\frac{\pi}{2} + 2j\pi}{6}, \quad j = 0, \ldots, 5. \tag{28.30}$$

The matrix A from (28.29) is in fact

$$A(\rho_1, \rho_2, \rho_3, \rho_4, \rho_5, \rho_6) = A(0, 0, 0, 0, 0, i),$$

where the notation is like in (25.70), therefore $\mu_k = 1,\ k = 1, \ldots, 5$.

Take $m = \left[\frac{N}{2}\right] = 3$. By (25.80), $\tau = -\frac{\beta}{\alpha}$ is a unit number, say

$$\tau = e^{i\vartheta}, \quad -\pi < \vartheta \leq \pi. \tag{28.31}$$

From (25.74) it follows that $G_0 = \begin{pmatrix} -1 & 0 & 0 \end{pmatrix}^T$ and we can use this computation and formula (25.78) to obtain that

$$B = \begin{pmatrix} 0 & 0 & e^{i\vartheta} \\ 1 & 0 & 0 \\ 0 & 1 & 0 \end{pmatrix}. \tag{28.32}$$

This matrix is also the companion matrix of the polynomial $\lambda^3 - e^{i\vartheta}$, therefore its eigenvalues are the order-3 roots of the complex number $\cos\vartheta + i\sin\vartheta$, which are known to be

$$\lambda_k = \cos\frac{\vartheta + 2k\pi}{3} + i\sin\frac{\vartheta + 2k\pi}{3}, \quad k = 0, 1, 2. \tag{28.33}$$

We will now compute the matrix C. By (25.75).

$$H_0 = \begin{pmatrix} 0 & 0 & -i \end{pmatrix},$$

and since $\tau^* = e^{-i\vartheta}$, we use formula (25.79) and we obtain that C is

$$C = \begin{pmatrix} 0 & 0 & ie^{-i\vartheta} \\ 1 & 0 & 0 \\ 0 & 1 & 0 \end{pmatrix} = \begin{pmatrix} 0 & 0 & e^{i(\frac{\pi}{2} - \vartheta)} \\ 1 & 0 & 0 \\ 0 & 1 & 0 \end{pmatrix}. \tag{28.34}$$

This matrix is also the companion matrix of the polynomial $\lambda^3 - e^{i(\frac{\pi}{2}-\vartheta)}$, therefore its eigenvalues are the order-3 roots of the complex number $\cos(\frac{\pi}{2}-\vartheta)+i\sin(\frac{\pi}{2}-\vartheta)$, which are known to be

$$\lambda_n = \cos\frac{(\frac{\pi}{2}-\vartheta)+2n\pi}{3} + i\sin\frac{(\frac{\pi}{2}-\vartheta)+2n\pi}{3}, \quad n = 0,1,2. \tag{28.35}$$

If we choose for instance $\beta = 1, \alpha = -1$, then in (28.31) $\vartheta = 0$, From (28.30), (28.33) and (28.35) it follows that the arguments of the 6 eigenvalues of A are

$$\frac{\pi}{12}, \quad \frac{5\pi}{12}, \quad \frac{9\pi}{12}, \quad \frac{13\pi}{12}, \quad \frac{17\pi}{12}, \quad \frac{21\pi}{12}, \tag{28.36}$$

while the arguments of the 6 eigenvalues of B and C combined are, respectively,

$$0, \quad \frac{2\pi}{3}, \quad \frac{4\pi}{3}, \quad \text{and} \quad \frac{\pi}{6}, \quad \frac{5\pi}{6}, \quad \frac{9\pi}{6}. \tag{28.37}$$

Therefore, there are no common eigenvalues and Theorem 26.1 and its corollaries apply.

Hence, in this case we can perform both the divide step and the conquer step. For the divide step, given the following Schur parameters and complementary parameters of A

$$\rho_0 = -1, \quad \rho_k = 0, \; \mu_k = 1, \; k = 1,\dots,5, \quad \rho_6 = 1,$$

use Algorithm 28.1 Step 1. to obtain the Schur parameters and complementary parameters of B,

$$\rho_0^B = -1, \quad \rho_1^B = \rho_2^B = 0, \quad \mu_1^B = \mu_2^B = 1, \quad \rho_3^B = 1$$

which give indeed

$$B = \begin{pmatrix} 0 & 0 & 1 \\ 1 & 0 & 0 \\ 0 & 1 & 0 \end{pmatrix}$$

as in (28.32), where we take $\vartheta = 0$.

Now use Algorithm 28.1, Step 2 to obtain the Schur parameters and complementary parameters of C,

$$\rho_0^C = -1, \quad \rho_1^C = \rho_2^C = 0, \quad \mu_1^C = \mu_2^C = 1, \quad \rho_3^C = i,$$

which give indeed

$$C = \begin{pmatrix} 0 & 0 & i \\ 1 & 0 & 0 \\ 0 & 1 & 0 \end{pmatrix}$$

as in (28.34), where we take $\vartheta = 0$. The divide step is accomplished.

Suppose now that we know, for instance from a previously conquer step, the eigendecompositions of B and C and then use Algorithm 28.8 to find the eigendecomposition of A. Namely, we are given the arguments $-\pi < \theta_j \leq \pi$ of the eigenvalues of B as in (28.37) $-\frac{2\pi}{3}$, 0, $\frac{2\pi}{3}$ and the unitary diagonalizing matrix

$$Q_B = \frac{1}{2\sqrt{3}} \begin{pmatrix} -2 & 2 & -2 \\ 1-\sqrt{3}i & 2 & 1+\sqrt{3}i \\ 1+\sqrt{3}i & 2 & 1-\sqrt{3}i \end{pmatrix},$$

and the arguments of the eigenvalues of C as in (28.37) $-\frac{3\pi}{6}$, $\frac{\pi}{6}$, $\frac{5\pi}{6}$ and the unitary diagonalizing matrix

$$Q_C = \frac{1}{2\sqrt{3}} \begin{pmatrix} 2 & 2i & 2i \\ 2i & 1+\sqrt{3}i & 1-\sqrt{3}i \\ -2 & \sqrt{3}+i & -\sqrt{3}+i \end{pmatrix}.$$

We will first compute the numbers ζ_j, $j = 1, \ldots, 6$ from Step 1 of Algorithm 28.8. Since the modulus of all the entries of the last row of Q_B and of the first row of Q_C is $\frac{1}{3}$ and since $\sigma = 1$, it follows that $\zeta_j = \frac{1}{6}$, $j = 1, \ldots, 6$. Therefore the function $\Phi(\psi)$ from Step 2 is

$$\begin{aligned} \Phi(\psi) = \frac{1}{6}\bigg(&\cot\left(\frac{\psi}{2} - \frac{5\pi}{12}\right) + \cot\left(\frac{\psi}{2} - \frac{\pi}{3}\right) + \cot\left(\frac{\psi}{2} - \frac{\pi}{12}\right) \\ &+ \cot\left(\frac{\psi}{2}\right) + \cot\left(\frac{\psi}{2} + \frac{\pi}{4}\right) + \cot\left(\frac{\psi}{2} + \frac{\pi}{3}\right)\bigg). \end{aligned} \tag{28.38}$$

It is a simple calculation to check that for the six angles from (28.36) this function vanishes and that those angles interlace with the 6 arguments of the eigenvalues of B and C combined. Let us list the angles $\frac{\psi_j}{2}$, $j = 1, \ldots, 6$ of the arguments in (28.36), seen as increasing angles between $-\pi$ and π:

$$\frac{-11\pi}{24}, \quad \frac{-7\pi}{24}, \quad \frac{-3\pi}{24}, \quad \frac{\pi}{24}, \quad \frac{5\pi}{24}, \quad \frac{9\pi}{24}.$$

We check, for instance, the third one in (28.38):

$$\frac{1}{6}\left(-\cot\frac{13\pi}{24} - \cot\frac{11\pi}{24} - \cot\frac{5\pi}{24} - \cot\frac{3\pi}{24} + \cot\frac{3\pi}{24} + \cot\frac{5\pi}{24}\right) = 0. \qquad \diamond$$

§28.5 Comments

The algorithm presented here was developed by M. Gu, R. Guzzo, X.-B. Chi and X.-Q. Cao in the paper [28]. We obtain this result as an application of our general method to the case of unitary upper Hessenberg matrices.

Part VII

Algorithms for QR Iterations and for Reduction to Hessenberg Form

Introduction to Part VII

The QR iteration method is a customary tool to compute all the eigenvectors of a matrix. In this part we develop fast QR iteration algorithms using quasiseparable representations of matrices. Unfortunately, in general the quasiseparable structure is not invariant under QR iterations. Thus one can obtain such algorithms only for special cases, such as Hermitian matrices, unitary matrices and their perturbations of small ranks.

In the first chapter we study explicit iteration methods. We show that the orders of lower quasiseparable generators of matrices do not increase in the iteration process. This allows one to obtain fast algorithms for the cases when one can reconstruct upper quasiseparable generators of a matrix in terms of lower ones and diagonal entries. This is valid for the aforementioned Hermitian and unitary matrices and their perturbations of small ranks.

An effective method for solving the eigenvalue problem is a preliminary reduction of a matrix to a unitary similar one in upper Hessenberg form. The subsequent QR iterations for upper Hessenberg matrices can be performed in an efficient way. The reduction to Hessenberg form method for matrices with quasiseparable representations is studied in the second chapter.

In the last chapter we study the implicit QR iteration methods for upper Hessenberg matrices obtained via the reduction step.

Chapter 29

The QR Iteration Method for Eigenvalues

Here we study the explicit QR iteration method for matrices with quasiseparable representations.

§29.1 The general description

The QR iteration method for the computation of eigenvalues of a matrix $A \in \mathbb{C}^{N \times N}$ consists of the following. One sets $A^{(0)} = A$ and for $k = 0, 1, 2, \ldots$ one computes the QR factorizations

$$A^{(k)} - \sigma_k I = Q^{(k)} R^{(k)}, \tag{29.1}$$

where $Q^{(k)}$ are unitary upper Hessenberg matrices, $R^{(k)}$ are upper triangular matrices, and σ_k are appropriate complex numbers, and next one computes the new iterates by the formulas

$$A^{(k+1)} = \sigma_k I + R^{(k)} Q^{(k)}. \tag{29.2}$$

The description of the choice of the numbers σ_k can be found in [35, 37]. The matrices $A^{(k)}$ are unitarily similar to the matrix A. Under some conditions (see [35, 37]), the sequence of complex matrices $A^{(k)}$ converges to an upper triangular matrix. Hence, the diagonal entries of the matrices $A^{(k)}$ approximate the eigenvalues of the matrix A. More details about the QR iteration method can be found in [18, 35, 37]. For a general matrix the complexity of the eigenvalue computation by QR iteration is $O(N^2)$ per iteration and $O(N^3)$ to compute all the eigenvalues. Using the quasiseparable structure the complexity may be reduced to $O(N)$ per iteration and $O(N^2)$ to compute all the set of eigenvalues.

§29.2 The QR iteration step via quasiseparable structure

We consider a step of the QR iteration algorithm for matrices with quasiseparable representations. In each iteration step, for a given matrix A and for a given number σ, the new iterate A_1 is obtained by the rule

$$\begin{cases} A - \sigma I = QR, \\ \quad A_1 = \sigma I + RQ, \end{cases}$$

where Q is a unitary matrix and R is an upper triangular matrix.

Here we present an algorithm to compute quasiseparable generators of the new iterate A_1. In this algorithm we obtain lower quasiseparable generators of the matrix A_1 with orders not greater than the orders of the corresponding generators of the matrix A. However, we will see that the orders of upper quasiseparable generators of the new iterate may increase.

Theorem 29.1. *Let $A = \{A_{ij}\}_{i,j=1}^N$ be a scalar matrix with lower quasiseparable generators $p(i)$ $(i = 2, \dots, N)$, $q(j)$ $(j = 1, \dots, N-1)$, $a(k)$ $(k = 2, \dots, N-1)$ of orders r_k^L $(k = 1, \dots, N-1)$, upper quasiseparable generators $g(i)$ $(i = 1, \dots, N-1)$, $h(j)$ $(j = 2, \dots, N)$, $b(k)$ $(k = 2, \dots, N-1)$ of orders r_k^U $(k = 1, \dots, N-1)$, and diagonal entries $d(k)$ $(k = 1, \dots, N)$, and let σ be a number. Let us define the numbers ρ_k via the recursion relations*

$$\rho_N = 0, \ \rho_{k-1} = \min\{1 + \rho_k, \ r_{k-1}^L\}, \ k = N, \dots, 2, \ \rho_0 = 0, \tag{29.3}$$

and the numbers

$$m_k = 1, n_k = 1, \nu_k = 1 + \rho_k - \rho_{k-1}, \ \rho'_k = r_k^U + \rho_k, \ k = 1, \dots, N. \tag{29.4}$$

Define the matrix A_1 by the rule

$$\begin{cases} A - \sigma I = QR, \\ \quad A_1 = \sigma I + RQ, \end{cases} \tag{29.5}$$

where Q is a unitary matrix and R is an upper triangular matrix.

Then lower quasiseparable generators $p^{(1)}(i)$ $(i = 2, \dots, N)$, $q^{(1)}(j)$ $(j = 1, \dots, N-1)$, $a^{(1)}(k)$ $(k = 2, \dots, N-1)$ of orders ρ_k $(k = 1, \dots, N-1)$, upper quasiseparable generators $g^{(1)}(i)$ $(i = 1, \dots, N-1)$, $h^{(1)}(j)$ $(j = 2, \dots, N)$, $b^{(1)}(k)$ $(k = 2, \dots, N-1)$ of orders ρ'_k $(k = 1, \dots, N-1)$, and diagonal entries $d^{(1)}(k)$ $(k = 1, \dots, N)$ of the matrix A_1 are obtained via the following algorithm.

1.1. *If $r_{N-1}^L > 0$, set*

$$C_N = p(N), \quad p_V(N) = 1, \quad h_R(N) = \begin{pmatrix} h(N) \\ d(N) - \sigma \end{pmatrix}, \tag{29.6}$$

$d_V(N)$ to be the 1×0 empty matrix, Δ_N to be the 0×1 empty matrix; if $r^L_{N-1}=0$, set C_N to be the 0×0 empty matrix, $p_V(N)$ to be the 1×0 empty matrix,

$$d_V(N)=1,\quad h_R(N)=h(N),\quad \Delta_N=d(N)-\sigma. \tag{29.7}$$

1.2. *For $k=N-1,\dots,2$ perform the following. Compute the QR factorization*

$$\begin{pmatrix} p(k) \\ C_{k+1}a(k) \end{pmatrix} = V_k \begin{pmatrix} C_k \\ 0_{\nu_k\times r^L_{k-1}} \end{pmatrix}, \tag{29.8}$$

where V_k is a unitary $(1+\rho_k)\times(1+\rho_k)$ matrix and C_k is a $\rho_{k-1}\times r^L_{k-1}$ matrix. Determine matrices $p_V(k)$, $a_V(k)$, $d_V(k)$, $q_V(k)$ of sizes $1\times\rho_{k-1}$, $\rho_k\times\rho_{k-1}$, $1\times\nu_k$, $\rho_k\times\nu_k$ from the partition

$$V_k=\begin{pmatrix} p_V(k) & d_V(k) \\ a_V(k) & q_V(k) \end{pmatrix}. \tag{29.9}$$

Compute

$$\begin{aligned} h'_k &= p^*_V(k)(d(k)-\sigma)+a^*_V(k)C_{k+1}q(k),\\ h_R(k) &= \begin{pmatrix} h(k) \\ h'_k \end{pmatrix},\\ b_R(k) &= \begin{pmatrix} b(k) & 0 \\ p^*_V(k)g(k) & a^*_V(k) \end{pmatrix},\\ \Theta_k &= \left[\begin{array}{cc} d^*_V(k)g(k) & q^*_V(k) \end{array}\right],\\ \Delta_k &= d^*_V(k)(d(k)-\sigma)+q^*_V(k)C_{k+1}q(k). \end{aligned} \tag{29.10}$$

1.3. *Set $V_1=I_{\nu_1}$ and define matrices $d_V(1)$, $q_V(1)$ of sizes $1\times\nu_1$, $\rho_1\times\nu_1$ from the partition*

$$V_1=\begin{pmatrix} d_V(1) \\ q_V(1) \end{pmatrix}; \tag{29.11}$$

compute

$$\Delta_1=\begin{pmatrix} d(1)-\sigma \\ C_2q(1) \end{pmatrix},\ \Theta_1=\begin{pmatrix} g(1) & 0 \\ 0 & I_{\rho_1} \end{pmatrix}. \tag{29.12}$$

Thus we have computed lower quasiseparable generators and diagonal entries of the matrix V and upper quasiseparable generators $b_R(k)$, $h_R(k)$ of the matrix R.

2.1. *Compute the QR factorization*

$$\Delta_1=U_1\left[\begin{array}{c} d_R(1) \\ 0_{\rho_1\times 1} \end{array}\right], \tag{29.13}$$

where U_1 is a unitary $\nu_1 \times \nu_1$ matrix and $d_R(1)$ is a number. Determine matrices $d_U(1)$, $g_U(1)$ of sizes $\nu_1 \times 1$, $\nu_1 \times \rho'_1$ from the partition

$$U_1 = \begin{pmatrix} d_U(1) & g_U(1) \end{pmatrix}. \tag{29.14}$$

Compute

$$g_R(1) = d_U^*(1)\Theta_1, \quad D_1 = g_U^*(1)\Theta_1. \tag{29.15}$$

2.2. *For $k = 2, \dots, N-1$ perform the following. Compute the QR factorization*

$$\begin{pmatrix} D_{k-1}h_R(k) \\ \Delta_k \end{pmatrix} = U_k \begin{pmatrix} d_R(k) \\ 0_{\rho_k \times 1} \end{pmatrix}, \tag{29.16}$$

where U_k is a unitary $(1+\rho_k) \times (1+\rho_k)$ matrix and $d_R(k)$ is a number. Determine matrices $d_U(k)$, $g_U(k)$, $h_U(k)$, $b_U(k)$ of sizes $(1+\rho_k-\rho_{k-1}) \times 1$, $(1+\rho_k-\rho_{k-1}) \times \rho_k$, $\rho_{k-1} \times 1$, $\rho_{k-1} \times \rho_k$ from the partition

$$U_k = \begin{pmatrix} h_U(k) & b_U(k) \\ d_U(k) & g_U(k) \end{pmatrix}. \tag{29.17}$$

Compute

$$\begin{aligned} g_R(k) &= h_U^*(k)D_{k-1}b_R(k) + d_U^*(k)\Theta_k, \\ D_k &= b_U^*(k)D_{k-1}b_R(k) + g_U^*(k)\Theta_k. \end{aligned} \tag{29.18}$$

2.3. *If $r^L_{N-1} > 0$, set $h_U(N) = 1$ and $d_U(N)$ to be the 0×1 empty matrix; if $r^L_{N-1} = 0$, set $d_U(N) = 1$ and $h_U(N)$ to be the 0×1 empty matrix; compute*

$$d_R(N) = \begin{pmatrix} D_{N-1}h_R(N) \\ \Delta_N \end{pmatrix}. \tag{29.19}$$

Thus we have computed generators and diagonal entries of the matrix U and generators $g_R(k)$ and diagonal entries $d_R(k)$ of the matrix R.

3. *Compute quasiseparable generators of the matrix $Q = VU$ as follows.*

3.1. *Determine the matrices $d_Q(1), g_Q(1), q_Q(1), \beta_1$ of the sizes $m_1 \times n_1, m_1 \times s_1, \rho_1 \times n_1, \rho_1 \times s_1$ from the partition*

$$U_1 = \begin{pmatrix} d_Q(1) & g_Q(1) \\ q_Q(1) & \beta_1 \end{pmatrix}. \tag{29.20}$$

3.2. *For $k = 2, \dots, N-1$ perform the following. Set*

$$\begin{aligned} p_Q(k) = p_V(k), \quad a_Q(k) = a_V(k), \\ h_Q(k) = h_U(k), \quad b_Q(k) = b_U(k). \end{aligned} \tag{29.21}$$

Compute the matrix

$$W_k = V_k \begin{pmatrix} \beta_{k-1} & 0 \\ 0 & I_{\nu_k} \end{pmatrix} U_k \tag{29.22}$$

and determine the matrices $d_Q(k), g_Q(k), q_Q(k), \beta_k$ *of the sizes* $m_k \times n_k, m_k \times \rho_k, \rho_k \times n_k, \rho_k \times \rho_k$ *from the partition*

$$W_k = \begin{pmatrix} d_Q(k) & g_Q(k) \\ q_Q(k) & \beta_k \end{pmatrix}. \tag{29.23}$$

3.3. *Set*

$$p_Q(N) = p_V(N), \ h_Q(N) = h_U(N). \tag{29.24}$$

Compute

$$d_Q(N) = d_V(N)d_U(N) + p_V(N)\beta_{N-1}h_U(N). \tag{29.25}$$

4. *Compute quasiseparable generators of the matrix* A_1 *as follows.*

4.1. *Compute*

$$\begin{gathered} \gamma_N = h_R(N)p_Q(N), \quad p^{(1)}(N) = d_R(N)p_Q(N), \\ d^{(1)}(N) = d_R(N)d_Q(N) + \sigma, \end{gathered} \tag{29.26}$$

$$h^{(1)}(N) = \begin{pmatrix} h_Q(N) \\ h_R(N)d_Q(N) \end{pmatrix}. \tag{29.27}$$

4.2. *For* $k = N-1, \dots, 2$ *perform the following. Set*

$$q^{(1)}(k) = q_Q(k), \quad a^{(1)}(k) = a_Q(k). \tag{29.28}$$

Compute the matrix

$$\Omega_k = \begin{pmatrix} h_R(k) & b_R(k) \\ d_R(k) & g_R(k) \end{pmatrix} \begin{pmatrix} 1 & 0 \\ 0 & \gamma_{k+1} \end{pmatrix} \begin{pmatrix} p_Q(k) & d_Q(k) \\ a_Q(k) & q_Q(k) \end{pmatrix} \tag{29.29}$$

and determine the matrices $\gamma_k, \tilde{h}(k), p^{(1)}(k), d''(k)$ *of the sizes* $\rho'_{k-1} \times \rho_{k-1}, \rho'_{k-1} \times m_k, m_k \times \rho_{k-1}, m_k \times m_k$ *from the partition*

$$\Omega_k = \begin{pmatrix} \gamma_k & \tilde{h}(k) \\ p^{(1)}(k) & d''(k) \end{pmatrix}. \tag{29.30}$$

Compute

$$d^{(1)}(k) = d''(k) + \sigma, \tag{29.31}$$

$$\begin{gathered} g^{(1)}(k) = \begin{pmatrix} d_R(k)g_Q(k) & g_R(k) \end{pmatrix}, \quad h^{(1)}(k) = \begin{pmatrix} h_Q(k) \\ \tilde{h}(k) \end{pmatrix}, \\ b^{(1)}(k) = \begin{pmatrix} b_Q(k) & 0 \\ h_R(k)g_Q(k) & b_R(k) \end{pmatrix}. \end{gathered} \tag{29.32}$$

4.3. *Set* $q^{(1)}(1) = q_Q(1)$. *Compute*

$$d^{(1)}(1) = d_R(1)d_Q(1) + g_R(1)\gamma_2 q_Q(1) + \sigma, \tag{29.33}$$

$$g^{(1)}(1) = \begin{pmatrix} d_R(1)g_Q(1) & g_R(1) \end{pmatrix}. \tag{29.34}$$

Proof. We start applying the algorithm from Theorem 20.10 to the matrix $A-\sigma I$, which has the same lower and upper quasiseparable generators as the matrix A and diagonal entries $d(k)-\sigma$ $(k=1,\dots,N)$. As a result we obtain the factorization $A-\sigma I = VUR$ with the block lower triangular unitary matrix V, block upper triangular unitary matrix U, and the upper triangular matrix R. Moreover, computing the quasiseparable generators of the matrices V, U, R according to the algorithm from Theorem 20.10, we justify Step 1 and Step 2 of the algorithm under consideration.

Next, on Step 3, we apply Theorem 17.4 to compute quasiseparable generators of the matrix $Q = VU$. The formula (17.10) with $k=1$ implies

$$\begin{pmatrix} d_Q(1) & g_Q(1) \\ q_Q(1) & \beta_1 \end{pmatrix} = \begin{pmatrix} d_V(1) \\ q_V(1) \end{pmatrix} \begin{pmatrix} d_U(1) & g_U(1) \end{pmatrix}.$$

Combining this with (29.11), (29.14) and using the fact that $V_1 = I$, we obtain the equality (29.20). For $k=2,\dots,N-1$ using the formulas (17.8) we obtain the formulas (29.21) and using the formula (17.10) and the representations (29.9), (29.17) we obtain the formulas (29.22), (29.23). For $k=N$ using (17.8) we obtain (29.24), and using (17.10) we obtain (29.25).

Finally, on Step 4, in order to compute quasiseparable generators of the matrix $A^{(1)} = RQ+\sigma I$, we apply Corollary 17.8 to the matrix RQ. Using (17.49) with $k=N$ we obtain (29.26) and the equality $\tilde{h}(N) = h_R(N)d_Q(N)$. Inserting the latter in (17.51) we obtain (29.27). For $k=N-1,\dots,2$, using (17.50) we obtain the formulas (29.28). Using the formula (17.49) we obtain the formulas (29.29), (29.30). The formula (29.31) is obvious. Next, using (17.51), (17.52) we obtain the relations (29.32). For $k=1$, using the first formula in (17.50) we obtain $q^{(1)}(1) = q_Q(1)$. Using (17.49) with $k=1$, we get (29.33). Finally, using the first formula from (17.51), we obtain (29.34). □

Remark. Using Corollary 20.6 or the definition of the numbers ρ_k in Theorem 29.1, we obtain the inequalities

$$\rho_k \le r_k^L, \quad k=1,\dots,N-1, \tag{29.35}$$

for the orders of lower quasiseparable generators of the matrices A and A_1. Thus, these orders do not increase in the iteration process.

As to the upper quasiseparable generators, their orders in general increase in the iteration process. Indeed, from formula (29.32) it follows that the orders $(r_k^U)^{(1)}$, $k=1,\dots,N-1$ of the upper quasiseparable generators of the matrix A_1 which emerge from the iteration process are

$$(r_k^U)^{(1)} = \rho_k + \rho_k' = 2\rho_k + r_k^U \le 2r_k^L + r_k^U, \; k=1,\dots,N-1,$$

and if for the initial matrix A we have $r_k^U \le r_k^L$, $k=1,\dots,N-1$, then the orders of the upper quasiseparable generators of the matrix A_1 satisfy

$$r_k^U < (r_k^U)^{(1)} \le 3r_k^L, \; k=1,\dots,N-1. \tag{29.36}$$

However, for some classes of matrices the iterates have also upper quasiseparable generators of non-increasing orders. In this case the formulas (29.27), (29.32), (29.34) are not suitable, since they yield generators with redundant orders. In the sequel we consider such classes of matrices and derive algorithms to compute upper quasiseparable generators of iterates with small orders.

§29.2.1 Complexity of the algorithm in Theorem 29.1

We will confine ourselves to the operations that are not performed just once, but about N times. These are the computations of the numbers $\rho_k = s_k, \nu_k$ and ρ'_k for $k = 1, \dots, N$, which comprise in total $5N$ operations of comparison or arithmetical addition, as well as the formulas in Steps 1.2, 2.2, 3.2 and 4.2 of the algorithm, in which each operation is performed $N-2$ times.

For the Step 1.2 in formula (29.8) we perform a matrix multiplication which requires $\rho_k r^L_k r^L_{k-1}$ arithmetical multiplications and $\rho_k(r^L_k - 1)r^L_{k-1}$ arithmetical additions and a QR decomposition of a $(1+\rho_k) \times r^L_{k-1}$ matrix. If we denote by $\mu(k,m)$ the number of arithmetical operations needed for the QR decomposition of a $k \times m$ matrix, then this formula requires less than $\mu(1+r^L_k, r^L_{k-1}) + 2(r^L_k)^2 r^L_{k-1}$ arithmetical operations.

In formulas (29.10) we perform complex conjugation of ρ_{k-1} plus $\rho_{k-1}\rho_k$ plus ν_k numbers and in the computation of h'_k three matrix multiplications, which require $\rho_{k-1} + \rho_{k-1}\rho_k r^L_k + \rho_k r^L_k$ arithmetical multiplications and $\rho_{k-1}(\rho_k - 1)r^L_k + \rho_k(r^L_k - 1)$ arithmetical additions. In the computation of $b_R(k)$ we perform a matrix multiplication, which requires $\rho_{k-1} r^U_k$ arithmetical multiplications and in the computations of Θ_k, Δ_k we perform four matrix multiplications, which require $\nu_k r^U_k + \nu_k + \nu_k \rho_k r^L_k + \rho_k r^L_k$ arithmetical multiplications and $\nu_k(\rho_k - 1)r^L_k + \rho_k(r^L_k - 1)$ arithmetical additions. In total, formulas (29.10) require less than

$$r^L_{k-1}(r^L_k + 3 + 2(r^L_k)^2 + r^U_k) + (r^L_k)^2 + \nu_k(2 + r^U_k + 2(r^L_k)^2) + 2(r^L_k)^2$$

arithmetical operations.

If $r_L = \max_{1 \le k \le N-1}(r^L_k)$ and $r_U = \max_{1 \le k \le N-1}(r^U_k)$, then Step 1.2 of the algorithm costs less than

$$(\mu(1 + r_L, r_L) + 6r_L^3 + 4r_L^2 + 5r_L + 2r_L r_U)N$$

arithmetical operations.

In Step 2.2 formula (29.16) performs two matrix multiplications which require $\rho_{k-1}\rho'_{k-1} + \rho_{k-1}\rho'_{k-1}\rho'_k$ arithmetical multiplication and less than this number additions. This step also performs a QR decomposition which requires $\mu(1+\rho_k, 1+\rho'_k)$ operations. In total, Step 2.2 costs less than

$$(\mu(1 + r_L, 1 + r_L + r_U) + 2r_L(r_L + r_U)(1 + r_L + r_U))N$$

arithmetical operations.

In Step 3.2 formula (29.22) performs two matrix multiplications which require $2(1+\rho_k)^3$ arithmetical multiplication and less than this number additions. In total, Step 3.2 costs less than

$$4(1+r_L)^3N$$

arithmetical operations.

In Step 4.2 formula (29.29) performs two matrix multiplications which require $(1+\rho'_{k-1})(1+\rho'_k)(1+\rho_k)+(1+\rho'_k)(1+\rho_k)(1+\rho_{k-1})$ arithmetical multiplication and less than this number additions. Finally, the formulas in (29.32) require $\rho_k+\rho'_k\rho_k$ arithmetical multiplications. In total, Step 4.2 costs less than

$$(2(1+r_L+r_U)(1+r_L)(2+2r_L+r_U)+r_L(1+r_L+r_U))N$$

arithmetical operations.

It follows that the overall complexity c_1 of the algorithm in Theorem 29.1 which builds one iteration satisfies

$$\begin{aligned} c_1 < \big(\mu(1+r_L,r_L)+\mu(1+r_L,1+r_L+r_U)+16r_L^3+10r_L^2r_U+4r_Lr_U^2+31r_L^2 \\ +15r_Lr_U+2r_U^2+30r_L+6r_U+9\big)N \end{aligned} \quad (29.37)$$

and if the orders of the upper quasiseparable generators of the initial matrix A are at most equal to the orders of its lower quasiseparable generators, then

$$c_1 < (\mu(1+r_L,r_L)+\mu(1+r_L,1+2r_L)+30r_L^3+48r_L^2+36r_L+9)N. \quad (29.38)$$

§29.3 Recovering of the upper triangular structure

In general, the upper quasiseparable structure is not invariant under QR iterations. We consider here classes of matrices for which the invariance of the upper quasiseparable structure takes place and derive algorithms to recover this structure.

§29.3.1 Hermitian matrices

The first case is that of Hermitian matrices. For a Hermitian matrix A with the given lower quasiseparable generators $p(i)$ $(i=2,\dots,N)$, $q(j)$ $(j=1,\dots,N)$, $a(k)$ $(k=2,\dots,N-1)$, the upper quasiseparable generators are obviously obtained by the formulas

$$\begin{aligned} g(i)&=q^*(i),\ i=1,\dots,N-1,\\ h(i)&=p^*(i),\ i=2,\dots,N,\\ b(i)&=a^*(i),\ i=2,\dots,N-1. \end{aligned} \quad (29.39)$$

In this case there are some changes in Step 1 of the algorithm in Theorem 29.1. In formulas (29.6) of substep 1.1, the part that computes $h_R(N)$, one must replace $h(N)$ by the $r^L_{N-1} \times 1$ column $p^*(N)$. In substep 1.2, formulas (29.10), the part that computes $h_R(k)$, one must replace $h(k)$ by the $r^L_{k-1} \times 1$ column $p^*(k)$. In the part that computes $b_R(k)$, one must replace $g(k)$ by the $1 \times r^L_k$ row $q^*(k)$, while $b(k)$ is to be replaced by the $r^L_{k-1} \times r^L_k$ matrix $a^*(k)$. Finally, in substep 1.3, formulas (29.12), the part that computes Θ_1, one must replace $g(1)$ by the $1 \times r^L_1$ row $q^*(1)$. Also in the Hermitian case the formulas (29.27), (29.32), (29.34) in Step 4 are not needed and $r_U = r_L$.

Thus, the total complexity of the algorithm in Theorem 29.1 is

$$c < (\mu(1 + r_L, r_L) + \mu(1 + r_L, 1 + 2r_L) + 30r_L^3 + 46r_L^2 + 35r_L + 9)N \quad (29.40)$$

arithmetical operations.

§29.3.2 Unitary matrices

To recover upper quasiseparable generators of a unitary matrix is a more complicated problem than in the Hermitian case. If the conditions of Theorem 12.4 hold one can apply the completion formulas from this theorem. In the general situation we use a compression method which is based on Theorem 7.5. Notice that from (7.22) it follows that for scalar matrices the numbers s_k and ρ_k used in Theorem 7.5 are equal.

Thus, in the case of a unitary matrix one can proceed as follows.

Algorithm 29.2. Let $A = \{A_{ij}\}_{i,j=1}^N$ be a scalar unitary matrix with lower quasiseparable generators $p(i)$ $(i = 2, \dots, N)$, $q(j)$ $(j = 1, \dots, N-1)$, $a(k)$ $(k = 2, \dots, N-1)$ of orders r^L_k $(k = 1, \dots, N-1)$, upper quasiseparable generators $g(i)$ $(i = 1, \dots, N-1)$, $h(j)$ $(j = 2, \dots, N)$, $b(k)$ $(k = 2, \dots, N-1)$ of orders r^U_k $(k = 1, \dots, N-1)$, and diagonal entries $d(k)$ $(k = 1, \dots, N)$, and let σ be a number. Then quasiseparable generators of the matrix A_1 from (29.5) are obtained as follows.

1. Using the algorithm from Theorem 12.4 determine lower quasiseparable generators $p^{(1)}(i)$ $(i = 2, \dots, N)$, $q^{(1)}(j)$ $(j = 1, \dots, N-1)$, $a^{(1)}(k)$ $(k = 2, \dots, N-1)$ of orders ρ_k $(k = 1, \dots, N-1)$ as defined in (29.3), upper quasiseparable generators $g^{(1)}(i)$ $(i = 1, \dots, N-1)$, $h^{(1)}(j)$ $(j = 2, \dots, N)$, $b^{(1)}(k)$ $(k = 2, \dots, N-1)$ of orders ρ'_k $(k = 1, \dots, N-1)$ as defined in (29.4), and diagonal entries $d^{(1)}(k)$ $(k = 1, \dots, N)$ of the matrix A_1.
2. Applying the algorithm from Theorem 7.5 to the matrix A_1, determine upper quasiseparable generators $g_s^{(1)}(i)$ $(i = 1, \dots, N-1)$, $h_s^{(1)}(j)$ $(j = 2, \dots, N)$, $b_s^{(1)}(k)$ $(k = 2, \dots, N-1)$ of orders ρ_k $(k = 1, \dots, N-1)$.

To calculate the complexity of Algorithm 29.2 one should compute the complexity of the algorithm in Theorem 7.5. We will constrain ourselves to the operations which are not performed just once, but about N times.

In formula (7.28) we perform a matrix multiplication which requires $\rho_k r_k^L r_{k-1}^L$ arithmetical multiplications and $\rho_k(r_k^L - 1)r_{k-1}^L$ arithmetical additions and a QR decomposition of an $(1+\rho_k)\times r_{k-1}^L$ matrix. If we denote by $\mu(k,m)$ the number of arithmetical operations needed for the QR decomposition of a $k \times m$ matrix, then this formula requires less than $\mu(1+r_k^L, r_{k-1}^L)+2(r_k^L)^2 r_{k-1}^L$ arithmetical operations.

In formula (7.29) we perform a complex conjugation of a $(\rho_k+1)\times(\rho_k+1)$ matrix and in the computation of Z_k three matrix multiplications. The multiplication by V_k^* requires $(\rho_k+1)^3$ arithmetical multiplications and $\rho_k(\rho_k+1)^2$ arithmetical additions, while the two other matrix multiplications require $\rho_k r_k^L + r_k^U s_k$ arithmetical multiplications $\rho_k(r_k^L - 1) + (r_k^U - 1)s_k$ arithmetical additions. In total, formula (7.29) requires less than

$$2(r_k^L + 1)^3 + 2(r_k^L)^2 + 2r_k^L r_k^U$$

arithmetical operations.

In formula (7.31) we perform a complex conjugation of an $(s_k+1)\times(s_k+1)$ matrix and we solve $1+\rho_k$ linear systems. If $\eta(k,m)$ is the complexity of solving a $k \times m$ linear system by a standard method, then the complexity of formula (7.31) is at most

$$(r_k^L + m_k)(r_k^L + 1 + \eta(\nu_k, n_k + r_k^L)).$$

In formula (7.33) we perform a complex conjugations of s_{k-1} plus $s_k s_{k-1}$ numbers, and in the computation of Y_k three matrix multiplications and in the computation of z_k another two matrix multiplications. These multiplications require $r_{k-1}^U s_{k-1} + r_{k-1}^U r_k^U s_k + r_k^U s_k s_{k-1}$ and respectively $\rho_{k-1}s_{k-1} + \rho_{k-1}s_k s_{k-1}$ arithmetical multiplications and $r_{k-1}^U(r_k^U - 1)s_k + r_k^U(s_k - 1)s_{k-1}$ and respectively $\rho_{k-1}(s_k - 1)s_{k-1}$ arithmetical additions. In total, if

$$r_L = \max_{k=1,\dots,N-1}(r_k^L), \quad r_U = \max_{k=1,\dots,N-1}(r_k^U)$$

then formula (7.33) requires less than

$$r_L + 2r_L^2 + 2r_L^3 + r_U r_L + 2r_U^2 r_L + 2r_U r_L^2$$

arithmetical operations.

Finally, formula (7.34) requires $r_{k-1}^U s_{k-1}$ arithmetical multiplications and $(r_{k-1}^U - 1)s_{k-1}$ arithmetical additions.

It follows that if

$$m = \max_{1\le k\le N} m_k, \quad n = \max_{1\le k\le N} n_k,$$

then the complexity c_2 of the algorithm in Theorem 7.5 satisfies

$$c_2 < ((m+r_L)\eta(m+r_L, n+r_L) + 4r_L^3 + 2r_L^2 r_U + 2r_L r_U^2 + 11r_L^2 + 5r_L r_U + 9r_L + 3)N$$

and since we apply this algorithm to the matrix A_1 that emerges from an iteration step using the algorithm in Theorem 29.1, (29.36) holds and $r_U \leq 3r_L$, so that

$$c_2 < ((m+r_L)\eta(m+r_L, n+r_L) + 28r_L^3 + 26r_L^2 + 9r_L + 3)N.$$

Using also formula (29.38) it follows that the cascade of the two algorithms used for unitary matrices has the complexity

$$c_1 + c_2 < (\mu(1+r_L, r_L) + \mu(1+r_L, 1+2r_L) + 58r_L^3 + 74r_L^2 + 45r_L + 12)N.$$

§29.4 QR iteration step for perturbations of Hermitian matrices

Let A be an $N \times N$ scalar matrix represented in the form

$$A = B - XY^*, \tag{29.41}$$

where B is a Hermitian matrix and $X = \operatorname{col}(x(i))_{i=1}^N$, $Y = \operatorname{col}(y(i))_{i=1}^N$ with m-dimensional rows $x(i), y(i)$ $(i = 1, \dots, N)$. To put it differently, A is an at most rank-m perturbation of a Hermitian matrix. The matrices X, Y are called *perturbation matrices* for the matrix A.

Assume that the matrix A has lower quasiseparable generators $p(i)$ $(i = 2, \dots, N)$, $q(j)$ $(j = 1, \dots, N-1)$, $a(k)$ $(k = 2, \dots, N-1)$, and diagonal entries $d(k)$ $(k = 1, \dots, N)$. One can easily check that upper quasiseparable generators $g(i)$ $(i = 1, \dots, N-1)$, $h(j)$ $(j = 2, \dots, N)$, $b(k)$ $(k = 2, \dots, N-1)$ of the matrix A are obtained in Steps 1–3, as follows.

1. Determine lower quasiseparable generators and diagonal entries of the matrix B via

$$\begin{aligned}
p_B(i) &= \begin{pmatrix} x(i) & p(i) \end{pmatrix}, & i &= 2, \dots, N, \\
q_B(j) &= \begin{pmatrix} y^*(j) \\ q(j) \end{pmatrix}, & j &= 1, \dots, N-1, \\
a_B(k) &= \begin{pmatrix} I_m & 0 \\ 0 & a(k) \end{pmatrix}, & k &= 2, \dots, N-1, \\
d_B(k) &= d(k) + x(k)y^*(k), & k &= 1, \dots, N.
\end{aligned}$$

2. Determine upper quasiseparable generators $g_B(i)$ $(i = 1, \dots, N-1)$, $h_B(j)$ $(j = 2, \dots, N)$, $b_B(k)$, $(k = 2, \dots, N-1)$ via the formulas (29.39)

$$\begin{aligned}
g_B(i) &= q_B^*(i) = \begin{pmatrix} y(j) & q^*(j) \end{pmatrix}, & i &= 1, \dots, N-1, \\
h_B(i) &= p_B^*(i) = \begin{pmatrix} x^*(i) \\ p^*(i) \end{pmatrix}, & i &= 2, \dots, N, \\
b_B(i) &= a_B^*(i) = \begin{pmatrix} I_m & 0 \\ 0 & a^*(k) \end{pmatrix}, & i &= 2, \dots, N-1.
\end{aligned}$$

3. Determine upper quasiseparable generators of the matrix A by the formulas

$$g(i) = \begin{pmatrix} -x(i) & g_B(i) \end{pmatrix}, \quad i = 1, \dots, N-1,$$
$$h(j) = \begin{pmatrix} y^*(j) \\ h_B(j) \end{pmatrix}, \quad j = 2, \dots, N-1,$$
$$b(k) = \begin{pmatrix} I_m & 0 \\ 0 & b_B(k) \end{pmatrix}, \quad k = 2, \dots, N-1.$$

From here one can easily get the formulas for upper generators of the matrix A when B is a Hermitian matrix:

$$g(i) = \begin{pmatrix} -x(i) & y(i) & q^*(i) \end{pmatrix}, \quad h(i) = \begin{pmatrix} y^*(i) \\ x^*(i) \\ p^*(i) \end{pmatrix}, \quad b(i) = \begin{pmatrix} I_{2m} & 0 \\ 0 & a^*(i) \end{pmatrix}. \tag{29.42}$$

Consider now a QR iteration step which transforms the matrix A into the matrix A_1 as in Theorem 29.1. Then there is a unitary matrix Q, as defined in the theorem, such that

$$A_1 = Q^*AQ = Q^*(B - XY^*)Q = Q^*BQ - (Q^*X)(Q^*Y)^* = B_1 - X_1Y_1^*$$

where $B_1 = Q^*BQ$, $X_1 = Q^*X$, $Y_1 = Q^*Y$. Thus, the new iteration matrix A_1 is an order at most m perturbation of a Hermitian matrix B_1 and this fact allows us to continue the iterations.

We will therefore find the diagonal entries and the lower quasiseparable generators of the matrix A_1 using Step 1 of Algorithm 29.3 below, which is the version of Theorem 29.1 for the matrix A. Like in Theorem 29.1, it follows that the orders of lower quasiseparable generators do not increase with a new iteration.

We will also find the new perturbation matrices $X_1 = Q^*X$, $Y_1 = Q^*Y$, which give the new matrix A_1 as

$$A_1 = B_1 - X_1Y_1^*,$$

using the algorithm of multiplication of a matrix by a vector 13.1 in Step 2 of Algorithm 29.3 below.

Algorithm 29.3. Let $A = \{A_{ij}\}_{i,j=1}^N$ be a scalar matrix which is a rank-m perturbation of a Hermitian matrix. Suppose that the matrix A has lower quasiseparable generators $p(i)$ $(i = 2, \dots, N)$, $q(j)$ $(j = 1, \dots, N-1)$, $a(k)$ $(k = 2, \dots, N-1)$ of orders r_k^L $(k = 1, \dots, N-1)$, diagonal entries $d(k)$ $(k = 1, \dots, N)$, and perturbation matrices $X = \mathrm{col}(x(i))_{i=1}^N$, $Y = \mathrm{col}(y(i))_{i=1}^N$ with m-dimensional columns $x(i), y(i)$ $(i = 1, \dots, N)$.

Define the matrix A_1 by the rule

$$\begin{cases} A - \sigma I = QR, \\ \quad A_1 = \sigma I + RQ, \end{cases}$$

where Q is a unitary matrix and R is an upper triangular matrix.

Lower quasiseparable generators $p^{(1)}(i)$ $(i = 2, \dots, N)$, $q^{(1)}(j)$ $(j = 1, \dots, N-1)$, $a^{(1)}(k)$ $(k = 2, \dots, N-1)$ of orders ρ_k $(k = 1, \dots, N-1)$, as defined in (29.3), and diagonal entries $d^{(1)}(k)$ $(k = 1, \dots, N)$ of the matrix A_1, as well as the new perturbation matrices $X_1 = \operatorname{col}(x^{(1)}(i))_{i=1}^N$, $Y_1 = \operatorname{col}(y^{(1)}(i))_{i=1}^N$ are obtained as follows.

1. Determine upper quasiseparable generators $g(i)$ $(i = 1, \dots, N-1)$, $h(j)$ $(j = 2, \dots, N)$, $b(k)$ $(k = 2, \dots, N-1)$ of the matrix A with orders
$$r_k^U = r_k^L + 2m, \quad k = 1, \dots, N-1,$$
by the formulas (29.48). Using the algorithm from Theorem 29.1, compute lower quasiseparable generators $p^{(1)}(i)$ $(i = 2, \dots, N)$, $q^{(1)}(j)$ $(j = 1, \dots, N-1)$, $a^{(1)}(k)$ $(k = 2, \dots, N-1)$ of orders ρ_k $(k = 1, \dots, N-1)$, and diagonal entries $d^{(1)}(k)$ $(k = 1, \dots, N)$ of the matrix A_1, and quasiseparable generators of the matrix Q.
2. Find the new matrices of perturbation
$$X_1 = \operatorname{col}(x^{(1)}(i))_{i=1}^N, \quad Y_1 = \operatorname{col}(y^{(1)}(i))_{i=1}^N.$$

2.1. Start with $x_L^{(1)}(0) = y_L^{(1)}(0) = 0$ and $\eta_1 = \zeta_1 = 0$ and define $b_Q(1) = 0$, and for $i = 2, \dots, N$ compute recursively the results of the multiplication of the lower part of Q^* with the vectors X and Y as follows:
$$\eta_i = b_Q^*(i-1)\eta_{i-1} + g_Q^*(i-1)x(i-1), \quad \zeta_i = b_Q^*(i-1)\zeta_{i-1} + g_Q^*(i-1)y(i-1)$$
and
$$x_L^{(1)}(i) = h_Q^*(i)\eta_i, \quad y_L^{(1)}(i) = h_Q^*(i)\zeta_i.$$

2.2. Compute for $i = 1, \dots, N$ recursively the results of the multiplication of the diagonal of Q^* with the vectors X and Y as follows:
$$x_D^{(1)}(i) = d_Q^*(i)x(i), \quad y_D^{(1)}(i) = d_Q^*(i)y(i).$$

2.3. Start with $x_U^{(N)}(0) = y_U^{(N)}(0) = 0$ and for $i = N-1, \dots, 1$ compute recursively the results of the multiplication of the upper part of Q^* with the vectors X and Y as follows:
$$x_U^{(1)}(i) = q_Q^*(i)\psi_{i+1}^*, \quad y_U^{(1)}(i) = q_Q^*(i)\chi_{i+1}^*.$$

2.4. Compute the vectors X_1 and Y_1 by
$$X_1 = x_L^{(1)} + x_D^{(1)} + x_U^{(1)}, \quad Y_1 = y_L^{(1)} + y_D^{(1)} + y_U^{(1)}.$$

The estimate for the complexity of Algorithm 29.3 may be obtained by substitution of the values $r_L = \max_{1 \le k \le N-1}$, $r_U = r_L + 2m$ in (29.37). However, notice that in the case under consideration the matrix $b_R(k)$ in (29.10) has the form

$$b_R(k) = \begin{pmatrix} I_m & 0 & 0 & 0 \\ 0 & I_m & 0 & 0 \\ 0 & 0 & a^*(k) & 0 \\ -p_V^*(k)x(k) & p_V^*(k)y(k) & p_V^*(k)q^*(k) & a_V^*(k) \end{pmatrix}. \tag{29.43}$$

This form may be used to reduce the number of operations in the formulas (29.18) and (29.29). More precisely, for the matrices D_{k-1} and γ_{k+1} we use the partitions

$$D_{k-1} = \begin{pmatrix} w_{k-1} & t_{k-1} & u_{k-1} & v_{k-1} \end{pmatrix}, \quad \gamma_{k+1} = \begin{pmatrix} \chi_{k+1} & \psi_{k+1} & \eta_{k+1} & \theta_{k+1} \end{pmatrix}^T,$$

which correspond to the form (29.43) of the matrix $b_R(k)$, and then compute the products $\Phi_k = D_{k-1}b_R(k)$ and $\Psi_k = b_R(k)\gamma_{k+1}$ by the formulas

$$\Phi_k = \begin{pmatrix} a_k & b_k & c_k & d_k \end{pmatrix}, \quad \Psi_k = \begin{pmatrix} \chi_{k+1} & \psi_{k+1} & \rho_k & \lambda_k \end{pmatrix}^T,$$

where

$$\begin{aligned} &a_k = w_{k-1} - v_{k-1}p_V^*(k)x(k), \ b_k = t_{k-1} + v_{k-1}p_V^*(k)y(k) \\ &c_k = u_{k-1}a^*(k) + v_{k-1}p_V^*(k)q^*(k), \ d_k = v_{k-1}a_V^*(k), \\ &\rho_k = a^*(k)\eta_{k+1}, \ \lambda_k = p_V^*(k)(-x(k)\chi_{k+1} + y(k)\eta_{k+1} + q^*(k)\eta_{k+1}) + a_V^*(k)\theta_{k+1}. \end{aligned}$$

Notice also that the formulas (29.32) are not used in Algorithm 29.3.

It follows that the overall complexity c_3 of the algorithm in Theorem 29.3 which builds one iteration satisfies

$$\begin{aligned} c_3 < \big(&\mu(1+r,r) + \mu(1+r,1+2r+2m) + 18r^3 \\ &+ 12r^2m + 33r^2 + 26rm + 18r + 2m + 17\big)N, \end{aligned} \tag{29.44}$$

and so if m is about the size of r, then this number is much smaller than c_1 in (29.37).

§29.5 QR iteration step for perturbations of unitary matrices

Let A be an $N \times N$ scalar matrix represented in the form

$$A = B - XY^*, \tag{29.45}$$

where B is a unitary matrix and $X = \mathrm{col}(x(i))_{i=1}^N$, $Y = \mathrm{col}(y(i))_{i=1}^N$ with m-dimensional rows $x(i), y(i)$ $(i = 1, \dots, N)$. To put it differently, A is an at most rank-m perturbation of a unitary matrix. The matrices X, Y are called as *perturbation matrices* for the matrix A.

Assume that the matrix A has lower quasiseparable generators $p(i)$ $(i = 2,\dots,N)$, $q(j)$ $(j = 1,\dots,N-1)$, $a(k)$ $(k = 2,\dots,N-1)$ of orders r_k^L, $k = 1,\dots,N-1$, and diagonal entries $d(k)$ $(k = 1,\dots,N)$. The formula (29.45) means that the matrix B has lower quasiseparable generators

$$\begin{aligned} p_B(i) &= \begin{pmatrix} x(i) & p(i) \end{pmatrix}, \quad & i = 2,\dots,N, \\ q_B(j) &= \begin{pmatrix} y^*(j) \\ q(j) \end{pmatrix}, & j = 1,\dots,N-1, \\ a_B(k) &= \begin{pmatrix} I_m & 0 \\ 0 & a(k) \end{pmatrix}, & k = 2,\dots,N-1, \end{aligned} \tag{29.46}$$

with orders $r_k^L + m$, $k = 1,\dots,N-1$, and diagonal entries

$$d_B(k) = d(k) + x(k)y^*(k), \quad k = 1,\dots,N. \tag{29.47}$$

Since B is a unitary scalar matrix, by Corollary 7.2 it has upper quasiseparable generators with the same orders. Let $g_B(i)$ $(i = 1,\dots,N-1)$, $h_B(j)$ $(j = 2,\dots,N)$, $b_B(k)$ $(k = 2,\dots,N-1)$ be such generators. Then upper quasiseparable generators $g(i)$ $(i = 1,\dots,N-1)$, $h(j)$ $(j = 2,\dots,N)$, $b(k)$ $(k = 2,\dots,N-1)$ of orders $r_k^L + 2m$, $k = 1,\dots,N-1$, of the matrix A are obtained as follows:

$$\begin{aligned} g(i) &= \begin{pmatrix} -x(i) & g_B(i) \end{pmatrix}, \quad & i = 1,\dots,N-1, \\ h(j) &= \begin{pmatrix} y^*(j) \\ h_B(j) \end{pmatrix}, & j = 2,\dots,N-1, \\ b(k) &= \begin{pmatrix} I_m & 0 \\ 0 & b_B(k) \end{pmatrix}, & k = 2,\dots,N-1. \end{aligned} \tag{29.48}$$

Consider now a QR iteration step which transforms the matrix A into the matrix A_1 as in Theorem 29.1. Then there is a unitary matrix Q, as defined in the theorem, such that

$$A_1 = Q^*AQ = Q^*(B - XY^*)Q = Q^*BQ - (Q^*X)(Q^*Y)^* = B_1 - X_1Y_1^*$$

where $B_1 = Q^*BQ$, $X_1 = Q^*X$, $Y_1 = Q^*Y$, and so the new iteration matrix A_1 is a rank at most m perturbation of a unitary matrix B_1, which allows us to continue the iterations.

We will therefore find the lower quasiseparable generators of the matrix A_1 in Step 1 of the algorithm in Theorem 29.4 below, which is the version of Theorem 29.1 for the matrix A. Like in Theorem 29.1, it follows that orders of lower quasiseparable generators do not increase with a new iteration.

We will also find the new perturbation vectors $X_1 = Q^*X$, $Y_1 = Q^*Y$ which give the new matrix A_1 as

$$A_1 = B_1 - X_1Y_1^*$$

using the Algorithm 13.1 of multiplication of a matrix by a vector in Step 2 of the algorithm in Theorem 29.4 below. The order m of the perturbation also remains the same.

We will find in Step 3 upper quasiseparable generators for the new perturbed unitary matrix B_1 and we will reduce their order in Step 4, so that the order will be at most as for the initial unitary perturbed matrix B.

Theorem 29.4. *Let $A = \{A_{ij}\}_{i,j=1}^N$ be a scalar matrix with the given lower quasiseparable generators $p(i)$ $(i = 2, \ldots, N)$, $q(j)$ $(j = 1, \ldots, N-1)$, $a(k)$ $(k = 2, \ldots, N-1)$ of orders r_k^L $(k = 1, \ldots, N-1)$, and diagonal entries $d(k)$ $(k = 1, \ldots, N)$, and let σ be a number. Assume that A is represented in the form*

$$A = B - XY^*, \tag{29.49}$$

where B is a unitary matrix with given upper quasiseparable generators $g_B(i)$ $(i = 1, \ldots, N-1)$, $h_B(j)$ $(j = 2, \ldots, N)$, $b_U(k)$ $(k = 2, \ldots, N-1)$ of orders $r_k^L + m$ $(k = 1, \ldots, N-1)$, and perturbation vectors $X = \mathrm{col}(x(i))_{i=1}^N$, $Y = \mathrm{col}(y(i))_{i=1}^N$ with $1 \times m$ rows $x(i), y(i)$. Define the numbers ρ_k via the recursion relations (29.3) and the numbers τ_k via

$$\tau_N = 0, \quad \tau_{k-1} = \min\{1 + \tau_k,\ m + \rho_{k-1}\},\ k = N, \ldots, 2.$$

Define the matrix A_1 by the rule

$$\begin{cases} A - \sigma I = QR, \\ \quad\quad A_1 = \sigma I + RQ, \end{cases} \tag{29.50}$$

where Q is a unitary matrix and R is an upper triangular matrix.

Then the matrix $A^{(1)}$ is represented in the form

$$A^{(1)} = B_1 - X_1 Y_1^*,$$

where B_1 is a unitary matrix and $X_1 = \mathrm{col}(x_1(i))_{i=1}^N$, $Y_1 = \mathrm{col}(y_1(i))_{i=1}^N$ are block columns with $1 \times m$ entries. Moreover, lower quasiseparable generators $p^{(1)}(i)$ $(i = 2, \ldots, N)$, $q^{(1)}(j)$ $(j = 1, \ldots, N-1)$, $a^{(1)}(k)$ $(k = 2, \ldots, N-1)$ of orders ρ_k $(k = 1, \ldots, N-1)$, and diagonal entries $d^{(1)}(k)$ $(k = 1, \ldots, N)$ of the matrix A_1, as well as upper quasiseparable generators $g_{B_1}(i)$ $(i = 1, \ldots, N-1)$, $h_{B_1}(j)$ $(j = 2, \ldots, N)$, $b_{B_1}(k)$ $(k = 2, \ldots, N-1)$ of orders τ_k and perturbation matrices $X_1 = \mathrm{col}(x_1(i))_{i=1}^N$, $Y_1 = \mathrm{col}(y_1(i))_{i=1}^N$, are obtained via the following algorithm.

1. *Determine upper quasiseparable generators $g(i)$ $(i = 1, \ldots, N-1)$, $h(j)$ $(j = 2, \ldots, N)$, $b(k)$ $(k = 2, \ldots, N-1)$ of the matrix A with orders*

   $$r_k^U = r_k^L + 2m, \quad k = 1, \ldots, N-1$$

   by the formulas (29.42). Using the algorithm from Theorem 29.1, compute lower quasiseparable generators $p^{(1)}(i)$ $(i = 2, \ldots, N)$, $q^{(1)}(j)$ $(j = 1, \ldots, N-$

1), $a^{(1)}(k)$ $(k = 2, \dots, N-1)$ of orders ρ_k $(k = 1, \dots, N-1)$, and diagonal entries $d^{(1)}(k)$ $(k = 1, \dots, N)$ of the matrix A_1, and quasiseparable generators of the matrices Q and R.

2. *Find the new perturbation matrices $X_1 = \mathrm{col}(x^{(1)}(i))_{i=1}^N$, $Y_1 = \mathrm{col}(y^{(1)}(i))_{i=1}^N$ as in Step 2 of Algorithm 29.3.*
3. *Determine upper quasiseparable generators $\hat{g}_B(i)$ $(i = 1, \dots, N-1)$, $\hat{h}_B(j)$ $(j = 2, \dots, N)$, $\hat{b}_B(k)$ $(k = 2, \dots, N-1)$ of the matrix B_1 with orders $3\rho_k + m$, $k = 1, \dots, m-1$ as follows.*

   3.1. *Determine quasiseparable generators of the matrix $W = Q^* B$.*
   Set

$$d_W(k) = d_R(k) + \sigma d_Q(k) + x^{(1)}(k) y^*(k), \ k = 1, \dots, N, \tag{29.51}$$

$$\begin{aligned} p_W(i) &= \begin{pmatrix} x^{(1)}(i) & h_Q^*(i) \end{pmatrix}, \ i = 2, \dots, N, \\ q_W(j) &= \begin{pmatrix} y^*(j) \\ \sigma g_Q^*(j) \end{pmatrix}, \qquad j = 1, \dots, N-1, \\ a_W(k) &= \begin{pmatrix} I_m & 0 \\ 0 & b_Q^*(k) \end{pmatrix}, \quad k = 2, \dots, N-1. \end{aligned} \tag{29.52}$$

Set

$$\tilde{g}(1) = 0_{1 \times (r_1^L + m)}, \quad \beta_1 = g_Q^*(1) g_B(1) \tag{29.53}$$

and for $k = 2, \dots, N-1$ compute

$$\tilde{g}(k) = h_Q^*(k) \beta_{k-1} b_B(k), \quad \beta_k = b_Q^*(k) \beta_{k-1} b_B(k) + g_Q^*(k) g_B(k). \tag{29.54}$$

Set

$$c_N = p_V^*(N) x(N), \quad \tilde{h}(N) = p_V^*(N) d(N) + c_N y^*(N) \tag{29.55}$$

and for $k = N-1, \dots, 2$ compute

$$\begin{aligned} c_k &= p_V^*(k) x(k) + a_V^*(k) c_{k+1} a(k), \\ \tilde{h}(k) &= p_V^*(k) d(k) + a_V^*(k) C_{k+1} a(k) + c_k y^*(k). \end{aligned} \tag{29.56}$$

Here the variables C_k $(k = N, \dots, 3)$ are obtained in Step 1 of the algorithm.
Set

$$\begin{aligned} g_W(i) &= \begin{pmatrix} d_V^*(i) g_B(i) + \tilde{g}(i) & q_V^*(i) \end{pmatrix}, \ i = 1, \dots, N-1, \\ h_W(j) &= \begin{pmatrix} h_B(j) \\ \tilde{h}(j) \end{pmatrix}, \qquad j = 2, \dots, N, \\ b_W(k) &= \begin{pmatrix} b_B(k) & 0 \\ p_V^*(k) g_B(k) & a_V^*(k) \end{pmatrix}, \quad k = 2, \dots, N-1. \end{aligned} \tag{29.57}$$

3.2. *Determine upper quasiseparable generators of the matrix* $B_1 = WQ$. *Set*

$$\hat{g}(1) = 0_{1\times\rho_1}, \quad \hat{\beta}_1 = q_W(1)g_Q(1), \tag{29.58}$$

and for $k = 2, \dots, N-1$ *compute*

$$\begin{aligned} \hat{g}(k) &= p_W(k)\hat{\beta}_{k-1}b_Q(k), \\ \hat{\beta}_k &= a_W(k)\hat{\beta}_{k-1}b_Q(k) + q_W(k)g_Q(k). \end{aligned} \tag{29.59}$$

Set

$$\hat{\gamma}_N = h_W(N)p_Q(N), \quad \hat{h}(N) = h_W(N)d_Q(N), \tag{29.60}$$

and for $k = N-1, \dots, 2$ *compute*

$$\begin{aligned} \hat{\gamma}_k &= h_W(k)p_Q(k) + b_W(k)\hat{\gamma}_{k+1}a_Q(k), \\ \hat{h}(k) &= h_W(k)d_Q(k) + b_W(k)\hat{\gamma}_{k+1}q_Q(k). \end{aligned} \tag{29.61}$$

Set

$$\begin{aligned} \hat{g}_B(i) &= \begin{pmatrix} d_W(i)g_Q(i) + \hat{g}(i) & g_W(i) \end{pmatrix}, \quad i = 1, \dots, N-1, \\ \hat{h}_B(j) &= \begin{pmatrix} h_Q(j) \\ \hat{h}(j) \end{pmatrix}, \quad j = 2, \dots, N, \\ \hat{b}_B(k) &= \begin{pmatrix} b_Q(k) & 0 \\ h_W(k)g_Q(k) & b_W(k) \end{pmatrix}, \quad k = 2, \dots, N-1. \end{aligned} \tag{29.62}$$

4. *Determine upper quasiseparable generators* $g_{B_1}(i)$ $(i = 1, \dots, N-1)$, $h_{B_1}(j)$ $(j = 2, \dots, N)$, $b_{B_1}(k)$ $(k = 2, \dots, N-1)$ *of orders* τ_k, $k = 1, \dots, N-1$ *of the matrix* B_1 *as follows.*

 4.1. *Determine lower quasiseparable generators of orders* $\rho_k + m$, $k = 1, \dots, N-1$ *and diagonal entries of the matrix* B_1 *via*

$$\begin{aligned} p_{B_1}(i) &= \begin{pmatrix} x^{(1)}(i) & p^{(1)}(i) \end{pmatrix}, \quad i = 2, \dots, N, \\ q_{B_1}(j) &= \begin{pmatrix} (y^{(1)}(j))^* \\ q^{(1)}(j) \end{pmatrix}, \quad j = 1, \dots, N-1, \\ a_{B_1}(k) &= \begin{pmatrix} I_m & 0 \\ 0 & a^{(1)}(k) \end{pmatrix}, \quad k = 2, \dots, N-1, \\ d_{B_1}(k) &= d^{(1)}(k) + x^{(1)}(k)(y^{(1)}(k))^*, \quad k = 1, \dots, N. \end{aligned}$$

 4.2. *Applying the algorithm from Theorem 7.5 to the unitary matrix* B_1 *with quasiseparable generators* $p_{B_1}(i)$ $(i = 2, \dots, N)$, $q_{B_1}(j)$ $(j = 1, \dots, N-1)$, $a_{B_1}(k)$; $(k = 2, \dots, N-1)$; $\hat{g}_B(i)(i = 1, \dots, N-1)$, $\hat{h}_B(j)$ $(j = 2, \dots, N)$, $\hat{b}_B(k)$; $(k = 2, \dots, N-1)$; $d_{B_1}(k)$ $(k = 1, \dots, N)$ *with orders* $(\rho_k + m, 3\rho_k + m)$ $k = 1, \dots, N-1$, *compute a set of upper quasiseparable generators* $g_{B_1}(i)$ $(i = 1, \dots, N-1)$, $h_{B_1}(j)$ $(j = 2, \dots, N)$, $b_{B_1}(k)$ $(k = 2, \dots, N-1)$ *of orders* τ_k, $k = 1, \dots, N-1$.

Proof. Step 1 of the algorithm follows from Theorem 29.1. Step 2 is a direct consequence of Algorithm 13.1.

We need to justify Step 3. Combining (29.49), the first relation in (29.50), and the equality $X_1 = Q^*X$, we get

$$W = Q^*B = Q^*A + (Q^*X)Y = R + \sigma Q^* + X_1Y^*.$$

From here we obtain (29.51) for the diagonal entries of the matrix W and also, since R is an upper triangular matrix, the formulas (29.52).

Next we apply Theorem 17.6 to compute upper quasiseparable generators of the matrix $W = Q^*B$. Notice that Q^* has quasiseparable generators $h_Q^*(i)$ $(i = 2,\dots,N)$, $g_Q^*(j)$ $(j = 1,\dots,N-1)$, $b_Q^*(k)$ $(k = 2,\dots,N-1)$, $q_Q^*(i)$ $(i = 1,\dots,N-1)$, $p_Q^*(j)$ $(j = 2,\dots,N)$, $a_Q^*(j)$ $(j = 2,\dots,N)$, and diagonal entries $d_Q^*(k)$ $(k = 1,\dots,N)$. Using the formulas (17.34) and (17.35) we obtain the formulas (29.57) with auxiliary variables $\tilde{g}(i)$ and $\tilde{h}(j)$, which are $(r_i^L + m)$-dimensional columns and ρ_{j-1}-dimensional columns. Moreover, using the relations (17.30) we get β_0 to be 0×0 empty matrix and

$$\begin{pmatrix} \tilde{g}(k) \\ \beta_k \end{pmatrix} = \begin{pmatrix} h_Q^*(k) & 0 \\ b_Q^*(k) & g_Q^*(k) \end{pmatrix} \begin{pmatrix} \beta_{k-1} & 0 \\ 0 & 1 \end{pmatrix} \begin{pmatrix} b_B(k) \\ g_B(k) \end{pmatrix}, \quad k = 1,\dots,N-1. \tag{29.63}$$

From here for $k = 1$, using the fact that $h_Q^*(1), b_Q^*(1), \beta_0, b_B(1)$ are empty matrices, we obtain (29.53). For $k = 2,\dots,N-1$ the relations (29.63) are equivalent to (29.54).

Using the relations (17.31) we get γ_{N+1} to be 0×0 empty matrix and

$$\begin{pmatrix} \gamma_k & \tilde{h}(k) \end{pmatrix} = \begin{pmatrix} p_Q^*(k) & a_Q^*(k) \end{pmatrix} \begin{pmatrix} 1 & 0 \\ 0 & \gamma_{k+1} \end{pmatrix} \begin{pmatrix} p_B(k) & d_B(k) \\ a_B(k) & q_B(k) \end{pmatrix}, \quad k = N,\dots,2. \tag{29.64}$$

Here γ_k, $k = N,\dots,2$ are auxiliary variables which are matrices of sizes $\rho_{k-1} \times (m + r_{k-1}^L)$. For these matrices we use the partitions

$$\gamma_k = \begin{pmatrix} c_k & C_k \end{pmatrix}, \quad k = N,\dots,2,$$

with matrices c_k, C_k of sizes $\rho_{k-1} \times m, \rho_{k-1} \times r_{k-1}^L$. Thus from (29.64), using also the formulas (29.21), (29.46) and the fact that $\gamma_{N+1}, a_Q^*(N), a_B(N), q_B(N)$ are empty matrices, we get

$$\begin{pmatrix} c_N & C_N & \tilde{h}(N) \end{pmatrix} = p_V^*(N) \begin{pmatrix} x(N) & p(N) & d(N) + x(N)y^*(N) \end{pmatrix},$$

$$\begin{pmatrix} c_k & C_k & \tilde{h}(k) \end{pmatrix} = \begin{pmatrix} p_V^*(k) & a_V^*(k) \end{pmatrix} \begin{pmatrix} x(k) & p(k) & d(k) + x(k)y^*(k) \\ c_{k+1} & C_{k+1}a(k) & c_{k+1}y^*(k) + C_{k+1}q(k) \end{pmatrix},$$

$$k = N-1,\dots,2.$$

It follows that

$$c_N = p_V^*(N)x(N), \quad c_k = p_V^*(k)x(k) + a_V^*(k)c_{k+1}, \qquad k = N-1,\dots,2, \tag{29.65}$$

$$C_N = p_V^*(N)p(N), \quad C_k = p_V^*(k)p(k) + a_V^*(k)C_{k+1}a(k), \quad k = N-1,\dots,2, \tag{29.66}$$

$$\begin{aligned}\tilde{h}(N) &= p_V^*(N)d(N) + (p_V^*(N)x(N))y^*(N),\\ \tilde{h}(k) &= p_V^*(k)d(k) + (p_V^*(k)x(k) + a_V^*(k)c_{k+1})y^*(k) + a_V^*(k)C_{k+1}q(k), \\ &k = N-1,\dots,2.\end{aligned} \tag{29.67}$$

Comparing (29.66) with (29.6), (29.7) and (29.8), (29.9) we conclude that the values C_k here coincide with ones obtained in Step 1 of the algorithm. Now notice that the formulas (29.65), (29.67) imply (29.55), (29.56), which completes the proof of Step 3.1.

On Step 3.2 we apply Theorem 17.6 to compute upper quasiseparable generators of the matrix $B_1 = WQ$. Using the formulas (17.34), (17.35) we obtain the formulas (29.62) with auxiliary variables $\hat{g}(i), \hat{h}(j)$, which are ρ_i-dimensional columns and $2\rho_{j-1}+m$-dimensional columns. Moreover, using the relations (17.30) we get $\hat{\beta}_0$ to be 0×0 empty matrix and

$$\begin{pmatrix} \hat{g}(k) \\ \hat{\beta}_k \end{pmatrix} = \begin{pmatrix} p_W(k) & 0 \\ a_W(k) & q_W(k) \end{pmatrix} \begin{pmatrix} \hat{\beta}_{k-1} & 0 \\ 0 & 1 \end{pmatrix} \begin{pmatrix} b_Q(k) \\ g_Q(k) \end{pmatrix}, \quad k = 1,\dots,N-1. \tag{29.68}$$

Here $\hat{\beta}_k$ are auxiliary variables which are matrices of sizes $(m+\rho_k)\times\rho_k$. From (29.68) for $k=1$, using the fact that $p_W(1), a_W(1), \beta_0, b_Q(1)$ are empty matrices, we obtain (29.58). For $k = 2,\dots,N-1$ the relations (29.68) are equivalent to (29.59).

Using the relations (17.31) we get $\hat{\gamma}_{N+1}$ to be 0×0 empty matrix and

$$\begin{pmatrix} \hat{\gamma}_k & \hat{h}(k) \end{pmatrix} = \begin{pmatrix} h_W(k) & b_W(k) \end{pmatrix} \begin{pmatrix} 1 & 0 \\ 0 & \hat{\gamma}_{k+1} \end{pmatrix} \begin{pmatrix} p_Q(k) & d_Q(k) \\ a_Q(k) & q_Q(k) \end{pmatrix}, \quad k = N,\dots,2. \tag{29.69}$$

Here $\hat{\gamma}_k$, $k = N,\dots,2$ are auxiliary variables, which are matrices of sizes $(2\rho_k + m)\times\rho_{k-1}$. From (29.69) for $k=N$, using the fact that $\hat{\gamma}_{N+1}, b_W(1), a_Q(N), q_Q(N)$ are empty matrices, we obtain (29.60). For $k = N-1,\dots,2$ the relations (29.69) are equivalent to (29.61).

Step 4.1 follows from the equality $B_1 = A_1 + X_1Y_1^*$. Step 4.2 is an application of Theorem 7.5 for the matrix B_1. □

We will now estimate the complexity of the computations required by Theorem 29.4. The complexity of Step 1 is c_1, which has been computed in (29.37), while the complexity of Step 2 is c_3, which has been computed in (29.44). The complexity of Step 3.1 is

$$\begin{aligned}4+m+\rho_k(r_k^L+m)+(2\rho_{k-1}-1)(r_k^L+m)+\rho_k(2r_k^L+2m-1)(r_k^L+m)\\ +2r_k^L(r_{k-1}^L+m)(r_k^L+m)+\rho_{k-1}(2m+1)+\rho_{k-1}(2\rho_k-1)m+\rho_{k-1}\\ +\rho_{k-1}(2\rho_k-1)+\rho_k(2m-1)+\rho_{k-1}(r_k^L+m).\end{aligned}$$

The complexity of Step 3.2 is

$$\begin{aligned}r_k^L+m+(r_k^L+2m)(2r_k^L+2m-1)(r_k^L+m)+2(r_k^L+m)^3\\ +2(r_k^L+2m)(2r_k^L+4m-1)(r_k^L+m)+2(r_k^L+m)^3\\ +(r_k^L+2m)(r_k^L+m+1)\end{aligned}$$

arithmetical operations. The complexity of Step 4.1 is $1+m$.

§29.6 The iteration process

The QR iteration procedure may be performed using the standard shift strategy (see for instance [35, p. 394]) with the use of the quasiseparable structure, as follows. In the first two iterations we take the shift parameter σ to be equal to zero. Next, for the iterate $A^{(s)}$ with lower quasiseparable generators $p^{(s)}(i)$ $(i=2,\dots,N)$, $q^{(s)}(j)$ $(j=1,\dots,N-1)$, $a^{(s)}(k)$ $(k=2,\dots,N-1)$ and diagonal entries $d^{(s)}(k)$ $(k=1,\dots,N)$ we check the condition

$$|d^{(s)}(N)-d^{(s-1)}(N)|\le 0.1|d^{(s)}(N)| \tag{29.70}$$

and we set $\sigma^{(s)}=d^{(s)}(N)$ if this is true, and $\sigma^{(s)}=0$ otherwise. We say that $d^{(s)}(N)=R^{(s)}(N,N)$ provides a numerical approximation of an eigenvalue λ of A whenever

$$\|p^{(s)}(N)\|\max\{\|a^{(s)}(N-1)\|,\|q^{(s)}(N-1)\|\}\le \upsilon|d^{(s)}(N)|,$$

where υ is the unit round-off error. If this condition is fulfilled, then we set $\lambda=d^{(s)}(N)$ and we deflate the matrix by restarting the process with the initial matrix $A^{(s)}(1:N-1,1:N-1)$, i.e., with the leading principal submatrix of $A^{(s)}$ of order $N-1$, and choosing the shifts in accordance with (29.70).

§29.7 Solution of the eigenproblem for a unitary Hessenberg matrix via reduction to a Hermitian matrix

Let U be an upper Hessenberg scalar unitary matrix. Define the linear-fractional transformation $F:\mathbb{C}\to\mathbb{C}$ by

$$F(z)=\frac{\delta z-\beta}{-\gamma z+\alpha},$$

with the coefficients $\alpha, \beta, \gamma, \delta$ satisfying the conditions (6.39). One can easily check that this is an inverse to the fractional transformation $\Phi(z)$ defined in (6.36). Assume that the matrix $-\gamma U + \alpha I$ is invertible. By Theorem 6.22, the matrix

$$F := F(U) = (\delta U - \beta I)(-\gamma U + \alpha I)^{-1} \tag{29.71}$$

is Hermitian.

Based on the formula (29.71) we obtain the following procedure to compute the Hermitian matrix F.

1. For the upper Hessenberg matrix $-\gamma U + \alpha I$, determine a unitary matrix Q and an upper triangular matrix R such that

$$-\gamma U + \alpha I = RQ. \tag{29.72}$$

2. Compute the product of matrices

$$G = (\delta U - \beta I)Q^*. \tag{29.73}$$

3. Determine the matrix F via

$$F = GR^{-1}. \tag{29.74}$$

Set

$$m_0 = m_1 = \cdots = m_{N-1} = 1,\ m_N = 0, \quad n_0 = 0,\ n_1 = n_2 = \cdots = n_N = 1. \tag{29.75}$$

As it was shown in Section §7.4, relative to this partition U is a block upper triangular unitary matrix $U = \{u_{ij}\}_{i,j=0}^{N}$ with upper quasiseparable generators

$$\begin{gathered} g(i) = -\rho_i^*,\ i = 0, \dots, N-1, \quad h(j) = \rho_j,\ j = 1, \dots, N, \\ b(k) = \mu_k,\ k = 1, \dots, N-1, \end{gathered} \tag{29.76}$$

and diagonal entries

$$d(k) = \mu_k,\ k = 1, \dots, N-1; \quad d(0) \text{ and } d(N) \text{ are } 1 \times 0 \text{ and } 0 \times 1 \text{ empty matrices.} \tag{29.77}$$

Here μ_k, ρ_k are numbers such that $\mu_k \geq 0$, $|\rho_k|^2 + \mu_k^2 = 1$ $(k = 1, \dots, N-1)$, $\rho_0 = -1$, $|\rho_N| = 1$.

Theorem 29.5. *Let U be a unitary upper Hessenberg matrix represented in the form (7.81) with the parameters ρ_k $(k = 0, \dots, N)$, μ_k $(k = 1, \dots, N-1)$.*

Then the Hermitian matrix $F = F(U)$ defined in (29.71) has lower quasiseparable generators with orders equal to one. Moreover, a set $p_F(i)$ $(i = 2, \dots, N)$, $q_F(j)$ $(j = 1, \dots, N-1)$, $a_F(k)$ $(k = 2, \dots, N-1)$ of such generators, and the diagonal entries $d_F(k)$ $(k = 1, \dots, N)$ are determined via the following algorithm.

1. *Set*

$$f_N = \rho_N, \; p_Q(N) = 1. \tag{29.78}$$

For $k = N-1, \dots, 1$ *perform the following. Determine a* 2×2 *unitary matrix* Q_k *and a number* t_k *from the condition*

$$\begin{pmatrix} \mu_k & \gamma\rho_k^* f_{k+1} + \alpha p_Q(k+1) \end{pmatrix} Q_k = \begin{pmatrix} 0 & t_k \end{pmatrix}. \tag{29.79}$$

Set $d_R(k+1) = t_k$. *Determine the numbers* $p_Q(k), d_Q(k), a_Q(k), q_Q(k)$ *from the partition*

$$Q_k = \begin{pmatrix} p_Q(k) & d_Q(k) \\ a_Q(k) & q_Q(k) \end{pmatrix}. \tag{29.80}$$

Compute

$$h_R(k+1) = \begin{pmatrix} \rho_k d_Q(k) + \mu_k f_{k+1} q_Q(k) \\ d_Q(k) \end{pmatrix}, \qquad f_k = \rho_k p_Q(k) + \mu_k f_{k+1} a_Q(k). \tag{29.81}$$

Set

$$d_R(1) = \gamma\rho_0^* f_1 + \alpha p_Q(1). \tag{29.82}$$

Set

$$g_R(k+1) = \begin{pmatrix} \gamma\rho_k^* & \alpha \end{pmatrix}, \; k = 0, \dots, N-2, \qquad b_R(k+1) = \begin{pmatrix} \mu_k & 0 \\ 0 & 0 \end{pmatrix}, \; k = 1, \dots, N-2. \tag{29.83}$$

2. *Set*

$$q_G(1) = 1, \quad q_G(k+1) = q_Q(k), \; a_G(k+1) = a_Q(k), \; k = 1, \dots, N-2. \tag{29.84}$$

Compute

$$d_G(1) = -\delta\rho_0^* f_1 - \beta p_Q(1). \tag{29.85}$$

For $k = 1, \dots, N-1$ *compute*

$$\begin{aligned} p_G(k+1) &= \mu_k p_Q(k) - (\delta\rho_k^* f_k + \beta p_Q(k+1)) a_Q(k), \\ d_G(k+1) &= \mu_k d_Q(k) - (\delta\rho_k^* f_k + \beta p_Q(k+1)) q_Q(k). \end{aligned} \tag{29.86}$$

3. *Set*

$$\begin{aligned} d'(k) &= (d_R(k))^{-1}, & k &= 1, \dots, N, \\ h'(k) &= h_R(k) d'(k), & k &= 2, \dots, N, \\ g'(k) &= -d'(k) g_R(k), & k &= 1, \dots, N-1, \\ b'(k) &= b_R(k) - h'(k) g_R(k), & k &= 2, \dots, N-1. \end{aligned} \tag{29.87}$$

Compute

$$d_F(1) = d_G(1) d'(1), \quad q_F(1) = q_Q(1) d'(1), \quad \beta_1 = q_Q(1) g'(1). \tag{29.88}$$

For $k = 2, \dots, N-1$ compute

$$\begin{pmatrix} d_F(k) & * \\ q_F(k) & \beta_k \end{pmatrix} = \begin{pmatrix} p_G(k) & d_G(k) \\ a_G(k) & q_G(k) \end{pmatrix} \begin{pmatrix} \beta_{k-1} & 0 \\ 0 & 1 \end{pmatrix} \begin{pmatrix} h'(k) & b'(k) \\ d'(k) & g'(k) \end{pmatrix}. \tag{29.89}$$

Compute

$$d_F(N) = p_G(N)\beta_{N-1}h'(N) + d_G(N)d'(N). \tag{29.90}$$

Set

$$p_F(k) = p_G(k),\ k = 2, \dots, N, \quad a_F(k) = a_G(k),\ k = 2, \dots, N-1. \tag{29.91}$$

Proof. Relative to the partition (29.75) one can treat the upper Hessenberg matrices $B = -\gamma U + \alpha I$ and $C = \delta U - \beta I$ as upper triangular matrices with upper quasiseparable generators

$$\begin{aligned} g_B(i) &= \begin{pmatrix} -\gamma g(i) & \alpha \end{pmatrix}, & i &= 0, \dots, N-1, \\ h_B(j) &= \begin{pmatrix} h(j) \\ 1 \end{pmatrix}, & j &= 1, \dots, N, \\ b_B(k) &= \begin{pmatrix} b(k) & 0 \\ 0 & 0 \end{pmatrix}, & k &= 1, \dots, N-1, \\ d_B(k) &= d(k), & k &= 0, \dots, N \end{aligned} \tag{29.92}$$

and

$$\begin{aligned} g_C(i) &= \begin{pmatrix} \delta g(i) & -\beta \end{pmatrix}, & i &= 0, \dots, N-1, \\ h_C(j) &= \begin{pmatrix} h(j) \\ 1 \end{pmatrix}, & j &= 1, \dots, N, \\ b_C(k) &= \begin{pmatrix} b(k) & 0 \\ 0 & 0 \end{pmatrix}, & k &= 1, \dots, N-1, \\ d_C(k) &= d(k), & k &= 0, \dots, N, \end{aligned} \tag{29.93}$$

respectively.

Here the generators $g(i), h(j), b(k), d(k)$ are defined in (29.76), (29.77).

Now set

$$Q^* = \tilde{Q}_N \tilde{Q}_{N-1} \cdots \tilde{Q}_1 \tilde{Q}_0, \tag{29.94}$$

where

$$\begin{aligned} \tilde{Q}_0 &= I_N = \operatorname{diag}\{Q_0, I_{N-1}\}, \\ \tilde{Q}_k &= \operatorname{diag}\{I_{k-1}, Q_k, I_{N-k-1}\}, \quad k = 1, \dots, N-1, \\ \tilde{Q}_N &= I_N = \operatorname{diag}\{I_{N-1}, Q_N\} \end{aligned} \tag{29.95}$$

with $Q_0 = Q_N = 1$ and 2×2 matrices Q_k $(k = 1, \dots, N-1)$ defined in (29.79). By Lemma 20.2, the matrix Q^*, treated as a block matrix with entries of sizes

$n_i \times m_j$, $i,j = 0,\dots,N$, is lower triangular with lower quasiseparable generators and diagonal entries defined by the formulas

$$q_Q(0) = Q_0 = 1, \quad Q_k = \begin{pmatrix} p_Q(k) & d_Q(k) \\ a_Q(k) & q_Q(k) \end{pmatrix}, \; k = 1,\dots,N-1, \quad Q_N = p_Q(N) = 1, \tag{29.96}$$

and with $d_Q(0)$ and $d_Q(N)$ being 0×1 and 1×0 empty matrices.

We treat $R = BQ^*$ as a block matrix with entries of sizes $m_i \times m_j$, $i,j = 0,\dots,N$. By Theorem 17.5 we obtain lower quasiseparable generators $\hat{p}_R(i)$ ($i = 1,\dots,N-1$), upper quasiseparable generators $\hat{g}_R(i)$ ($i = 0,\dots,N-1$), $\hat{h}_R(j)$ ($j = 1,\dots,N-1$), $\hat{b}_R(k)$ ($k = 1,\dots,N-1$), and diagonal entries $\hat{d}_R(k)$ ($k = 0,\dots,N-1$) of the block matrix R via relations

$$\hat{g}_R(i) = g_B(i), \; i = 0,\dots,N-2, \quad \hat{b}_R(i) = b_B(i), \; i = 1,\dots,N-2, \tag{29.97}$$

and

$$\begin{gathered} \hat{f}_N = h_B(N)p_Q(N), \\ \begin{pmatrix} \hat{f}_k & \hat{h}_R(k) \\ \hat{p}_R(k) & \hat{d}_R(k) \end{pmatrix} = \begin{pmatrix} h_B(k) & b_B(k)\hat{f}_{k+1} \\ d_B(k) & g_B(k)\hat{f}_{k+1} \end{pmatrix} \begin{pmatrix} p_Q(k) & d_Q(k) \\ a_Q(k) & q_Q(k) \end{pmatrix}, \\ k = N-1,\dots,1, \\ \hat{d}_R(0) = g_B(0)\hat{f}_1 q_Q(0). \end{gathered} \tag{29.98}$$

Here $\hat{f}_k$ ($k = N,\dots,1$) are auxiliary variables, which are two-dimensional columns, i.e., $\hat{f}_k = \begin{pmatrix} f_k \\ \phi_k \end{pmatrix}$.

Now we treat R as a usual scalar matrix with quasiseparable generators $p_R(i)$ ($i = 2,\dots,N$), $q_R(j)$ ($j = 1,\dots,N-1$), $a_R(k)$ ($k = 2,\dots,N-1$), $g_R(i)$ ($i = 1,\dots,N-1$), $h_R(j)$ ($j = 2,\dots,N$), $b_R(k)$ ($k = 2,\dots,N-1$), and diagonal entries $d_R(k)$ ($k = 1,\dots,N$). We have

$$\begin{pmatrix} \hat{f}_k & \hat{h}_R(k) \\ \hat{p}_R(k) & \hat{d}_R(k) \end{pmatrix} = \begin{pmatrix} \hat{f}_k & h_R(k+1) \\ p_R(k+1) & d_R(k+1) \end{pmatrix}, \; k = N-1,\dots,1, \quad \hat{d}_R(0) = d_R(1). \tag{29.99}$$

Next, using the formulas (29.97), (29.92) and (29.76) we obtain the formulas (29.83). Using (29.92) and (29.76), (29.77) we obtain the formulas

$$\begin{gathered} h_B(N) = \begin{pmatrix} \rho_N \\ 1 \end{pmatrix}, \\ \begin{pmatrix} h_B(k) & b_B(k)\hat{f}_{k+1} \\ d_B(k) & g_B(k)\hat{f}_{k+1} \end{pmatrix} = \begin{pmatrix} \rho_k & \mu_k f_{k+1} \\ 1 & 0 \\ \mu_k & \gamma\rho_k^* f_{k+1} + \alpha\phi_{k+1} \end{pmatrix}, \\ k = N-1,\dots,1, \\ g_B(0)\hat{f}_1 = \gamma\rho_0^* f_1 + \alpha\phi_1. \end{gathered} \tag{29.100}$$

Inserting (29.99), (29.100) in (29.98) we obtain the relations

$$\hat{f}_N = \begin{pmatrix} \rho_N \\ 1 \end{pmatrix} p_Q(N), \tag{29.101}$$

$$\begin{pmatrix} \hat{f}_k & h_R(k+1) \end{pmatrix} = \begin{pmatrix} \rho_k p_Q(k) + \mu_k f_{k+1} a_Q(k) & \rho_k d_Q(k) + \mu_k f_{k+1} q_Q(k) \\ p_Q(k) & d_Q(k) \end{pmatrix},$$
$$k = N-1, \dots, 1, \tag{29.102}$$

$$\begin{pmatrix} p_R(k+1) & d_R(k+1) \end{pmatrix} = \begin{pmatrix} d(k) & \gamma \rho_k^* f_{k+1} + \alpha \phi_{k+1} \end{pmatrix} Q_k,$$
$$k = N-1, \dots, 1, \tag{29.103}$$

$$d_R(1) = -\gamma \rho_0^* f_1 + \alpha \phi_1. \tag{29.104}$$

Using (29.101) and $p_Q(N) = 1$ we obtain (29.78). The relations (29.102) imply (29.81). Moreover the equalities (29.101), (29.102) and $\hat{f}_k = \begin{pmatrix} f_k \\ \phi_k \end{pmatrix}$ imply, in particular, that $\phi_k = p_Q(k)$, $k = N, \dots, 1$. Inserting this in (29.103) and (29.104) we obtain the equalities

$$\begin{pmatrix} p_R(k+1) & d_R(k+1) \end{pmatrix} = \begin{pmatrix} d(k) & \gamma \rho_k^* f_{k+1} + \alpha p_Q(k+1) \end{pmatrix} Q_k, \quad k = N-1, \dots, 1 \tag{29.105}$$

and the equality (29.82). Notice that the matrices Q_k $(k = N-1, \dots, 1)$ were determined in such a way that the relations (29.79) hold. Hence, comparing (29.105) with (29.79) we conclude that $d_R(k+1) = t_k$, $k = N-1, \dots, 1$ and $p_R(k+1) = 0$, $k = N-1, \dots, 1$. The latter implies that R is an upper triangular matrix.

Next we apply Theorem 17.5 to compute the product $G = CQ^*$ of the block upper triangular matrix C and the block lower triangular matrix Q^*. We obtain quasiseparable generators of C, treated as a block matrix with entries of sizes $m_i \times m_j$, $i, j = 0, \dots, N$ via the relations

$$\begin{aligned} \hat{q}_G(k) &= q_Q(k), \quad k = 0, \dots, N-1, \\ \hat{a}_G(k) &= a_Q(k), \quad k = 1, \dots, N-1, \end{aligned} \tag{29.106}$$

and

$$\hat{f}_N = h_B(N) p_Q(N),$$
$$\begin{pmatrix} \hat{f}_k & * \\ \hat{p}_G(k) & \hat{d}_G(k) \end{pmatrix} = \begin{pmatrix} h_C(k) & b_C(k)\hat{f}_{k+1} \\ d_C(k) & g_C(k)\hat{f}_{k+1} \end{pmatrix} \begin{pmatrix} p_Q(k) & d_Q(k) \\ a_Q(k) & q_Q(k) \end{pmatrix}, \tag{29.107}$$
$$k = N-1, \dots, 1,$$
$$\hat{d}_G(0) = g_C(0) \hat{f}_1 q_Q(0).$$

Comparing (29.92) and (29.93), we get

$$h_C(k) = h_B(k), \ k = 1, \dots, N, \quad b_C(k) = b_B(k), \ k = 1, \dots, N-1,$$

and therefore the values

$$\hat{f}_k = \begin{pmatrix} f_k \\ p_Q(k) \end{pmatrix}, \quad k = 1, \dots, N \tag{29.108}$$

are the same as in the previous step.

Now we treat the matrix G as an ordinary scalar matrix with lower quasiseparable generators $p_G(i)$ $(i = 2, \dots, N)$, $q_G(j)$ $(j = 1, \dots, N-1)$, $a_G(k)$ $(k = 2, \dots, N-1)$, and diagonal entries $d_G(k)$ $(k = 1, \dots, N)$. We have

$$\begin{aligned} q_G(k+1) &= \hat{q}_G(k), \quad k = 0, \dots, N-2, \\ a_G(k+1) &= \hat{a}_G(k), \quad k = 0, \dots, N-2 \end{aligned} \tag{29.109}$$

and

$$\begin{aligned} \begin{pmatrix} \hat{p}_G(k) & \hat{d}_G(k) \end{pmatrix} &= \begin{pmatrix} p_G(k+1) & d_G(k+1) \end{pmatrix}, \\ & k = N-1, \dots, 1, \quad \hat{d}_G(0) = d_G(1). \end{aligned} \tag{29.110}$$

From (29.109) using $q_Q(0) = 1$ we obtain (29.84). Next using the formulas (29.93), (29.76), (29.77), and also (29.108), we obtain

$$\begin{aligned} d_C(k) &= \mu_k, & k &= 1, \dots, N-1, \\ g_C(k)\hat{f}_{k+1} &= -\delta\rho^*_{k-1} f_{k+1} - \beta p_Q(k+1), & k &= N-1, \dots, 0. \end{aligned} \tag{29.111}$$

Thus combining (29.107), (29.110), (29.111) we obtain the relations (29.85), (29.86) to determine lower quasiseparable generators $p_G(i)$ and diagonal entries $d_G(k)$ of the matrix G.

Next, by Theorem 13.13, R^{-1} is an upper triangular matrix with upper quasiseparable generators $g'(i)$ $(i = 1, \dots, N-1)$, $h'(j)$ $(j = 2, \dots, N)$, $b'(k)$ $(k = 2, \dots, N-1)$, and diagonal entries $d'(k)$ $(k = 1, \dots, N)$ defined in (29.87). We need to determine lower quasiseparable generators and diagonal entries of the Hermitian matrix $F = GR^{-1}$. To this end, notice that one can represent the matrix G in the form $G = G_L + G_U$, with a lower triangular matrix G_L and a strictly upper triangular matrix G_U. We have $GR^{-1} = G_L R^{-1} + G_U R^{-1}$. Since R^{-1} is upper triangular, the matrix $G_U R^{-1}$ is strictly upper triangular. Hence, in order to determine the lower triangular part of the matrix F it is enough to compute the matrix $G_L R^{-1}$. Applying Theorem 17.4 to the product $G_L R^{-1}$ we obtain the formulas (29.88)–(29.91) for computing lower quasiseparable generators and diagonal entries of the matrix F. Moreover, one can see easily that the orders of these quasiseparable generators equal to one. □

§29.8 Comments

The study of explicit QR iteration methods via quasiseparable representations was initiated by D. Bini, L. Gemignani and V. Pan in the papers [10, 11]. In [11] the matrices with lower quasiseparable order one which are rank one perturbations of Hermitian matrices were studied. In the paper [23] the general algorithm for computation of lower quasiseparable generators and diagonal entries of a new iterate was obtained and the QR iteration method for Hermitian quasiseparable matrices of arbitrary orders was developed.

Chapter 30

The Reduction to Hessenberg Form

One of the methods used in the solution of the eigenvalue problem is to reduce at first a matrix to a unitarily similar one in the upper Hessenberg form. The subsequent QR iterations for an upper Hessenberg matrix can be performed in an efficient way. The complexity of reduction for a non-structured $N \times N$ matrix to the upper Hessenberg form is $O(N^3)$ operations.

In this chapter we present an algorithm of reduction to the upper Hessenberg form for matrices with quasiseparable representations. We show that for unitary and Hermitian matrices and for their perturbations of small ranks one can reduce the complexity of algorithm to $O(N^2)$ operations. The crucial role is played by the factorization obtained in Theorem 20.5.

§30.1 The general description

Let A be an $N \times N$ scalar matrix. Determine a unitary $(N-1) \times (N-1)$ matrix $V^{(0)}$ from the condition

$$(V^{(0)})^* A(2:N,1) = \begin{pmatrix} \sigma_1 & 0 & \dots & 0 \end{pmatrix}^T$$

as described for instance in Section 7.1, and define the $N \times N$ unitary matrix $\hat{V}^{(0)}$ via

$$\hat{V}^{(0)} = I_1 \oplus V^{(0)}.$$

Consider the matrix $A_1 = (\hat{V}^{(0)})^* A \hat{V}^{(0)}$ unitarily similar to A. One can see easily that all the entries in the first column of A_1, except for the first two, are zeros, i.e.,

$$A_1(:,1) = \begin{pmatrix} \delta_1 & \sigma_1 & 0 & \dots & 0 \end{pmatrix}^T.$$

Next we apply a similar procedure to the matrix A_1. We determine a unitary $(N-2) \times (N-2)$ matrix $V^{(1)}$ from the condition

$$(V^{(1)})^* A_1(3:N,2) = \begin{pmatrix} \sigma_2 & 0 & \dots & 0 \end{pmatrix}^T$$

and define the $N \times N$ unitary matrix $\hat{V}^{(1)}$ by

$$\hat{V}^{(1)} = I_2 \oplus V^{(1)}.$$

Determine the unitarily similar to A matrix $A_2 = (\hat{V}^{(1)})^* A_1 \hat{V}^{(1)}$. All the entries below the first subdiagonal in the first two columns in the matrix A_2 are zeros.

We go on and obtain the sequence of matrices

$$A_0 = A, A_1, A_2, \dots, A_{N-2} \tag{30.1}$$

unitarily similar to A, such that

$$A_k = (\hat{V}^{(k-1)})^* A_{k-1} \hat{V}^{(k-1)}, \quad k = 1, \dots, N-2, \tag{30.2}$$

where

$$\hat{V}^{(k-1)} = I_k \oplus V^{(k-1)}, \quad k = 1, \dots, N-2, \tag{30.3}$$

with $(N-k) \times (N-k)$ matrices $V^{(k-1)}$ determined from the conditions

$$(V^{(k-1)})^* A_{k-1}(k+1 : N, k) = \begin{pmatrix} \sigma_k & 0 & \dots & 0 \end{pmatrix}^T, \quad k = 1, \dots, N-2. \tag{30.4}$$

Here each matrix A_k $(k = 1, \dots, N-2)$ has the partition

$$A_k = \begin{pmatrix} \Delta_k & * \\ C_k & A^{(k)} \end{pmatrix}, \tag{30.5}$$

where

$$\Delta_k = \begin{pmatrix} \delta_1 & * & * & \dots & * & * \\ \sigma_1 & \delta_2 & * & \dots & * & * \\ 0 & \sigma_2 & \delta_3 & \dots & * & * \\ \vdots & \vdots & \vdots & \ddots & \vdots & \vdots \\ 0 & 0 & 0 & \dots & \delta_{k-1} & * \\ 0 & 0 & 0 & \dots & \sigma_{k-1} & \delta_k \end{pmatrix}, \quad C_k = \begin{pmatrix} 0 & \dots & \sigma_k \\ \vdots & \ddots & \vdots \\ 0 & \dots & 0 \end{pmatrix}, \tag{30.6}$$

$$A^{(k)} = A_k(k+1 : N, k+1 : N). \tag{30.7}$$

In the final step, for $k = N-2$, we obtain a matrix A_{N-2} similar to A in the upper Hessenberg form.

The complexity of such an algorithm for non-structured matrices is $O(N^2)$ per step and $O(N^3)$ for the total algorithm.

§30.2 The reduction steps via quasiseparable generators

In this section we describe the steps of the algorithm for reduction to Hessenberg form via quasiseparable structure of a matrix.

We will show that the orders of lower quasiseparable generators do not increase in the reduction process. However, this is not true in general for upper quasiseparable generators. Fortunately, for some classes of matrices such as Hermitian matrices, unitary matrices and their perturbations of small rank, the orders of upper quasiseparable generators depend on the ones for lower generators. For these cases one can obtain upper quasiseparable generators with non-increasing orders in the reduction process. We will consider these classes of matrices in the subsequent sections.

§30.2.1 The general theorem

In Theorem 20.5 above we have obtained a factorization of a block matrix A in the form $A = VT$ with a block lower triangular unitary matrix V and a block upper triangular matrix T. It was shown, in particular, that the orders of lower quasiseparable generators of the matrix V are not greater than the corresponding orders of the matrix A. It turns out that using V as a matrix of unitary similarity transformation of the matrix A we obtain the matrix $A_1 = V^*AV$ with the same orders of lower quasiseparable generators as V. Moreover, one can choose V in such a way that all the entries of the matrix A_1 in the first column below the first two blocks will be equal to zero. This result lies at the basis of a fast algorithm for reducing a matrix with quasiseparable generators to the upper Hessenberg form.

Theorem 30.1. *Let $A = \{A_{ij}\}_{i,j=1}^N$ be a block matrix with entries of sizes $m_i \times m_j$ $(i,j = 1,\dots,N)$, with quasiseparable generators $p(i)$ $(i = 2,\dots,N)$, $q(j)$ $(j = 1,\dots,N-1)$, $a(k)$ $(k = 2,\dots,N-1)$, $g(i)$ $(i = 1,\dots,N-1)$, $h(j)$ $(j = 2,\dots,N)$, $b(k)$ $(k = 2,\dots,N-1)$, $d(k)$ $(k = 1,\dots,N)$ of orders (r_k^L, r_k^U) $(k = 1,\dots,N-1)$.*

Set

$$\begin{aligned} {}_N = 0, \quad \rho_{k-1} &= \min\{m_k + \rho_k,\ r_{k-1}^L\}, \quad && k = N,\dots,2,\ \rho_0 = 0, \\ \nu_k &= m_k + \rho_k - \rho_{k-1}, && k = 1,\dots,N, \end{aligned} \tag{30.8}$$

and using lower quasiseparable generators $p(i)$ $(i = 2,\dots,N)$, $a(k)$ $(k = 2,\dots,N-1)$ determine the unitary $(m_k + \rho_k) \times (m_k + \rho_k)$ matrices V_k, $k = N,\dots,2$ by using the QR *factorizations*

$$p(N) = V_N \begin{pmatrix} X_N \\ 0_{\nu_N \times r_{N-1}^L} \end{pmatrix}, \tag{30.9}$$

$$\begin{pmatrix} p(k) \\ X_{k+1}a(k) \end{pmatrix} = V_k \begin{pmatrix} X_k \\ 0_{\nu_k \times r_{k-1}^L} \end{pmatrix}, \quad k = N-1,\dots,2, \tag{30.10}$$

with matrices X_k of sizes $\rho_{k-1} \times r_{k-1}^L$. Next, take a unitary $\rho_1 \times \rho_1$ matrix Q_1 and set

$$V_1 = \begin{pmatrix} I_{m_1} & 0 \\ 0 & Q_1 \end{pmatrix}. \tag{30.11}$$

Using the matrices V_k $(k = 1, \dots, N)$ *define the unitary matrices* $\tilde{V}_k$ $(k = 1, \dots, N)$ *by*

$$\tilde{V}_1 = \mathrm{diag}\{V_1, I_{\phi_1}\}; \ \tilde{V}_k = \mathrm{diag}\{I_{\eta_k}, V_k, I_{\phi_k}\}, \\ k = 2, \dots, N-1; \ \tilde{V}_N = \mathrm{diag}\{I_{\eta_N}, V_N\}, \tag{30.12}$$

where

$$\eta_k = \sum_{i=1}^{k-1} m_i, \quad \phi_k = \sum_{i=k+1}^{N} \nu_i,$$

and next define the unitary matrix V *via*

$$V = \tilde{V}_N \cdots \tilde{V}_2 \tilde{V}_1. \tag{30.13}$$

Then the matrix $B = V^* A V$*, treated as a block matrix with entries of sizes* $\nu_i \times \nu_j$ $(i, j = 1, \dots, N)$*, satisfies the condition*

$$B(3 : N, 1) = 0. \tag{30.14}$$

Moreover lower quasiseparable generators $\hat{p}_B(i)$ $(i = 2, \dots, N)$, $\hat{q}_B(j)$ $(j = 1, \dots, N-1)$, $\hat{a}_B(k)$ $(k = 2, \dots, N-1)$ *of orders* ρ_k $(k = 1, \dots, N-1)$*, and the diagonal entries* $\hat{d}_B(k)$ $(k = 1, \dots, N)$ *of the matrix* B *are determined as follows.*

1. *Determine the matrices* $p_V(N)$, $d_V(N)$ *of sizes* $m_N \times \rho_{N-1}$, $m_N \times \nu_N$ *from the partition*

$$V_N = \begin{pmatrix} p_V(N) & d_V(N) \end{pmatrix}. \tag{30.15}$$

Set

$$\begin{aligned} Y_N &= h(N) p_V(N), & \gamma_N &= p_V^*(N) d(N) p_V(N), \\ \hat{p}_B(N) &= d_V^*(N) d(N) p_V(N), & \hat{d}_B(N) &= d_V^*(N) d(N) d_V(N), \\ \hat{h}(N) &= h(N) d_V(N), & h'(N) &= p_V^*(N) d(N) d_V(N). \end{aligned} \tag{30.16}$$

2. *For* $k = N-1, \dots, 2$ *perform the following. Determine matrices* $p_V(k)$, $a_V(k)$, $d_V(k)$, $q_V(k)$ *of sizes* $m_k \times \rho_{k-1}$, $\rho_k \times \rho_{k-1}$, $m_k \times \nu_k$, $\rho_k \times \nu_k$ *from the partition*

$$V_k = \begin{pmatrix} p_V(k) & d_V(k) \\ a_V(k) & q_V(k) \end{pmatrix}. \tag{30.17}$$

Set

$$\hat{q}_B(k) = q_V(k), \quad \hat{a}_B(k) = a_V(k). \tag{30.18}$$

Compute the matrix

$$\Psi_k = \begin{pmatrix} h(k) & b(k) Y_{k+1} \end{pmatrix} V_k \tag{30.19}$$

and determine the matrices $Y_k, \hat{h}(k)$ *of the sizes* $r_{k-1}^U \times \rho_{k-1}, r_{k-1}^U \times \nu_k$ *from the partition*

$$\Psi_k = \begin{pmatrix} Y_k & \hat{h}(k) \end{pmatrix}. \tag{30.20}$$

Compute the matrix

$$\Omega_k = V_k^* \begin{pmatrix} d(k) & g(k)Y_{k+1} \\ X_{k+1}q(k) & \gamma_{k+1} \end{pmatrix} V_k \tag{30.21}$$

and determine the matrices γ_k, $h'(k)$, $\hat{p}_B(k)$, $\hat{d}_B(k)$ of the sizes $\rho_{k-1} \times \rho_{k-1}$, $\rho_{k-1} \times \nu_k$, $\nu_k \times \rho_{k-1}$, $\nu_k \times \nu_k$ from the partition

$$\Omega_k = \begin{pmatrix} \gamma_k & h'(k) \\ \hat{p}_B(k) & \hat{d}_B(k) \end{pmatrix}. \tag{30.22}$$

3. *Compute*

$$\hat{d}_B(1) = \begin{pmatrix} d(1) & g(1)Y_2Q_1 \\ Q_1^*X_2q(1) & Q_1^*\gamma_2Q_1 \end{pmatrix} \tag{30.23}$$

and set

$$\hat{q}_B(1) = \begin{pmatrix} 0_{\rho_1 \times m_1} & Q_1 \end{pmatrix}. \tag{30.24}$$

Proof. By Theorem 20.5, V is a unitary block lower triangular matrix with entries of sizes $m_i \times \nu_j$ $(i,j = 1,\dots,N)$ and $T = V^*A$ is a block upper triangular matrix with entries of sizes $\nu_i \times m_j$ $(i,j = 1,\dots,N)$. Moreover, lower quasiseparable generators $p_V(i)$ $(i = 2,\dots,N)$, $q_V(j)$ $(j = 2,\dots,N)$, $a_V(k)$ $(k = 2,\dots,N)$ of orders ρ_k $(k = 1,\dots,N-1)$ and diagonal entries $d_V(k)$ $(k = 2,\dots,N)$ of the matrix V are obtained for $k = 2,\dots,N$ via the partitions (30.15), (30.17), and for $k = 1$ via the partition

$$V_1 = \begin{pmatrix} d_V(1) \\ q_V(1) \end{pmatrix},$$

with matrices $d_V(1), q_V(1)$ of sizes $m_1 \times \nu_1, \rho_1 \times \nu_1$. The latter together with (30.11) means that

$$d_V(1) = \begin{pmatrix} I_{m_1} & 0_{m_1 \times \rho_1} \end{pmatrix}, \quad q_V(1) = \begin{pmatrix} 0_{\rho_1 \times m_1} & Q_1 \end{pmatrix}. \tag{30.25}$$

Furthermore, one can easily check that the formulas (20.19), (20.20), (20.21) to compute upper quasiseparable generators and diagonal entries of the matrix T may recast as

$$\begin{pmatrix} h_T(N) \\ d_T(N) \end{pmatrix} = \begin{pmatrix} I & 0 \\ 0 & p_V^*(N) \\ 0 & d_V^*(N) \end{pmatrix} \begin{pmatrix} h(N) \\ d(N) \end{pmatrix}, \tag{30.26}$$

$$\begin{pmatrix} h_T(k) & b_T(k) \\ d_T(k) & g_T(k) \end{pmatrix} = \begin{pmatrix} I & 0 & 0 \\ 0 & p_V^*(k) & a_V^*(k) \\ 0 & d_V^*(k) & q_V^*(k) \end{pmatrix} \begin{pmatrix} h(k) & b(k) & 0 \\ d(k) & g(k) & 0 \\ X_{k+1}q(k) & 0 & I \end{pmatrix}, \tag{30.27}$$

$$k = N-1,\dots,2,$$

$$\begin{pmatrix} d_T(1) & g_T(1) \end{pmatrix} = \begin{pmatrix} d_V^*(1) & q_V^*(1) \end{pmatrix} \begin{pmatrix} d(1) & g(1) & 0 \\ X_2q(1) & 0 & I \end{pmatrix}. \tag{30.28}$$

Next we consider the product $B = TV$ of the block upper triangular matrix T and the block lower triangular matrix V. B is obtained as a block matrix with entries of sizes $\nu_i \times \nu_j$, $i, j = 1, \dots, N$. We apply Theorem 17.5 to compute quasiseparable generators $\hat{p}_B(i)$ $(i = 2, \dots, N)$, $\hat{q}_B(j)$ $(j = 1, \dots, N-1)$, $\hat{a}_B(k)$ $(k = 2, \dots, N-1)$, $\hat{g}_B(i)$ $(i = 1, \dots, N-1)$, $\hat{h}_B(j)$ $(j = 2, \dots, N)$, $\hat{b}_B(k)$ $(k = 2, \dots, N-1)$, $\hat{d}_B(k)$ $(k = 1, \dots, N)$ of the matrix B relative to this partition. Using (17.20), (17.21), and the partitions (30.17) we obtain

$$\begin{pmatrix} \Gamma_N & \hat{h}_B(N) \\ \hat{p}_B(N) & \hat{d}_B(N) \end{pmatrix} = \begin{pmatrix} h_T(N) \\ d_T(N) \end{pmatrix} \begin{pmatrix} p_V(N) & d_V(N) \end{pmatrix}, \tag{30.29}$$

$$\begin{pmatrix} \Gamma_k & \hat{h}_B(k) \\ \hat{p}_B(k) & \hat{d}_B(k) \end{pmatrix} = \begin{pmatrix} h_T(k) & b_T(k) \\ d_T(k) & g_T(k) \end{pmatrix} \begin{pmatrix} I & 0 \\ 0 & \Gamma_{k+1} \end{pmatrix} V_k, \quad k = N-1, \dots, 2, \tag{30.30}$$

$$\hat{d}_B(1) = \begin{pmatrix} d_T(1) & g_T(1) \end{pmatrix} \begin{pmatrix} I & 0 \\ 0 & \Gamma_2 \end{pmatrix} \begin{pmatrix} d_V(1) \\ q_V(1) \end{pmatrix}. \tag{30.31}$$

Here Γ_k $(k = 2, \dots, N)$ are auxiliary variables, which are $(r^U_{k-1} + \rho_{k-1}) \times \rho_{k-1}$ matrices. We use the representations

$$\Gamma_k = \begin{pmatrix} Y_k \\ \gamma_k \end{pmatrix}, \quad \hat{h}_B(k) = \begin{pmatrix} \hat{h}(k) \\ h'(k) \end{pmatrix}, \quad k = 2, \dots, N \tag{30.32}$$

with matrices $Y_k, \gamma_k, \hat{h}(k), h'(k)$ of sizes $r^U_{k-1} \times \rho_{k-1}, \rho_{k-1} \times \rho_{k-1}, r^U_{k-1} \times \nu_k, \rho_{k-1} \times \nu_k$, respectively.

Inserting (30.32) in (30.29) we get

$$\begin{pmatrix} Y_N & \hat{h}(N) \\ \gamma_N & h'(N) \\ \hat{p}_B(N) & \hat{d}_B(N) \end{pmatrix} = \begin{pmatrix} h_T(N) \\ d_T(N) \end{pmatrix} \begin{pmatrix} p_V(N) & d_V(N) \end{pmatrix}. \tag{30.33}$$

Combining (30.33) and (30.26) we get

$$\begin{pmatrix} Y_N & \hat{h}(N) \\ \gamma_N & h'(N) \\ \hat{p}_B(N) & \hat{d}_B(N) \end{pmatrix} = \begin{pmatrix} I & 0 \\ 0 & p^*_V(N) \\ 0 & d^*_V(N) \end{pmatrix} \begin{pmatrix} h(N) \\ d(N) \end{pmatrix} \begin{pmatrix} p_V(N) & d_V(N) \end{pmatrix},$$

which implies (30.16).

Inserting (30.32) in (30.30) and using (30.17) we get

$$\begin{pmatrix} Y_k & \hat{h}(k) \\ \gamma_k & h'(k) \\ \hat{p}_B(k) & \hat{d}_B(k) \end{pmatrix} = \begin{pmatrix} h_T(k) & b_T(k) \\ d_T(k) & g_T(k) \end{pmatrix} \begin{pmatrix} I & 0 \\ 0 & Y_{k+1} \\ 0 & \gamma_{k+1} \end{pmatrix} V_k, \quad k = N-1, \dots, 2. \tag{30.34}$$

Combining this with (30.27) and using also (30.17) we get

$$\begin{pmatrix} Y_k & \hat{h}(k) \\ \gamma_k & h'(k) \\ p_B(k) & d_B(k) \end{pmatrix} = \begin{pmatrix} I & 0 \\ 0 & V_k^* \end{pmatrix} \begin{pmatrix} h(k) & b(k)Y_{k+1} \\ d(k) & g(k)Y_{k+1} \\ X_{k+1}q(k) & \gamma_{k+1} \end{pmatrix} V_k,$$
$$k = N-1, \dots, 2,$$

which implies the relations (30.19), (30.20) and (30.21), (30.22).

Combining (30.31) and (30.28) we get

$$\hat{d}_B(1) = \begin{pmatrix} d_V^*(1) & q_V^*(1) \end{pmatrix} \begin{pmatrix} d(1) & g(1) & 0 \\ X_2 q(1) & 0 & I \end{pmatrix} \begin{pmatrix} I & 0 \\ 0 & Y_2 \\ 0 & \gamma_2 \end{pmatrix} \begin{pmatrix} d_V(1) \\ q_V(1) \end{pmatrix},$$

and using (30.25) we obtain

$$\hat{d}_B(1) = \begin{pmatrix} I_{m_1} & 0 \\ 0 & Q_1^* \end{pmatrix} \begin{pmatrix} d(1) & g(1)Y_2 \\ X_2 q(1) & \gamma_2 \end{pmatrix} \begin{pmatrix} I_{m_1} & 0 \\ 0 & Q_1 \end{pmatrix}.$$

From here we obtain formula (30.23).

Next, using (17.19) we obtain the formulas (30.18) and the equality

$$\hat{q}_B(1) = q_V(1). \tag{30.35}$$

Combining (30.35) and (30.25) we obtain formula (30.24), which in turn yields (30.14) follows. □

Corollary 30.2. *Under the conditions of Theorem 30.1, assume that A is a scalar matrix, i.e., $m_k = 1$, $k = 1, \dots, N$. Set*

$$\begin{gathered} \lambda_k = \sum_{i=1}^{k-1} \nu_k, \quad k = 1, \dots, N, \\ r_{\lambda_k + i} = \rho_{k-1} + i, \; i = 1, \dots, \nu_k - 1, \quad r_{\lambda_{k+1}} = \rho_k, \quad k = 1, \dots, N. \end{gathered} \tag{30.36}$$

Determine the unitary $\rho_1 \times \rho_1$ matrix Q_1 and the number σ_1 from the condition

$$Q_1^*(X_2 q(1)) = \begin{pmatrix} \sigma_1 & 0_{1 \times (\rho_1 - 1)} \end{pmatrix}^T. \tag{30.37}$$

Then B, treated as a scalar matrix, satisfies the condition

$$B(:,1) = \begin{pmatrix} d(1) \\ \sigma_1 \\ 0_{(N-2)\times 1} \end{pmatrix}. \tag{30.38}$$

Moreover, lower quasiseparable generators $p_B(i)$ $(i = 2, \dots, N)$, $q_B(j)$ $(j = 1, \dots, N-1)$, $a_B(k)$ $(k = 2, \dots, N-1)$ of orders r_i $(i = 1, \dots, N-1)$ and *the diagonal entries $d(k)$ $(k = 1, \dots, N)$ of the scalar matrix B are determined as follows.*

1. *Set*

$$D = Q_1^* \gamma_2 Q_1, \tag{30.39}$$

$$d_B(1) = d(1), \quad d_B(i) = D(i-1, i-1),\ i = 2, \dots, \nu_1, \tag{30.40}$$

$$p_B(2) = \sigma_1,\ q_B(1) = 1,\ a_B(2) = 0,\ q_B(2) = 1, \tag{30.41}$$

$$p_B(i+1) = D(i, 1:i-1), \quad i = 2, \dots, \rho_1, \tag{30.42}$$

$$a_B(i) = \begin{pmatrix} I_{i-2} \\ 0_{1\times(i-2)} \end{pmatrix},\ q_B(i) = \begin{pmatrix} 0_{(i-2)\times 1} \\ 1 \end{pmatrix}, \quad i = 3, \dots, \nu_1 - 1, \tag{30.43}$$

$$a_B(\nu_1) = Q_1(:, 1:\rho_1 - 1), \quad q_B(\nu_1) = Q_1(:, \rho_1). \tag{30.44}$$

2. *For $k = 2, \dots, N-1$, $\nu_k \neq 0$ set $\hat{p}_k = \hat{p}_B(k)$, $\hat{q}_k = \hat{q}_B(k)$, $\hat{d}_k = \hat{d}_B(k)$ and*

$$d_B(\lambda_k + i) = \hat{d}_k(i, i), \quad i = 1, \dots, \nu_k, \tag{30.45}$$

$$p_B(\lambda_k + i) = \begin{bmatrix} \hat{p}_k(i, :) & \hat{d}_k(i, 1:i-1) \end{bmatrix},\ i = 1, \dots, \nu_k, \tag{30.46}$$

$$q_B(\lambda_k + i) = \begin{pmatrix} 0_{(\rho_{k-1}+i-1)\times 1} \\ 1 \end{pmatrix},\ a_B(\lambda_k + i) = \begin{pmatrix} I_{\rho_{k-1}+i-1} \\ 0_{1\times(\rho_{k-1}+i-1)} \end{pmatrix}, \quad i = 1, \dots, \nu_k - 1, \tag{30.47}$$

$$a_B(\lambda_{k+1}) = \begin{pmatrix} \hat{a}_B(k) & \hat{q}_k(:, 1:\nu_k - 1) \end{pmatrix}, \quad q_B(\lambda_{k+1}) = \hat{q}_k(:, \nu_k). \tag{30.48}$$

3. *If $\nu_N \neq 0$, set*

$$p_B(N) = \hat{p}_B(N), \tag{30.49}$$

$$d_B(N) = \hat{d}_B(N). \tag{30.50}$$

Proof. From (30.23), using the condition (30.37) we obtain the formula

$$\hat{d}_B(1) = \begin{bmatrix} d(1) & g(1)Y_2Q_1 \\ \hat{\sigma}_1 & Q_1^*\gamma_2 Q_1 \end{bmatrix}, \tag{30.51}$$

with $\hat{\sigma}_1 = \begin{pmatrix} \sigma_1 & 0_{1\times(\rho_1-1)} \end{pmatrix}^T$. Combining this with (30.14) we obtain (30.38).

Using (30.23) and the fact that B was obtained as a block $\nu_i \times \nu_j$ matrix we obtain the formulas (30.40), (30.45) for diagonal entries. Next, we check that the relations

$$B(i, 1:i-1) = p_B(i) Q_{i-1}^B, \qquad i = 2, \dots, N, \tag{30.52}$$

with

$$Q_j^B = \mathrm{row}((a_B)^{>}_{j+1,i} q_B(i))_{i=1}^{j}, \quad j = 1, \dots, N-1,$$

hold. By Lemma 5.3, this implies that $p_B(i)$ $(i = 2, \dots, N)$, $q_B(j)$ $(j = 1, \dots, N-1)$, $a_B(j)$ $(j = 2, \dots, N-1)$ are lower quasiseparable generators of B, treated as a scalar matrix. Applying (30.23) and Lemma 5.2 to B. treated as a block matrix, we have

$$B(1 : \lambda_2, 1 : \lambda_2) = \hat{d}_B(1) \tag{30.53}$$

and

$$B(\lambda_k + 1 : \lambda_{k+1}, 1 : \lambda_{k+1}) = \left[\begin{array}{cc} \hat{p}_B(k)\hat{Q}^B_{k-1} & \hat{d}_B(k) \end{array} \right], \quad k = 2, \dots, N, \tag{30.54}$$

with

$$\hat{Q}^B_j = \operatorname{row}((\hat{a}_B)^>_{j+1,i}\hat{q}_B(i))^j_{i=1}, \quad j = 1, \dots, N-1.$$

To get from here the relations (30.52) one needs to derive some relations for the matrices Q^B_i $(i = 1, \dots, N-1)$. Using the recursion (5.3), we have

$$Q^B_1 = q_B(1), \quad Q^B_j = \left(\begin{array}{cc} a(j)Q^B_{j-1} & q(j) \end{array} \right) = \left(\begin{array}{cc} a(j) & q(j) \end{array} \right) \left(\begin{array}{cc} Q^B_{j-1} & 0 \\ 0 & 1 \end{array} \right), \tag{30.55}$$
$$i = 2, \dots, N-1.$$

From here using (30.41) we have

$$Q^B_1 = 1, \quad Q^B_2 = \left(\begin{array}{cc} 0 & 1 \end{array} \right). \tag{30.56}$$

The relations (30.43) imply

$$\left(\begin{array}{cc} a_B(i) & q_B(i) \end{array} \right) = I_{i-1}, \quad i = 3, \dots, \nu_1 - 1.$$

Hence, using (30.55) we get

$$Q^B_i = \left(\begin{array}{cc} 0_{(i-1)\times 1} & I_{i-1} \end{array} \right), \quad i = 3, \dots, \nu_1 - 1. \tag{30.57}$$

The relations (30.44) imply

$$\left(\begin{array}{cc} a_B(\nu_1) & q_B(\nu_1) \end{array} \right) = Q_1,$$

so that using (30.55) and (30.57) with $i = \nu_1 - 1 = \rho_1$ we get

$$Q^B_{\nu_1} = Q_1 \left(\begin{array}{cc} 0_{\rho_1\times 1} & I_{\rho_1} \end{array} \right) = \left(\begin{array}{cc} 0_{\rho_1\times 1} & Q_1 \end{array} \right).$$

Comparing this with (30.24) we get

$$Q^B_{\nu_1} = Q^B_{\lambda_2} = \hat{q}_B(1) = \hat{q}^B_1. \tag{30.58}$$

Next, we prove by induction that

$$\hat{Q}^B_k = Q^B_{\lambda_{k+1}}, \quad k = 1, \dots, N-1, \tag{30.59}$$

and the relations

$$Q^B_{\lambda_k+i} = \begin{pmatrix} \hat{Q}^B_{k-1} & 0 \\ 0 & I_i \end{pmatrix}, \quad i = 1, \dots, \nu_1 - 1 \tag{30.60}$$

hold. The equality (30.58) implies (30.59) with $k = 1$. Assume that for some k with $2 \le k \le N-2$

$$\hat{Q}^B_{k-1} = Q^B_{\lambda_k}. \tag{30.61}$$

Using (30.47) we have

$$\begin{pmatrix} a_B(\lambda_k + i) & q_B(\lambda_k + i) \end{pmatrix} = I_{\rho_{k-1}+i}, \quad i = 1, \dots, \nu_k - 1,$$

whence, using (30.55),

$$Q^B_{\lambda_k+i} = \begin{pmatrix} Q^B_{\lambda_k} & 0 \\ 0 & I_i \end{pmatrix}, \quad i = 1, \dots, \nu_k - 1, \tag{30.62}$$

which together with (30.61) implies (30.60). The equalities (30.48) imply

$$\begin{pmatrix} a_B(\lambda_{k+1}) & q_B(\lambda_{k+1}) \end{pmatrix} = \begin{pmatrix} \hat{a}_B(k) & \hat{q}_B(k) \end{pmatrix}$$

and therefore using (30.55) and (30.62) with $i = \nu_k - 1$ we get

$$Q^B_{\lambda_{k+1}} = \begin{pmatrix} \hat{a}_B(k) & \hat{q}_B(k) \end{pmatrix} \begin{pmatrix} \hat{Q}^B_{k-1} & 0 \\ 0 & I_{\nu_k} \end{pmatrix}$$

which implies (30.59).

Now using (30.53) and (30.23) we get

$$B(2,1) = \sigma_1, \quad B(i, 1:i-1) = \begin{pmatrix} 0 & D(i-1, 1:i-2) \end{pmatrix}, \quad i = 3, \dots, \nu_1,$$

which together with (30.41), (30.42) and (30.56), (30.57) imply (30.52) with $i = 2, \dots, \nu_1$. Using (30.54), we have

$$\begin{gathered} B(\lambda_k + 1, 1 : \lambda_k) = \hat{p}_k(1,:)\hat{Q}^B_{k-1}, \\ B(\lambda_k + i, 1 : \lambda_k + i - 1) = \begin{bmatrix} \hat{p}_k(i,:)\hat{Q}^B_{k-1} & \hat{d}_k(i, 1:i-1) \end{bmatrix}, \\ i = 2, \dots, \nu_k, \ k = 2, \dots, N-1. \end{gathered} \tag{30.63}$$

Hence using (30.59), (30.62) and (30.46) we obtain (30.52) with $i = \nu_1 + 1, \dots, \lambda_N$.

Finally from (30.8) with $k = N$ it follows that either $\nu_N = 0$ and the proof is finished, or $\nu_N = 1$, $\lambda_N = N - 1$. In the second case using (30.54) with $k = N$, (30.59) with $k = N - 1$, and (30.49), we get (30.52) with $k = N$ and the formula (30.50) for the last diagonal entry. □

§30.3 The reduction step for special cases

Next, based on Theorem 30.1, we derive algorithms for a step of reduction to upper Hessenberg form for Hermitian matrices, unitary matrices and their perturbations of small ranks. For these cases one can obtain also upper quasiseparable generators with non-increasing orders.

§30.3.1 The reduction step for Hermitian and perturbations of Hermitian matrices

We start the description of the above-mentioned important cases with Hermitian matrices. For this case a matrix is defined completely by its lower quasiseparable generators and diagonal entries.

Theorem 30.3. *Let $A = \{A_{ij}\}_{i,j=1}^N$ be a Hermitian scalar matrix with lower quasiseparable generators $p(i)$ $(i = 2,\dots,N)$, $q(j)$ $(j = 1,\dots,N-1)$, $a(k)$ $(k = 2,\dots,N-1)$ of orders r_k^L $(k = 1,\dots,N-1)$ and diagonal entries $d(k)$ $(k = 1,\dots,N)$. Set $m_k = 1$, $k = 1,\dots,N$ and define the numbers ρ_k $(k = 0,\dots,N)$, ν_k $(k = 1,\dots,N)$ via* (30.8) *and the numbers λ_k $(k = 1,\dots,N)$, r_i $(i = 1,\dots,N-1)$ via* (30.36)*. Determine the $(m_k+\rho_k)\times(m_k+\rho_k)$ unitary matrices V_k, $k = N,\dots,2$ and auxiliary $\rho_{k-1}\times r_{k-1}^L$ matrices X_k via* (30.9), (30.10)*, and the $\rho_1\times\rho_1$ unitary matrix Q_1 and $\nu_1\times\nu_1$ unitary matrix V_1 via* (30.37), (30.11)*. Next define the unitary $N\times N$ matrix V via* (30.12), (30.13) *and the Hermitian matrix $B = V^*AV$. By Corollary* 30.2*, the matrix B satisfies the condition* (30.38)*.*

Then lower quasiseparable generators $p_B(i)$ $(i = 2,\dots,N)$, $q_B(j)$ $(j = 1,\dots,N-1)$, $a_B(k)$ $(k = 2,\dots,N-1)$ of orders r_i $(i = 1,\dots,N-1)$ of the matrix B as well as its diagonal entries $d_B(k)$ $(k = 1,\dots,N)$ are determined using the following algorithm.

1.1. *Determine the matrices $p_V(N)$, $d_V(N)$ of sizes $m_N\times\rho_{N-1}$, $m_N\times\nu_N$ from the partition*

$$V_N = \begin{pmatrix} p_V(N) & d_V(N) \end{pmatrix}. \tag{30.64}$$

Set

$$\gamma_N = p_V^*(N)d(N)p_V(N), \quad \hat{p}_B(N) = d_V^*(N)d(N)p_V(N), \\ \hat{d}_B(N) = d_V^*(N)d(N)d_V(N). \tag{30.65}$$

1.2. *For $k = N-1,\dots,2$ perform the following. Determine the matrices $p_V(k)$, $a_V(k)$, $d_V(k)$, $q_V(k)$ of sizes $m_k\times\rho_{k-1}$, $\rho_k\times\rho_{k-1}$, $m_k\times\nu_k$, $\rho_k\times\nu_k$ from the partition*

$$V_k = \begin{pmatrix} p_V(k) & d_V(k) \\ a_V(k) & q_V(k) \end{pmatrix}. \tag{30.66}$$

Set

$$\hat{q}_B(k) = q_V(k), \quad \hat{a}_B(k) = a_V(k). \tag{30.67}$$

Compute the matrix

$$\Omega_k = V_k^* \begin{pmatrix} d(k) & q^*(k)X_{k+1}^* \\ X_{k+1}q(k) & \gamma_{k+1} \end{pmatrix} V_k \tag{30.68}$$

and determine the matrices $\gamma_k, h'(k), \hat{p}_B(k), \hat{d}_B(k)$ *of sizes* $\rho_{k-1} \times \rho_{k-1}$, $\rho_{k-1} \times \nu_k$, $\nu_k \times \rho_{k-1}$, $\nu_k \times \nu_k$ *from the partition and determine the matrices* γ_k, $h'(k)$, $p_B(k+1)$, $d_B(k+1)$ *of the sizes from the partition*

$$\Omega_k = \begin{pmatrix} \gamma_k & h'(k) \\ \hat{p}_B(k) & \hat{d}_B(k) \end{pmatrix}. \tag{30.69}$$

2. *Using Corollary* 30.2 *determine lower quasiseparable generators and diagonal entries of the matrix* B, *treated as a scalar matrix.*

Proof. Upper quasiseparable generators of the Hermitian matrix A are $q^*(i)$ $(i = 1, \dots, N-1)$, $p^*(j)$ $(j = 2, \dots, N)$, $a^*(k)$ $(k = 2, \dots, N-1)$. To apply Theorem 30.1 to the matrix A we only need to check that the relations (30.68), (30.69) hold. To this end, we check that the parameters X_k, Y_k $(k = 2, \dots, N)$ used therein satisfy the equalities

$$Y_k = X_k^*, \quad k = 2, \dots, N. \tag{30.70}$$

Indeed, for $k = N$ we have

$$Y_N = h(N) = p^*(N) = X_N^*.$$

Using (30.19), (30.20) we get

$$\begin{pmatrix} Y_k & \hat{h}(k) \end{pmatrix} = \begin{pmatrix} p^*(k) & a^*(k)Y_{k+1} \end{pmatrix} V_k,$$

which means that

$$V_k^* \begin{pmatrix} p(k) \\ Y_{k+1}^* a(k) \end{pmatrix} = \begin{pmatrix} Y_k^* \\ (\hat{h}(k))^* \end{pmatrix}. \tag{30.71}$$

Thus, comparing (30.71) with (30.10) we obtain (30.70) and also the equalities $\hat{h}(k) = 0$ $(k = N, \dots, 2)$. Inserting (30.70) and $g(k) = q^*(k)$ in (30.21), and using (30.22) we obtain the equalities (30.68) and (30.69).

Step 2 of the algorithm is obvious. □

The complexity of Step 1.2 as performed in formula (30.68) is

$$4(m_k + \rho_k)^3 + 2(\rho_k + r_k^L).$$

Next we consider a generalization to perturbations of Hermitian matrices.

Theorem 30.4. *Let $A = \{A_{ij}\}_{i,j=1}^{N}$ be a scalar matrix with lower quasiseparable generators $p(i)$ $(i = 2,\dots,N)$, $q(j)$ $(j = 1,\dots,N-1)$, $a(k)$ $(k = 2,\dots,N-1)$ and diagonal entries $d(k)$ $(k = 1,\dots,N)$. Assume that A is represented in the form*

$$A = H - XY^*, \tag{30.72}$$

where H is a Hermitian matrix and $X = \operatorname{col}(x(i))_{i=1}^{N}$, $Y = \operatorname{col}(y(i))_{i=1}^{N}$, with m-dimensional rows $x(i), y(i)$. Set $m_k = 1$, $k = 1,\dots,N$, and define the numbers ρ_k $(k = 0,\dots,N)$, ν_k $(k = 1,\dots,N)$ via (30.8) and the numbers λ_k $(k = 1,\dots,N)$, r_i $(i = 1,\dots,N-1)$ via (30.36). Determine the $(m_k+\rho_k)\times(m_k+\rho_k)$ unitary matrices V_k, $k = N,\dots,2$ and auxiliary $\rho_{k-1}\times r^L_{k-1}$ matrices X_k via (30.9), (30.10), and the $\rho_1\times\rho_1$ unitary matrix Q_1 and $\nu_1\times\nu_1$ unitary matrix V_1 via (30.37), (30.11). Next define the unitary $N\times N$ matrix V via (30.12), (30.13) and the matrix $B = V^*AV$. By Corollary 30.2, the matrix B satisfies the condition (30.38).

Then the matrix $B = V^*AV$ is represented in the form

$$B = H_B - X_B Y_B^*, \tag{30.73}$$

where H_B is a Hermitian matrix and $X_B = \operatorname{col}(x_B(i))_{i=1}^{N}$, $Y_B = \operatorname{col}(y_B(i))_{i=1}^{N}$ are block columns with $1\times m$ entries. Moreover, lower quasiseparable generators $p_B(i)$ $(i = 2,\dots,N)$, $q_B(j)$ $(j = 1,\dots,N-1)$, $a_B(k)$ $(k = 2,\dots,N-1)$ of orders r_i $(i = 1,\dots,N-1)$ of the matrix B as well as its diagonal entries $d_B(k)$ $(k = 1,\dots,N)$, and the perturbation matrices X_B, Y_B are determined using the following algorithm.

1.1. *Determine the matrices $p_V(N)$, $d_V(N)$ of sizes $m_N\times\rho_{N-1}$, $m_N\times\nu_N$ from the partition*

$$V_N = \begin{pmatrix} p_V(N) & d_V(N) \end{pmatrix}. \tag{30.74}$$

Compute

$$\begin{gathered}\gamma_N = p_V^*(N)d(N)p_V(N),\quad \hat{p}_B(N) = d_V^*(N)d(N)p_V(N),\\ \hat{d}_B(N) = d_V^*(N)d(N)d_V(N),\end{gathered} \tag{30.75}$$

$$\begin{aligned} c_N &= p_V^*(N)x(N), & \hat{x}_B(N) &= d_V^*(N)x(N),\\ z_N &= p_V^*(N)y(N), & \hat{y}_B(N) &= d_V^*(N)y(N).\end{aligned} \tag{30.76}$$

1.2. *For $k = N-1,\dots,2$ perform the following. Determine the matrices $p_V(k)$, $a_V(k)$, $d_V(k)$, $q_V(k)$ of sizes $m_k\times\rho_{k-1}$, $\rho_k\times\rho_{k-1}$, $m_k\times\nu_k$, $\rho_k\times\nu_k$ from the partition*

$$V_k = \begin{pmatrix} p_V(k) & d_V(k)\\ a_V(k) & q_V(k)\end{pmatrix}. \tag{30.77}$$

Set

$$\hat{q}_B(k) = q_V(k),\quad \hat{a}_B(k) = a_V(k). \tag{30.78}$$

Compute

$$
\begin{aligned}
c_k &= p_V^*(k)x(k) + a_V^*(k)c_{k+1}, \\
\hat{x}_B(k) &= d_V^*(k)x(k) + q_V^*(k)c_{k+1},
\end{aligned}
\tag{30.79}
$$

$$
\begin{aligned}
z_k &= p_V^*(k)y(k) + a_V^*(k)z_{k+1}, \\
\hat{y}_B(k) &= d_V^*(k)y(k) + q_V^*(k)z_{k+1},
\end{aligned}
\tag{30.80}
$$

$$
\hat{Y}_k = -x(k)z_{k+1}^* + y(k)c_{k+1}^* + q^*(k)X_{k+1}^*. \tag{30.81}
$$

Compute the matrix

$$
\Omega_k = V_k^* \begin{pmatrix} d(k) & \hat{Y}_k \\ X_{k+1}q(k) & \gamma_{k+1} \end{pmatrix} V_k \tag{30.82}
$$

and determine the matrices $\gamma_k, h'(k), \hat{p}_B(k), \hat{d}_B(k)$ *of the sizes* $\rho_{k-1} \times \rho_{k-1}, \rho_{k-1} \times \nu_k, \nu_k \times \rho_{k-1}, \mu_{k+1} \times \mu_{k+1}$ *from the partition*

$$
\Omega_k = \begin{pmatrix} \gamma_k & h'(k) \\ \hat{p}_B(k) & \hat{d}_B(k) \end{pmatrix}. \tag{30.83}
$$

1.3. *Compute*

$$
\hat{x}_B(1) = \begin{pmatrix} x(1) \\ Q_1^* c_2 \end{pmatrix}, \quad \hat{y}_B(1) = \begin{pmatrix} y(1) \\ Q_1^* z_2 \end{pmatrix}. \tag{30.84}
$$

2. *Using Corollary* 30.2 *determine lower quasiseparable generators and diagonal entries of* B*, treated as a scalar matrix.*
   Determine the vector rows $x_B(i), y_B(i),\ i = 1, \dots, N$ *as follows.*
   For $k = 1, \dots, N, \quad \nu_k \neq 0$*, set* $\hat{x}_k = \hat{x}_B(k), \hat{y}_k = \hat{y}_B(k)$ *and*

$$
x_B(\lambda_k + i) = \hat{x}_k(i,:),\ y_B(\lambda_k + i) = \hat{y}_k(i,:), \quad i = 1, \dots, \nu_k. \tag{30.85}
$$

Proof. We obviously get (30.73) with the Hermitian matrix $H_B = V^*HV$ and $X_B = V^*X,\ Y_B = V^*Y$. We use the representations $X_B = \mathrm{col}(\hat{x}_B(i))_{i=1}^N$, $Y_B = \mathrm{col}(\hat{y}_B(i))_{i=1}^N$ with $\nu_i \times m$ matrices $\hat{x}_B(i), \hat{y}_B(i)$. The formulas (30.76), (30.79), (30.80) and (30.84) to compute the components $\hat{x}_B(i), \hat{y}_B(i)$ follow directly from the factorization (30.13) and the partitions (30.74), (30.77), and (30.11).

Next we apply Theorem 30.1 to compute lower quasiseparable generators and diagonal entries of B, treated as a block matrix. Upper quasiseparable generators of the matrix A are obtained by the formulas (29.42):

$$
g(i) = \begin{pmatrix} -x(i) & y(i) & q^*(i) \end{pmatrix},\ i = 1, \dots, N-1, \tag{30.86}
$$

$$
\begin{aligned}
h(j) &= \begin{pmatrix} y^*(j) \\ x^*(j) \\ p^*(j) \end{pmatrix}, \qquad j = 2, \dots, N, \\
b(k) &= \begin{pmatrix} I_{2m} & 0 \\ 0 & a^*(k) \end{pmatrix},\ k = 2, \dots, N-1.
\end{aligned}
\tag{30.87}
$$

We need to check that the relations (30.81), (30.82), (30.83) hold. For the variables $Y_k, \hat{h}(k)$ from Theorem 30.1 we use the partitions

$$Y_k = \begin{pmatrix} z'_k \\ c'_k \\ \theta_k \end{pmatrix}, \quad \hat{h}(k) = \begin{pmatrix} y'_k \\ x'_k \\ w_k \end{pmatrix}. \tag{30.88}$$

Using the first equality in (30.16) we get

$$z'_N = y^*(N)p_V(N), \quad c'_N = x^*(N)p_V(N), \quad \theta_N = p^*(N)p_V(N).$$

Inserting (30.88) in (30.19), (30.20) and using (30.87) we get

$$\begin{pmatrix} z'_k & y'_k \\ c'_k & x'_k \\ \theta_k & w_k \end{pmatrix} = \begin{pmatrix} y^*(k) & z'_{k+1} \\ x^*(k) & c'_{k+1} \\ p^*(k) & a^*(k)\theta_{k+1} \end{pmatrix} V_k, \quad k = N-1, \dots, 2.$$

From here, using the partitions (30.77), we obtain the equalities

$$\begin{aligned} (z'_N)^* &= p^*_V(N)y(N), \\ (z'_k)^* &= p^*_V(k)y(k) + a^*_V(k)(z'_{k+1})^*, \quad k = N-1, \dots, 2, \end{aligned} \tag{30.89}$$

$$\begin{aligned} (c'_N)^* &= p^*_V(N)x(N), \\ (c'_k)^* &= p^*_V(k)x(k) + a^*_V(k)(c'_{k+1})^*, \quad k = N-1, \dots, 2, \end{aligned} \tag{30.90}$$

$$\begin{aligned} \theta^*_N &= p^*_V(N)p(N), \\ \begin{pmatrix} \theta^*_k \\ w^*_k \end{pmatrix} &= V^*_k \begin{pmatrix} p(k) \\ \theta^*_{k+1}a(k) \end{pmatrix}, \quad k = N-1, \dots, 2. \end{aligned} \tag{30.91}$$

Comparing (30.89), (30.90) with (30.76), (30.79), (30.80) and the equality (30.91) with (30.9), (30.10), we conclude that

$$z'_k = z^*_k, \quad c'_k = c^*_k, \quad \theta_k = X^*_k, \quad k = 2, \dots, N.$$

Inserting this in (30.88) we get

$$Y_k = \begin{pmatrix} z^*_k \\ c^*_k \\ X^*_k \end{pmatrix}, \quad k = 2, \dots, N,$$

which in conjunction with formula (30.86) yields

$$\hat{Y}_k := g(k)Y_{k+1} = -x(k)z^*_{k+1} + y(k)c^*_{k+1} + q^*(k)X^*_{k+1}, \quad k = 1, \dots, N-1.$$

Hence, using (30.21), (30.22) we obtain the equalities (30.82), (30.83).

Step 2 to determine quasiseparable generators of B, treated as a scalar matrix, and the vector rows $x_B(i), y_B(i)$, $i = 1, \dots, N$, is obvious. □

The complexity of Step 1.2 which is executed N times (formulas (30.79)–(30.82)) is less than

$$2\big(\rho_{k-1}(2m_k-1)m+\rho_{k-1}(2\rho_k-1)m+\nu_k(2m_k-1)m+\nu_k(2\rho_k-1)m+m_k(2m-1)\rho_k\big) \\ +2m_k(2r_k^L-1)\rho_k+4(\rho_k+m_k)^3$$

arithmetical multiplications and additions.

§30.4 The reduction step for unitary matrices

Here we consider the reduction step for a unitary matrix. In this case we obtain upper quasiseparable generators of the resulting matrix with the same orders as for the lower quasiseparable generators.

Theorem 30.5. *Let $A=\{A_{ij}\}_{i,j=1}^N$ be a unitary scalar matrix with quasiseparable generators $p(i)$ $(i=2,\dots,N)$, $q(j)$ $(j=1,\dots,N-1)$, $a(k)$ $(k=2,\dots,N-1)$, $g(i)$ $(i=1,\dots,N-1)$, $h(j)$ $(j=2,\dots,N)$, $b(k)$ $(k=2,\dots,N-1)$ of orders r_k^L,r_k^U $(k=1,\dots,N-1)$, and diagonal entries $d(k)$ $(k=1,\dots,N)$. Set $m_k=1$, $k=1,\dots,N$ and define the numbers ρ_k $(k=0,\dots,N)$, ν_k $(k=1,\dots,N)$ via (30.8) and the numbers λ_k $(k=1,\dots,N)$, r_i $(i=1,\dots,N-1)$ via (30.36). Determine the $(m_k+\rho_k)\times(m_k+\rho_k)$ unitary matrices V_k, $k=N,\dots,2$ and auxiliary $\rho_{k-1}\times r_{k-1}^L$ matrices X_k via (30.9), (30.10), and the $\rho_1\times\rho_1$ unitary matrix Q and $\nu_1\times\nu_1$ unitary matrix V_1 via (30.37), (30.11). Next define the unitary $N\times N$ matrix V via (30.12), (30.13) and the unitary matrix $B=V^*AV$. By Theorem 30.1, the matrix B satisfies the condition (30.38).*

Then the matrix B has quasiseparable generators $p_B(i)$; $(i=2,\dots,N)$, $q_B(j)$ $(j=1,\dots,N-1)$, $a_B(k)$ $(k=2,\dots,N-1)$; $g_B(i)$ $(i=1,\dots,N-1)$, $h_B(j)$ $(j=2,\dots,N)$, $b_B(k)$ $(k=2,\dots,N-1)$ of orders r_k,r_k $(k=1,\dots,N-1)$, and diagonal entries $d_B(k)$ $(k=1,\dots,N)$. These quasiseparable generators are determined using the following algorithm.

1.1. *Determine the matrices $p_V(N)$, $d_V(N)$ of sizes $m_N\times\rho_{N-1}$, $m_N\times\nu_N$ from the partition*

$$V_N=\begin{bmatrix} p_V(N) & d_V(N)\end{bmatrix}.$$

Compute

$$\gamma_N=p_V^*(N)d(N)p_V(N),\quad \hat{p}_B(N)=d_V^*(N)d(N)p_V(N),$$
$$\hat{d}_B(N)=d_V^*(N)d(N)d_V(N),\quad f_N=h(N),$$
$$\tilde{h}(N)=p_V^*(N)d(N),\quad d_W(N)=d_V^*(N)d(N).$$

The matrix $d_W(N)$ has orthonormal rows and one can determine the $m_N\times m_N$ unitary matrix F_N from the condition

$$d_W(N)F_N^*=\begin{pmatrix} 0_{\nu_N\times\rho_{N-1}} & I_{\nu_N}\end{pmatrix}.$$

Determine the matrices $h_s(N), d_F(N)$ *of sizes* $\rho_{N-1} \times m_N, \nu_N \times m_N$ *from the partition*

$$F_N = \begin{pmatrix} h_s(N) \\ d_F(N) \end{pmatrix}.$$

Compute

$$u_N = h(N)h_s^*(N), \qquad v_N = \tilde{h}(N)h_s^*(N),$$
$$\alpha_N = h_s(N)p_V(N), \quad \hat{h}_B(N) = h_s(N)d_V(N).$$

Set

$$\hat{g}_B(N-1) = g_s(N-1).$$

1.2. *For* $k = N-1, \dots, 2$ *perform the following. Determine the matrices* $p_V(k)$, $a_V(k)$, $d_V(k)$, $q_V(k)$ *of sizes* $m_k \times \rho_{k-1}$, $\rho_k \times \rho_{k-1}$, $m_k \times \nu_k$, $\rho_k \times \nu_k$ *from the partition*

$$V_k = \begin{pmatrix} p_V(k) & d_V(k) \\ a_V(k) & q_V(k) \end{pmatrix}.$$

Compute

$$f_k = h(k)p_V(k) + b(k)f_{k+1}a_V(k),$$
$$g_s(k) = d_V^*(k)g(k)u_{k+1} + q_V^*(k)v_{k+1}.$$

Compute the matrix

$$\Omega_k = V_k^* \begin{pmatrix} d(k) & g(k)f_{k+1} \\ X_{k+1}q(k) & \gamma_{k+1} \end{pmatrix} V_k$$

and determine the matrices $\gamma_k, h'(k), \hat{p}_B(k), \hat{d}_B(k)$ *of sizes* $\rho_{k-1}\times\rho_{k-1}, \rho_{k-1}\times \nu_k, \nu_k \times \rho_{k-1}, \nu_k \times \nu_k$ *from the partition*

$$\Omega_k = \begin{bmatrix} \gamma_k & h'(k) \\ \hat{p}_B(k) & \hat{d}_B(k) \end{bmatrix}.$$

Compute

$$\tilde{h}(k) = p_V^*(k)d(k) + a_V^*(k)X_{k+1}q(k),$$
$$d_W(k) = d_V^*(k)d(k) + q_V^*(k)X_{k+1}q(k).$$

The matrix $\begin{pmatrix} d_W(k) & g_s(k) \end{pmatrix}$ *has orthonormal rows and one can determine the* $(m_k + \rho_k) \times (m_k + \rho_k)$ *unitary matrix* F_k *from the condition*

$$\begin{pmatrix} d_W(k) & g_s(k) \end{pmatrix} F_k^* = \begin{pmatrix} 0_{\nu_k \times \rho_{k-1}} & I_{\nu_k} \end{pmatrix}.$$

Determine the matrices $h_s(k)$, $b_s(k)$, $d_F(k)$, $g_F(k)$ *of sizes* $\rho_{k-1}\times m_k$, $\rho_{k-1}\times \rho_k$, $\nu_k \times m_k$, $\nu_k \times \rho_k$ *from the partition*

$$F_k = \begin{pmatrix} h_s(k) & b_s(k) \\ d_F(k) & g_F(k) \end{pmatrix}.$$

Compute

$$\alpha''_k = p^*_V(k)g(k)u_{k+1} + a^*_V(k)v_{k+1},$$
$$u_k = h(k)h^*_s(k) + b(k)u_{k+1}b^*_s(k),$$
$$v_k = \tilde{h}(k)h^*_s(k) + \alpha''_k b^*_s(k).$$

Set

$$\hat{q}_B(k) = q_V(k), \quad \hat{a}_B(k) = a_V(k),$$
$$\hat{b}_B(k) = b_s(k), \quad \hat{g}_B(k) = g_s(k)$$

and compute

$$\alpha_k = h_s(k)p_V(k) + b_s(k)\alpha_{k+1}a_V(k),$$
$$\hat{h}_B(k) = h_s(k)d_V(k) + b_s(k)\alpha_{k+1}q_V(k).$$

1.3. *Compute*

$$g_s(1) = \begin{pmatrix} g(1)f_2 \\ v_2 \end{pmatrix}, \quad \hat{d}_B(1) = \begin{pmatrix} d(1) & g(1)f_2Q_1 \\ Q_1^*X_2q(1) & Q_1^*\gamma_2Q_1 \end{pmatrix}$$

and set

$$\hat{q}_B(1) = \begin{pmatrix} 0_{\rho_1\times m_1} & Q_1 \end{pmatrix}, \quad \hat{g}_B(1) = g_s(1).$$

2.1. *Using Corollary* 30.2 *determine the subdiagonal entry* σ_1, *lower quasiseparable generators and diagonal entries of* B, *treated as a scalar matrix.*

2.2. *Determine upper quasiseparable generators and diagonal entries of* B, *treated as a scalar matrix, as follows.*

2.2.1. *Set* $\hat{g}_1 = \hat{g}_B(1)$ *and* $g_B(1) = 1$,

$$h_B(i) = D_1(1:i-1, i-1), \quad i = 2,\dots,\nu_1,$$
$$b_B(i) = \begin{pmatrix} I_{i-1} & 0_{(i-1)\times 1} \end{pmatrix}, \ g_B(i) = \begin{pmatrix} 0_{1\times(i-1)} & 1 \end{pmatrix}, \quad i = 2,\dots,\nu_1-1,$$
$$b_B(\nu_1) = \hat{g}_1(1:\rho_1,:), \quad g_B(\nu_1) = \hat{g}_1(\nu_1,:).$$

2.2.2. *For* $k = 2,\dots,N-1$, $\nu_k \neq 0$ *set* $\hat{h}_k = \hat{h}_B(k)$, $\hat{g}_k = \hat{g}_B(k)$ *and*

$$h_B(\lambda_k + i) = \begin{pmatrix} \hat{h}_k(:,i) \\ \hat{d}_k(1:i-1,i) \end{pmatrix}, \ i = 1,\dots,\nu_k,$$
$$g_B(\lambda_k + i) = \begin{pmatrix} 0_{(\rho_{k-1}+i-1)\times 1} \\ 1 \end{pmatrix}, \ b_B(\lambda_k + i) = \begin{pmatrix} I_{\rho_{k-1}+i-1} \\ 0_{1\times(\rho_{k-1}+i-1)} \end{pmatrix},$$
$$i = 1,\dots,\nu_k - 1,$$
$$b_B(\lambda_{k+1}) = \begin{pmatrix} \hat{b}_B(k) \\ \hat{g}_k(1:\rho_k,:) \end{pmatrix}, \quad g_B(\lambda_{k+1}) = \hat{g}_k(\rho_k+1,:).$$

2.2.3. *If* $\nu_N \neq 0$, *set*

$$h_B(N) = \hat{h}_B(N).$$

The proof follows from the more general Theorem 30.7 given below.

Corollary 30.6. *Under the conditions of Theorem* 30.5, *there exists a* 2×2 *unitary matrix* W *such that the relation*

$$W^*\begin{pmatrix} d(1) \\ \sigma_1 \end{pmatrix} = \begin{pmatrix} 1 \\ 0 \end{pmatrix} \tag{30.92}$$

holds. Moreover the matrix B *admits the factorization*

$$B = \begin{pmatrix} W & 0 \\ 0 & I_{N-2} \end{pmatrix}\begin{pmatrix} I_{m_1} & 0 \\ 0 & U \end{pmatrix}, \tag{30.93}$$

where U *is an* $(N-1)\times(N-1)$ *unitary matrix. The matrix* U *has lower quasiseparable generators* $p_B(i+1)$ $(i=2,\dots,N-1)$, $q_B(j+1)$ $(j=1,\dots,N-2)$, $a_B(k+1)$ $(k=2,\dots,N-2)$, *upper quasiseparable generators* $g_U(1)$, $g_B(i+1)$ $(i=2,\dots,N-2)$, $h_B(j+1)$ $(j=2,\dots,N-1)$, $b_B(k+1)$ $(k=2,\dots,N-2)$, *and diagonal entries* $d_U(1)$, $d_B(k+1)$ $(k=1,\dots,N-1)$, *with the matrices* $g_U(1), d_U(1)$ *defined via the equalities*

$$g_U(1) = w_1 g_B(1) b_B(2) + w_2 g_B(2), \quad d_U(1) = w_1 g_B(1) h_B(2) + w_2 d_B(2), \tag{30.94}$$

where the numbers w_1, w_2 *are determined from the partition*

$$W^* = \begin{pmatrix} * & * \\ w_1 & w_2 \end{pmatrix}.$$

Proof. Applying Corollary 5.5 to the matrix B we get

$$B(k,k+1:N) = g_B(k)H^B_{k+1}, \quad k=1,\dots,N-1, \tag{30.95}$$

where the matrices H^B_k, $k=2,\dots,N$ are defined via upper quasiseparable generators $b_B(k), h_B(k)$ by the formulas (5.6). We have

$$B = \begin{pmatrix} d(1) & g_B(1)h_B(2) & g_B(1)b_B(2)H^B_3 \\ \sigma_1 & d_B(2) & g_B(2)H^B_3 \\ 0 & B(3:N,2) & B(3:N,3:N) \end{pmatrix}.$$

From here, since B is a unitary matrix, we conclude that there exists a 2×2 unitary matrix W such that the relation (30.92) holds and moreover the representation (30.93) is valid with the unitary matrix

$$U = \begin{pmatrix} d_U(1) & g_U(1)H^B_3 \\ B(3:N,2) & B(3:N,3:N) \end{pmatrix}, \tag{30.96}$$

with the matrices $d_U(1), g_U(1)$ defined in (30.94). From (30.96) it follows that the matrix U has lower quasiseparable generators which are equal to the corresponding generators of the matrix B, i.e., $p_B(i+1)$ $(i=2,\dots,N-1)$, $q_B(j+1)$ $(j=

$1,\dots,N-2$), $a_B(k+1)$ $(k=2,\dots,N-2)$, and diagonal entries $d_U(1)$, $d_B(k+1)$ $(k=2,\dots,N-1)$. Moreover, using (30.96) we get

$$U(2,3:N)=g_U(1)H_3^B,\quad U(k,k+1:N)=B(k,k+1:N),\quad k=2,\dots,N-1. \tag{30.97}$$

Combining (30.95) with (30.97) and using Lemma 5.6, we conclude that $g(1)$, $g_B(i+1)$ $(i=2,\dots,N-2)$, $h_B(j+1)$ $(j=2,\dots,N-1)$, $b_B(k+1)$ $(k=2,\dots,N-2)$ are upper quasiseparable generators of the matrix U. □

The complexity of the algorithm in the above theorem is given mostly by Step 1.2. It is less than

$$2r_k^U\rho_{k-1}+r_k^U(2r_{k+1}^U-1)\rho_k+r_{k+1}^U(2\rho_k-1)\rho_{k-1}$$
$$\nu_kr_k^U+(2r_k^U-1)\rho_k+\nu_k\rho_k+\nu_k(2\nu_k-1)\rho_k+4(1+\rho_k)^3+\rho_k(2r_k-1)$$

and, after Ω_k is partitioned, another

$$2\rho_{k-1}+\rho_{k-1}(2\rho_k-1)r_k^L+\rho_k(2r_k^L-1)+2\nu_k+\nu_k(2\rho_k-1)r_k^L+\nu_k(2r_k^L-1)$$

arithmetical operations until the operations needed for computing the matrix F_k, and then

$$\rho_{k-1}r_k^L+(2r_k^L-1)\rho_k+2\rho_{k-1}\rho_k^2+2r_k^U\rho_{k-1}+r_k^U(2r_{k-1}^U-1)\rho_k$$
$$r_k^U(2\rho_k-1)\rho_{k-1}+2\rho_{k-1}^2+\rho_{k-1}^2(2\rho_k-1)+2\rho_{k-1}^2$$
$$\rho_{k-1}(2\rho_k-1)\rho_k+\rho_k(2\rho_k-1)+2\rho_{k-1}\nu_k$$
$$2\rho_{k-1}\nu_k+\rho_{k-1}(2\rho_k-1)\rho_k+\rho_k(2\rho_k-1)\nu_k.$$

§30.5 The reduction step for perturbations of unitary matrices

The last case we consider is a reduction step for a small rank perturbation of a unitary matrix.

Theorem 30.7. *Let $A=\{A_{ij}\}_{i,j=1}^N$ be a block matrix with entries of sizes $m_i\times m_j$, $(i,j=1,\dots,N)$, with lower quasiseparable generators $p(i)$ $(i=2,\dots,N)$, $q(j)$ $(j=1,\dots,N-1)$, $a(k)$ $(k=2,\dots,N-1)$, and diagonal entries $d(k)$ $(k=1,\dots,N)$. Assume that A is represented in the form*

$$A=U-XY^*, \tag{30.98}$$

where U is a unitary matrix with the given upper quasiseparable generators $g_U(i)$ $(i=1,\dots,N-1)$, $h_U(j)$ $(j=2,\dots,N)$, $b_U(k)$ $(k=2,\dots,N-1)$ of orders r_k^U $(k=1,\dots,N-1)$, and $X=\mathrm{col}(x(i))_{i=1}^N$, $Y=\mathrm{col}(y(i))_{i=1}^N$ with $m_i\times m$ entries

$x(i)$, $y(i)$. Define the numbers ρ_k $(k=0,\dots,N)$, ν_k $(k=1,\dots,N)$ via (30.8) and consider the numbers

$$\tau_N=0, \quad \tau_{k-1}=\min\{\nu_k+\tau_k,\ m\},\ k=N,\dots,2, \qquad \tau_0=0, \tag{30.99}$$

$$t_k=\nu_k+\tau_k-\tau_{k-1}, \quad k=1,\dots,N, \tag{30.100}$$

$$s_k=\rho_k+\tau_k, \qquad k=1,\dots,N-1, \quad s_N=0. \tag{30.101}$$

Determine the $(m_k+\rho_k)\times(m_k+\rho_k)$ unitary matrices V_k, $k=N,\dots,2$ and auxiliary $\rho_{k-1}\times r^L_{k-1}$ matrices X_k via (30.9), (30.10), and then take a unitary $\rho_1\times\rho_1$ matrix Q_1 and define a $\nu_1\times\nu_1$ unitary matrix V_1 via (30.11). Next define the unitary $N\times N$ matrix V via (30.12), (30.13) and the matrix $B=V^*AV$, treated as a block matrix with entries of sizes $\nu_i\times\nu_j$, $(i,j=1,\dots,N)$. By Theorem 30.1, the matrix B satisfies the condition (30.14).

Then the matrix $B=V^*AV$ is represented in the form

$$B=U_B-X_BY_B^*, \tag{30.102}$$

where U_B is a unitary matrix and $X_B=\mathrm{col}(\hat{x}_B(i))_{i=1}^N$, $Y_B=\mathrm{col}(\hat{y}_B(i))_{i=1}^N$ are block columns with $\nu_i\times m$ entries. Moreover, lower quasiseparable generators $\hat{p}_B(i)$ $(i=2,\dots,N)$, $\hat{q}_B(j)$ $(j=1,\dots,N-1)$, $\hat{a}_B(k)$ $(k=2,\dots,N-1)$ of orders ρ_k $(k=1,\dots,N-1)$ and diagonal entries $\hat{d}_B(k)$ $(k=1,\dots,N)$ of the matrix B as well as upper quasiseparable generators $\hat{g}_B(i)$ $(i=1,\dots,N-1)$, $\hat{h}_B(j)$ $(j=2,\dots,N)$, $\hat{b}_B(k)$ $(k=2,\dots,N-1)$ of orders s_k $(k=1,\dots,N-1)$ of the matrix U_B and the matrices X_B,Y_B are determined using the following algorithm.

1.1. *Determine the matrices $p_V(N)$, $d_V(N)$ of sizes $m_N\times\rho_{N-1}$, $m_N\times\nu_N$ from the partition*

$$V_N=\begin{pmatrix} p_V(N) & d_V(N)\end{pmatrix}. \tag{30.103}$$

Compute

$$\begin{gathered}\gamma_N=p_V^*(N)d(N)p_V(N), \quad \hat{p}_B(N)=d_V^*(N)d(N)p_V(N),\\ \hat{d}_B(N)=d_V^*(N)d(N)d_V(N),\end{gathered} \tag{30.104}$$

$$\begin{gathered}c_N=p_V^*(N)x(N), \quad \hat{x}_B(N)=d_V^*(N)x(N),\\ z_N=p_V^*(N)y(N), \quad \hat{y}_B(N)=d_V^*(N)y(N),\end{gathered} \tag{30.105}$$

$$f_N=h_U(N), \tag{30.106}$$

$$\begin{gathered}\tilde{h}(N)=p_V^*(N)d(N)+c_Ny^*(N),\\ d_W(N)=d_V^*(N)d(N)+\hat{x}_B(N)y^*(N).\end{gathered} \tag{30.107}$$

Determine the $\nu_N\times\nu_N$ unitary matrix Φ_N and the $\tau_{N-1}\times m$ matrix χ_N from the condition

$$\Phi_N^*\hat{x}_B(N)=\begin{pmatrix}\chi_N\\ 0_{t_N\times m}\end{pmatrix}. \tag{30.108}$$

Compute the $\nu_N \times m_N$ *matrix*

$$Z_N = \Phi_N^* d_W(N) \tag{30.109}$$

and determine the matrices f'_N, Δ_N *of sizes* $\tau_{N-1} \times m_N, t_N \times m_N$ *from the partition*

$$Z_N = \begin{pmatrix} f'_N \\ \Delta_N \end{pmatrix}. \tag{30.110}$$

The submatrix Δ_N *has orthonormal rows and one can determine the* $m_N \times m_N$ *unitary matrix* F_N *from the condition*

$$\Delta_N F_N^* = \begin{pmatrix} 0_{t_N \times s_{N-1}} & I_{t_N} \end{pmatrix}. \tag{30.111}$$

Determine the matrices $h_s(N), d_F(N)$ *of sizes* $s_{N-1} \times m_N, t_N \times m_N$ *from the partition*

$$F_N = \begin{pmatrix} h_s(N) \\ d_F(N) \end{pmatrix}. \tag{30.112}$$

Compute

$$u_N = h_U(N)h_s^*(N), \qquad v_N = \tilde{h}(N)h_s^*(N), \quad \beta_N = f'_N h_s^*(N), \tag{30.113}$$

$$\alpha_N = h_F(N)p_V(N), \quad \hat{h}_B(N) = h_s(N)d_V(N). \tag{30.114}$$

1.2. *For* $k = N-1, \dots, 2$ *perform the following. Determine the matrices* $p_V(k)$, $a_V(k)$, $d_V(k)$, $q_V(k)$ *of sizes* $m_k \times \rho_{k-1}$, $\rho_k \times \rho_{k-1}$, $m_k \times \nu_k$, $\rho_k \times \nu_k$ *from the partition*

$$V_k = \begin{pmatrix} p_V(k) & d_V(k) \\ a_V(k) & q_V(k) \end{pmatrix}. \tag{30.115}$$

Compute

$$f_k = h_U(k)p_V(k) + b_U(k)f_{k+1}a_V(k), \tag{30.116}$$

$$g_s(k) = d_V^*(k)g_U(k)u_{k+1} + q_V^*(k)v_{k+1}, \tag{30.117}$$

$$\begin{aligned} c_k &= p_V^*(k)x(k) + a_V^*(k)c_{k+1}, \\ \hat{x}_B(k) &= d_V^*(k)x(k) + q_V^*(k)c_{k+1}, \end{aligned} \tag{30.118}$$

$$\begin{aligned} z_k &= p_V^*(k)y(k) + a_V^*(k)z_{k+1}, \\ \hat{y}_B(k) &= d_V^*(k)y(k) + q_V^*(k)z_{k+1}. \end{aligned} \tag{30.119}$$

Compute the matrix

$$\Omega_k = V_k^* \begin{pmatrix} d(k) & -x(k)z_{k+1}^* + g_U(k)f_{k+1} \\ X_{k+1}q(k) & \gamma_{k+1} \end{pmatrix} V_k \tag{30.120}$$

and determine the matrices $\gamma_k, h'(k), \hat{p}_B(k), \hat{d}_B(k)$ *of sizes* $\rho_{k-1} \times \rho_{k-1}, \rho_{k-1} \times \nu_k, \nu_k \times \rho_{k-1}, \nu_k \times \nu_k$ *from the partition*

$$\Omega_k = \begin{pmatrix} \gamma_k & h'(k) \\ \hat{p}_B(k) & \hat{d}_B(k) \end{pmatrix}. \tag{30.121}$$

Compute

$$\tilde{h}(k) = p_V^*(k)d(k) + a_V^*(k)X_{k+1}q(k) + c_k y^*(k), \tag{30.122}$$
$$d_W(k) = d_V^*(k)d(k) + q_V^*(k)X_{k+1}q(k) + \hat{x}_B(k)y^*(k). \tag{30.123}$$

Determine the unitary $(\nu_k + \tau_k) \times (\nu_k + \tau_k)$ *matrix* Φ_k *and the* $\tau_{k-1} \times m$ *matrix* χ_k *from the condition*

$$\begin{pmatrix} \hat{x}_B(k) \\ \chi_{k+1} \end{pmatrix} = \Phi_k \begin{pmatrix} \chi_k \\ 0_{t_k \times m} \end{pmatrix}. \tag{30.124}$$

Compute the $(\nu_k + \tau_k) \times (m_k + s_k)$ *matrix*

$$Z_k = \Phi_k^* \begin{pmatrix} d_W(k) & g_s(k) \\ \chi_{k+1}y^*(k) & \beta_k \end{pmatrix} \tag{30.125}$$

and determine the matrices $f_k', f_k'', \Delta_k, \Theta_k$ *of sizes* $\tau_{k-1} \times m_k, \tau_{k-1} \times s_k, t_k \times m_k, t_k \times s_k$ *from the partition*

$$Z_k = \begin{pmatrix} f_k' & f_k'' \\ \Delta_k & \Theta_k \end{pmatrix}. \tag{30.126}$$

The submatrix $\begin{pmatrix} \Delta_k & \Theta_k \end{pmatrix}$ *has orthonormal rows and one can determine the* $(m_k + s_k) \times (m_k + s_k)$ *unitary matrix* F_k *from the condition*

$$\begin{pmatrix} \Delta_k & \Theta_k \end{pmatrix} F_k^* = \begin{pmatrix} 0_{t_k \times s_{k-1}} & I_{t_k} \end{pmatrix}. \tag{30.127}$$

Determine the matrices $h_s(k), b_s(k), d_F(k), g_F(k)$ *of sizes* $s_{k-1} \times m_k, s_{k-1} \times s_k, t_k \times m_k, t_k \times s_k$ *from the partition*

$$F_k = \begin{pmatrix} h_s(k) & b_s(k) \\ d_F(k) & g_F(k) \end{pmatrix}. \tag{30.128}$$

Compute

$$\begin{gathered} \alpha_k'' = p_V^*(k)g_U(k)u_{k+1} + a_V^*(k)v_{k+1}, \\ u_k = h_U(k)h_s^*(k) + b_U(k)u_{k+1}b_s^*(k), \quad v_k = \tilde{h}(k)h_s^*(k) + \alpha_k''b_s^*(k), \end{gathered} \tag{30.129}$$

$$\beta_k = f_k'h_s^*(k) + f_k''b_s^*(k), \tag{30.130}$$

Set

$$\hat{q}_B(k) = q_V(k), \quad \hat{a}_B(k) = a_V(k), \tag{30.131}$$
$$\hat{b}_B(k) = b_s(k), \quad \hat{g}_B(k) = g_s(k) \tag{30.132}$$

and compute

$$\begin{aligned} \alpha_k &= h_s(k)p_V(k) + b_s(k)\alpha_{k+1}a_V(k), \\ \hat{h}_B(k) &= h_s(k)d_V(k) + b_s(k)\alpha_{k+1}q_V(k). \end{aligned} \tag{30.133}$$

1.3. *Compute*

$$g_s(1) = \begin{pmatrix} g_U(1)u_2 \\ Q_1^* v_2 \end{pmatrix}, \tag{30.134}$$

$$\hat{d}_B(1) = \begin{pmatrix} d(1) & g_U(1)f_2Q_1 - x(1)z_2^* \\ Q_1^* X_2 q(1) & Q_1^* \gamma_2 Q_1 \end{pmatrix} \tag{30.135}$$

and set

$$\hat{q}_B(1) = \begin{pmatrix} 0_{\rho_1 \times m_1} & Q_1 \end{pmatrix}, \tag{30.136}$$

$$\hat{x}_B(1) = \begin{pmatrix} x(1) \\ Q_1^* c_2 \end{pmatrix}, \quad \hat{y}_B(1) = \begin{pmatrix} y(1) \\ Q_1^* z_2 \end{pmatrix}. \tag{30.137}$$

Finally, set

$$\hat{g}_B(1) = g_s(1). \tag{30.138}$$

Proof. We obviously get (30.102), with the unitary matrix $H_B = V^*HV$ and $X_B = V^*X$, $Y_B = V^*Y$. We use the representations $X_B = \text{col}(\hat{x}_B(i))_{i=1}^N$, $Y_B = \text{col}(\hat{y}_B(i))_{i=1}^N$ with $\nu_i \times m$ matrices $\hat{x}_B(i), \hat{y}_B(i)$. The formulas (30.105), (30.118), (30.119) and (30.137) to compute the components $\hat{x}_B(i), \hat{y}_B(i)$ follow directly from the factorization (30.13) and the partitions (30.103), (30.115). (30.11).

Using the formula (30.98) we obtain upper quasiseparable generators of the matrix A in the form

$$g(i) = \begin{pmatrix} -x(i) & g_U(i) \end{pmatrix}, \quad i = 1, \dots, N-1, \tag{30.139}$$

$$h(j) = \begin{pmatrix} y^*(j) \\ h_U(j) \end{pmatrix}, \quad j = 2, \dots, N,$$
$$b(k) = \begin{pmatrix} I_m & 0 \\ 0 & b_U(k) \end{pmatrix}, \quad k = 2, \dots, N-1, \tag{30.140}$$

with orders $r_k^U + m$, $k = 1, \dots, N-1$. We apply Theorem 29.1 to compute lower quasiseparable generators $p_B(i)$ $(i = 2, \dots, N)$, $q_B(j)$ $(j = 1, \dots, N-1)$, $a_B(k)$ $(k = 2, \dots, N-1)$ and diagonal entries $d_B(k)$ $(k = 1, \dots, N)$ of the matrix B.

The formulas (30.104), (30.131) and (30.136) are the same as in Theorem 29.1. For the auxiliary variables Y_k from Theorem 29.1, which are $(r_{k-1}^U + m) \times \rho_{k-1}$ matrices, we use the partitions

$$Y_k = \begin{pmatrix} w_k \\ f_k \end{pmatrix}, \quad k = N, \dots, 2, \tag{30.141}$$

with matrices w_k, f_k of sizes $m \times \rho_{k-1}, r_{k-1}^U \times \rho_{k-1}$, respectively. Inserting (30.140), (30.141) in the first equality in (30.16) and in (30.19), (30.20) we obtain (30.106), the equality $w_N = y^*(N)p_V(N)$, and the relations

$$\begin{pmatrix} w_k & * \\ f_k & * \end{pmatrix} = \begin{pmatrix} y^*(k) & w_{k+1} \\ h_U(k) & b_U(k)f_{k+1} \end{pmatrix} V_k, \quad k = N-1, \dots, 2. \tag{30.142}$$

From (30.142) using the partitions (30.115) we obtain (30.116) and the equalities

$$w_k = y^*(k)p_V(k) + w_{k+1}a_V(k), \quad k = N-1,\dots,2. \tag{30.143}$$

Comparing the equalities $w_N = y^*(N)p_V(N)$ and (30.143) with (30.105), (30.119), we conclude that

$$w_k = z_k, \quad k = N,\dots,2. \tag{30.144}$$

Combining (30.139), (30.141) and (30.144) we get

$$g(k)Y_{k+1} = -x(k)z^*_{k+1} + g_U(k)f_{k+1}, \quad k = N-1,\dots,1. \tag{30.145}$$

Inserting this in (30.21), (30.22) we obtain the formulas (30.120), (30.121). Inserting (30.145) with $k=1$ in (30.23) we obtain (30.135).

Next we compute the matrix $W = V^*U$. Using (30.98) we obtain lower quasiseparable generators of orders $r^L_k + m$ ($k = 1,\dots,N-1$) and the diagonal entries of the matrix U in the form

$$\begin{aligned}
p_U(i) &= \begin{pmatrix} x(i) & p(i) \end{pmatrix}, && i = 2,\dots,N,\\
q_U(j) &= \begin{pmatrix} y^*(j) \\ q(j) \end{pmatrix}, && j = 2,\dots,N,\\
a_U(k) &= \begin{pmatrix} I_m & 0 \\ 0 & a(k) \end{pmatrix}, && k = 2,\dots,N-1,\\
d_U(k) &= d(k) + x(k)y^*(k), && k = 1,\dots,N.
\end{aligned} \tag{30.146}$$

We treat W as a block matrix with entries of sizes $\nu_i \times m_j$, $i,j = 1,\dots,N$. Using (30.98) we have

$$W = V^*(A + XY^*) = T + X_BY^*,$$

with $T = V^*A$, $X_B = V^*X$. By Theorem 20.5, T is a block upper triangular matrix. Hence, lower quasiseparable generators of the matrix W are obtained via the matrix X_BY^*, i.e.,

$$\begin{aligned}
&p_W(i) = \hat{x}_B(i),\ i = 2,\dots,N, \quad q_W(j) = y^*(j),\ j = 1,\dots,N-1,\\
&a_W(k) = I_m, \quad k = 2,\dots,N-1.
\end{aligned} \tag{30.147}$$

Next we apply Corollary 17.8 to compute upper quasiseparable generators and diagonal entries of the matrix $W = V^*U$. V^* is a block upper triangular matrix with upper quasiseparable generators $q^*_V(i)$ ($i = 1,\dots,N-1$), $p^*_V(j)$ ($j = 2,\dots,N$), $a^*_V(k)$ ($k = 2,\dots,N-1$), and diagonal entries $d^*_V(k)$ ($k = 1,\dots,N$). Hence, using (17.49) and (30.146) we get

$$\begin{aligned}
&\begin{pmatrix} \Gamma_N & \tilde{h}(N) \\ * & d_W(N) \end{pmatrix} = \begin{pmatrix} p^*_V(N) \\ d^*_V(N) \end{pmatrix} \begin{pmatrix} x(N) & p(N) & d(N) + x(N)y^*(N) \end{pmatrix},\\
&\begin{pmatrix} \Gamma_k & \tilde{h}(k) \\ * & d_W(k) \end{pmatrix} = \begin{pmatrix} p^*_V(k) & a^*_V(k) \\ d^*_V(k) & q^*_V(k) \end{pmatrix} \begin{pmatrix} I & 0 \\ 0 & \Gamma_{k+1} \end{pmatrix} \begin{pmatrix} x(k) & p(k) & d(k) + x(k)y^*(k) \\ I_m & 0 & y^*(k) \\ 0 & a(k) & q(k) \end{pmatrix},\\
&\qquad k = N-1,\dots,2.
\end{aligned} \tag{30.148}$$

Here Γ_k are auxiliary variables which are $\rho_{k-1} \times (m + r^L_{k-1})$ matrices. We can check that

$$\Gamma_k = \begin{pmatrix} c_k & X_k \end{pmatrix}, \quad k = N, \dots, 2, \tag{30.149}$$

with c_k as in (30.105), (30.118) and X_k as in (30.9), (30.10). Indeed, introduce the partitions

$$\Gamma_k = \begin{pmatrix} c'_k & X'_k \end{pmatrix}, \quad k = N, \dots, 2.$$

with matrices c'_k, X'_k of sizes $\rho_{k-1} \times m, \rho_{k-1} \times r^L_{k-1}$ respectively. Using (30.148) we get

$$c'_N = p^*_V(N)x(N), \quad X'_N = p^*_V(N)p(N) \tag{30.150}$$

and

$$\begin{pmatrix} c'_k & X'_k \end{pmatrix} = \begin{pmatrix} p^*_V(k) & a^*_V(k)c'_{k+1} & a^*_V(k)X'_{k+1} \end{pmatrix} \begin{pmatrix} x(k) & p(k) \\ I_m & 0 \\ 0 & a(k) \end{pmatrix},$$
$$k = N-1, \dots, 2,$$

which implies

$$\begin{aligned} c'_k &= p^*_V(k)x(k) + a^*_V(k)c'_{k+1}, \\ X'_k &= p^*_V(k)p(k) + a^*_V(k)X'_{k+1}a(k), \quad k = N-1, \dots, 2. \end{aligned} \tag{30.151}$$

Comparing (30.150), (30.151) with (30.9), (30.10) and (30.105), (30.118) we obtain (30.149). Now inserting (30.149) in (30.148) we obtain the equalities

$$\begin{aligned} \tilde{h}(N) &= p^*_V(N)d(N) + p^*_V(N)x(N)y^*(N), \\ d_W(N) &= d^*_V(N)d(N) + d^*_V(N)x(N)y^*(N), \\ \tilde{h}(k) &= p^*_V(k)d(k) + (p^*_V(k)x(k) + a^*_V(k)c_{k+1})y^*(k) + a^*_V(k)X_{k+1}q(k), \\ d_W(k) &= d^*_V(k)d(k) + (d^*_V(k)x(k) + q^*_V(k)c_{k+1})y^*(k) + q^*_V(k)X_{k+1}q(k), \\ k &= N-1, \dots, 2. \end{aligned}$$

From here using (30.105), (30.118) we obtain (30.107), (30.122), (30.123). Next, using the formulas (17.51), (17.52) and (30.25) we get

$$\begin{aligned} g_W(1) &= \begin{pmatrix} g_U(1) & 0 \\ 0 & Q^*_1 \end{pmatrix}, \\ g_W(k) &= \begin{pmatrix} d^*_V(k)g_U(k) & q^*_V(k) \end{pmatrix}, \; k = 2, \dots, N-1 \end{aligned} \tag{30.152}$$

and

$$\begin{aligned} h_W(k) &= \begin{pmatrix} h_U(k) \\ \tilde{h}(k) \end{pmatrix}, & k &= 2, \dots, N, \\ b_W(k) &= \begin{pmatrix} b_U(k) & 0 \\ p^*_V(k)g_U(k) & a^*_V(k) \end{pmatrix}, & k &= 2, \dots, N-1. \end{aligned} \tag{30.153}$$

Next we apply the compression algorithm from Theorem 7.5 to the matrix W in order to obtain upper quasiseparable generators $g_s(i)$ $(i = 1, \dots, N-1)$, $h_s(j)$ $(j = 2, \dots, N)$, $b_s(k)$ $(k = 2, \dots, N-1)$ with orders s_k $(k = 1, \dots, N-1)$. Following (7.22), we define the numbers τ_k $(k = 0, \dots, N)$, t_k $(k = 1, \dots, N)$ via (30.99), (30.100), and the numbers s_k $(k = 1, \dots, N)$ via

$$s_N = 0, \quad s_{k-1} = m_k + s_k - t_k, \quad k = N, \dots, 2. \tag{30.154}$$

One can easily check by induction that (30.154) means (30.101). Using (7.28) we obtain the relations (30.108) and (30.124). The formula (7.29) implies (30.109) and (30.125) with auxiliary variables β_k, which are $\tau_{k-1} \times s_{k-1}$ matrices. The partition (7.30) implies the partitions (30.110), (30.126). Following (7.31), we obtain the formulas (30.111), (30.127) to determine unitary matrices F_k $(k = N, \dots, 2)$. Using the partitions (7.32) we obtain (30.112), (30.128). Following (7.33), we introduce auxiliary $(r^U_{k-1} + \rho_{k-1}) \times s_{k-1}$ matrices T_k via the recursion relations

$$\begin{aligned} T_N &= h_W(N)h_s^*(N), \\ T_k &= h_W(k)h_s^*(k) + b_W(k)T_{k+1}b_s^*(k), \ k = N-1, \dots, 2. \end{aligned} \tag{30.155}$$

Moreover, the second equality in (7.33) implies $\beta_N = f'_N h_s^*(N)$ and (30.130). We introduce the partition

$$T_k = \begin{pmatrix} u_k \\ v_k \end{pmatrix}, \quad k = N, \dots, 2, \tag{30.156}$$

with the matrices u_k, v_k of sizes $r^U_{k-1} \times s_{k-1}, \rho_{k-1} \times s_{k-1}$. Inserting this in (30.155) and using (30.153) we obtain the relations (30.113), (30.129). Next, using (7.34) we obtain

$$g_s(k-1) = g_W(k-1)T_k, \quad k = N, \dots, 2. \tag{30.157}$$

From (30.157) using (30.152) and (30.156) we obtain (30.117) and (30.134).

Finally we obtain upper quasiseparable generators of the matrix $U_B = WV$ by means of Corollary 17.9. Using the relations (17.53) we get

$$\begin{pmatrix} \alpha_k & \hat{h}_B(k) \end{pmatrix} = \begin{pmatrix} h_s(k) & b_s(k) \end{pmatrix} \begin{pmatrix} I & 0 \\ 0 & \alpha_{k+1} \end{pmatrix} \begin{pmatrix} p_V(k) & d_V(k) \\ a_V(k) & q_V(k) \end{pmatrix}, \quad k = N, \dots, 2. \tag{30.158}$$

From here we obtain (30.114), (30.133). Using (17.54) we get (30.132) and (30.138). □

Corollary 30.8. *Under the conditions of Theorem 30.7, assume that* $m_2 = m_3 = \dots = m_N = 1$.

Then one can determine upper quasiseparable generators of U_B *treated as a matrix with* $m_i \times m_j$ $(i, j = 1, \dots, N)$ *entries, as well as the* $m_i \times m$ *matrices* $x_B(i), y_B(i)$ *in the partitions* $X_B = \mathrm{col}(x_B(i))_{i=1}^N$, $Y_B = \mathrm{col}(y_B(i))_{i=1}^N$, *as follows.*

1. *Set*

$$s_0 = r_1^U, \quad \lambda_2 = 1+\rho_1, \quad \lambda_k = \lambda_{k-1}+\nu_{k-1},\ k=3,\dots,N$$

and

$$\hat{g}_1 = \begin{pmatrix} u_2 \\ Q_1^* v_2 \end{pmatrix}, \tag{30.159}$$

$$g_B(1) = g_U(1), \tag{30.160}$$

$$\hat{D} = Q_1^*\gamma_2 Q_1 + Q_1^* c_2 z_2^* Q_1, \quad d = f_2 Q_1, \tag{30.161}$$

$$D_1 = \begin{pmatrix} d \\ \hat{D} \end{pmatrix}, \tag{30.162}$$

$$x_B(1) = x(1),\ \hat{x}_1 = Q_1^* c_2, \quad y_B(1) = y(1),\ \hat{y}_1 = Q_1^* z_2, \tag{30.163}$$

$$h_B(i) = D_1(1:i-1, i-1), \quad i = 2,\dots,\rho_1+1, \tag{30.164}$$

$$\begin{gathered} b_B(i) = \begin{pmatrix} I_{s_0+i-1} & 0_{(s_0+i-1)\times 1} \end{pmatrix},\ g_B(i) = \begin{pmatrix} 0_{1\times(s_0+i-1)} & 1 \end{pmatrix}, \\ i = 2,\dots,\rho_1, \end{gathered} \tag{30.165}$$

$$b_B(\nu_1) = \hat{g}_1(1:\rho_1,:), \quad g_B(\nu_1) = \hat{g}_1(\rho_1+1,:), \tag{30.166}$$

$$x_B(i+1) = \hat{x}_1(i),\ y_B(i+1) = \hat{y}_1(i), \quad i = 1,\dots,\rho_1. \tag{30.167}$$

2. *For* $k = 2,\dots,N-1,\quad \nu_k \neq 0,$ *set* $\hat{h}_k = \hat{h}_B(k),\ \hat{g}_k = \hat{g}_B(k),\ \hat{x}_k = \hat{x}_B(k),\ \hat{y}_k = \hat{y}_B(k)$ *and*

$$h_B(\lambda_k+i) = \begin{pmatrix} \hat{h}_k(:,i) \\ \hat{d}_k(1:i-1,i) \end{pmatrix},\ i = 1,\dots,\nu_k, \tag{30.168}$$

$$\begin{gathered} g_B(\lambda_k+i) = \begin{pmatrix} 0_{1\times(s_{k-1}+i-1)} & 1 \end{pmatrix}, \\ b_B(\lambda_k+i) = \begin{pmatrix} I_{s_{k-1}+i-1} & 0_{(s_{k-1}+i-1)\times 1} \end{pmatrix}, \quad i = 1,\dots,\nu_k-1, \end{gathered} \tag{30.169}$$

$$b_B(\lambda_{k+1}) = \begin{pmatrix} \hat{b}_B(k) \\ \hat{g}_k(1:\rho_k+\tau_{k-1},:) \end{pmatrix}, \quad g_B(\lambda_{k+1}) = \hat{g}_k(\nu_k+s_{k-1},:), \tag{30.170}$$

$$x_B(\lambda_k+i) = \hat{x}_k(i,:),\ ,y_B(\lambda_k+i) = \hat{y}_k(i,:), \quad i = 1,\dots,\nu_k. \tag{30.171}$$

3. *If* $\nu_N \neq 0$, *set*

$$h_B(N) = \hat{h}_B(N), \quad x_B(N) = \hat{x}_B(N), \quad y_B(N) = \hat{y}_B(N). \tag{30.172}$$

Proof. First, by (30.102), we obtain the representation of the first block diagonal entry $\tilde{d}_B(1) = U_B(1,1)$ in the form

$$\tilde{d}_B(1) = \hat{d}_B(1) + \hat{x}_B(1)(\hat{y}_B(1))^*.$$

From here using (30.23) and (30.137) we get

$$\tilde{d}_B(1) = \begin{pmatrix} d(1)+x(1)y^*(1) & g_U(1)d \\ * & D \end{pmatrix}, \tag{30.173}$$

with d, D defined in (30.161).

We check that the relations

$$U_B(1:i-1,i) = G^B_{i-1} h_B(i), \quad i=2,\dots,N \tag{30.174}$$

with

$$G^B_j = \mathrm{col}(g_B(i)(b_B)^{>}_{i,j+1})^j_{i=1}, \quad j=1,\dots,N-1$$

hold. By Lemma 5.6, this implies that $g_B(i)$ $(i=1,\dots,N-1)$, $h_B(j)$ $(j=2,\dots,N)$, $b_B(j)$ $(j=2,\dots,N-1)$ are upper quasiseparable generators of U_B, treated as matrix with $\nu_i \times \nu_j$ $(i,j=1,\dots,N)$ a scalar matrix. Applying (30.173) and Lemma 5.5 to U_B, treated as a block matrix, we have

$$U_B(1:\lambda_2, 2:\lambda_2) = \begin{pmatrix} g_U(1)d \\ D \end{pmatrix} = \begin{pmatrix} g_U(1) & 0 \\ 0 & I_{\rho_1} \end{pmatrix} D_1 \tag{30.175}$$

with D_1 as in (30.162) and

$$U_B(1:\lambda_{k+1}, \lambda_k+1:\lambda_{k+1}) = \begin{pmatrix} \hat{G}^B_{k-1}\hat{h}_B(k) \\ \tilde{d}_B(k) \end{pmatrix}, \quad k=2,\dots,N, \tag{30.176}$$

with

$$\hat{G}^B_j = \mathrm{col}(\hat{g}_B(i)(\hat{b}_B)^{>}_{i,j+1})^j_{i=1}, \quad j=1,\dots,N-1.$$

To get from here the relations (30.174) one needs to derive first some relations for the matrices G^B_i $(i=1,\dots,N-1)$. Using the recursion (5.7) we have

$$G^B_1 = g_B(1), \quad G^B_j = \begin{pmatrix} G^B_{j-1} b_B(j) \\ g_B(j) \end{pmatrix} = \begin{pmatrix} G^B_{j-1} & 0 \\ 0 & 1 \end{pmatrix} \begin{pmatrix} b_B(j) \\ g_B(j) \end{pmatrix}, \quad i=2,\dots,N-1, \tag{30.177}$$

whence, using (30.160) and (30.177),

$$G^B_1 = g_U(1). \tag{30.178}$$

The relations (30.165) imply

$$\begin{pmatrix} b_B(i) \\ g_B(i) \end{pmatrix} = I_{s_0+i-1}, \quad i=2,\dots,\rho_1.$$

Hence, using (30.177), we get

$$G^B_i = \begin{pmatrix} g_U(1) & 0 \\ 0 & I_{i-1} \end{pmatrix}, \quad i=2,\dots,\rho_1. \tag{30.179}$$

The relations (30.166) imply

$$\begin{pmatrix} b_B(\rho_1+1) \\ g_B(\rho_1+1) \end{pmatrix} = \hat{g}_1,$$

so using (30.177), (30.179) with $i = \rho_1$, (30.159) and (30.134), (30.138) we get

$$G^B_{\nu_1} = G^B_{\lambda_2} = \hat{g}_B(1) = \hat{g}^B_1. \tag{30.180}$$

Next we prove by induction that

$$\hat{G}^B_k = G^B_{\lambda_{k+1}}, \quad k = 1, \dots, N-1, \tag{30.181}$$

and also that the relations

$$G^B_{\lambda_k+i} = \begin{pmatrix} \hat{G}^B_{k-1} & 0 \\ 0 & I_i \end{pmatrix}, \quad i = 1, \dots, \nu_1 - 1 \tag{30.182}$$

hold. The equality (30.180) implies (30.181) with $k = 1$. Assume that for some k with $2 \le k \le N-2$,

$$\hat{G}^B_{k-1} = G^B_{\lambda_k}. \tag{30.183}$$

Using (30.169) we have

$$\begin{pmatrix} b_B(\lambda_k + i) \\ g_B(\lambda_k + i) \end{pmatrix} = I_{s_{k-1}+i}, \quad i = 1, \dots, \nu_k - 1.$$

From here, using (30.177), we get

$$G^B_{\lambda_k+i} = \begin{pmatrix} G^B_{\lambda_k} & 0 \\ 0 & I_i \end{pmatrix}, \quad i = 1, \dots, \nu_k - 1, \tag{30.184}$$

which together with (30.183) implies (30.182). The equalities (30.170) imply

$$\begin{pmatrix} b_B(\lambda_{k+1}) \\ g_B(\lambda_{k+1}) \end{pmatrix} = \begin{pmatrix} \hat{b}_B(k) \\ \hat{g}_B(k) \end{pmatrix},$$

and therefore using (30.177) and (30.184) with $i = \nu_k - 1$ we get

$$G^B_{\lambda_{k+1}} = \begin{pmatrix} \hat{G}^B_{k-1} & 0 \\ 0 & I_{\nu_k} \end{pmatrix} \begin{pmatrix} \hat{b}_B(k) \\ \hat{g}_B(k) \end{pmatrix}$$

which yields (30.181).

Now using (30.175) we get

$$U_B(1 : i-1, i) = \begin{pmatrix} g_U(1) & 0 \\ 0 & I_{i-2} \end{pmatrix} D_1(1 : i-1, i-1), \quad i = 2, \dots, \rho_1 + 1,$$

which together with (30.164) and (30.178), (30.179) imply (30.174) with $i = 2, \dots, \rho_1 + 1$. Using (30.176) we have

$$U_B(1 : \lambda_k, \lambda_k + 1) = \hat{G}^B_{k-1} \hat{h}_k(:, 1), \tag{30.185}$$

$$U_B(1 : \lambda_k + i - 1, \lambda_k + i) = \begin{bmatrix} \hat{G}^B_{k-1} \hat{h}_k(:, i) \\ \hat{d}_k(1 : i-1, i) \end{bmatrix}, \; i = 2, \dots, \nu_k, \; k = 2, \dots, N-1.$$

Hence, using (30.181), (30.184) and (30.168) we obtain (30.174) with $i = \nu_1 + 1, \dots, \lambda_N$.

Finally, from (30.8) with $k = N$ it follows that either $\nu_N = 0$ and the proof is finished, or $\nu_N = 1$, $\lambda_N = N-1$. In the second case, using (30.176) with $k = N$, (30.181) with $k = N-1$ and (30.172), we get (30.174) with $k = N$. □

We will now compute the complexity of Step 1.2 only, since this step is performed about N times. We start with formulas (30.116)–(30.119) which use

$$2r^U_{k-1}m_k\rho_{k-1} + r^U_{k-1}(2r^U_k - 1)\rho_k + r^U_k(2\rho_k - 1)\rho_{k-1},$$

$$\nu_k(2m_k - 1)r^U_k + m_k(2r^U_k - 1)s_k + 2\nu_k\rho_k s_k,$$

$$2\rho_{k-1}m_k m + \rho_{k-1}(2\rho_k - 1)m + \rho_k(2m-1)s_k + 2\nu_k m_k m + \nu_k(2\rho_k - 1)m$$

and

$$2\rho_{k-1}m_k m + \rho_{k-1}(2\rho_k - 1)m + \rho_k(2m-1)s_k + 2\nu_k m_k m + \nu_k(2\rho_k - 1)m$$

arithmetic operations, respectively. Further, one uses less than

$$\begin{aligned}4(\rho_{k-1}+\nu_k)^3 &+ \rho_k(2r^L_k - 1)m_k + m_k(2m-1)\rho_k + m_k(2r^U_k - 1)\rho_k\\ &+ 2\rho_{k-1}m_k^2 + \rho_{k-1}(2\rho_k - 1)r^L_k + \rho_k(2r^L_k - 1)m_k\\ &+ 2\nu_k m_k^2 + \nu_k(2\rho_k - 1)r^L_k + \rho_k(2r^L_k - 1)m_k\end{aligned}$$

arithmetic operations up to the computation of Φ_k. Then less than

$$\begin{aligned}2(\nu_k + \tau_k)^2(m_k + s_k) &+ 2\rho_{k-1}m_k r^U_k + 2m_k r^U_k s_k + 2\rho_{k-1}\rho_k s_k + 2r^U_k m_k s_{k-1}\\ &+ 2r^U_k r^U_{k+1}s_k + 2r^U_{k+1}s_k s_{k-1} + 2\rho_{k-1}(m_k + s_k)s_{k-1} + 2s_{k-1}m_k\rho_{k-1}\\ &+ 2s_{k-1}s_k\rho_k + 2s_k\rho_k\rho_{k-1} + 2s_{k-1}m_k\nu_k + 2s_{k-1}s_k\rho_k + 2s_k\rho_k\nu_k\end{aligned}$$

arithmetical operations are used to finish Step 1.2.

§30.6 The reduction process

Here we consider the complete reduction process described in Section §30.1, to be applied to the classes of matrices presented in Section §30.3. In every such case, on each step using quasiseparable generators of the matrix $A^{(k)}$ $(k = 0, \dots, N-3)$ from (30.7), we determine by Theorem 30.1 the unitary $(N-k-1) \times (N-k-1)$ matrix $V^{(k)}$ such that the matrix

$$B^{(k)} = \begin{pmatrix} 1 & 0 \\ 0 & (V^{(k)})^* \end{pmatrix} A^{(k)} \begin{pmatrix} 1 & 0 \\ 0 & V^{(k)} \end{pmatrix} \tag{30.186}$$

satisfies the condition

$$B^{(k)}(:,1) = \begin{pmatrix} d^{(k)}_B(1) \\ \sigma^{(k)}_1 \\ 0 \end{pmatrix}. \tag{30.187}$$

This means that the matrices $V^{(k)}$ $(k = 0, \dots, N-3)$ satisfy the conditions (30.4).

Let us give the concrete implementation of the reduction procedure for different classes of matrices.

§30.6.1 The case of Hermitian matrices

Let A be a Hermitian scalar matrix with lower quasiseparable generators $p(i)$ $(i = 2,\dots,N)$, $q(j)$ $(j = 1,\dots,N-1)$, $a(k)$ $(k = 2,\dots,N-1)$ of orders r_k^L $(k = 1,\dots,N-1)$ and diagonal entries $d(k)$ $(k = 1,\dots,N)$.

In the sequence (30.1) defined via (30.2), (30.3), (30.4) all the matrices A_k $(k = 0,\dots,N-2)$ are Hermitian. Moreover, in the relations (30.5), (30.6), (30.7) the matrices Δ_k have the form

$$\Delta_k = \begin{pmatrix} \delta_1 & \sigma_1^* & 0 & \dots & 0 & 0 \\ \sigma_1 & \delta_2 & \sigma_2^* & \dots & 0 & 0 \\ 0 & \sigma_2 & \delta_3 & \dots & 0 & 0 \\ \vdots & \vdots & \vdots & \ddots & \vdots & \vdots \\ 0 & 0 & 0 & \dots & \delta_{k-1} & \sigma_{k-1}^* \\ 0 & 0 & 0 & \dots & \sigma_{k-1} & \delta_k \end{pmatrix}$$

and the matrices $A^{(k)} = A_k(k+1:N, k+1:N)$ are Hermitian.

As a result of the reduction process we obtain a Hermitian tridiagonal matrix A_H with diagonal entries δ_k $(k = 1,\dots,N)$ and subdiagonal entries σ_k $(k = 1,\dots,N-1)$. The numbers δ_k $(k = 1,\dots,N)$, σ_k $(k = 1,\dots,N-1)$ are computed as follows.

Algorithm 30.9.

1. Set

$$p^{(0)}(i) = p(i),\ i = 2,\dots,N, \qquad q^{(0)}(j) = q(j),\ j = 1,\dots,N-1,$$
$$a^{(0)}(i) = a(i),\ i = 2,\dots,N-1, \quad d^{(0)}(i) = d(i),\ i = 1,\dots,N.$$

2. For $k = 0,\dots,N-3$ perform the following.
   Applying Theorem 30.3 to the Hermitian matrix $A^{(k)} = A_k(k+1:N, k+1:N)$ determine the unitary matrix $V^{(k)}$, the diagonal entries $d_B^{(k)}(j)$, $j = 1,\dots,N-k$, the subdiagonal entry $\sigma_1^{(k)}$ and lower quasiseparable generators $p_B^{(k)}(j)$ $(j = 3,\dots,N-k)$, $q_B^{(k)}(j)$ $(j = 2,\dots,N-k-1)$, $a_B^{(k)}(j)$ $(j = 3,\dots,N-k-1)$ of the matrix $B^{(k)}$ of the form (30.186) such that (30.187) holds.
   Set

$$\delta_{k+1} = d_B^{(k)}(1), \quad \sigma_{k+1} = \sigma_1^{(k)}.$$

   Determine lower quasiseparable generators and diagonal entries of the matrix $A^{(k+1)} = A_{k+1}(k+2:N, k+2:N)$ by setting

$$p^{(k+1)}(i) = p_B^{(k)}(i+1), \quad i = 2,\dots,N-k-1,$$
$$q^{(k+1)}(j) = q_B^{(k)}(j+1), \quad j = 1,\dots,N-k-2,$$

$$a^{(k+1)}(i) = a_B^{(k)}(i+1), \quad i = 2, \ldots, N-k-2,$$
$$d^{(k+1)}(i) = d_B^{(k)}(i+1), \quad i = 1, \ldots, N-k-1.$$

3. Set

$$\sigma_{N-1} = \sigma_1^{(N-2)}, \quad \delta_{N-1} = d_B^{(N-2)}(1), \quad \delta_N = d_B^{(N-2)}(2).$$

§30.6.2 The case of perturbations of Hermitian matrices

Let $A = \{A_{ij}\}_{i,j=1}^N$ be a scalar matrix with lower quasiseparable generators $p(i)$ $(i = 2, \ldots, N)$, $q(1) = 1$, $q(j)$ $(j = 2, \ldots, N-1)$, $a(k)$ $(k = 2, \ldots, N-1)$ of orders $r_1^L = 1$, $r_k^L = r$ $(k = 2, \ldots, N-1)$, and diagonal entries $d(k)$ $(k = 1, \ldots, N)$, represented in the form

$$A = E - XY^*, \tag{30.188}$$

where E is a Hermitian matrix and $X = \mathrm{col}(x(i))_{i=1}^N$, $Y = \mathrm{col}(y(i))_{i=1}^N$ are $N \times m$ matrices. From (30.2), (30.3) it follows that the matrices A_k in (30.1) have the form $A_k = E_k - X_k Y_k^*$, where E_k is a Hermitian matrix and X_k, Y_k are $N \times m$ matrices. For the matrices X_k, Y_k we use the partitions

$$X_k = \begin{pmatrix} X_H^{(k)} \\ x^{(k)} \end{pmatrix}, \quad Y_k = \begin{pmatrix} Y_H^{(k)} \\ y^{(k)} \end{pmatrix}, \quad k = 0, \ldots, N-3,$$

with submatrices $X_H^{(k)}, Y_H^{(k)}$ and $x^{(k)}, y^{(k)}$ of sizes $(k+1) \times m$ and $(N-k-1) \times m$, respectively. Hence for the matrices $A^{(k)} = A_k(k+1 : N, k+1 : N)$ we have the representations

$$A^{(k)} = E^{(k)} - x^{(k)}(y^{(k)})^*, \quad k = 0, 1, \ldots, N-2,$$

where $E^{(k)} = E_k(k+1 : N, k+1 : N)$ is a Hermitian matrix. Hence on each step one can apply Theorem 30.4 to the matrix $A^{(k)}$.

As a result of the reduction procedure we obtain the upper Hessenberg matrix A_H in the form

$$A_H = E_H - X_H Y_H^*,$$

where E_H is a Hermitian matrix and $X_H = \mathrm{col}(x_H(i))_{i=1}^N$, $Y_H = \mathrm{col}(y_H(i))_{i=1}^N$ are $N \times m$ matrices. The diagonal and subdiagonal entries of the matrix A_H and the perturbation matrices X_H, Y_H are obtained as follows.

Algorithm 30.10.

1. Set

$$p^{(0)}(i) = p(i),\ i = 2, \ldots, N, \qquad q^{(0)}(j) = q(j),\ j = 1, \ldots, N-1,$$
$$a^{(0)}(i) = a(i),\ i = 2, \ldots, N-1, \quad d^{(0)}(i) = d(i),\ i = 1, \ldots, N,$$
$$x^{(0)}(i) = x(i),\ i = 1, \ldots, N \qquad y^{(0)}(i) = y(i),\ i = 1, \ldots, N.$$

2. For $k = 0, \dots, N-3$ perform the following.
Applying Theorem 30.4 to the matrix $A^{(k)} = A_k(k+1:N, k+1:N)$ determine the unitary matrix $V^{(k)}$, the diagonal entries $d_B^{(k)}(j)$ $(j = 1, \dots, N-k)$, the subdiagonal entry $\sigma_1^{(k)}$ and lower quasiseparable generators $p_B^{(k)}(j)$ $(j = 3, \dots, N-k)$, $q_B^{(k)}(j)$ $(j = 2, \dots, N-k-1)$, $a_B^{(k)}(j)$ $(j = 3, \dots, N-k-1)$ of the matrix $B^{(k)}$ of the form (30.186) such that (30.187) holds, and the perturbation matrices $x_B^{(k)} = V^{(k)}x^{(k)} = \mathrm{col}(x_B^{(k)}(i))_{i=1}^{N-k}$, $y_B^{(k)} = V^{(k)}y^{(k)} = \mathrm{col}(y_B^{(k)}(i))_{i=1}^{N-k}$.
Set
$$\delta_{k+1} = d_B^{(k)}(1), \qquad \sigma_{k+1} = \sigma_1^{(k)},$$
$$x_H(k+1) = x_B^{(k)}(1), \quad y_H(k+1) = y_B^{(k)}(1).$$
Determine lower quasiseparable generators and diagonal entries of the matrix $A^{(k+1)} = A_{k+1}(k+2:N, k+2:N)$ and perturbation matrices $x^{(k+1)}, y^{(k+1)}$ by setting
$$\begin{aligned}
p^{(k+1)}(i) &= p_B^{(k)}(i+1), \ i = 2, \dots, N-k-1,\\
q^{(k+1)}(j) &= q_B^{(k)}(j+1), j = 1, \dots, N-k-2,\\
a^{(k+1)}(i) &= a_B^{(k)}(i+1), \ i = 2, \dots, N-k-2,\\
d^{(k+1)}(i) &= d_B^{(k)}(i+1), \ i = 1, \dots, N-k-1,\\
x^{(k+1)}(i) &= x_B^{(k)}(i+1), \ i = 1, \dots, N-k-1,\\
y^{(k+1)}(i) &= y_B^{(k)}(i+1), \ i = 1, \dots, N-k-1.
\end{aligned}$$

3. Set
$$\sigma_{N-1} = \sigma_1^{(N-2)}, \quad \delta_{N-1} = d^{(N-2)}(1), \quad \delta_N = d^{(N-2)}(2),$$
$$x_H(N-1) = x^{(N-2)}(1), \quad x_H(N) = x^{(N-2)}(2),$$
$$y_H(N-1) = y^{(N-2)}(1), \quad y_H(N) = y^{(N-2)}(2).$$

§30.6.3 The case of unitary matrices

Let $A = \{A_{ij}\}_{i,j=1}^N$ be a unitary scalar matrix with quasiseparable generators $p(i)$ $(i = 2, \dots, N)$, $q(j)$ $(j = 2, \dots, N-1)$, $a(k)$ $(k = 2, \dots, N-1)$, $g(i)$ $(i = 1, \dots, N-1)$, $h(j)$ $(j = 2, \dots, N)$, $b(k)$ $(k = 2, \dots, N-1)$; $d(k)$ $(k = 1, \dots, N)$ of orders (r_k^L, r_k^L) $(k = 1, \dots, N-1)$. In the sequence (30.1) defined via (30.2), (30.3), (30.4) all the matrices A_k $(k = 0, \dots, N-2)$ are unitary. As a result of the reduction process we obtain a unitary upper Hessenberg matrix A_H. Using Theorem 30.5 one can obtain an algorithm to compute diagonal and subdiagonal entries, and upper quasiseparable generators of the matrix A_H. Such an approach is realized in a more general situation in the next subsection. Here using Corollary 30.6 we obtain the resulting unitary upper Hessenberg matrix A_H represented as in (7.78), i.e., as a product

$$A_H = \tilde{W}_1 \tilde{W}_2 \cdots \tilde{W}_{N-1}, \tag{30.189}$$

with

$$\tilde{W}_k = \text{diag}\{I_{k-1}, W_k, I_{N-k-1}\}, \quad k = 1, \dots, N-1, \tag{30.190}$$

where W_k are 2×2 unitary matrices.

More precisely, on the first step, applying Corollary 30.6 we obtain the representation

$$A_1 = \begin{pmatrix} W_1 & 0 \\ 0 & I_{N-2} \end{pmatrix} \begin{pmatrix} 1 & 0 \\ 0 & U_1 \end{pmatrix}$$

with the 2×2 unitary matrix W_1 and the $(N-1) \times (N-1)$ unitary matrix U_1. Moreover, the matrix W_1 and quasiseparable generators $p_U^{(1)}(i)$ $(i = 2, \dots, N-1)$, $q_U^{(1)}(j)$ $(j = 1, \dots, N-2)$, $a_U^{(1)}(k)$ $(k = 2, \dots, N-2)$, $g_U^{(1)}(i)$ $(i = 2, \dots, N-2)$, $h_U^{(1)}(j)$ $(j = 2, \dots, N-1)$, $b_U^{(1)}(k)$ $(k = 2, \dots, N-2)$, and diagonal entries $d_U^{(1)}(k)$ $(k = 2, \dots, N-1)$ of the matrix U_1 are obtained via the algorithm from Theorem 30.5 and the relations (30.92), (30.94).

Next using (30.2) we have

$$A_2 = (\hat{V}^{(1)})^* A_1 \hat{V}^{(1)} = \begin{pmatrix} W_1 & 0 \\ 0 & I_{N-2} \end{pmatrix} \begin{pmatrix} 1 & 0 \\ 0 & (V^{(1)})^* U_1 V^{(1)} \end{pmatrix}. \tag{30.191}$$

Applying Theorem 30.5 and Corollary 30.6 to the matrix U_1 we obtain the representation

$$U_1 = \begin{pmatrix} W_2 & 0 \\ 0 & I_{N-3} \end{pmatrix} \begin{pmatrix} 1 & 0 \\ 0 & U_2 \end{pmatrix} \tag{30.192}$$

with the 2×2 unitary matrix W_2 and the $(N-2) \times (N-2)$ unitary matrix U_2. Moreover, the matrix W_2 and the quasiseparable generators $p_U^{(2)}(i)$ $(i = 2, \dots, N-2)$, $q_U^{(2)}(j)$ $(j = 1, \dots, N-3)$, $a_U^{(2)}(k)$ $(k = 2, \dots, N-3)$, $g_U^{(2)}(i)$ $(i = 1, \dots, N-3)$, $h_U^{(2)}(j)$ $(j = 2, \dots, N-2)$, $b_U^{(2)}(k)$ $(k = 2, \dots, N-3)$, and diagonal entries $d_U^{(2)}(k)$ $(k = 1, \dots, N-2)$ of the matrix U_2 are obtained via the algorithm from Theorem 30.5 and the relations (30.92), (30.94). Inserting (30.192) in (30.191) we get

$$A_2 = \begin{pmatrix} W_1 & 0 \\ 0 & I_{N-2} \end{pmatrix} \begin{pmatrix} 1 & 0 & 0 \\ 0 & W_2 & 0 \\ 0 & 0 & I_{N-3} \end{pmatrix} \begin{pmatrix} I_2 & 0 \\ 0 & U_2 \end{pmatrix}.$$

We continue in the same way. After $N-2$ steps, we get

$$A_{N-2} = A_H = \tilde{W}_1 \tilde{W}_2 \cdots \tilde{W}_{N-2} \begin{pmatrix} I_{N-2} & 0 \\ 0 & U_{N-1} \end{pmatrix},$$

where U_{N-1} is a 2×2 unitary matrix. Hence setting $W_{N-1} = U_{N-1}$ we obtain (30.189).

To compute the unitary matrices W_k $(k = 1, \dots, N-1)$ we obtain the following algorithm.

Algorithm 30.11.

1. Set $U_0 = A$, i.e., determine upper quasiseparable generators of the matrix U_0 via

$$\begin{array}{ll} p^{(0)}(i) = p(i), \ i = 2, \dots, N, & q^{(0)}(j) = q(j), j = 1, \dots, N-1, \\ a^{(0)}(i) = a(i), \ i = 2, \dots, N-1, & g^{(0)}(i) = g(i), \ i = 1, \dots, N-1, \\ h^{(0)}(j) = h(j), j = 2, \dots, N, & b^{(0)}(i) = b(i), \ i = 2, \dots, N-1, \\ d^{(0)}(i) = d(i), \ i = 1, \dots, N. & \end{array}$$

2. For $k = 0, \dots, N-3$ perform the following.
   Applying Theorem 30.5 and Corollary 30.6 to the unitary matrix U_k compute the 2×2 unitary matrix W_{k+1} and quasiseparable generators of $p^{(k+1)}(i) = p(i)$ $(i = 2, \dots, N-k-1)$, $q^{(k+1)}(j)$, $(j = 1, \dots, N-k-2)$, $a^{(k+1)}(i)$ $(i = 2, \dots, N-k-2)$, $g^{(k+1)}(i)$ $(i = 1, \dots, N-k-2)$, $h^{(k+1)}(j)$ $(j = 2, \dots, N-k-1)$, $b^{(k+1)}(i)$ $(i = 2, \dots, N-k-2)$, and diagonal entries $d^{(0)}(i)$ $(i = 1, \dots, N-k-1)$ of the unitary matrix U_{k+1}.
3. Set
$$W_{N-1} = \begin{pmatrix} d_U^{(N-2)}(1) & g_U^{(N-2)}(1) h_B^{(N-2)}(2) \\ p_B^{(N-2)}(2) q_B^{(N-2)}(1) & d_B^{(N-2)}(2) \end{pmatrix}.$$

§30.6.4 The case of perturbations of unitary matrices

Let $A = \{A_{ij}\}_{i,j=1}^N$ be a scalar matrix with lower quasiseparable generators $p(i)$ $(i = 2, \dots, N)$, $q(j)$ $(j = 1, \dots, N-1)$, $a(k)$ $(k = 2, \dots, N-1)$ of orders r_k^L $(k = 1, \dots, N-1)$ and diagonal entries $d(k)$ $(k = 1, \dots, N)$. Assume that A is represented in the form

$$A = U - XY^*, \tag{30.193}$$

where U is a unitary matrix with the given upper quasiseparable generators $g_U(i)$ $(i = 1, \dots, N-1)$, $h_U(j)$ $(j = 2, \dots, N)$, $b_U(k)$ $(k = 2, \dots, N-1)$ of orders r_k^U $(k = 1, \dots, N-1)$, and $X = \mathrm{col}(x(i))_{i=1}^N$, $Y = \mathrm{col}(y(i))_{i=1}^N$, with $1 \times m$ matrices $x(i), y(i)$.

In the reduction process we obtain the sequence of matrices

$$A_0 = A, A_1, \dots, A_{N-3}, A_{N-2} = A_H$$

in the form

$$A_k = U_k - \hat{X}^{(k)}(\hat{Y}^{(k)})^*, \quad k = 0, \dots, N-2, \tag{30.194}$$

with unitary matrices U_k defined via the recursion relations

$$U_0 = U, \quad U_{k+1} = (\hat{V}^{(k)})^* U_k \hat{V}^{(k)}$$

and $N \times m$ matrices $\hat{X}^{(k)}, \hat{Y}^{(k)}$ defined via

$$\hat{X}^{(0)} = X,\ \hat{Y}^{(0)} = Y, \quad \hat{X}^{(k+1)} = (\hat{V}^{(k)})^* \hat{X}^{(k)},$$
$$\hat{Y}^{(k+1)} = (\hat{V}^{(k)})^* \hat{Y}^{(k)}, \qquad k = 0, \dots, N-2.$$

On every step we have lower quasiseparable generators $p^{(k)}(i)$ $(i = 2, \dots, N-k)$, $q^{(k)}(j)$ $(j = 1, \dots, N-k-1)$, $a^{(k)}(j)$ $(j = 2, \dots, N-k-1)$ and diagonal entries $d^{(k)}(j)$ $(j = 1, \dots, N-k)$ of the matrix $A^{(k)}$, upper quasiseparable generators $g^{(k)}(i)$ $(i = 1, \dots, N-1)$, $h^{(k)}(i)$ $(i = 2, \dots, N)$, $b^{(k)}(j)$ $(j = 2, \dots, N-2)$ of the matrix U_k and the perturbation matrices $\hat{X}^{(k)} = \mathrm{col}(x^{(k)}(i))_{i=1}^N, \hat{Y}^{(k)} = \mathrm{col}(y^{(k)}(i))_{i=1}^N$.

We treat A_k and U_k as block matrices with $m_i \times m_j$, blocks $i, j = 1, \dots, N-k$, with

$$m_1 = k+1, \quad m_2 = m_3 = \cdots = m_{N-k} = 1. \tag{30.195}$$

Relative to this partition the matrix A_k has lower quasiseparable generators

$$\begin{aligned} \tilde{p}^{(k)}(i) &= p^{(k)}(i),\ i = 2, \dots, N-k, \\ \tilde{q}^{(k)}(1) = \begin{pmatrix} 0_{1\times k} & q^{(k)}(1) \end{pmatrix},\ \tilde{q}^{(k)}(j) &= q^{(k)}(j),\ j = 2, \dots, N-k-1, \\ \tilde{a}^{(k)}(j) &= a^{(k)}(j),\ j = 2, \dots, N-k-1, \end{aligned} \tag{30.196}$$

and diagonal entries

$$\tilde{d}^{(k)}(1) = A_k(1:k+1, 1:k+1), \quad \tilde{d}^{(k)}(j) = d^{(k)}(j),\ j = 2, \dots, N-k, \tag{30.197}$$

and the matrix U_k has upper quasiseparable generators

$$\begin{aligned} g_U^{(k)}(1) &= G_{k+1}^{(k)} = \mathrm{col}(g^{(k)}(i)(b^{(k)})^{>}_{i,k+2})_{i=1}^{k+1}, \\ g_U^{(k)}(i) &= g^{(k)}(i+k), \quad i = 2, \dots, N-k-1, \\ h_U^{(k)}(j) &= h^{(k)}(j+k), \quad j = 2, \dots, N-k, \\ b_U^{(k)}(j) &= b^{(k)}(j+k), \quad j = 2, \dots, N-k-1. \end{aligned}$$

For the perturbation matrices we use the partitions $\hat{X}^{(k)} = \mathrm{col}(\tilde{x}^{(k)}(i))_{i=1}^{N-k}, \hat{Y}^{(k)} = \mathrm{col}(\tilde{y}^{(k)}(i))_{i=1}^{N-k}$ with $m_i \times m$ matrices $\tilde{x}^{(k)}(i)$, $\tilde{y}^{(k)}(i)$ defined via

$$\tilde{x}^{(k)}(1) = \mathrm{col}(\tilde{x}^{(k)}(i))_{i=1}^{k+1}, \quad \tilde{x}^{(k)}(j) = x^{(k)}(j+k),\ j = 2, \dots, N-k,$$
$$\tilde{y}^{(k)}(1) = \mathrm{col}(\tilde{y}^{(k)}(i))_{i=1}^{k+1}, \quad \tilde{y}^{(k)}(j) = y^{(k)}(j+k),\ j = 2, \dots, N-k.$$

Next, we apply Theorem 30.7 (with $N-k$ instead of N) to the matrix A_k partitioned in the form (30.195). This allows us to determine the numbers

ρ_i $(i = 0, \dots, N-k)$, ν_i $(i = 1, \dots, N-k)$ via (30.8), compute $(m_i+\rho_i)\times(m_i+\rho_i)$ matrices V_i, $i = N-k, \dots, 2$, via (30.9), (30.10), take a $\rho_1 \times \rho_1$ matrix Q_1, and determine the $(m_1+\rho_1)\times(m_1+\rho_1)$ matrix V_1 via (30.11). Then we define the unitary $N \times N$ matrix $\hat{V}^{(k)}$ via (30.12), (30.13). The matrix $B_k = (\hat{V}^{(k)})^* A_k \hat{V}^{(k)}$ treated as a block one with $\nu_i \times \nu_j$ entries, $i, j = 1, \dots, N-k$, satisfies the condition (30.14). Moreover, the algorithm from Theorem 30.7 yields lower quasiseparable generators $\hat{p}_B^{(k)}(i)$ $(i = 2, \dots, N-k)$, $\hat{q}_B^{(k)}(j)$ $(i = 1, \dots, N-k-1)$, $\hat{a}_B^{(k)}(j)$ $(j = 2, \dots, N-k-1)$ and diagonal entries $\hat{d}_B^{(k)}(i)$ $(i = 1, \dots, N-k)$ of the matrix B_k, upper quasiseparable generators $\hat{g}_B^{(k)}(i)$ $(i = 1, \dots, N-k-1)$, $\hat{h}_B^{(k)}(j)$ $(j = 2, \dots, N-k-1)$, $\hat{b}_B^{(k)}(j)$ $(j = 2, \dots, N-k-1)$ of the unitary matrix $U_{k+1} = (\hat{V}^{(k)})^* U_k \hat{V}^{(k)}$, and block rows $\hat{x}_B^{(k)}(j)$, $j = 2, \dots, N-k$, $\hat{y}_B^{(k)}(j)$, $j = 2, \dots, N-k$ of the perturbation matrices $\hat{X}^{(k+1)} = (\hat{V}^{(k)})^* \hat{X}^{(k)}$, $\hat{Y}^{(k+1)} = (\hat{V}^{(k)})^* \hat{Y}^{(k)}$. Here B_k, U_k are treated as block matrices with $\nu_i \times \nu_j$ entries $(i, j = 1, \dots, N-k)$ and $\hat{X}^{(k+1)} = \mathrm{col}(\hat{x}_B^{(k)}(i))_{i=1}^{N-k}$, $\hat{Y}^{(k+1)} = \mathrm{col}(\hat{y}_B^{(k)}(i))_{i=1}^{N-k}$ as columns with $\nu_i \times m$ $(i = 1, \dots, N-k)$ entries $\hat{x}_B^{(k)}(i)), \hat{y}_B^{(k)}(i)$. Notice that the relations (30.12), (30.13) and (30.11) imply

$$\hat{V}^{(k)} = I_{m_1} \oplus V^{(k)} = I_{k+1} \oplus V^{(k)}.$$

Hence, using (30.5) we get

$$A_k = \begin{pmatrix} \Delta_k & * \\ C_k & B^{(k)} \end{pmatrix},$$

with $B^{(k)}$ defined in (30.186). Notice that the matrix

$$\begin{pmatrix} 1 & 0 \\ 0 & V^{(k)} \end{pmatrix}$$

coincides with the unitary matrix V in Theorem 30.1, to be applied to the scalar matrix $A^{(k)}$. Moreover, lower quasiseparable generators $\hat{p}_B^{(k)}(i)$ $(i = 2, \dots, N-k)$, $\hat{q}_B^{(k)}(j)$ $(j = 2, \dots, N-k-1)$, $\hat{a}_B^{(k)}(j)$ $(j = 2, \dots, N-k-1)$ and diagonal entries $\hat{d}_B^{(k)}(i)$ $(i = 2, \dots, N-k)$ are the same as the ones obtained via the algorithm from Theorem 30.1. Notice that using the formulas (30.135), (30.136) and the formulas (30.196), (30.197) with $j = 1$ we get

$$\hat{d}_B^{(k)}(1) = \begin{pmatrix} A_k(1:k, 1:k) & A_k(1:k, k+1) & * \\ A_k(k+1, 1:k) & d^{(k)}(1) & * \\ 0_{\rho_1 \times k} & Q_1^* X_2 q^{(k)}(1) & Q_1^* \gamma_2 Q_1 \end{pmatrix},$$
$$\hat{q}_B^{(k)}(1) = \begin{pmatrix} 0_{\rho_1 \times k} & 0_{\rho_1 \times 1} & Q_1 \end{pmatrix}.$$

It follows that the first diagonal entry $\boldsymbol{d}_B^{(k)}(1)$ and the generator $\boldsymbol{q}_B^{(k)}(1)$ of the matrix $B^{(k)}$ are obtained by the formulas

$$\boldsymbol{d}_B^{(k)}(1) = \begin{pmatrix} d^{(k)}(1) & * \\ Q_1^* X_2 q^{(k)}(1) & Q_1^*\gamma_2 Q_1 \end{pmatrix},$$

$$\boldsymbol{q}_B^{(k)}(1) = \begin{pmatrix} 0_{\rho_1\times 1} & Q_1 \end{pmatrix}.$$

Thus we have obtained lower quasiseparable generators $\hat{p}_B^{(k)}(i)$ $(i = 2,\dots,N-k)$, $\boldsymbol{q}_B^{(k)}(1)$, $\hat{q}_B^{(k)}(j)$ $(j = 2,\dots,N-k-1)$, $\hat{a}_B^{(k)}(j)$ $(j = 2,\dots,N-k-1)$ and diagonal entries $\boldsymbol{d}_B^{(k)}(1)$, $\hat{d}_B^{(k)}(i)$ $(i = 2,\dots,N-k)$ of the matrix $B^{(k)}$, treated as a block one with entries of sizes $\nu'_i \times \nu'_j$, $i,j = 1,\dots,N-k$, where

$$\nu'_1 = \rho_1 + 1, \quad \nu'_i = \nu_i,\ i = 2,\dots,N-k, \tag{30.198}$$

in the same way as in Theorem 30.1. Hence, we can apply Corollary 30.2. Taking the unitary matrix Q_1 as in this corollary we satisfy the condition (30.187). Moreover, applying Corollary 30.2 we obtain lower quasiseparable generators $p_B^{(k)}(i)$ $(i = 2,\dots,N-k)$, $q_B^{(k)}(j)$ $(j = 1,\dots,N-k-1)$, $a_B^{(k)}(j)$ $(j = 2,\dots,N-k-1)$ and diagonal entries $d_B^{(k)}(i)$ $(i = 1,\dots,N-k)$ of $B^{(k)}$, now treated as a usual scalar matrix.

Using Corollary 30.8 we obtain upper quasiseparable generators $g_B^{(k)}(i)$ $(i = 1,\dots,N-k-1)$, $h_B^{(k)}(j)$ $(j = 2,\dots,N-k-1)$, $b_B^{(k)}(j)$ $(j = 2,\dots,N-k-1)$ of the matrix U_{k+1} and the entries $x_B^{(k)}(j), y_B^{(k)}(j)$, $j = 2,\dots,N-k$ of the perturbation matrices $\hat{X}^{(k+1)}, \hat{Y}^{(k+1)}$ with respect to the partition (30.195). From here we obtain upper quasiseparable generators of U_{k+1}, treated as a usual scalar matrix, and the rows of the matrices $\hat{X}^{(k+1)}, \hat{Y}^{(k+1)}$ by the formulas

$$g^{(k+1)}(i) = g^{(k)}(i),\ i = 1,\dots,k+1,\ \ g^{(k+1)}(i) = g_B^{(k)}(i-k),\ i = k+2,\dots,N-1,$$

$$h^{(k+1)}(i) = h^{(k)}(i),\ i = 2,\dots,k+1,\ \ h^{(k+1)}(i) = h_B^{(k)}(i-k),\ i = k+2,\dots,N,$$

$$b^{(k+1)}(i) = b^{(k)}(i),\ i = 2,\dots,k+1,\ \ b^{(k+1)}(i) = b_B^{(k)}(i-k),\ i = k+2,\dots,N-1,$$

$$x^{(k+1)}(i) = x^{(k)}(i),\ i = 1,\dots,k+1,\ \ x^{(k+1)}(i) = x_B^{(k)}(i-k),\ i = k+2,\dots,N,$$

$$y^{(k+1)}(i) = y^{(k)}(i),\ i = 1,\dots,k+1,\ \ y^{(k+1)}(i) = y_B^{(k)}(i-k),\ i = k+2,\dots,N.$$

As a result of the reduction procedure we obtain the upper Hessenberg matrix A_H in the form

$$A_H = U_H - X_H Y_H^*,$$

where U_H is a unitary matrix and $X_H = \mathrm{col}(x_H(i))_{i=1}^N$, $Y_H = \mathrm{col}(y_H(i))_{i=1}^N$ are $N \times m$ matrices.

The diagonal and subdiagonal entries of the matrix A_H, the perturbation matrices X_H, Y_H and upper quasiseparable generators of the matrix U_H are obtained as follows.

Algorithm 30.12.

1. Set

$$
\begin{array}{lll}
p^{(0)}(i) = p(i), & i = 2, \dots, N, & \\
q^{(0)}(j) = q(j), & j = 1, \dots, N-1, & \\
a^{(0)}(i) = a(i), & i = 2, \dots, N-1, & \\
d^{(0)}(i) = d(i), & i = 1, \dots, N, & \delta_1 = d(1), \\
g_U^{(0)}(i) = g_U(i), & i = 1, \dots, N-1, & g_H(1) = g_U(1), \\
h_U^{(0)}(j) = h_U(j), & j = 2, \dots, N, & \\
b_U^{(0)}(i) = b_U(i), & i = 2, \dots, N-1, & \\
x^{(0)}(i) = x(i), & i = 1, \dots, N, & \\
y^{(0)}(i) = y(i), & i = 1, \dots, N, & \\
x_H(1) = x(1), \quad y_H(1) = y(1). & &
\end{array}
$$

2. For $k = 0, \dots, N-3$ perform the following.
   2.1. Applying Theorem 30.7 to the matrix A_k, treated as a block one with $m_i \times m_j$, $i, j = 1, \dots, N-k$, entries, where $m_1 = k+1$, $m_2 = \cdots = m_{N-k} = 1$, determine the parameters ν_i, ρ_i $(i = 1, \dots, N)$, unitary matrix $V^{(k)}$ and lower quasiseparable generators $\hat{p}_B^{(k)}(i)$ $(i = 2, \dots, N-k)$, $\hat{q}_B^{(k)}(j)$ $(j = 2, \dots, N-k-1)$, $\hat{a}_B^{(k)}(j)$ $(j = 2, \dots, N-k-1)$ and diagonal entries $\hat{d}_B^{(k)}(i)$ $(i = 2, \dots, N-k)$ of the matrix B_k such that $B_k = (\hat{V}^{(k)})^* A_k \hat{V}^{(k)}$ satisfies the condition (30.14), and then upper quasiseparable generators $\hat{g}_B^{(k)}(i)$ $(i = 1, \dots, N-k-1)$, $\hat{h}_B^{(k)}(j)$ $(j = 2, \dots, N-k-1)$, $\hat{b}_B^{(k)}(j)$ $(j = 2, \dots, N-k-1)$ of the unitary matrix $U_{k+1} = (\hat{V}^{(k)})^* U_k \hat{V}^{(k)}$ and the perturbation matrices $x_B^{(k)} = \hat{V}^{(k)} x^{(k)} = \mathrm{col}(\hat{x}_B^{(k)}(i))_{i=1}^{N-k}$, $y_B^{(k)} = \hat{V}^{(k)} y^{(k)} = \mathrm{col}(\hat{y}_B^{(k)}(i))_{i=1}^{N-k}$. Here B_k, U_k are treated as block matrices with $\nu_i \times \nu_j$ $(i, j = 1, \dots, N-k)$ entries and $\hat{X}^{(k+1)}, \hat{Y}^{(k+1)}$ as columns with $\nu_i \times m$ $(i = 1, \dots, N-k)$ entries. The elements $\hat{p}_B^{(k)}(i)$ $(i = 2, \dots, N-k)$, $\hat{q}_B^{(k)}(j)$ $(j = 2, \dots, N-k-1)$, $\hat{a}_B^{(k)}(j)$ $(j = 2, \dots, N-k-1)$ and $\hat{d}_B^{(k)}(i)$ $(i = 2, \dots, N-k)$ of the matrix $B^{(k)}$ of the form (30.186), treated as a block one with $\nu'_i \times \nu'_j$, $i, j = 1, \dots, N-k$, entries, where the numbers ν'_i are defined in (30.198).
   2.2. Applying Corollary 30.2 determine the unitary matrix Q_1 such that the condition (30.38) for the matrix $B^{(k)}$ holds, diagonal entries $d_B^{(k)}(j)$ $(j = 1, \dots, N-k)$, the subdiagonal entry $\sigma_1^{(k)}$ and lower quasiseparable generators $p_B^{(k)}(j)$ $(j = 3, \dots, N-k)$, $q_B^{(k)}(j)$ $(j = 2, \dots, N-k-1)$, $a_B^{(k)}(j)$ $(j = 3, \dots, N-k-1)$ of $B^{(k)}$, treated as a scalar matrix.

2.3. Using Corollary 30.8 determine upper quasiseparable generators $g_B^{(k)}(i)$ $(i = 2, \dots, N-k-1)$, $h_B^{(k)}(j)$ $(j = 2, \dots, N-k-1)$, $b_B^{(k)}(j)$ $(j = 2, \dots, N-k-1)$ of the matrix U_{k+1} and the rows $x_B^{(k)}(j)$, $y_B^{(k)}(j)$, $j = 2, \dots, N-k$ of the perturbation matrices $\hat{X}^{(k+1)}$, $\hat{Y}^{(k+1)}$ with respect to the partition (30.195).
Set

$$\delta_{k+1} = d_B^{(k)}(1), \qquad \sigma_{k+1} = \sigma_1^{(k)},$$
$$x_H(k+2) = x_B^{(k)}(2), \quad y_H(k+2) = y_B^{(k)}(2).$$

Determine lower quasiseparable generators and diagonal entries of the matrix $A^{(k+1)} = A_{k+1}(k+2:N, k+2:N)$ and perturbation matrices $x^{(k+1)}, y^{(k+1)}$ by setting

$$p^{(k+1)}(i) = p_B^{(k)}(i+1), \; i = 2, \dots, N-k-1,$$
$$q^{(k+1)}(j) = q_B^{(k)}(j+1), \, j = 1, \dots, N-k-2,$$
$$a^{(k+1)}(i) = a_B^{(k)}(i+1), \; i = 2, \dots, N-k-2,$$
$$d^{(k+1)}(i) = d_B^{(k)}(i+1), \; i = 1, \dots, N-k-1,$$
$$x^{(k+1)}(i) = x_B^{(k)}(i-k), \; i = k+2, \dots, N,$$
$$y^{(k+1)}(i) = y_B^{(k)}(i-k), \; i = k+2, \dots, N.$$

Determine upper quasiseparable generators of the matrix U_{k+1} by setting

$$g_U^{(k+1)}(i) = g_B^{(k)}(i-k), \; i = k+2, \dots, N-1,$$
$$h_U^{(k+1)}(j) = h_B^{(k)}(j-k), \, j = k+2, \dots, N,$$
$$b_U^{(k+1)}(i) = b_B^{(k)}(i-k), \; i = k+2, \dots, N-1.$$

Set

$$g_H(k+2) = g_B^{(k)}(2), \quad h_H(k+2) = h_B^{(k)}(2), \quad b_H(k+2) = b_B^{(k)}(2).$$

3. Set

$$\sigma_{N-1} = \sigma_1^{(N-2)}, \quad \delta_{N-1} = d^{(N-2)}(1), \quad \delta_N = d^{(N-2)}(2),$$
$$x_H(N-1) = x^{(N-2)}(N-1), \quad x_H(N) = x^{(N-2)}(N),$$
$$y_H(N-1) = y^{(N-2)}(N-1), \quad y_H(N) = y^{(N-2)}(N),$$
$$h_H(N) = h^{(N-2)}(N).$$

§30.6.5 Application to block companion matrices

For the monic matrix polynomial $P(x) = x^N + \alpha_{N-1}x^{N-1} + \cdots + \alpha_1 x + \alpha_0$ with $n \times n$ matrix coefficients α_k $(k = 0, \dots, N-1)$, the block companion $N \times N$ matrix associated to it is

$$C' = \begin{pmatrix} 0 & 0 & \dots & 0 & -\alpha_0 \\ I_n & 0 & \dots & 0 & -\alpha_1 \\ \vdots & \ddots & \ddots & & \vdots \\ 0 & \dots & I_n & 0 & -\alpha_{N-2} \\ 0 & 0 & \dots & I_n & -\alpha_{N-1} \end{pmatrix}. \tag{30.199}$$

We will regard C' as an $nN \times nN$ scalar matrix. Using the formulas (5.37), (5.38), (5.39) we obtain a set $p(i)$ $(i = 2, \dots, Nn)$, $q(j)$ $(j = 1, \dots, Nn-1)$, $a(k)$ $(k = 2, \dots, Nn-1)$ of lower quasiseparable generators of the matrix C' of orders not greater than n. The diagonal entries of C' are

$$\begin{aligned} d(k) &= 0,\ k = 1, \dots, n(N-1), \\ d(k) = -\alpha_{N-1}(k - n(N-1), k - n(N-1)),\ k &= n(N-1)+1, \dots, Nn. \end{aligned}$$

Next one can see that the matrix C' has the form (30.193), i.e., $A = U - XY^*$ with the unitary matrix

$$U = \begin{pmatrix} 0 & 0 & \dots & 0 & 0 \\ I_n & 0 & \dots & 0 & 0 \\ \vdots & \ddots & \ddots & & \vdots \\ 0 & \dots & I_n & 0 & 0 \\ 0 & 0 & \dots & I_n & 0 \end{pmatrix}$$

and the perturbation matrices

$$X = \begin{pmatrix} \alpha_0 \\ \alpha_1 \\ \vdots \\ \alpha_{N-1} \end{pmatrix}, \quad Y = \begin{pmatrix} 0_{n(N-1)} \\ I_n \end{pmatrix}.$$

The upper quasiseparable generators of the matrix U are

$$\begin{aligned} g_U(i) &= \begin{pmatrix} 0_{1\times(i-1)} & 1 & 0_{1\times(n-i)} \end{pmatrix},\quad i = 1, \dots, n, \\ g_U(i) &= 0_{1\times n},\ i = n+1, \dots, Nn-1, \\ h_U(j) &= 0_{n\times 1},\ j = 2, \dots, N(n-1), \\ h_U(j + N(n-1)) &= \begin{pmatrix} 0_{1\times(i-1)} & 1 & 0_{1\times(n-i)} \end{pmatrix}^T, \\ b_U(k) &= I_n, \qquad\qquad k = 2, \dots, Nn-1. \end{aligned}$$

Thus one can apply Algorithm 30.12 to reduce the matrix C' to the upper Hessenberg form.

§30.7 Comments

The methods for reduction to tridiagonal form of symmetric banded plus semiseparable and diagonal plus semiseparable of the first order matrices were studied by S. Chandrasekaran and M. Gu in [16] and by N. Mastronardi, S. Chandrasekaran and S. Van Huffel in [33]. The reduction method presented here and its implementation for Hermitian matrices and their small rank perturbations were developed in the paper [21]. The reduction to upper Hessenberg form methods for rank structured matrices based on Givens-weight representations were suggested by S. Delvaux and M. Van Barel for Hermitian matrices and their small rank perturbations in [19] and for unitary matrices in [20]. The complete algorithm for reduction to Hessenberg form for small rank perturbations of unitary matrices appears here for the first time.

Chapter 31

The Implicit QR Iteration Method for Eigenvalues of Upper Hessenberg Matrices

In this chapter we consider the implicit QR iteration method for upper Hessenberg matrices obtained via the algorithms presented in the previous chapter. The first section is a general description of the QR iteration method for the cases of the single shift and the double shift. In the second section we derive some auxiliary results for products of matrices with quasiseparable generators and derive general algorithms for the QR iteration step for upper Hessenberg matrices with single shift and with double shift. Here we treat upper Hessenberg matrices as upper triangular ones with block entries. In such a treatment the diagonal of a scalar matrix is included in the quasiseparable structure and the corresponding quasiseparable generators are called upper triangular. In the last two sections we apply the general method to upper Hessenberg matrices which are small rank perturbations of Hermitian and unitary matrices.

§31.1 The general description of the implicit QR iteration method

As it was mentioned in Section §29.1, the QR iteration method for determining eigenvalues of an $N \times N$ matrix consists of the computation of the sequence of matrices

$$A^{(0)} = A, \quad A^{(k+1)} = (Q^{(k)})^* A^{(k)} Q^{(k)}, \quad k = 0, 1, 2, \dots \tag{31.1}$$

with unitary matrices $Q^{(k)}$ such that the relations (29.1) and (29.2) hold.

The method usually employed for upper Hessenberg matrices is the implicit QR algorithm. In this method one computes the matrices $Q^{(k)}$ and the iterates

$A^{(k+1)} = (Q^{(k)})^* A^{(k)} Q^{(k)}$ without forming the factors $R^{(k)}$ in (29.1), (29.2) explicitly.

Under some conditions (see [35, 37]), the sequence of complex matrices $A^{(k)}$ converges to an upper triangular matrix. Hence, the diagonal entries of the matrices $A^{(k)}$ approximate eigenvalues of the matrix A. In the case of a real matrix A one can choose a sequence of orthogonal matrices $Q^{(k)}$ such that the sequence (31.1) converges to a block upper Hessenberg real matrix with 1×1 or 2×2 blocks on the main diagonal. One can find more details about the implicit QR iteration method in [18, 35, 37].

§31.1.1 An implicit iteration step with a single shift

Let A be an upper Hessenberg matrix. Consider a single step of the QR iteration for this matrix with the shift α. The parameter α is usually chosen equal to $A(N,N)$. The implicit step consists of the computation of the unitary matrix Q of the form

$$Q = \tilde{Q}_1 \tilde{Q}_2 \cdots \tilde{Q}_{N-1}, \tag{31.2}$$

where

$$\tilde{Q}_i = I_{i-1} \oplus Q_i \oplus I_{N-i-1} \tag{31.3}$$

and Q_i, $i = 1, \dots, N-1$, are complex Givens rotation matrices, and of the matrix $A^{(1)} = Q^* A Q$ such that

$$(Q^*(A - \alpha I))(2,1) = 0 \quad \text{and } A^{(1)} \text{ is an upper Hessenberg matrix.} \tag{31.4}$$

§31.1.2 An implicit iteration step with a double shift

Let $A = (a_{ij})_{i,j=1}^N$ be a real upper Hessenberg matrix with possibly complex eigenvalues. A single QR iteration step for such matrix consists of the computation of a new iterate as follows. One is given two real parameters s, t, usually chosen as $s = \alpha_1 + \alpha_2$, $t = \alpha_1 \alpha_2$, where α_1, α_2 are the eigenvalues of the matrix $\begin{pmatrix} a_{N-1,N-1} & a_{N-1,N} \\ a_{N,N-1} & a_{N,N} \end{pmatrix}$, i.e.,

$$s = a_{N-1,N-1} + a_{N,N}, \quad t = \det \begin{pmatrix} a_{N-1,N-1} & a_{N-1,N} \\ a_{N,N-1} & a_{N,N} \end{pmatrix}.$$

The implicit step with double shift consists of the computation of the unitary matrix Q of the form

$$Q = \tilde{Q}_1 \tilde{Q}_2 \cdots \tilde{Q}_{N-2} \tilde{Q}_{N-1}, \tag{31.5}$$

where

$$\tilde{Q}_i = I_{i-1} \oplus Q_i \oplus I_{N-i-2}, \; i = 1, \dots, N-2, \quad \tilde{Q}_{N-1} = I_{N-2} \oplus Q_{N-1}, \tag{31.6}$$

and Q_i, $i = 1, \dots, N-2$ are 3×3 orthogonal matrices, Q_{N-1} is a Givens rotation, and of the matrix $A^{(1)} = Q^*AQ$, such that

$$(Q^*(A^2 - sA + tI))(2:3,1) = \begin{pmatrix} 0 \\ 0 \end{pmatrix} \quad \text{and } A^{(1)} \text{ is an upper Hessenberg matrix.} \tag{31.7}$$

§31.1.3 Deflation

Deflation amounts to setting a small subdiagonal element of the Hessenberg matrix to zero. Suppose that after k iterations of the QR algorithm the subdiagonal entry $A^{(k)}(N-s+1, N-s)$ of $A^{(k)}$ becomes small enough to be considered negligible. A customary criterion requires that such an entry be small compared with its diagonal neighbors, namely

$$|A^{(k)}(N-s+1, N-s)| \leq \upsilon \cdot (|A^{(k)}(N-s, N-s)| + |A^{(k)}(N-s+1, N-s+1)|),$$

where υ is the unit round-off error. If this condition is satisfied, then $A^{(k)}(N-s+1, N-s)$ is set to zero, which makes $A^{(k)}$ a block upper triangular matrix:

$$A^{(k)} = \begin{pmatrix} A_1^{(k)} & \star \\ 0 & A_2^{(k)} \end{pmatrix},$$

where $A_1^{(k)} \in \mathbb{C}^{(N-s)\times(N-s)}$ and $A_2^{(k)} \in \mathbb{C}^{s \times s}$. Now $A_1^{(k)}$ and $A_2^{(k)}$ can be reduced to upper triangular form separately. The process is called *deflation*, and continuing in this fashion, by operating on smaller and smaller matrices, we can approximate all the eigenvalues of A. Specifically, if $s = 1$ or $s = 2$, then the computed eigenvalues of $A_2^{(k)}$ may be taken to be eigenvalues of the initial matrix A.

§31.2 The basic theorems

In this section we derive algorithms to compute the unitary matrix Q and the new iterate $A^{(1)} = Q^*AQ$ in the implicit QR iteration process. In fact in these algorithms we follow the bulge chasing method presented for instance in [18, pp. 168–170], [35, pp. 389–398]. However we obtain our results independently, using algorithms of matrix multiplication developed in Chapter 17.

§31.2.1 The multiplication lemma

Based on the results of Chapter 17 we derive an algorithm for multiplication of block triangular matrices. This result is used essentially in our proofs.

Lemma 31.1. *Let A be a block matrix with entries of sizes $m_i \times n_j$, $i,j = 1,\dots,N$ and with quasiseparable generators $p_A(i)$ $(i = 2,\dots,N)$, $q_A(j)$ $(j = 1,\dots,N-1)$, $a_A(k)$ $(k = 2,\dots,N-1)$; $g_A(i)$ $(i = 1,\dots,N-1)$, $h_A(j)$ $(j = 2,\dots,N)$, $b_A(k)$ $(k = 2,\dots,N-1)$; $d_A(k)$ $(k = 1,\dots,N)$ of orders ρ_k^A, r_k^A $(k = 1,\dots,N-1)$. Let S be a block lower triangular matrix with entries of sizes $\tau_i \times m_j$, $i,j = 1,\dots,N$, represented in the factorized form*

$$S = \tilde{S}_N \tilde{S}_{N-1} \cdots \tilde{S}_1 \tag{31.8}$$

with

$$\tilde{S}_1 = \operatorname{diag}\{S_1, I_{\gamma_1}\}, \quad \tilde{S}_k = \operatorname{diag}\{I_{\eta_k}, S_k, I_{\gamma_k}\},$$
$$k = 2,\dots,N-1, \quad \tilde{S}_N = \operatorname{diag}\{I_{\eta_N}, S_N\},$$

where $\eta_k = \sum_{i=1}^{k-1} \tau_i$, $\gamma_k = \sum_{i=k+1}^{N} m_i$, and with matrices S_1, S_k $(k = 2,\dots,N-1)$, and S_N of sizes $(\tau_1 + r_1^S) \times m_1$, $(\tau_k + r_k^S) \times (m_k + r_{k-1}^S)$ $(k = 2,\dots,N-1)$, and $\tau_N \times (m_N + r_{N-1}^S)$, respectively. Let T be a block upper triangular matrix with entries of sizes $n_i \times \nu_j$, $i,j = 1,\dots,N$ represented in the factorized form

$$T = \tilde{T}_1 \tilde{T}_2 \cdots \tilde{T}_N$$

with

$$\tilde{T}_1 = \operatorname{diag}\{T_1, I_{\lambda_1}\}, \quad \tilde{T}_k = \operatorname{diag}\{I_{\chi_k}, T_k, I_{\lambda_k}\},$$
$$k = 2,\dots,N-1, \quad \tilde{T}_N = \operatorname{diag}\{I_{\chi_N}, A_N\},$$

where $\chi_k = \sum_{i=1}^{k-1} \nu_i$, $\lambda_k = \sum_{i=k+1}^{N} n_i$, and with matrices T_1, T_k $(k = 2,\dots,N-1)$, and T_N of sizes $n_1 \times (\nu_1 + r_1^T)$, $(n_k + r_{k-1}^T) \times (\nu_k + r_k^T)$ $(k = 2,\dots,N-1)$, and $(n_N + r_{N-1}^T) \times \nu_N$, respectively.

Then the product $F = SAT$ is a block matrix with entries of sizes $\tau_i \times \nu_j$ $(i,j = 1,\dots,N)$ with quasiseparable generators $p_F(i)$ $(i = 2,\dots,N)$, $q_F(j)$ $(j = 1,\dots,N-1)$, $a_F(k)$ $(k = 2,\dots,N-1)$, $g_F(i)$ $(i = 1,\dots,N-1)$, $h_F(j)$ $(j = 2,\dots,N)$, $b_F(k)$ $(k = 2,\dots,N-1)$, $d_F(k)$ $(k = 1,\dots,N)$ of orders $\rho_k^A + r_k^S$, $r_k^A + r_k^T$ $(k = 1,\dots,N-1)$. These generators are determined via the following algorithm.

1. *Compute the matrix $(\tau_1 + r_1^S + \rho_1^A) \times (\nu_1 + r_1^A + r_1^T)$ matrix Δ_1 by the formula*

$$\Delta_1 = \begin{pmatrix} S_1 & 0 \\ 0 & I_{\rho_1^A} \end{pmatrix} \begin{pmatrix} d_A(1) & g_A(1) \\ q_A(1) & 0 \end{pmatrix} \begin{pmatrix} T_1 & 0 \\ 0 & I_{r_1^A} \end{pmatrix} \tag{31.9}$$

and determine the matrices $d_F(1), g_F(1), q_F(1), \Gamma_1, f_1, \phi_1, \alpha_1$ of sizes $\tau_1 \times \nu_1, \tau_1 \times (r_1^A + r_1^T), (r_1^S + \rho_1^A) \times \nu_1, (r_1^S + \rho_1^A) \times (r_1^A + r_1^T), r_1^S \times r_1^T, r_1^S \times r_1^A, \rho_1^A \times r_1^T$, respectively, from the partitions

$$\Delta_1 = \begin{pmatrix} d_F(1) & g_F(1) \\ q_F(1) & \Gamma_1 \end{pmatrix} \tag{31.10}$$

and

$$\Gamma_1 = \begin{bmatrix} f_1 & \phi_1 \\ \alpha_1 & 0 \end{bmatrix}. \tag{31.11}$$

2. *For $k = 2, \dots, N-1$, compute the $(r^T_{k-1} + r^A_{k-1} + \tau_k + r^S_k + \rho^A_k) \times (\nu_k + r^A_k + r^T_k)$ matrix Δ_k by the formula*

$$\Delta_k = \begin{pmatrix} I_{r^T_{k-1}+r^A_{k-1}} & 0 & 0 \\ 0 & S_k & 0 \\ 0 & 0 & I_{\rho^A_k} \end{pmatrix} \begin{pmatrix} I_{r^T_{k-1}} & 0 & 0 \\ 0 & h_A(k) & b_A(k) \\ f_{k-1} & \phi_{k-1}h_A(k) & \phi_{k-1}b_A(k) \\ p_A(k)\alpha_{k-1} & d_A(k) & g_A(k) \\ a_A(k)\alpha_{k-1} & q_A(k) & 0 \end{pmatrix} \begin{pmatrix} T_k & 0 \\ 0 & I_{r^A_k} \end{pmatrix} \tag{31.12}$$

and determine the matrices $h_F(k)$, $b_F(k)$, $d_F(k)$, $g_F(k)$, $q_F(k)$, $\Gamma_k, f_k, \phi_k, \alpha_k$ of sizes $(r^T_{k-1} + r^A_{k-1}) \times \nu_k$, $(r^T_{k-1} + r^A_{k-1}) \times (r^A_k + r^T_k)$, $\tau_k \times \nu_k$, $\tau_k \times (r^A_k + r^T_k)$, $(r^S_k + \rho^A_k) \times \nu_k$, $(r^S_k + \rho^A_k) \times (r^A_k + r^T_k)$, $r^S_k \times r^T_k, r^S_k \times r^A_k$, $\rho^A_k \times r^T_k$, respectively, from the partitions

$$\Delta_k = \begin{pmatrix} h_F(k) & b_F(k) \\ d_F(k) & g_F(k) \\ q_F(k) & \Gamma_k \end{pmatrix} \tag{31.13}$$

and

$$\Gamma_k = \begin{pmatrix} f_k & \phi_k \\ \alpha_k & 0 \end{pmatrix}. \tag{31.14}$$

Compute the matrix

$$\Phi_k = \begin{pmatrix} S_k & 0 \\ 0 & I_{\rho^A_k} \end{pmatrix} \begin{pmatrix} I_{r^S_{k-1}} & 0 \\ 0 & p_A(k) \\ 0 & a_A(k) \end{pmatrix} \tag{31.15}$$

and determine the matrices $p_F(k), a_F(k)$ of sizes $\tau_k \times (r^S_{k-1} + \rho^A_{k-1}), (r^S_k + \rho^A_k) \times (r^S_{k-1} + \rho^A_{k-1})$ from the partition

$$\Phi_k = \begin{pmatrix} p_F(k) \\ a_F(k) \end{pmatrix}. \tag{31.16}$$

3. *Compute the $(r^T_{N-1} + r^A_{N-1} + \tau_N) \times \nu_N$ matrix Δ_N by the formula*

$$\Delta_N = \begin{pmatrix} I_{r^T_{N-1}+r^A_{N-1}} & 0 \\ 0 & S_N \end{pmatrix} \begin{pmatrix} I_{r^T_{N-1}} & 0 \\ 0 & h_A(N) \\ f_{N-1} & \phi_{N-1}h_A(N) \\ p_A(N)\alpha_{N-1} & d_A(N) \end{pmatrix} T_N \tag{31.17}$$

and determine the matrices $h_F(N), d_F(N)$ of sizes $(r^T_{N-1} + r^A_{N-1}) \times \nu_N, \tau_N \times \nu_N$, respectively, from the partitions

$$\Delta_N = \begin{pmatrix} h_F(N) \\ d_F(N) \end{pmatrix}. \tag{31.18}$$

Compute the matrix

$$p_F(N) = S_N \begin{pmatrix} I_{r^S_{N-1}} & 0 \\ 0 & p_A(N) \end{pmatrix}. \tag{31.19}$$

Proof. Consider partitions of the matrices T_k in the form

$$\begin{aligned} T_1 &= \begin{pmatrix} d_T(1) & g_T(1) \end{pmatrix}; \quad T_k = \begin{pmatrix} h_T(k) & b_T(k) \\ d_T(k) & g_T(k) \end{pmatrix}, \\ k &= 2,\dots,N-1; \qquad T_N = \begin{pmatrix} h_T(N) \\ d_T(N) \end{pmatrix} \end{aligned} \tag{31.20}$$

with submatrices $d_T(k)$ $(k = 1,\dots,N)$, $g_T(i)$ $(i = 1,\dots,N-1)$, $h_T(j)$ $(j = 2,\dots,N)$, $b_T(k)$ $(k = 2,\dots,N-1)$ of sizes $n_k \times \nu_k$, $n_i \times r^T_i$, $r^T_{j-1} \times \nu_j$, $r^T_{k-1} \times r^T_k$, respectively. By Lemma 20.4, the matrices $g_T(i)$ $(i = 1,\dots,N-1)$, $h_T(j)$ $(j = 2,\dots,N)$, $b_T(k)$ $(k = 2,\dots,N-1)$ and $d_T(k)$ $(k = 1,\dots,N)$ are upper quasiseparable generators of orders r^T_k $(k = 1,\dots,N-1)$, and diagonal entries of the matrix T.

Similarly consider partitions of the matrices S_k in the form

$$\begin{aligned} S_1 &= \begin{pmatrix} d_S(1) \\ q_S(1) \end{pmatrix}, \quad S_k = \begin{pmatrix} p_S(k) & d_S(k) \\ a_S(k) & q_S(k) \end{pmatrix}, \\ k &= 2,\dots,N-1, \quad S_N = \begin{pmatrix} p_S(N) & d_S(N) \end{pmatrix} \end{aligned} \tag{31.21}$$

with submatrices $d_S(k)$ $(k = 1,\dots,N)$, $p_S(i)$ $(i = 2,\dots,N)$, $q_S(j)$ $(j = 1,\dots,N-1)$, $a_S(k)$ $(k = 2,\dots,N-1)$ of sizes $\tau_k \times m_k$, $\tau_i \times r^S_{i-1}$, $r^S_j \times m_j$, $r^S_k \times r^S_{k-1}$, respectively. By Lemma 20.2, the matrices $p_S(i)$ $(i = 2,\dots,N)$, $q_S(j)$ $(j = 1,\dots,N-1)$, $a_S(k)$ $(k = 2,\dots,N-1)$ and $d_S(k)$ $(k = 1,\dots,N)$ are lower quasiseparable generators of orders r^S_k $(k = 1,\dots,N-1)$, and diagonal entries of the matrix S.

Consider the matrix $G = AT$. By Corollary 17.11, we obtain quasiseparable generators of the matrix G with orders $\rho^A_k, r^A_k + r^T_k$, $k = 1,\dots,N-1$, via the relations

$$\begin{pmatrix} d_G(1) & \tilde{g}(1) \\ q_G(1) & \alpha_1 \end{pmatrix} = \begin{pmatrix} d_A(1) \\ q_A(1) \end{pmatrix} \begin{pmatrix} d_T(1) & g_T(1) \end{pmatrix}, \tag{31.22}$$

$$\begin{pmatrix} d_G(k) & \tilde{g}(k) \\ q_G(k) & \beta_k \end{pmatrix} = \begin{pmatrix} p_A(k) & d_A(k) \\ a_A(k) & q_A(k) \end{pmatrix} \begin{pmatrix} \alpha_{k-1} & 0 \\ 0 & I_{n_k} \end{pmatrix} \begin{pmatrix} h_T(k) & b_T(k) \\ d_T(k) & g_T(k) \end{pmatrix}, \quad k = 2,\dots,N-1, \tag{31.23}$$

$$d_G(N) = \begin{pmatrix} p_A(N) & d_A(N) \end{pmatrix} \begin{pmatrix} \alpha_{N-1} & 0 \\ 0 & I_{n_N} \end{pmatrix} \begin{pmatrix} h_T(N) \\ d_T(N) \end{pmatrix}, \tag{31.24}$$

with auxiliary variables α_k, which are $\rho_k^A \times r_k^T$ matrices,

$$g_G(k) = \begin{pmatrix} \tilde{g}(k) & g_A(k) \end{pmatrix}, \quad k = 1, \dots, N-1, \tag{31.25}$$

$$p_G(k) = p_A(k), \quad k = 2, \dots, N, \tag{31.26}$$

$$a_G(k) = a_A(k), \quad k = 2, \dots, N-1, \tag{31.27}$$

and

$$h_G(k) = \begin{pmatrix} h_T(k) \\ h_A(k) d_T(k) \end{pmatrix}, \ b_G(k) = \begin{pmatrix} b_T(k) & 0 \\ h_A(k) g_T(k) & b_A(k) \end{pmatrix}, \quad k = 2, \dots, N-1, \tag{31.28}$$

$$h_G(N) = \begin{pmatrix} h_T(N) \\ h_A(N) d_T(N) \end{pmatrix}.$$

Using (31.20) and setting

$$\tilde{\alpha}_k = \begin{pmatrix} \alpha_k & 0_{\rho_k^A \times r_k^A} \end{pmatrix}, \quad k = 1, \dots, N-1, \tag{31.29}$$

one can rewrite the relations (31.22), (§31.2.1), (31.25), (31.28) and (31.24) in the form

$$\begin{pmatrix} d_G(1) & g_G(1) \\ q_G(1) & \tilde{\alpha}_1 \end{pmatrix} = \begin{pmatrix} d_A(1) & g_A(1) \\ q_A(1) & 0 \end{pmatrix} \begin{pmatrix} T_1 & 0 \\ 0 & I_{r_1^A} \end{pmatrix}, \tag{31.30}$$

$$\begin{pmatrix} h_G(k) & b_G(k) \\ d_G(k) & g_G(k) \\ q_G(k) & \tilde{\alpha}_k \end{pmatrix} = \begin{pmatrix} I_{r_{k-1}^T} & 0 & 0 \\ 0 & h_A(k) & b_A(k) \\ p_A(k)\alpha_{k-1} & d_A(k) & g_A(k) \\ a_A(k)\alpha_{k-1} & q_A(k) & 0 \end{pmatrix} \begin{pmatrix} T_k & 0 \\ 0 & I_{r_k^A} \end{pmatrix}, \quad k = 2, \dots, N-1, \tag{31.31}$$

$$\begin{pmatrix} h_G(N) \\ d_G(N) \end{pmatrix} = \begin{pmatrix} I_{r_{N-1}^T} & 0 \\ 0 & h_A(N) \\ p_A(N)\alpha_{N-1} & d_A(N) \end{pmatrix} T_N. \tag{31.32}$$

Next, by Corollary 17.10, we obtain quasiseparable generators of the matrix $F = SG = SAT$ with orders $\rho_k^A + r_k^S, r_k^A + r_k^T$, $k = 1, \dots, N-1$, via the relations

$$\begin{pmatrix} d_F(1) & g_F(1) \\ \tilde{q}(1) & \beta_1 \end{pmatrix} = \begin{pmatrix} d_S(1) \\ q_S(1) \end{pmatrix} \begin{pmatrix} d_G(1) & g_G(1) \end{pmatrix}, \tag{31.33}$$

$$\begin{pmatrix} d_F(k) & g_F(k) \\ \tilde{q}(k) & \beta_k \end{pmatrix} = \begin{pmatrix} p_S(k) & d_S(k) \\ a_S(k) & q_S(k) \end{pmatrix} \begin{pmatrix} \beta_{k-1} & 0 \\ 0 & I_{m_k} \end{pmatrix} \begin{pmatrix} h_G(k) & b_G(k) \\ d_G(k) & g_G(k) \end{pmatrix}, \quad k = 2, \dots, N-1, \tag{31.34}$$

$$d_F(N) = \begin{pmatrix} p_S(N) & d_S(N) \end{pmatrix} \begin{pmatrix} \beta_{N-1} & 0 \\ 0 & I_{m_N} \end{pmatrix} \begin{pmatrix} h_G(N) \\ d_G(N) \end{pmatrix}, \tag{31.35}$$

with auxiliary variables β_k, which are $r_k^S \times (r_k^A + r_k^T)$ matrices, $k = 1, \dots, N-1$,

$$\begin{aligned} h_F(k) &= h_G(k), \qquad && k = 2, \dots, N, \\ b_F(k) &= b_G(k), && k = 2, \dots, N-1, \end{aligned} \tag{31.36}$$

$$q_F(k) = \begin{pmatrix} \tilde{q}(k) \\ q_G(k) \end{pmatrix}, \qquad k = 1, \dots, N-1, \tag{31.37}$$

$$\begin{aligned} p_F(k) &= \begin{pmatrix} p_S(k) & d_S(k)p_G(k) \end{pmatrix}, \quad k = 2, \dots, N, \\ a_F(k) &= \begin{pmatrix} a_S(k) & q_S(k)p_G(k) \\ 0 & a_G(k) \end{pmatrix}, k = 2, \dots, N-1. \end{aligned} \tag{31.38}$$

Using (31.21) and the partitions above one can rewrite the relations (31.33)–(31.37) in the form

$$\begin{pmatrix} d_F(1) & g_F(1) \\ \tilde{q}(1) & \beta_1 \end{pmatrix} = S_1 \begin{pmatrix} d_G(1) & g_G(1) \end{pmatrix}, \tag{31.39}$$

$$\begin{pmatrix} h_F(k) & b_F(k) \\ d_F(k) & g_F(k) \\ \tilde{q}(k) & \beta_k \end{pmatrix} = \begin{pmatrix} I_{r_{k-1}^A + r_{k-1}^T} & 0 \\ 0 & S_k \end{pmatrix} \begin{pmatrix} I_{r_{k-1}^A + r_{k-1}^T} & 0 \\ \beta_{k-1} & 0 \\ 0 & I_{m_k} \end{pmatrix} \begin{pmatrix} h_G(k) & b_G(k) \\ d_G(k) & g_G(k) \end{pmatrix}, \quad k = 2, \dots, N-1, \tag{31.40}$$

$$\begin{pmatrix} h_F(N) \\ d_F(N) \end{pmatrix} = S_N \begin{pmatrix} \beta_{N-1} & 0 \\ 0 & I_{m_N} \end{pmatrix} \begin{pmatrix} h_G(N) \\ d_G(N) \end{pmatrix}, \tag{31.41}$$

and the relations (31.38) in the form

$$\begin{aligned} \begin{pmatrix} p_F(k) \\ a_F(k) \end{pmatrix} &= \begin{pmatrix} S_k & 0 \\ 0 & \rho_k^A \end{pmatrix} \begin{pmatrix} I_{r_{k-1}^S} & 0 \\ 0 & p_G(k) \\ 0 & a_G(k) \end{pmatrix}, \quad k = 2, \dots, N-1, \\ p_F(N) &= S_N \begin{pmatrix} I_{r_{N-1}^S} & 0 \\ 0 & p_G(N) \end{pmatrix}. \end{aligned} \tag{31.42}$$

Next we introduce the partitions

$$\beta_k = \begin{pmatrix} f_k & \phi_k \end{pmatrix}, \quad k = 1, \dots, N-1, \tag{31.43}$$

with the matrices f_k, ϕ_k of sizes $r_k^S \times r_k^T, r_k^S \times r_k^A$, respectively, and define the matrices Γ_k, $k = 1, \dots, N-1$, via (31.11), (31.14). The relations (31.29), (31.43) imply

$$\Gamma_k = \begin{pmatrix} \beta_k \\ \tilde{\alpha}_k \end{pmatrix}, \quad k = 1, \dots, N-1.$$

Using (31.37) we have

$$\begin{pmatrix} q_F(k) & \Gamma_k \end{pmatrix} = \begin{pmatrix} \tilde{q}(k) & \beta_k \\ q_G(k) & \tilde{\alpha}_k \end{pmatrix}, \quad k = 1, \dots, N-1. \tag{31.44}$$

Combining (31.39) and (31.44) with $k = 1$ we get

$$\begin{pmatrix} d_F(1) & g_F(1) \\ q_F(1) & \Gamma_1 \end{pmatrix} \begin{pmatrix} d_F(1) & g_F(1) \\ \tilde{q}(1) & \beta_1 \\ q_G(1) & \tilde{\alpha}_1 \end{pmatrix} = \begin{pmatrix} S_1 & 0 \\ 0 & I_{\rho_1^A} \end{pmatrix} \begin{pmatrix} d_G(1) & g_G(1) \\ q_G(1) & \tilde{\alpha}_1 \end{pmatrix}.$$

Combining this with (31.30) we obtain (31.9), (31.11).

Combining (31.40) and (31.44) together we get

$$\begin{pmatrix} h_F(k) & b_F(k) \\ d_F(k) & g_F(k) \\ q_F(k) & \Gamma_k \end{pmatrix}$$

$$= \begin{pmatrix} I_{r_{k-1}^A + r_{k-1}^T} & 0 & 0 \\ 0 & S_k & 0 \\ 0 & 0 & I_{\rho_k^A} \end{pmatrix} \begin{pmatrix} I_{r_{k-1}^A + r_{k-1}^T} & 0 & 0 \\ \beta_{k-1} & 0 & 0 \\ 0 & I_{m_k} & 0 \\ 0 & 0 & I_{\rho_k^A} \end{pmatrix} \begin{pmatrix} h_G(k) & b_G(k) \\ d_G(k) & g_G(k) \\ q_G(k) & \tilde{\alpha}_k \end{pmatrix},$$

$$k = 2, \dots, N-1. \tag{31.45}$$

Using the partitions (31.43) we have

$$\begin{pmatrix} I_{r_{k-1}^A + r_{k-1}^T} \\ \beta_{k-1} \end{pmatrix} = \begin{pmatrix} I_{r_{k-1}^T} & 0 \\ 0 & I_{r_{k-1}^A} \\ f_{k-1} & \phi_{k-1} \end{pmatrix}, \quad k = 2, \dots, N. \tag{31.46}$$

Hence combining (31.45), (31.31) and (31.46) we obtain (31.12), (31.13). Then, combining (31.42) and (31.26) we obtain (31.15), (31.16) and (31.19).

Finally combining (31.41), (31.43) and (31.32) we obtain (31.17), (31.18). □

Corollary 31.2. *Under the condition of Lemma* 31.1 *assume that A is a block upper triangular matrix with upper quasiseparable generators $g_A(i)$ ($i = 1, \dots, N-1$), $h_A(j)$ ($j = 2, \dots, N$), $b_A(k)$ ($k = 2, \dots, N-1$) of orders r_k^A ($k = 1, \dots, N-1$) and diagonal entries $d_A(k)$ ($k = 1, \dots, N$).*

Then lower quasiseparable generators $\tilde{q}(j)$ ($j = 1, \dots, N-1$), upper quasiseparable generators $g_F(i)$ ($i = 1, \dots, N-1$), $h_F(j)$ ($j = 2, \dots, N$), $b_F(k)$ ($k = 2, \dots, N-1$) and diagonal entries $d_F(k)$ ($k = 1, \dots, N$) of the product $F = SAT$ are determined via the following algorithm.

1. *Compute the $(\tau_1 + r_1^S) \times (\nu_1 + r_1^A + r_1^T)$ matrix Δ_1 by the formula*

$$\Delta_1 = S_1 \begin{pmatrix} d_A(1) & g_A(1) \end{pmatrix} \begin{pmatrix} T_1 & 0 \\ 0 & I_{r_1^A} \end{pmatrix}, \tag{31.47}$$

and determine the matrices $d_F(1), g_F(1), \tilde{q}(1), \beta_1, f_1, \phi_1$ of sizes $\tau_1 \times \nu_1, \tau_1 \times (r_1^A + r_1^T), r_1^S \times \nu_1, r_1^S \times (r_1^A + r_1^T), r_1^S \times r_1^T, r_1^S \times r_1^A$, respectively, from the partitions

$$\Delta_1 = \begin{pmatrix} d_F(1) & g_F(1) \\ \tilde{q}(1) & \beta_1 \end{pmatrix} \tag{31.48}$$

and

$$\beta_1 = \begin{pmatrix} f_1 & \phi_1 \end{pmatrix}. \tag{31.49}$$

2. *For* $k = 2, \dots, N-1$ *compute the* $(\tau_k + r_k^S) \times (\nu_k + r_k^A + r_k^T)$ *matrix* Δ_k *by the formula*

$$\Delta_k = \begin{pmatrix} I_{r^T_{k-1}+r^A_{k-1}} & 0 \\ 0 & S_k \end{pmatrix} \begin{pmatrix} I_{r^T_{k-1}} & 0 & 0 \\ 0 & h_A(k) & b_A(k) \\ f_{k-1} & \phi_{k-1}h_A(k) & \phi_{k-1}b_A(k) \\ 0 & d_A(k) & g_A(k) \end{pmatrix} \begin{pmatrix} T_k & 0 \\ 0 & I_{r^A_k} \end{pmatrix}. \tag{31.50}$$

Determine the matrices $h_F(k), b_F(k), d_F(k), g_F(k), \tilde{q}(k), \beta_k, f_k, \phi_k$ *of sizes* $(r^A_{k-1} + r^T_{k-1}) \times \nu_k, (r^A_{k-1} + r^T_{k-1}) \times (r^A_k + r^T_k), \tau_k \times \nu_k, \tau_k \times (r^A_k + r^T_k), r^S_k \times \nu_k, r^S_k \times (r^A_k + r^T_k), r^S_k \times r^T_k, r^S_k \times r^A_k$, *respectively, from the partitions*

$$\Delta_k = \begin{pmatrix} h_F(k) & b_F(k) \\ d_F(k) & g_F(k) \\ \tilde{q}(k) & \beta_k \end{pmatrix} \tag{31.51}$$

and

$$\beta_k = \begin{pmatrix} f_k & \phi_k \end{pmatrix}. \tag{31.52}$$

3. *Compute the* $(r^T_{N-1} + r^A_{N-1} + \tau_N) \times \nu_N$ *matrix* Δ_N *by the formula*

$$\Delta_N = \begin{pmatrix} I_{r^T_{N-1}+r^A_{N-1}} & 0 \\ 0 & S_N \end{pmatrix} \begin{pmatrix} I_{r^T_{N-1}} & 0 \\ 0 & h_A(N) \\ f_{N-1} & \phi_{N-1}h_A(N) \\ 0 & d_A(N) \end{pmatrix} T_N \tag{31.53}$$

and determine the matrices $h_F(N), d_F(N)$ *of sizes* $(r^T_{N-1} + r^A_{N-1}) \times \nu_N, \tau_N \times \nu_N$, *respectively, from the partition*

$$\Delta_N = \begin{pmatrix} h_F(N) \\ d_F(N) \end{pmatrix}. \tag{31.54}$$

Proof. Since A is upper triangular, we have

$$p_A(i) = 0, \ i = 2, \dots, N, \quad q_A(j) = 0, \ j = 1, \dots, N-1,$$
$$a_A(k) = 0, \ k = 2, \dots, N-1.$$

Inserting this in the relations of Lemma 31.1 we get

$$\alpha_k = 0, \ k = 1, \dots, N-1 \tag{31.55}$$

and

$$\begin{aligned} p_F(i) &= \begin{pmatrix} \tilde{p}(i) & 0 \end{pmatrix}, & i &= 2, \dots, N, \\ q_F(j) &= \begin{pmatrix} \tilde{q}(j) \\ 0 \end{pmatrix}, & j &= 1, \dots, N-1, \\ a_F(k) &= \begin{pmatrix} \tilde{a}(k) & 0 \\ 0 & 0 \end{pmatrix}, & k &= 2, \dots, N-1, \end{aligned} \tag{31.56}$$

with the matrices $\tilde{p}(i), \tilde{q}(j), \tilde{a}(k)$ of sizes $\tau_i \times r^S_{i-1}, r^S_j \times \nu_j, r^S_k \times r^S_{k-1}$, respectively. The relations (31.56) imply that the matrices $\tilde{p}(i)$ $(i = 2, \dots, N), \tilde{q}(j)$ $(j = 1, \dots, N-1)$, $\tilde{a}(k)$ $(k = 2, \dots, N-1)$ are lower quasiseparable generators of the matrix F. Next, inserting (31.55) in (31.9), (31.10), (31.12), (31.13) and (31.17), (31.18) we obtain the formulas for lower quasiseparable generators $\tilde{q}(j)$ $(j = 1, \dots, N-1)$, upper quasiseparable generators $g_F(i)$ $(i = 1, \dots, N-1)$, $h_F(j)$ $(j = 2, \dots, N)$, $b_F(k)$ $(k = 2, \dots, N-1)$ and diagonal entries $d_F(k)$ $(k = 1, \dots, N)$ of the matrix $F = SAT$. □

§31.2.2 Upper triangular generators

For upper Hessenberg and related matrices we need to introduce quasiseparable representations which include the main diagonal of a matrix. Assume that A is an $N \times N$ scalar matrix with the entries in the upper triangular part represented in the form

$$A(i,j) = g(i) b^{<}_{i-1,j} h(j), \quad 1 \le i \le j \le N, \tag{31.57}$$

with matrices $g(i), h(i)$ $(i = 1, \dots, N)$, $b(k)$ $(k = 1, \dots, N-1)$ of sizes $1 \times r_i, r_i \times 1, r_k \times r_{k+1}$. The elements $g(i), h(i)$ $(i = 1, \dots, N)$, $b(k)$ $(k = 1, \dots, N-1)$ are called *upper triangular generators* of the matrix A with orders r_k $(k = 1, \dots, N)$.

Upper triangular and upper quasiseparable generators of a scalar matrix are connected as follows.

Lemma 31.3. *Let A be a scalar matrix with upper quasiseparable generators $g(i)$ $(i = 1, \dots, N-1)$, $h(j)$ $(j = 2, \dots, N)$, $b(k)$ $(k = 2, \dots, N-1)$ of orders r_k $(k = 1, \dots, N-1)$, and diagonal entries $d(k)$ $(k = 1, \dots, N)$. Then the formulas*

$$g_u(i) = \begin{pmatrix} d(i) & g(i) \end{pmatrix}, \qquad i = 1, \dots, N-1, \qquad g_u(N) = d(N), \tag{31.58}$$

$$h_u(j) = \begin{pmatrix} 1 \\ 0_{r_j \times 1} \end{pmatrix}, \qquad j = 1, \dots, N-1, \qquad h_u(N) = 1, \tag{31.59}$$

$$b_u(k) = \begin{pmatrix} 0 & 0_{1 \times r_{k+1}} \\ h(k+1) & b(k+1) \end{pmatrix}, \ k = 1, \dots, N-2, \ b_u(N-1) = \begin{pmatrix} 0 \\ h(N) \end{pmatrix} \tag{31.60}$$

yield upper triangular generators of the matrix A with orders ρ_k $(k = 1, \dots, N)$, where

$$\rho_k = r_k + 1, \ k = 1, \dots, N-1, \quad \rho_N = 1. \tag{31.61}$$

In the reverse direction, assume that a scalar matrix A has upper triangular generators $g_u(i), h_u(i)$ $(i = 1, \dots, N)$, $b_u(k)$ $(k = 1, \dots, N-1)$ with orders ρ_k $(k = 1, \dots, N)$. Then the formulas

$$\begin{aligned} g(i) &= g_u(i) b_u(i) & i &= 1, \dots, N-1, \\ h(j) &= h_u(j), & j &= 2, \dots, N, \\ b(k) &= b_u(k), & k &= 2, \dots, N-1, \end{aligned} \tag{31.62}$$

and

$$d(k) = g_u(k)h_u(k), \quad k = 1, \dots, N, \tag{31.63}$$

yield upper quasiseparable generators with orders ρ_k $(k = 1, \dots, N-1)$ *and diagonal entries of* A.

Proof. Assume that $g(i)$ $(i = 1, \dots, N-1)$, $h(j)$ $(j = 2, \dots, N)$, $b(k)$ $(k = 2, \dots, N-1)$ and $d(k)$ $(k = 1, \dots, N)$ are upper quasiseparable generators of orders r_k $(k = 1, \dots, N-1)$ and diagonal entries of A. Using (31.58), (31.59) with $i = j$, we get

$$g_u(i)h_u(i) = d(i), \quad i = 1, \dots, N. \tag{31.64}$$

Next, (31.59) and (31.60) yield

$$(b_u)^{<}_{i-1,j} h_u(j) = (b_u)^{<}_{i-1,j-1} b_u(j-1) h_u(j) = \begin{pmatrix} 0 & 0 \\ * & b^{<}_{ij} \end{pmatrix} \begin{pmatrix} 0 \\ h(j) \end{pmatrix}, \quad 1 \le i < j \le N,$$

whence, here using (31.58), we obtain

$$g_u(i)(b_u)^{<}_{i-1,j} h_u(j) = g(i) b^{<}_{ij} h(j), \quad 1 \le i < j \le N. \tag{31.65}$$

Now from the relations (31.64), (31.65) we obtain the relations (31.57). It follows that $g_u(i), h_u(i)$ $(i = 1, \dots, N)$, $b_u(k)$ $(k = 1, \dots, N-1)$ are upper triangular generators of the matrix A with the orders given by (31.61).

Assume that $g_u(i), h_u(i)$ $(i = 1, \dots, N)$, $b_u(k)$ $(k = 1, \dots, N-1)$ are upper triangular generators of the matrix A with orders ρ_k $(k = 1, \dots, N-1)$. Using (31.57) with $j = i$ we get $A(i,i) = g_u(i)h_u(i)$, $i = 1, \dots, N$. Hence the numbers $d(i)$ $(i = 1, \dots, N)$ defined in (31.63) are the diagonal entries of the matrix A. Using (31.57) with $j > i$ we get

$$A(i,j) = (g_u(i)b_u(i))(b_u)^{<}_{i,j} h(j), \quad 1 \le i < j \le N.$$

We conclude that the formulas (31.62) yield upper quasiseparable generators of the matrix A with orders $g_u(i), h_u(i)$ $(i = 1, \dots, N)$, $b_u(k)$ $(k = 1, \dots, N-1)$ with orders ρ_k $(k = 1, \dots, N)$. □

§31.2.3 Single shift

Here we apply Lemma 31.2 to compute the matrix Q defined in Subsection §31.1.1 and quasiseparable generators of the new iterate $A^{(1)} = Q^* A Q$.

Theorem 31.4. *Let* A *be an upper Hessenberg matrix with subdiagonal entries* σ_k $(k = 1, \dots, N-1)$ *and upper triangular generators* $g(i), h(i)$ $(i = 1, \dots, N)$, $b(k)$ $(k = 1, \dots, N-1)$ *of orders* r_k $(k = 1, \dots, N)$, *and let* α *be a number.*

Then the Givens rotation matrices $Q_k,\ k=1,\dots,N-1$ *such that the unitary upper Hessenberg matrix*

$$Q = \tilde{Q}_1\tilde{Q}_2\cdots\tilde{Q}_{N-2}\tilde{Q}_{N-1} \tag{31.66}$$

with

$$\tilde{Q}_i = I_{i-1}\oplus Q_i \oplus I_{N-i-1},\ i=1,\dots,N-1, \tag{31.67}$$

satisfies the condition (31.4), *and the subdiagonal entries* $\sigma_k^{(1)}$ $(k=1,\dots,N-1)$ *and upper triangular generators* $g^{(1)}(i), h^{(1)}(i)$ $(i=1,\dots,N)$, $b^{(1)}(k)$ $(k=1,\dots,N-1)$ *with orders* s_k $(k=1,\dots,N)$, *where*

$$s_k = 1 + r_{k+1},\ k=1,\dots,N-1, \quad s_N = 1,$$

of the upper Hessenberg matrix $A^{(1)} = Q^*AQ$, *are determined as follows.*

1. *Determine the complex Givens transformation matrix* Q_1 *from the condition*

$$Q_1^*\begin{pmatrix} g(1)h(1)-\alpha \\ \sigma_1 \end{pmatrix} = \begin{pmatrix} * \\ 0 \end{pmatrix} \tag{31.68}$$

and compute

$$\begin{pmatrix} g^{(1)}(1) \\ \beta_2 \end{pmatrix} = Q_1^*\begin{pmatrix} g(1)h(1) & g(1)b(1) \\ \sigma_1 & g(2) \end{pmatrix}. \tag{31.69}$$

Determine the number f_2 *and the row vector* ϕ_2 *from the partition*

$$\beta_2 = \begin{pmatrix} f_2 & \phi_2 \end{pmatrix}. \tag{31.70}$$

2. *For* $k=1,\dots,N-2$ *perform the following. Compute the* 2×2 *matrix* Ω_k *by the formula*

$$\Omega_k = \begin{pmatrix} f_{k+1} & \phi_{k+1}h(k+1) \\ 0 & \sigma_{k+1} \end{pmatrix} Q_k. \tag{31.71}$$

Determine the complex Givens rotation matrix Q_{k+1} *and the number* $\sigma_k^{(1)}$ *such that*

$$Q_{k+1}^*\Omega_k(:,1) = \begin{pmatrix} \sigma_k^{(1)} \\ 0 \end{pmatrix}. \tag{31.72}$$

Compute

$$\begin{pmatrix} g^{(1)}(k+1) \\ \beta_{k+2} \end{pmatrix} = Q_{k+1}^*\begin{pmatrix} \Omega_k(1,2) & \phi_{k+1}b(k+1) \\ \Omega_k(2,2) & g(k+2) \end{pmatrix}. \tag{31.73}$$

Determine the number f_{k+2} *and the row vector* ϕ_{k+2} *from the partition*

$$\beta_{k+2} = \begin{pmatrix} f_{k+2} & \phi_{k+2} \end{pmatrix}. \tag{31.74}$$

Compute

$$\begin{pmatrix} h^{(1)}(k) & b^{(1)}(k) \end{pmatrix} = \begin{pmatrix} 1 & 0 & 0 \\ 0 & h(k+1) & b(k+1) \end{pmatrix} \begin{pmatrix} Q_k & 0 \\ 0 & I_{r_{k+2}} \end{pmatrix}. \tag{31.75}$$

3. *Compute*

$$\begin{pmatrix} \sigma_{N-1}^{(1)} & g^{(1)}(N) \end{pmatrix} = \begin{pmatrix} f_N & \phi_N h(N) \end{pmatrix} Q_{N-1} \tag{31.76}$$

and

$$\begin{pmatrix} h^{(1)}(N-1) & b^{(1)}(N-1) \end{pmatrix} = \begin{pmatrix} 1 & 0 \\ 0 & h(N) \end{pmatrix} Q_{N-1}, \quad h^{(1)}(N) = 1. \tag{31.77}$$

Proof. We treat A as a block matrix with entries of sizes $m_i \times n_j$, $i, j = 1, \dots, N+2$, where

$$\begin{gathered} m_1 = \cdots = m_N = 1, \quad m_{N+1} = m_{N+2} = 0, \\ n_1 = 0, \quad n_2 = \cdots = n_{N+1} = 1, \; n_{N+2} = 0. \end{gathered} \tag{31.78}$$

Relative to this partition, the matrix A is a block upper triangular matrix with the diagonal entries given by

$$\begin{gathered} d_A(1) \text{ is the } 1 \times 0 \text{ empty matrix,} \\ d_A(k) = \sigma_{k-1}, \; k = 2, \dots, N, \\ d_A(N+1), d_A(N+2) \text{ are the } 0 \times 1, 0 \times 0 \text{ empty matrices.} \end{gathered} \tag{31.79}$$

Moreover, the equalities (31.57) imply that the formulas

$$\begin{array}{lll} g_A(k) = g(k), & k = 1, \dots, N, & g_A(N+1) \text{ is the } 0 \times 0 \text{ empty matrix,} \\ h_A(k) = h(k-1), & k = 2, \dots, N+1, & h_A(N+2) \text{ is the } 0 \times 0 \text{ empty matrix,} \\ b_A(k) = b(k-1), & k = 2, \dots, N, & b_A(N+1) \text{ is the } r_N \times 0 \text{ empty matrix} \end{array} \tag{31.80}$$

yield upper quasiseparable generators of the matrix A with orders

$$r_k^A = r_k, \; k = 1, \dots, N, \quad r_{N+1}^A = 0 \tag{31.81}$$

relative to the partition (31.78).

Using (31.66) and setting

$$\begin{gathered} \tilde{S}_1 = I_N, \quad \tilde{S}_i = \tilde{Q}_{i-1}^*, \; i = 2, \dots, N, \quad \tilde{S}_{N+1} = \tilde{S}_{N+2} = I_N, \\ \tilde{T}_1 = \tilde{T}_2 = I_N, \quad \tilde{T}_i = \tilde{Q}_{i-2}, \; i = 3, \dots, N+1, \quad \tilde{T}_{N+2} = I_N, \end{gathered} \tag{31.82}$$

we get

$$Q^* = \tilde{S}_{N+2} \cdots \tilde{S}_1, \quad Q = \tilde{T}_1 \cdots \tilde{T}_{N+2}. \tag{31.83}$$

We have

$$\tilde{S}_1 = \mathrm{diag}\{S_1, I_{N-1}\},$$
$$\tilde{S}_k = \mathrm{diag}\{I_{k-2}, S_k, I_{N-k}\},\ k = 2, \dots, N, \quad \tilde{S}_{N+1} = \mathrm{diag}\{I_{N-1}, S_{N+1}\},$$
$$\tilde{S}_{N+2} = \mathrm{diag}\{I_N, S_{N+2}\},$$

with

$$S_1 = 1, \quad S_k = Q^*_{k-1},\ k = 2, \dots, N, \quad S_{N+1} = 1, \qquad (31.84)$$
$$S_{N+2} \text{ is the } 0 \times 0 \text{ empty matrix,}$$

and

$$\tilde{T}_1 = \mathrm{diag}\{T_1, I_N\}, \quad \tilde{T}_2 = \mathrm{diag}\{T_2, I_{N-1}\};$$
$$\tilde{T}_k = \mathrm{diag}\{I_{k-3}, T_k, I_{N-k+1}\},\ k = 3, \dots, N+1;$$
$$\tilde{T}_{N+2} = \mathrm{diag}\{I_{N-1}, T_{N+2}\}$$

with

$$T_1 \text{ is the } 0 \times 0 \text{ empty matrix,} \qquad (31.85)$$
$$T_2 = 1, \quad T_k = Q_{k-2},\ k = 3, \dots, N+1, \quad T_{N+2} = 1.$$

We treat the lower Hessenberg matrix Q^* as a block matrix with entries of sizes $\tau_i \times m_j,\ i, j = 1, \dots, N+2$, where

$$\tau_1 = 0, \quad \tau_2 = \dots = \tau_{N+1} = 1, \quad \tau_{N+2} = 0. \qquad (31.86)$$

The matrix Q^* has the representation (31.83) with the matrices S_k $(k = 1, \dots, N+2)$ in (31.84) of sizes $(\tau_1 + r^S_1) \times m_1$, $(\tau_k + r^S_k) \times (m_k + r^S_{k-1})$ $(k = 2, \dots, N+1)$, $\tau_{N+2} \times (m_{N+2} + r^S_{N+1})$, where $r^S_k = 1$ $(k = 1, \dots, N)$, $r^S_{N+1} = 0$. We treat the upper Hessenberg matrix Q as a block matrix with entries of sizes $n_i \times \nu_j,\ i, j = 1, \dots, N+2$, where

$$\nu_1 = \nu_2 = 0, \quad \nu_3 = \dots = \nu_{N+2} = 1. \qquad (31.87)$$

The matrix Q has the representation (31.83) with the matrices T_k $(k = 1, \dots, N+2)$ in (31.85) of sizes $n_1 \times (\nu_1 + r^T_1)$, $(n_k + r^T_{k-1}) \times (\nu_k + r^T_k)$ $(k = 2, \dots, N+1)$, $(n_{N+2} + r^T_{N+1}) \times \nu_{N+2}$, where $r^T_1 = 0$, $r^T_k = 1$ $(k = 2, \dots, N+1)$.

Now we are in the position to apply Corollary 31.2 to the product $A^{(1)} = Q^*AQ$. We treat $A^{(1)}$ as a block matrix with entries of sizes $\tau_i \times \nu_j,\ i, j = 1, \dots, N+2$, and obtain quasiseparable generators $\tilde{q}(j)$ $(j = 1, \dots, N+1)$, $g_F(i)$ $(i = 1, \dots, N+1)$, $h_F(j)$ $(j = 2, \dots, N+2)$, $b_F(k)$ $(k = 2, \dots, N+1)$, $d_F(k)$ $(k = 1, \dots, N+2)$ relative to this partition. By (31.86) and (31.87), the elements $A^{(1)}_B(i, j)$ of the block representation of the matrix $A^{(1)}$ are connected with the elements $A^{(1)}(i, j)$ of $A^{(1)}$, treated as an ordinary scalar matrix, via the relations

$$A^{(1)}(i, j) = A^{(1)}_B(i+1, j+2), \quad 1 \le i, j \le N.$$

Hence, we have

$$A^{(1)}(k+1,k) = d_F(k+2), \quad k = 1,\dots,N-1, \tag{31.88}$$

$$g^{(1)}(k) = g_F(k+1), \quad k = 1,\dots,N, \tag{31.89}$$

$$\begin{aligned} h^{(1)}(k) &= h_F(k+2), \quad k = 1,\dots,N, \\ b^{(1)}(k) &= b_F(k+2), \quad k = 1,\dots,N-1. \end{aligned} \tag{31.90}$$

Using (31.47) with empty matrices $d_A(1), T_1$ and $g_A(1) = g(1)$, $S_1 = 1$, we get $\Delta_1 = g(1)$. Hence using (31.48), (31.49) we get that $\phi_1 = g(1)$ and that f_1 is the 1×0 empty matrix. Inserting this in (31.50) and using (31.79), (31.80), (31.84), (31.85) with $k = 2$ we get

$$\begin{pmatrix} g_F(2) \\ \beta_2 \end{pmatrix} = Q_1^* \begin{pmatrix} g(1)h(1) & g(1)b(1) \\ \sigma_1 & g(2) \end{pmatrix}.$$

From here using $g_F(2) = g^{(1)}(1)$ we obtain (31.69). Using (31.70) we determine the variables f_2, ϕ_2

Next for $k = 3,\dots,N$, combining (31.50), (31.51) with (31.84), (31.85), (31.79) and (31.80), we get

$$\begin{pmatrix} d_F(k) & g_F(k) \\ \tilde{q}(k) & \beta_k \end{pmatrix} = Q_{k-1}^* \begin{pmatrix} f_{k-1} & \phi_{k-1}h(k-1) & \phi_{k-1}b(k-1) \\ 0 & \sigma_{k-1} & g(k) \end{pmatrix} \begin{pmatrix} Q_{k-2} & 0 \\ 0 & I_{r_k} \end{pmatrix} \tag{31.91}$$

and

$$\begin{pmatrix} h_F(k) & b_F(k) \end{pmatrix} = \begin{pmatrix} 1 & 0 & 0 \\ 0 & h(k+1) & b(k+1) \end{pmatrix} \begin{pmatrix} Q_{k-2} & 0 \\ 0 & I_{r_k} \end{pmatrix}. \tag{31.92}$$

From (31.91) using (31.71) we obtain

$$\begin{pmatrix} d_F(k+2) & g_F(k+2) \\ \tilde{q}(k+2) & \beta_{k+2} \end{pmatrix} = Q_{k+1}^* \begin{pmatrix} \Omega_k(1,1) & \Omega_k(1,2) & \phi_{k+1}b(k+1) \\ \Omega_k(2,1) & \Omega_k(2,2) & g(k+2) \end{pmatrix}, \quad k = 1,\dots,N-2. \tag{31.93}$$

Hence, using (31.72) we have

$$d_F(k+2) = \sigma_k^{(1)}, \quad k = 1,\dots,N-2 \tag{31.94}$$

and

$$\tilde{q}(k+2) = 0, \qquad k = 1,\dots,N-2. \tag{31.95}$$

The formulas (31.94) and (31.88) mean that $\sigma_k^{(1)}$, $k = 1,\dots,N-2$, are subdiagonal entries of the matrix $A^{(1)}$. The equalities (31.95) imply that $A^{(1)}$ is an upper Hessenberg matrix. The latter together with (31.68) implies that (31.4) holds.

Next, (31.93) and (31.89) yield (31.73). Using (31.52) we obtain the partitions (31.74). From (31.92) using (31.90), we obtain (31.75).

Further, (31.50),(31.51) and (31.84), (31.85), (31.79), (31.80) with $k = N+1$ yield

$$\begin{pmatrix} d_F(N+1) & g_F(N+1) \end{pmatrix} = \begin{pmatrix} f_N & \phi_N h_A(N+1) \end{pmatrix} Q_{N-1}. \tag{31.96}$$

and

$$\begin{pmatrix} h_F(N+1) & b_F(N+1) \end{pmatrix} = \begin{pmatrix} 1 & 0 \\ 0 & h(N) \end{pmatrix} Q_{N-1}. \tag{31.97}$$

Using (31.53), (31.54) and (31.80), (31.85) we get

$$h_F(N+2) = 1. \tag{31.98}$$

From (31.96), using (31.80), (31.88) and (31.89) with $k = N$, we obtain (31.76). Finally, from (31.97) and (31.98), using (31.90), we obtain (31.77). □

§31.2.4 Double shift

Applying Lemma 31.2 to matrices obtained via an implicit step with a double shift as in Subsection §31.1.2, we obtain the following algorithm.

Theorem 31.5. *Let A be an upper Hessenberg real matrix with subdiagonal entries σ_k $(k = 1, \dots, N-1)$ and upper triangular generators $g(i), h(i)$ $(i = 1, \dots, N)$, $b(k)$ $(k = 1, \dots, N-1)$ of orders r_k $(k = 1, \dots, N)$, and let s, t be real numbers.*

Then the 3×3 orthogonal matrices Q_k, $k = 1, \dots, N-2$, and the 2×2 orthogonal matrix Q_{N-1} such that the orthogonal matrix

$$Q = \tilde{Q}_1 \tilde{Q}_2 \cdots \tilde{Q}_{N-2} \tilde{Q}_{N-1}, \tag{31.99}$$

with

$$\tilde{Q}_i = I_{i-1} \oplus Q_i \oplus I_{N-i-2},\ i = 1, \dots, N-2, \quad \tilde{Q}_{N-1} = I_{N-2} \oplus Q_{N-1}, \tag{31.100}$$

satisfies the condition (31.7), and subdiagonal entries $\sigma_k^{(1)}$ $(k = 1, \dots, N-1)$ and upper triangular generators $g^{(1)}(i), h^{(1)}(i)$ $(i = 1, \dots, N)$, $b^{(1)}(k)$ $(k = 1, \dots, N-1)$ with orders s_k $(k = 1, \dots, N)$, where

$$s_k = 2 + r_{k+2},\ k = 1, \dots, N-2, \quad s_{N-1} = 2, \quad s_N = 1,$$

of the upper Hessenberg matrix $A^{(1)} = Q^* A Q$, are determined as follows.

1. *Set*

$$a_{11} = g(1)h(1),\ a_{12} = g(1)b(1)h(2),\ a_{21} = \sigma_1,\ a_{22} = g(2)h(2),\ a_{32} = \sigma_2 \tag{31.101}$$

and determine the 3×3 *orthogonal matrix* Q_1 *from the condition*

$$Q_1^* \begin{pmatrix} a_{11}^2 + a_{12}a_{21} - sa_{11} + t & a_{21}(a_{11} + a_{22} - s) & a_{32}a_{21} \end{pmatrix}^T = \begin{pmatrix} * & 0 & 0 \end{pmatrix}^T. \tag{31.102}$$

Compute the $3 \times (2 + r_3)$ *matrix* Δ_3 *by the formula*

$$\Delta_3 = Q_1^* \begin{pmatrix} g(1)h(1) & g(1)b(1)h(2) & g(1)b(1)b(2) \\ \sigma_1 & g(2)h(2) & g(2)b(2) \\ 0 & \sigma_2 & g(3) \end{pmatrix} \tag{31.103}$$

and determine the matrices $g^{(1)}(1), \beta_3, f_3, \phi_3$ *of sizes* $1 \times (2 + r_3), 2 \times (2 + r_3), 2 \times 2, 2 \times r_3$, *respectively, from the partitions*

$$\Delta_3 = \begin{pmatrix} g^{(1)}(1) \\ \beta_3 \end{pmatrix}, \quad \beta_3 = \begin{pmatrix} f_3 & \phi_3 \end{pmatrix}. \tag{31.104}$$

2. *For* $k = 1, \ldots, N - 3$, *perform the following.*
   Compute the 3×3 *matrix* Ω_k *by the formula*

$$\Omega_k = \begin{pmatrix} f_{k+2} & \phi_{k+2}h(k+2) \\ 0_{1\times 2} & \sigma_{k+2} \end{pmatrix} Q_k. \tag{31.105}$$

Determine the 3×3 *orthogonal matrix* Q_{k+1} *and the number* $\sigma_k^{(1)}$ *such that*

$$Q_{k+1}^* \Omega_k(:, 1) = \begin{pmatrix} \sigma_k^{(1)} & 0 & 0 \end{pmatrix}^T. \tag{31.106}$$

Compute the matrix Δ_{k+3} *by the formula*

$$\Delta_{k+3} = Q_{k+1}^* \begin{pmatrix} \Omega_k(1:2, 2:3) & \phi_{k+2}b(k+2) \\ \Omega_k(3, 2:3) & g(k+3) \end{pmatrix} \tag{31.107}$$

and determine the matrices $g^{(1)}(k+1), \beta_{k+3}, f_{k+3}, \phi_{k+3}$ *of sizes* $1 \times (2 + r_{k+3}), 2 \times (2 + r_{k+3})$, $2 \times 2, 2 \times r_{k+3}$, *respectively, from the partitions*

$$\Delta_{k+3} = \begin{pmatrix} g^{(1)}(k+1) \\ \beta_{k+3} \end{pmatrix}, \tag{31.108}$$

$$\beta_{k+3} = \begin{pmatrix} f_{k+3} & \phi_{k+3} \end{pmatrix}. \tag{31.109}$$

Compute

$$\begin{pmatrix} h^{(1)}(k) & b^{(1)}(k) \end{pmatrix} = \begin{pmatrix} I_2 & 0 & 0 \\ 0 & h(k+2) & b(k+2) \end{pmatrix} \begin{pmatrix} Q_k & 0 \\ 0 & I_{r_{k+3}} \end{pmatrix}. \tag{31.110}$$

3. *Compute the* 2×3 *matrix* Ω_{N-2} *by the formula*

$$\Omega_{N-2} = \begin{pmatrix} f_N & \phi_N h(N) \end{pmatrix} Q_{N-2}. \tag{31.111}$$

Determine the 2×2 *orthogonal matrix* Q_{N-1} *and the number* $\sigma^{(1)}_{N-2}$ *such that*

$$Q^*_{N-1}\Omega_{N-2}(:,1) = \begin{pmatrix} \sigma^{(1)}_{N-2} & 0 \end{pmatrix}^T. \tag{31.112}$$

Compute the matrix Δ_{N+1} *by the formula*

$$\Delta_{N+1} = Q^*_{N-1}\Omega_{N-2}(:,2:3) \tag{31.113}$$

and determine the two-dimensional row vectors $g^{(1)}(N-1), \beta_{N+1}$ *from the partition*

$$\Delta_{N+1} = \begin{pmatrix} g^{(1)}(N-1) \\ \beta_{N+1} \end{pmatrix}. \tag{31.114}$$

Compute

$$\begin{pmatrix} h^{(1)}(N-2) & b^{(1)}(N-2) \end{pmatrix} = \begin{pmatrix} I_2 & 0 \\ 0 & h(N) \end{pmatrix} Q_{N-2}. \tag{31.115}$$

4. *Compute*

$$\begin{pmatrix} \sigma^{(1)}_{N-1} & g^{(1)}(N) \end{pmatrix} = \beta_{N+1}Q_{N-1}. \tag{31.116}$$

Determine the two-dimensional column vectors $h_Q(N-1), b_Q(N-1)$ *from the partition*

$$Q_{N-1} = \begin{pmatrix} h_Q(N-1) & b_Q(N-1) \end{pmatrix}. \tag{31.117}$$

Set

$$h^{(1)}(N-1) = h_Q(N-1), \quad b^{(1)}(N-1) = b_Q(N-1), \quad h^{(1)}(N) = 1. \tag{31.118}$$

Proof. We treat A as a block matrix with entries of sizes $m_i \times n_j,\ i,j = 1,\dots,N+3$, where

$$\begin{aligned} &m_1 = \cdots = m_N = 1,\ m_{N+1} = m_{N+2} = m_{N+3} = 0, \\ &n_1 = 0,\ n_2 = \cdots = n_{N+1} = 1,\ n_{N+2} = n_{N+3} = 0. \end{aligned} \tag{31.119}$$

With respect to this partition, A is a block upper triangular matrix with the following diagonal entries:

$$\begin{aligned} &d_A(1) \text{ is the } 1 \times 0 \text{ empty matrix,} \\ &d_A(k) = \sigma_{k-1},\ k = 2,\dots,N, \\ &d_A(N+1), d_A(N+2), d_A(N+3) \text{ are} \\ &\quad \text{the } 0 \times 1, 0 \times 0, 0 \times 0 \text{ empty matrices.} \end{aligned} \tag{31.120}$$

Moreover, the equalities (31.57) imply that the formulas

$$\begin{aligned} g_A(i) &= g(i),\ i=1,\dots,N, \quad g_A(N+1), g_A(N+2) \\ &\quad \text{are the } 0\times 0 \text{ empty matrices}, \\ h_A(j) &= h(j-1),\ j=2,\dots,N+1, \quad h_A(N+2), h_A(N+3) \\ &\quad \text{are the } 0\times 0 \text{ empty matrices}, \\ b_A(k) &= b(k-1),\ k=2,\dots,N, b_A(N+1), b_A(N+2), \\ &\quad \text{are the } r_N \times 0, 0\times 0 \text{ empty matrices}, \end{aligned} \tag{31.121}$$

yield upper quasiseparable generators of the matrix A with orders $r_k^A = r_k,\ k = 1,\dots,N,\ r_{N+1}^A = r_{N+2}^A = 0$ with respect to partition (31.119).

Using (31.99) and setting

$$\begin{aligned} &\tilde{S}_1 = \tilde{S}_2 = I_N, \ \ \tilde{S}_i = \tilde{Q}_{i-2}^*,\ i=3,\dots,N+1, \ \ \tilde{S}_{N+2} = \tilde{S}_{N+3} = I_N, \\ &\tilde{T}_1 = \tilde{T}_2 = \tilde{T}_3 = I_N, \quad \tilde{T}_i = \tilde{Q}_{i-3},\ i=4,\dots,N+2, \quad \tilde{T}_{N+3} = I_N, \end{aligned} \tag{31.122}$$

we get

$$Q^* = \tilde{S}_{N+3}\cdots\tilde{S}_1, \quad Q = \tilde{T}_1\cdots\tilde{T}_{N+3}. \tag{31.123}$$

We have

$$\begin{aligned} \tilde{S}_1 &= \operatorname{diag}\{S_1, I_{N-1}\}, & \tilde{S}_2 &= \operatorname{diag}\{S_2, I_{N-2}\}, \\ \tilde{S}_k &= \operatorname{diag}\{I_{k-3}, S_k, I_{N-k}\},\ k=3,\dots,N, & \tilde{S}_{N+1} &= \operatorname{diag}\{I_{N-2}, S_{N+1}\}, \\ \tilde{S}_{N+2} &= \operatorname{diag}\{I_{N-1}, S_{N+2}\}, & \tilde{S}_{N+3} &= \operatorname{diag}\{I_N, S_{N+3}\} \end{aligned}$$

with

$$\begin{aligned} &S_1 = 1, S_2 = I_2, \quad S_i = Q_{i-2}^*,\ i=3,\dots,N+1, \quad S_{N+2} = 1, \\ &\qquad S_{N+3} \text{ is the } 0\times 0 \text{ empty matrix}, \end{aligned} \tag{31.124}$$

and

$$\begin{aligned} \tilde{T}_1 &= \operatorname{diag}\{T_1, I_N\}, \quad \tilde{T}_2 = \operatorname{diag}\{T_2, I_{N-1}\}, \quad \tilde{T}_3 = \operatorname{diag}\{T_3, I_{N-2}\}, \\ \tilde{T}_k &= \operatorname{diag}\{I_{k-4}, T_k, I_{N-k+1}\},\ k=4,\dots,N+2, \\ \tilde{T}_{N+3} &= \operatorname{diag}\{I_{N-1}, T_{N+3}\}, \end{aligned}$$

with

$$\begin{aligned} &T_1 \text{ is the } 0\times 0 \text{ empty matrix}, \\ &T_2 = 1,\ T_3 = I_2, \quad T_i = Q_{i-3},\ i=4,\dots,N+2, \quad T_{N+3} = 1. \end{aligned} \tag{31.125}$$

We treat Q^* as a block matrix with entries of sizes $\tau_i \times m_j,\ i,j = 1,\dots,N+3$, where

$$\tau_1 = \tau_2 = 0, \quad \tau_3 = \cdots = \tau_{N+2} = 1,\ \tau_{N+3} = 0. \tag{31.126}$$

The matrix Q^* has the representation (31.123) with the matrices S_k ($k = 1,\dots, N+2$) in (31.124) of sizes $(\tau_1+r_1^S)\times m_1$, $(\tau_k+r_k^S)\times(m_k+r_{k-1}^S)$ ($k = 2,\dots,N+2$), $\tau_{N+3}\times(m_{N+2}+r_{N+1}^S)$, where

$$r_k^S = 1, \quad r_k^S = 2,\ k = 2,\dots,N, \quad r_{N+1}^S = 1,\ r_{N+2}^S = 0.$$

We treat Q as a block matrix with entries of sizes $n_i \times \nu_j$, $i,j = 1,\dots,N+3$, where

$$\nu_1 = \nu_2 = \nu_3 = 0, \quad \nu_4 = \cdots = \nu_{N+3} = 1. \tag{31.127}$$

The matrix Q has the representation (31.123) with the matrices T_k ($k = 1,\dots,N+2$) in (31.125) of sizes $n_1 \times (\nu_1 + r_1^T)$, $(n_k + r_{k-1}^T) \times (\nu_k + r_k^T)$ ($k = 2,\dots,N+2$), $(n_{N+3} + r_{N+2}^T) \times \nu_{N+3}$, where

$$r_1^T = 0,\ r_2^T = 1, \quad r_k^T = 2,\ k = 3,\dots,N+1, \quad r_{N+2}^T = 1. \tag{31.128}$$

Now we are in the position to apply Corollary 31.2 to the product $A^{(1)} = Q^*AQ$. We treat $A^{(1)}$ as a block matrix with entries of sizes $\tau_i\times\nu_j$, $i,j = 1,\dots,N+3$ and obtain quasiseparable generators $\tilde{q}(j)$ ($j = 1,\dots,N+2$), $g_F(i)$ ($i = 1,\dots,N+2$), $h_F(j)$ ($j = 2,\dots,N+3$), $b_F(k)$ ($k = 2,\dots,N+2$), $d_F(k)$ ($k = 1,\dots,N+3$) relative to this partition. By (31.126) and (31.127) the elements $A_B^{(1)}(i,j)$ of the block representation of the matrix $A^{(1)}$ are connected with the elements $A^{(1)}(i,j)$ of $A^{(1)}$, treated as an ordinary scalar matrix, by the relations

$$A^{(1)}(i,j) = A_B^{(1)}(i+2, j+3), \quad 1 \le i,j \le N.$$

Hence,

$$A^{(1)}(k+1,k) = d_F(k+3), \quad k = 1,\dots,N-1, \tag{31.129}$$

$$g^{(1)}(k) = g_F(k+2), \quad k = 1,\dots,N, \tag{31.130}$$

$$\begin{aligned} h^{(1)}(k) &= h_F(k+3), \quad k = 1,\dots,N, \\ b^{(1)}(k) &= b_F(k+3), \quad k = 1,\dots,N-1. \end{aligned} \tag{31.131}$$

Using (31.47) with empty matrices $d_A(1), T_1$ and $S_1 = 1$, $g_A(1) = g(1)$, we get that $\phi_1 = g(1)$ and f_1 is the 1×0 empty matrix. Inserting this in (31.50), (31.51) and using (31.120), (31.121), (31.124), (31.125) with $k = 2$, we get

$$\Delta_2 = \begin{pmatrix} g(1)h(1) & g(1)b(1) \\ \sigma_1 & g(2) \end{pmatrix}.$$

Using (31.52) with $k = 2$ we get

$$f_2 = \begin{pmatrix} g(1)h(1) \\ \sigma_1 \end{pmatrix}, \quad \phi_2 = \begin{pmatrix} g(1)b(1) \\ g(2) \end{pmatrix}.$$

Inserting this in (31.50) and using (31.120), (31.121), (31.124), (31.125) with $k = 3$ we obtain the equality (31.103). Using (31.51), (31.52) with $k = 3$ and $g_F(3) = g^{(1)}(1)$ we obtain (31.104).

Next, for $k = 4, \dots, N+1$, combining (31.50), (31.51) with (31.124), (31.125), (31.120) and (31.121), we get

$$\begin{pmatrix} d_F(k) & g_F(k) \\ \tilde{q}(k) & \beta_k \end{pmatrix} = Q_{k-2}^* \begin{pmatrix} f_{k-1} & \phi_{k-1}h(k-1) & \phi_{k-1}b(k-1) \\ 0_{1\times 2} & \sigma_{k-1} & g(k) \end{pmatrix} \begin{pmatrix} Q_{k-3} & 0 \\ 0 & I_{r_k} \end{pmatrix},$$
$$k = 4, \dots, N, \tag{31.132}$$
$$\begin{pmatrix} d_F(N+1) & g_F(N+1) \\ \tilde{q}(N+1) & \beta_{N+1} \end{pmatrix} = Q_{N-1}^* \begin{pmatrix} f_N & \varphi_N h(N) \end{pmatrix} Q_{N-2}$$

and

$$\begin{pmatrix} h_F(k) & b_F(k) \end{pmatrix} = \begin{pmatrix} I_2 & 0 & 0 \\ 0 & h(k-1) & b(k-1) \end{pmatrix} \begin{pmatrix} Q_{k-3} & 0 \\ 0 & I_{r_k} \end{pmatrix}, \quad k = 4, \dots, N,$$
$$\begin{pmatrix} h_F(N+1) & b_F(N+1) \end{pmatrix} = \begin{pmatrix} I_2 & 0 \\ 0 & h(N) \end{pmatrix} Q_{N-2}. \tag{31.133}$$

From (31.132) using (31.105) and (31.111) we obtain

$$\begin{pmatrix} d_F(k+3) & g_F(k+3) \\ \tilde{q}(k+3) & \beta_{k+3} \end{pmatrix} = Q_{k+1}^* \begin{pmatrix} \Omega_k(1:2, 1:2) & \Omega_k(1:2, 3) & \phi_{k+2}b(k+2) \\ \Omega_k(3, 1:2) & \Omega_k(3,3) & g(k+3) \end{pmatrix},$$
$$k = 1, \dots, N-3, \tag{31.134}$$
$$\begin{pmatrix} d_F(N+1) & g_F(N+1) \\ \tilde{q}(N+1) & \beta_{N+1} \end{pmatrix} = Q_{N-1}^* \Omega_{N-2}.$$

Hence, using (31.106), (31.112) we have

$$d_F(k+3) = \sigma_k^{(1)}, \quad k = 1, \dots, N-2 \tag{31.135}$$

and

$$\tilde{q}(k+3) = 0, \qquad k = 1, \dots, N-2. \tag{31.136}$$

The formulas (31.135) and (31.129) mean that $\sigma_k^{(1)}$, $k = 1, \dots, N-2$ are subdiagonal entries of the matrix $A^{(1)}$. The equalities (31.136) imply that $A^{(1)}$ is an upper Hessenberg matrix. The latter together with (31.102) imply that (31.7) holds. Next, using (§31.2.4) and (31.130) we obtain the equalities (31.107), (31.108) and (31.113), (31.114). Using (31.52) we obtain the partitions (31.109) and the equality $f_{N+1} = \beta_{N+1}$. From (§31.2.4), using (31.131), we obtain (31.110) and (31.115).

Using (31.50), (31.51) and (31.124), (31.125), (31.120), (31.121) with $k = N+2$ we get

$$\begin{pmatrix} d_F(N+2) & g_F(N+2) \end{pmatrix} = \beta_{N+1} Q_{N-1} \tag{31.137}$$

and

$$\begin{pmatrix} h_F(N+2) & b_F(N+2) \end{pmatrix} = Q_{N-1}. \tag{31.138}$$

From (31.137) using (31.129), (31.130) with $k = N$ we obtain (31.116). Then (31.53), (31.54) and (31.121), (31.125) yield

$$h_F(N+3) = 1. \tag{31.139}$$

Finally, from (31.138), (31.139) using (31.131) we obtain (31.117), (31.118). □

§31.3 Perturbations of Hermitian matrices

We denote by $\mathcal{F}_{N,m}$ the class of upper Hessenberg $N \times N$ matrices A which are rank m perturbations of Hermitian matrices, i.e.,

$$A = H - XY^*,$$

where H is a Hermitian matrix and $X = \text{col}(x(i))_{i=1}^N$, $Y = \text{col}(y(i))_{i=1}^N$, with m-dimensional rows $x(i), y(i)$.

Every matrix A from the class $\mathcal{F}_{N,m}$ is completely defined by the following parameters:

1. the diagonal entries $d(k)$ $(k = 1, \dots, N)$ of A;
2. the subdiagonal entries σ_k $(k = 1, \dots, N-1)$ of A;
3. the perturbation matrices $X = \text{col}(x(i))_{i=1}^N$, $Y = \text{col}(y(i))_{i=1}^N$.

Let Q be a unitary matrix such that $A^{(1)} = Q^*AQ$ is an upper Hessenberg matrix. We obviously have

$$A^{(1)} = H^{(1)} - X^{(1)}(Y^{(1)})^*,$$

with the Hermitian matrix $H^{(1)} = Q^*HQ$ and the perturbation matrices $X^{(1)} = Q^*X$, $Y^{(1)} = Q^*Y$. It follows that the matrix $A^{(1)}$ also belongs to $\mathcal{F}_{N,m}$.

§31.3.1 Single shift

Here we apply Theorem 31.4 to matrices from the class $\mathcal{F}_{N,m}$.

Theorem 31.6. *Let A be a matrix from the class $\mathcal{F}_{N,m}$ with diagonal entries $d(k)$ $(k = 1, \dots, N)$, subdiagonal entries σ_k $(k = 1, \dots, N-1)$ and perturbation matrices $X = \text{col}(x(i))_{i=1}^N$, $Y = \text{col}(y(i))_{i=1}^N$ and let α be a number.*

Then the Givens rotation matrices Q_k, $k = 1, \dots, N-1$, such that the unitary matrix

$$Q = \tilde{Q}_1\tilde{Q}_2 \cdots \tilde{Q}_{N-2}\tilde{Q}_{N-1}, \tag{31.140}$$

with

$$\tilde{Q}_i = I_{i-1} \oplus Q_i \oplus I_{N-i-1},\ i = 1, \dots, N-1, \tag{31.141}$$

satisfies the condition (31.4), and diagonal entries $d^{(1)}(k)$ $(k = 1, \dots, N)$, *subdiagonal entries* $\sigma_k^{(1)}$ $(k = 1, \dots, N-1)$ *and matrices of perturbation* $X^{(1)} = \mathrm{col}(x^{(1)}(i))_{i=1}^N$, $Y^{(1)} = \mathrm{col}(y^{(1)}(i))_{i=1}^N$ *of the upper Hessenberg matrix* $A^{(1)} = Q^*AQ$ *are determined as follows.*

1. *Determine the complex Givens transformation matrix* Q_1 *from the condition*

$$Q_1^* \begin{pmatrix} d(1) - \alpha \\ \sigma_1 \end{pmatrix} = \begin{pmatrix} * \\ 0 \end{pmatrix} \tag{31.142}$$

and compute

$$\varepsilon_1 = -x(1)y^*(2) + y(1)x^*(2) + \sigma_1^*. \tag{31.143}$$

Determine the numbers $h_Q(1), b_Q(1), d_Q(1), g_Q(1)$ *from the partition*

$$Q_1 = \begin{pmatrix} h_Q(1) & b_Q(1) \\ d_Q(1) & g_Q(1) \end{pmatrix} \tag{31.144}$$

and compute

$$\gamma_1 = h_Q^*(1)d(1) + d_Q^*(1)\sigma_1, \quad \delta_1 = h_Q^*(1)\varepsilon_1 + d_Q^*(1)d(2), \tag{31.145}$$

$$f_2 = b_Q^*(1)d(1) + g_Q^*(1)\sigma_1, \quad \varphi_2 = b_Q^*(1)\varepsilon_1 + g_Q^*(1)d(2), \quad \rho_2 = g_Q^*(1)\sigma_2^*, \tag{31.146}$$

$$x^{(1)}(1) = h_Q^*(1)x(1) + d_Q^*(1)x(2), \quad c_2 = b_Q^*(1)x(1) + g_Q^*(1)x(2),$$
$$y^{(1)}(1) = h_Q^*(1)y(1) + d_Q^*(1)y(2), \quad z_2 = b_Q^*(1)y(1) + g_Q^*(1)y(2), \tag{31.147}$$

$$d^{(1)}(1) = \gamma_1 h_Q(1) + \delta_1 d_Q(1). \tag{31.148}$$

Set $\sigma_N = 0$.

2. *For* $k = 1, \dots, N-2$, *perform the following. Compute the* 2×2 *matrix* Ω_k *by the formula*

$$\Omega_k = \begin{pmatrix} f_{k+1} & \varphi_{k+1} \\ 0 & \sigma_{k+1} \end{pmatrix} Q_k. \tag{31.149}$$

Determine the complex Givens rotation matrix Q_{k+1} *and the number* $\sigma_k^{(1)}$ *such that*

$$Q_{k+1}^* \Omega_k(:,1) = \begin{pmatrix} \sigma_k^{(1)} \\ 0 \end{pmatrix}. \tag{31.150}$$

Compute

$$\varepsilon_{k+1} = -c_{k+1}y^*(k+2) + z_{k+1}x^*(k+2) + \rho_{k+1}. \tag{31.151}$$

Determine the numbers $h_Q(k+1), b_Q(k+1), d_Q(k+1), g_Q(k+1)$ *from the partition*

$$Q_{k+1} = \begin{pmatrix} h_Q(k+1) & b_Q(k+1) \\ d_Q(k+1) & g_Q(k+1) \end{pmatrix} \tag{31.152}$$

and compute

$$\begin{aligned}\gamma_{k+1} &= h_Q^*(k+1)\Omega_k(1,2) + d_Q^*(k+1)\Omega_k(2,2),\\ \delta_{k+1} &= h_Q^*(k+1)\varepsilon_{k+1} + d_Q^*(k+1)d(k+2),\end{aligned} \tag{31.153}$$

$$\begin{aligned}f_{k+2} &= b_Q^*(k+1)\Omega_k(1,2) + g_Q^*(k+1)\Omega_k(2,2),\\ \varphi_{k+2} &= b_Q^*(k+1)\varepsilon_{k+1} + g_Q^*(k+1)d(k+2), \quad \rho_{k+2} = g_Q^*(k+1)\sigma_{k+2}^*,\end{aligned} \tag{31.154}$$

$$\begin{aligned}x^{(1)}(k+1) &= h_Q^*(k+1)c_{k+1} + d_Q^*(k+1)x(k+2),\\ c_{k+2} &= b_Q^*(k+1)c_{k+1} + g_Q^*(k+1)x(k+2),\\ y^{(1)}(k+1) &= h_Q^*(k+1)z_{k+1} + d_Q^*(k+1)y(k+2),\\ z_{k+2} &= b_Q^*(k+1)z_{k+1} + g_Q^*(k+1)y(k+2),\end{aligned} \tag{31.155}$$

$$d^{(1)}(k+1) = \gamma_{k+1}h_Q(k+1) + \delta_{k+1}d_Q(k+1). \tag{31.156}$$

3. *Compute*

$$\begin{pmatrix} \sigma_{N-1}^{(1)} & d^{(1)}(N) \end{pmatrix} = \begin{pmatrix} f_N & \varphi_N \end{pmatrix} Q_{N-1}. \tag{31.157}$$

Set

$$x^{(1)}(N) = c_N, \quad y^{(1)}(N) = z_N. \tag{31.158}$$

Proof. The upper Hessenberg matrix A has lower quasiseparable generators

$$p(i) = 1,\ i = 2,\dots,N, \quad q(j) = \sigma_j,\ j = 1,\dots,N-1, \quad a(k) = 0,\ k = 2,\dots,N-1.$$

Using the formulas (29.42) we obtain upper quasiseparable generators $g_u(i)$ ($i = 1,\dots,N-1$), $h_u(j)$ ($j = 2,\dots,N$), $b_u(k)$ ($k = 2,\dots,N-1$) of A by the formulas

$$\begin{aligned} g_u(i) &= \begin{pmatrix} -x(i) & y(i) & \sigma_i^* \end{pmatrix}, \quad i = 1,\dots,N-1,\\ h_u(j) &= \begin{pmatrix} y^*(j) \\ x^*(j) \\ 1 \end{pmatrix}, \qquad j = 2,\dots,N,\\ b_u(k) &= \begin{pmatrix} I_{2m} & 0 \\ 0 & 0 \end{pmatrix}, \qquad k = 2,\dots,N-1.\end{aligned} \tag{31.159}$$

By Lemma 31.3, the relations

$$g(i) = \begin{pmatrix} d(i) & g_u(i) \end{pmatrix}, \qquad i = 1,\dots,N-1, \quad g(N) = d(N), \tag{31.160}$$

$$h(j) = \begin{pmatrix} 1 \\ 0_{(2m+1)\times 1} \end{pmatrix}, \qquad j = 1,\dots,N-1, \quad h(N) = 1, \tag{31.161}$$

$$b(k) = \begin{pmatrix} 0_{1\times 1} & 0_{1\times(2m+1)} \\ h_u(k+1) & b_u(k+1) \end{pmatrix}, k = 1,\dots,N-2, \quad b(N-1) = \begin{pmatrix} 0 \\ h_u(N) \end{pmatrix} \tag{31.162}$$

yield upper triangular generators of the matrix A of orders $r_k = 2m+2,\ k = 1,\dots,N-1,\ r_N = 1$.

We apply Theorem 31.4 to A. For the row vectors $g^{(1)}(k), \beta_k$ used in this theorem we introduce the partitions

$$\begin{gathered} g^{(1)}(k) = \begin{pmatrix} \gamma_k & \delta_k & -x^{(1)}(k) & y^{(1)}(k) & \epsilon_k \end{pmatrix},\ k = 1,\dots,N-2, \\ g^{(1)}(N-1) = \begin{pmatrix} \gamma_{N-1} & \delta_{N-1} \end{pmatrix}, \quad g^{(1)}(N) = \gamma_N, \end{gathered} \tag{31.163}$$

$$\begin{gathered} \beta_k = \begin{pmatrix} f_k & \phi_k \end{pmatrix}, \quad \phi_k = \begin{pmatrix} \varphi_k & -c_k & z_k & \rho_k \end{pmatrix}, \\ k = 2,\dots,N-1, \quad \beta_N = \begin{bmatrix} f_N & \varphi_N \end{bmatrix} \end{gathered} \tag{31.164}$$

with numbers $\gamma_k, \delta_k, \epsilon_k, f_k, \varphi_k, \rho_k$ and m-dimensional rows $x^{(1)}(k), y^{(1)}(k), c_k, z_k$.

Using (31.159)–(31.162) with $k = 1, 2$ we have

$$\begin{gathered} g(1)h(1) = d(1), \quad g(1)b(1) = \begin{pmatrix} g_u(1)h_u(2) & g_u(1)b_u(2) \end{pmatrix} = \begin{pmatrix} \varepsilon_1 & -x(1) & y(1) & 0 \end{pmatrix}, \\ g(2) = \begin{pmatrix} d(2) & -x(2) & y(2) & \sigma_2^* \end{pmatrix} \end{gathered}$$

with the number ε_1 as in (31.143). Inserting this in (31.68) and (31.69) and using (31.163), (31.164) we obtain the condition (31.142) and the equality

$$\begin{pmatrix} \gamma_1 & \delta_1 & -x^{(1)}(1) & y^{(1)}(1) & \epsilon_1 \\ f_2 & \varphi_2 & -c_2 & z_2 & \rho_2 \end{pmatrix} = Q_1^* \begin{pmatrix} d(1) & \varepsilon_1 & -x(1) & y(1) & 0 \\ \sigma_1 & d(2) & -x(2) & y(2) & \sigma_2^* \end{pmatrix}.$$

From here using the partition (31.144) we obtain (31.145), (31.146) and (31.147).

Next, using (31.164) and (31.161), (31.162) we get

$$\phi_{k+1}h(k+1) = \varphi_{k+1}, \quad k = 1,\dots,N-1, \tag{31.165}$$

and

$$\begin{aligned} \phi_{k+1}b(k+1) &= \begin{pmatrix} -c_{k+1} & z_{k+1} & \rho_{k+1} \end{pmatrix} \begin{bmatrix} h_u(k+2) & b_u(k+2) \end{bmatrix} \\ &= \begin{bmatrix} \varepsilon_{k+1} & -c_{k+1} & z_{k+1} & 0 \end{bmatrix},\ k = 1,\dots,N-3, \\ \phi_{N-1}b(N-1) &= \varepsilon_{N-1}, \end{aligned} \tag{31.166}$$

with ε_{k+1} as in (31.151).

Thus for $k = 1,\dots,N-2$ we have the following. Inserting (31.165) in (31.71) we obtain (31.149). Next, the condition (31.72) means the condition (31.150). Inserting (31.166) and (31.160) in (31.73) and using (31.163), (31.164) we obtain the equalities

$$\begin{aligned} &\begin{pmatrix} \gamma_{k+1} & \delta_{k+1} & -x^{(1)}(k+1) & y^{(1)}(k+1) & \epsilon_{k+1} \\ f_{k+2} & \varphi_{k+2} & -c_{k+2} & z_{k+2} & \rho_{k+2} \end{pmatrix} \\ &= Q_{k+1}^* \begin{pmatrix} \Omega_k(1,2) & \varepsilon_{k+1} & -c_{k+1} & z_{k+1} & 0 \\ \Omega_k(2,2) & d(k+2) & -x(k+2) & y(k+2) & \sigma_{k+2}^* \end{pmatrix}, \quad k = 1,\dots,N-3, \\ &\begin{pmatrix} \gamma_{N-1} & \delta_{N-1} \\ f_N & \varphi_N \end{pmatrix} = Q_{N-1}^* \begin{pmatrix} \Omega_{N-2}(1,2) & \varepsilon_{N-1} \\ \Omega_{N-2}(2,2) & d(N) \end{pmatrix}. \end{aligned} \tag{31.167}$$

From (31.167) using the partitions (31.152) we obtain the formulas (31.153), (31.154) and (31.155).

Further, using (31.57) we obtain the formulas

$$d^{(1)}(k) = g^{(1)}(k)h^{(1)}(k), \quad k = 1, \dots, N. \tag{31.168}$$

Using the formulas (31.75), (31.77), (31.161) and the partitions (31.144), (31.152), we get

$$h^{(1)}(k) = \begin{pmatrix} h_Q(k) & d_Q(k) & 0_{1\times(2m+1)} \end{pmatrix}^T, k = 1, \dots, N-2,$$
$$h^{(1)}(N-1) = \begin{pmatrix} h_Q(N-1) & d_Q(N-1) \end{pmatrix}^T, \quad h^{(1)}(N) = 1.$$

Combining this with (31.163), (31.168) we obtain the formulas (31.148), (31.156) and $\gamma_N = d^{(1)}(N)$.

Finally, using (31.76) and (31.163) with $k = N$ and (31.165) with $k = N-1$, we get

$$\begin{pmatrix} \sigma^{(1)}_{N-1} & \gamma_N \end{pmatrix} = \begin{pmatrix} f_N & \varphi_N \end{pmatrix} Q_{N-1}.$$

From here, since $\gamma_N = d^{(1)}(N)$ we obtain (31.157).

From the relations (31.140), (31.141) and the partitions (31.144), (31.152) it follows that the formulas (31.147), (31.155), (31.158) yield the rows of the perturbation matrices $X^{(1)} = Q^*X$ and $Y^{(1)} = Q^*Y$. □

The complexity of the operations in Step 2 of Theorem 31.6, which are performed almost N times, is $14m + 38$.

§31.3.2 Double shift

Here we apply Theorem 31.5 on the algorithm for the double shift case to matrices from the class $\mathcal{F}_{N,m}$.

Theorem 31.7. *Let A be a real matrix from the class $\mathcal{F}_{N,m}$ with diagonal entries $d(k)$ $(k = 1, \dots, N)$, subdiagonal entries σ_k $(k = 1, \dots, N-1)$, and perturbation matrices $X = \mathrm{col}(x(i))_{i=1}^N$, $Y = \mathrm{col}(y(i))_{i=1}^N$ and let s, t be real numbers.*

Then the 3×3 orthogonal matrices Q_k, $k = 1, \dots, N-2$ and the 2×2 orthogonal matrix Q_{N-1}, such that the orthogonal matrix

$$Q = \tilde{Q}_1\tilde{Q}_2\cdots\tilde{Q}_{N-2}\tilde{Q}_{N-1}, \tag{31.169}$$

with

$$\tilde{Q}_i = I_{i-1} \oplus Q_i \oplus I_{N-i-2},\ i = 1, \dots, N-2, \quad \tilde{Q}_{N-1} = I_{N-2} \oplus Q_{N-1}, \tag{31.170}$$

satisfies the condition (31.7), and the diagonal entries $d^{(1)}(k)$ $(k = 1, \dots, N)$, subdiagonal entries $\sigma^{(1)}_k$ $(k = 1, \dots, N-1)$ and perturbation matrices $X^{(1)} =$

$\operatorname{col}(x^{(1)}(i))_{i=1}^N$, $Y^{(1)} = \operatorname{col}(y^{(1)}(i))_{i=1}^N$ *of the upper Hessenberg matrix* $A^{(1)} = Q^*AQ$ *are determined as follows.*

1. *Set*

$$c_2 = \begin{pmatrix} x(1) \\ x(2) \end{pmatrix}, \quad z_2 = \begin{pmatrix} y(1) \\ y(1) \end{pmatrix}. \tag{31.171}$$

Compute

$$\begin{aligned} \varepsilon_1' &= -x(1)y^*(2) + y(1)x^*(2) + \sigma_1^*, \\ \chi_1 &= -x(1)y^*(3) + y(1)x^*(3), \\ \varepsilon_2' &= -x(2)y^*(3) + y(2)x^*(3) + \sigma_2^* \end{aligned} \tag{31.172}$$

and set

$$\Psi_1 = \begin{pmatrix} d(1) & \varepsilon_1' \\ \sigma_1 & d(2) \end{pmatrix}, \quad w_1 = \begin{pmatrix} 0 & \sigma_2 \end{pmatrix}, \quad \varepsilon_2 = \begin{pmatrix} \chi_1 \\ \varepsilon_2' \end{pmatrix}. \tag{31.173}$$

Set

$$a_{11} = d(1),\ a_{12} = \varepsilon_1',\ a_{21} = \sigma_1,\ a_{22} = d(2),\ a_{32} = \sigma_2 \tag{31.174}$$

and determine the 3×3 *orthogonal matrix* Q_1 *from the condition*

$$Q_1^* \begin{pmatrix} a_{11}^2 + a_{12}a_{21} - sa_{21} + t & a_{21}(a_{11} + a_{22} - s) & a_{32}a_{21} \end{pmatrix}^T = \begin{pmatrix} * & 0 & 0 \end{pmatrix}^T. \tag{31.175}$$

Determine the matrices $h_Q(1), b_Q(1), d_Q(1), g_Q(1)$ *of sizes* $2 \times 1, 2 \times 2, 1 \times 1, 1 \times 2$ *from the partition*

$$Q_1 = \begin{pmatrix} h_Q(1) & b_Q(1) \\ d_Q(1) & g_Q(1) \end{pmatrix} \tag{31.176}$$

and compute

$$\gamma_1 = h_Q^*(1)\Psi_1 + d_Q^*(1)w_1, \quad \delta_1 = h_Q^*(1)\varepsilon_2 + d_Q^*(1)d(3), \tag{31.177}$$

$$\begin{gathered} f_3 = b_Q^*(1)\Psi_1 + g_Q^*(1)w_1, \quad \varphi_3 = b_Q^*(1)\Psi_1 + g_Q^*(1)d(3), \\ \rho_3 = g_Q^*(1)\sigma_3^*, \end{gathered} \tag{31.178}$$

$$\begin{aligned} x^{(1)}(1) &= h_Q^*(1)c_2 + d_Q^*(1)x(3), \quad c_3 = b_Q^*(1)c_2 + g_Q^*(1)x(3), \\ y^{(1)}(1) &= h_Q^*(1)z_2 + d_Q^*(1)y(3), \quad z_3 = b_Q^*(1)z_2 + g_Q^*(1)y(3), \end{aligned} \tag{31.179}$$

$$d^{(1)}(1) = \gamma_1 h_Q(1) + \delta_1 d_Q(1). \tag{31.180}$$

Set $\sigma_N = 0$.

2. *For* $k = 1, \dots, N-3$, *perform the following.*

Compute the 3×3 *matrix* Ω_k *by the formula*

$$\Omega_k = \begin{pmatrix} f_{k+2} & \varphi_{k+2} \\ 0 & \sigma_{k+1} \end{pmatrix} Q_k. \tag{31.181}$$

Determine the complex Givens rotation matrix Q_{k+1} *and the number* $\sigma_k^{(1)}$ *such that*

$$Q_{k+1}^*\Omega_k(:,1) = \begin{pmatrix} \sigma_k^{(1)} & 0 & 0 \end{pmatrix}^T. \tag{31.182}$$

Compute

$$\varepsilon_{k+2} = -c_{k+2}y^*(k+2) + z_{k+2}x^*(k+2) + \rho_{k+2}. \tag{31.183}$$

Determine the matrices $h_Q(k+1), b_Q(k+1), d_Q(k+1), g_Q(k+1)$ *of sizes* $2\times 1, 2\times 2, 1\times 1, 1\times 2$ *from the partition*

$$Q_{k+1} = \begin{pmatrix} h_Q(k+1) & b_Q(k+1) \\ d_Q(k+1) & g_Q(k+1) \end{pmatrix}, \tag{31.184}$$

and compute

$$\begin{aligned} \gamma_{k+1} &= h_Q^*(k+1)\Omega_k(1:2,2:3) + d_Q^*(k+1)\Omega_k(3,2:3), \\ \delta_{k+1} &= h_Q^*(k+1)\varepsilon_{k+2} + d_Q^*(k+1)d(k+3), \end{aligned} \tag{31.185}$$

$$\begin{aligned} f_{k+3} &= b_Q^*(k+1)\Omega_k(1:2,2:3) + g_Q^*(k+1)\Omega_k(3,2:3), \\ \varphi_{k+3} &= b_Q^*(k+1)\varepsilon_{k+2} + g_Q^*(k+1)d(k+3), \quad \rho_{k+3} = g_Q^*(k+1)\sigma_{k+2}^*, \end{aligned} \tag{31.186}$$

$$\begin{aligned} x^{(1)}(k+1) &= h_Q^*(k+1)c_{k+2} + d_Q^*(k+1)x(k+3), \\ c_{k+3} &= b_Q^*(k+1)c_{k+2} + g_Q^*(k+1)x(k+3), \\ y^{(1)}(k+1) &= h_Q^*(k+1)z_{k+2} + d_Q^*(k+1)y(k+3), \\ z_{k+3} &= b_Q^*(k+1)z_{k+2} + g_Q^*(k+1)y(k+3), \end{aligned} \tag{31.187}$$

$$d^{(1)}(k+1) = \gamma_{k+1}h_Q(k+1) + \delta_{k+1}d_Q(k+1). \tag{31.188}$$

3. *Compute the* 2×3 *matrix* Ω_{N-2} *by the formula*

$$\Omega_{N-2} = \begin{pmatrix} f_N & \varphi_N \end{pmatrix} Q_{N-2}. \tag{31.189}$$

Determine the 2×2 *orthogonal matrix* Q_{N-1} *and the number* $\sigma_{N-2}^{(1)}$ *such that*

$$Q_{N-1}^*\Omega_{N-2}(:,1) = \begin{pmatrix} \sigma_{N-2}^{(1)} & 0 \end{pmatrix}^T. \tag{31.190}$$

Determine the column vectors $h_Q(N-1), b_Q(N-1)$ *from the partition*

$$Q_{N-1} = \begin{pmatrix} h_Q(N-1) & b_Q(N-1) \end{pmatrix}. \tag{31.191}$$

Compute

$$d^{(1)}(N-1) = h_Q^*(N-1)\Omega(:,2:3)h_Q(N-1), \tag{31.192}$$

$$\beta_{N+1} = b_Q^*(N-1)\Omega(:,2:3), \tag{31.193}$$

$$\sigma_{N-1}^{(1)} = \beta_{N+1}h_Q(N-1), \tag{31.194}$$

$$d^{(1)}(N) = \beta_{N+1}b_Q(N-1), \tag{31.195}$$

and

$$\begin{pmatrix} x^{(1)}(N-1) \\ x^{(1)}(N) \end{pmatrix} = Q^*_{N-1}c_N, \quad \begin{pmatrix} y^{(1)}(N-1) \\ y^{(1)}(N) \end{pmatrix} = Q^*_{N-1}z_N. \tag{31.196}$$

Proof. In the same way as in the proof of Theorem 31.6 we obtain the formulas (31.171), (31.159), (31.160), (31.161), (31.162) for determining upper triangular generators of the matrix A.

We apply Theorem 31.5 to the matrix A. For the row vectors $g^{(1)}(k)$ and the matrices β_k used in this theorem we introduce the partitions

$$\begin{aligned} g^{(1)}(k) &= \begin{pmatrix} \gamma_k & \delta_k & -x^{(1)}(k) & y^{(1)}(k) & \epsilon_k \end{pmatrix}, \; k = 1, \dots, N-3, \\ g^{(1)}(N-2) &= \begin{pmatrix} \gamma_{N-2} & \delta_{N-2} \end{pmatrix}, \end{aligned} \tag{31.197}$$

$$\begin{aligned} \beta_k = \begin{pmatrix} f_k & \phi_k \end{pmatrix}, \quad \phi_k = \begin{pmatrix} \varphi_k & -c_k & z_k & \rho_k \end{pmatrix}, \\ k = 3, \dots, N-1, \quad \beta_N = \begin{pmatrix} f_N & \varphi_N \end{pmatrix} \end{aligned} \tag{31.198}$$

with matrices $\gamma_k, \delta_k, x^{(1)}(k), y^{(1)}(k), \epsilon_k, f_k, \varphi_k, c_k, z_k, \rho_k$ of sizes $1 \times 2, 1 \times 1, 1 \times m, 1 \times m, 1 \times 1, 2 \times 2, 2 \times 1, 2 \times m, 2 \times m, 2 \times 1$, respectively. Using (31.159)–(31.162) with $k = 1, 2, 3$ we have

$$\begin{aligned} g(1)h(1) = d(1),\; g(1)b(1) &= \begin{pmatrix} \varepsilon'_1 & -x(1) & y(1) & 0 \end{pmatrix},\; g(1)b(1)h(2) = \varepsilon'_1, \\ g(1)b(1)b(2) &= \begin{pmatrix} \chi_1 & -x(1) & y(1) & 0 \end{pmatrix},\; g(2)h(2) = d(2), \\ g(2)b(2) &= \begin{pmatrix} \varepsilon'_2 & -x(2) & y(2) & 0 \end{pmatrix},\; g(3) = \begin{pmatrix} d(3) & -x(3) & y(3) & \sigma^*_3 \end{pmatrix} \end{aligned}$$

with the numbers $\varepsilon'_1, \chi_1, \varepsilon'_2$ as in (31.172). Inserting this in (31.101), (31.102) and (31.103) and using (31.197), (31.198), we obtain (31.175) and the equality

$$\begin{pmatrix} \gamma_1 & \delta_1 & -x^{(1)}(1) & y^{(1)}(1) & \epsilon_1 \\ f_3 & \varphi_3 & -c_3 & z_3 & \rho_3 \end{pmatrix} = Q^*_1 \begin{pmatrix} d(1) & \varepsilon'_1 & \chi_1 & -x(1) & y(1) & 0 \\ \sigma_1 & d(2) & \varepsilon'_2 & -x(2) & y(2) & 0 \\ 0 & \sigma_2 & d(3) & -x(3) & y(3) & \sigma^*_3 \end{pmatrix}.$$

From here, using the partition (31.176), we obtain the formulas (31.177), (31.178) and (31.179). Next, using (31.198) and (31.161), (31.162) we get

$$\phi_{k+2}h(k+2) = \varphi_{k+2}, \quad k = 1, \dots, N-2 \tag{31.199}$$

and

$$\begin{aligned} \phi_{k+2}b(k+2) = \begin{pmatrix} \varepsilon_{k+2} & -c_{k+2} & z_{k+2} & 0 \end{pmatrix},\; k = 1, \dots, N-4, \\ \phi_{N-1}b(N-1) = \varepsilon_{N-1}, \end{aligned} \tag{31.200}$$

with ε_{k+2} as in (31.183).

Thus, for $k = 1, \ldots, N-3$ we have the following. Inserting (31.199) in (31.105) we obtain the equality (31.181). Next, the condition (31.106) is equivalent to (31.182). Inserting (31.200) and (31.160) in (31.107) and using the partitions (31.108), (31.109) and (31.197), (31.198), we obtain the equalities

$$\begin{pmatrix} \gamma_{k+1} & \delta_{k+1} & -x^{(1)}(k+1) & y^{(1)}(k+1) & \epsilon_{k+1} \\ f_{k+3} & \varphi_{k+3} & -c_{k+3} & z_{k+3} & \rho_{k+3} \end{pmatrix}$$
$$= Q_{k+1}^* \begin{pmatrix} \Omega_k(1:2,2:3) & \varepsilon_{k+2} & -c_{k+2} & z_{k+2} & 0_{2\times 1} \\ \Omega_k(3,2:3) & d(k+3) & -x(k+3) & z(k+3) & \sigma_{k+3}^* \end{pmatrix},$$
$$k = 1, \ldots, N-4, \tag{31.201}$$
$$\begin{pmatrix} \gamma_{N-2} & \delta_{N-2} \\ f_N & \varphi_N \end{pmatrix} = Q_{N-2}^* \begin{pmatrix} \Omega_{N-2}(1:2,2:3) & \varepsilon_{N-1} \\ \Omega_{N-2}(3,2:3) & d(N) \end{pmatrix},$$

which, using the partitions (31.184) yields (31.185), (31.186) and (31.187).

Next, using (31.111) and (31.199) with $k = N-2$ we obtain the formula (31.189). The condition (31.112) is equivalent to (31.190). Using (31.113), (31.114) and the partition (31.191), we obtain (31.193) and the formula

$$g^{(1)}(N-1) = h_Q^*(N-1)\Omega_{N-2}(:,2:3). \tag{31.202}$$

Using (31.116) and the partition (31.191), we obtain (31.194) and the formula

$$g^{(1)}(N) = \beta_{N+1} b_Q(N-1). \tag{31.203}$$

Further, (31.57) yields

$$d^{(1)}(k) = g^{(1)}(k)h^{(1)}(k), \quad k = 1, \ldots, N. \tag{31.204}$$

Using the formulas (31.110), (31.115), (31.118) and (31.161) we get

$$h^{(1)}(k) = \begin{pmatrix} h_Q(k) \\ d_Q(k) \\ 0_{(2m+1)\times 1} \end{pmatrix}, \ k = 1, \ldots, N-2, \ h^{(1)}(N-1) = h_Q(N-1), \ h^{(1)}(N) = 1.$$

Combining this with (31.197), (31.202), (31.203) and (31.204) we obtain the formulas (31.180), (31.188), (31.192) and (31.195) for the diagonal entries of the matrix $A^{(1)}$.

From the relations (31.169), (31.170) and the partitions (31.176), (31.184) it follows that the formulas (31.171), (31.179), (31.187) and (31.196) yield the rows of the perturbation matrices $X^{(1)} = Q^*X$ and $Y^{(1)} = Q^*Y$. □

The complexity of the operations in Step 2 of Theorem 31.7, which are performed almost N times, is $28m + 85$.

§31.4 Perturbations of unitary matrices

We denote by $\mathcal{H}_{N,m}$ the class of upper Hessenberg $N \times N$ matrices A which are rank m perturbations of unitary matrices, i.e.,

$$A = U - XY^*, \tag{31.205}$$

where U is a unitary matrix and $X = \mathrm{col}(x(i))_{i=1}^N$, $Y = \mathrm{col}(y(i))_{i=1}^N$ with m-dimensional rows $x(i), y(i)$.

Set

$$m_1 = \cdots = m_N = 1,\ m_{N+1} = 0, \quad n_1 = 0,\ n_2 = \cdots = n_{N+1} = 1. \tag{31.206}$$

One can treat A and U as block matrices with entries of sizes $m_i \times n_j$, $i, j = 1, \dots, N+1$. Using (7.8) we obtain the formulas

$$\rho_k^L + 1 = \rho_k^U, \quad k = 1, \dots, N$$

for the lower and upper rank numbers ρ_k^L and ρ_k^U of the unitary matrix U relative to the partition (31.206). Since relative to this partition the matrix A is upper triangular, we get $\rho_k^L \leq m$, $k = 1, \dots, N$, and therefore $\rho_k^U \leq m+1$, $k = 1, \dots, N$. Hence it follows that the matrix U treated as a scalar matrix has upper triangular generators with orders not greater than $m+1$.

Thus every matrix A from the class $\mathcal{H}_{N,m}$ is completely defined by the following parameters:

1. the subdiagonal entries σ_k $(k = 1, \dots, N-1)$ of the matrix A;
2. the perturbation matrices $X = \mathrm{col}(x(i))_{i=1}^N$, $Y = \mathrm{col}(y(i))_{i=1}^N$;
3. upper triangular generators $g_U(i)$, $h_U(i)$ $(i = 1, \dots, N)$, $b_U(k)$ $(k = 1, \dots, N-1)$ of the matrix U.

Let Q be a unitary matrix such that $A^{(1)} = Q^* A Q$ is an upper Hessenberg matrix. We obviously have

$$A^{(1)} = U^{(1)} - X^{(1)}(Y^{(1)})^* \tag{31.207}$$

with the unitary matrix $U^{(1)} = Q^* U Q$ and the perturbation matrices $X^{(1)} = Q^* X$, $Y^{(1)} = Q^* Y$. Hence, the matrix $A^{(1)}$ belongs to the class $\mathcal{H}_{N,m}$.

§31.4.1 Single shift

Based on general Theorem 31.4 and Theorem 7.5 for quasiseparable generators of unitary matrices, we obtain the following algorithm for matrices from the class $\mathcal{H}_{N,m}$.

Theorem 31.8. *Let A be a matrix from the class $\mathcal{H}_{N,m}$ with subdiagonal entries σ_k $(k = 1, \dots, N-1)$, perturbation matrices $X = \mathrm{col}(x(i))_{i=1}^N$, $Y = \mathrm{col}(y(i))_{i=1}^N$, upper triangular generators $g_U(i), h_U(i)$ $(i = 1, \dots, N)$, $b_U(k)$ $(k = 1, \dots, N-1)$ of orders r_k $(k = 1, \dots, N)$ of the matrix U, and let α be a number.*

Then the Givens rotation matrices Q_k, $k = 1, \dots, N-1$, such that the unitary matrix

$$Q = \tilde{Q}_1\tilde{Q}_2\cdots\tilde{Q}_{N-2}\tilde{Q}_{N-1}, \tag{31.208}$$

with

$$\tilde{Q}_i = I_{i-1} \oplus Q_i \oplus I_{N-i-1},\ i = 1, \dots, N-1, \tag{31.209}$$

satisfies the condition (31.4), and the diagonal entries $d^{(1)}(k)$ $(k = 1, \dots, N)$, subdiagonal entries $\sigma_k^{(1)}$ $(k = 1, \dots, N-1)$, of the upper Hessenberg matrix $A^{(1)} = Q^*AQ$, the perturbation matrices $X^{(1)} = \mathrm{col}(x^{(1)}(i))_{i=1}^N$, $Y^{(1)} = \mathrm{col}(y^{(1)}(i))_{i=1}^N$, upper triangular generators $g_U^{(1)}(i), h_U^{(1)}(i)$ $(i = 1, \dots, N)$, $b_U^{(1)}(k)$ $(k = 1, \dots, N-1)$ of the unitary matrix $U^{(1)}$ from (31.207) with orders not greater than $m+1$ are determined as follows:

1.1. *Compute*

$$\sigma_1^U = \sigma_1 + x(2)y^*(1), \tag{31.210}$$

$$\varepsilon_1 = g_U(1)h_U(1) - x(1)y^*(1), \quad \theta_1 = y(1). \tag{31.211}$$

Determine the complex Givens transformation matrix Q_1 from the condition

$$Q_1^* \begin{pmatrix} \varepsilon_1 - \alpha \\ \sigma_1 \end{pmatrix} = \begin{pmatrix} * \\ 0 \end{pmatrix}. \tag{31.212}$$

Compute

$$\Lambda_2 = Q_1^* \begin{pmatrix} g_U(1)h_U(1) & g_U(1)b_U(1) \\ \sigma_1^U & g_U(2) \end{pmatrix} \tag{31.213}$$

and determine the matrices $g_B(2), \beta_2^U, f_2^U, \phi_2^U$ of sizes $1 \times 1, 1 \times (r_2+1), 1 \times (r_2+1), 1 \times 1, 1 \times r_2$ from the partitions

$$\Lambda_2 = \begin{pmatrix} g_B(2) \\ \beta_2^U \end{pmatrix}, \quad \beta_2^U = \begin{pmatrix} f_2^U & \phi_2^U \end{pmatrix}. \tag{31.214}$$

Compute

$$\begin{pmatrix} x^{(1)}(1) \\ c_2 \end{pmatrix} = Q_1^* \begin{pmatrix} x(1) \\ x(2) \end{pmatrix}, \tag{31.215}$$

$$\begin{pmatrix} y^{(1)}(1) \\ \theta_2 \end{pmatrix} = Q_1^* \begin{pmatrix} y(1) \\ y(2) \end{pmatrix}, \tag{31.216}$$

with the m-dimensional rows $x^{(1)}(1), c_2, y^{(1)}(1), \theta_2$.

Compute

$$f_2 = f_2^U - c_2 y^*(1), \quad \varphi_2 = \phi_2^U. \tag{31.217}$$

1.2. *For $k = 1, \dots, N-2$, perform the following. Compute*

$$\sigma_{k+1}^U = \sigma_{k+1} + x(k+2)y^*(k+1), \tag{31.218}$$
$$\varepsilon_{k+1} = \varphi_{k+1} h_U(k+1) - c_{k+1} y^*(k+1). \tag{31.219}$$

Compute the 2×2 matrix Ω_k by the formula

$$\Omega_k = \begin{pmatrix} f_{k+1} & \varepsilon_{k+1} \\ 0 & \sigma_{k+1} \end{pmatrix} Q_k \tag{31.220}$$

and determine the complex Givens rotation matrix Q_{k+1} and the number $\sigma_k^{(1)}$ such that

$$Q_{k+1}^* \Omega_k(:,1) = \begin{pmatrix} \sigma_k^{(1)} \\ 0 \end{pmatrix}. \tag{31.221}$$

Compute

$$\Lambda_{k+2} = Q_{k+1}^* \begin{pmatrix} f_{k+1}^U & \phi_{k+1}^U h_U(k+1) & \phi_{k+1}^U b_U(k+1) \\ x(k+2)\theta_k^* & \sigma_{k+1}^U & g_U(k+2) \end{pmatrix} \begin{pmatrix} Q_k & 0 \\ 0 & I_{r_{k+2}} \end{pmatrix}, \tag{31.222}$$

$$B_{k+2} = \begin{pmatrix} 1 & 0 & 0 \\ 0 & h_U(k+1) & b_U(k+1) \end{pmatrix} \begin{pmatrix} Q_k & 0 \\ 0 & I_{r_{k+2}} \end{pmatrix}, \tag{31.223}$$

and determine the matrices $d_B(k+2), g_B(k+2), \beta_{k+2}^U, f_{k+2}^U, \phi_{k+2}^U$ of sizes $1\times 1, 1\times (r_{k+2}+1), 1\times(r_{k+2}+1), 1\times 1, 1\times r_{k+2}$ from the partitions

$$\Lambda_{k+2} = \begin{pmatrix} d_B(k+2) & g_B(k+2) \\ * & \beta_{k+2}^U \end{pmatrix}, \quad \beta_{k+2}^U = \begin{pmatrix} f_{k+2}^U & \phi_{k+2}^U \end{pmatrix}. \tag{31.224}$$

Compute

$$\begin{pmatrix} x^{(1)}(k+1) \\ c_{k+2} \end{pmatrix} = Q_{k+1}^* \begin{pmatrix} c_{k+1} \\ x(k+2) \end{pmatrix}, \tag{31.225}$$

$$\begin{pmatrix} y^{(1)}(k+1) \\ \theta_{k+2} \end{pmatrix} = Q_{k+1}^* \begin{pmatrix} \theta_{k+1} \\ y(k+2) \end{pmatrix}, \tag{31.226}$$

with the m-dimensional rows $x^{(1)}(k+1), c_{k+2}, y^{(k+1)}(1), \theta_{k+2}$.

Compute

$$f_{k+2} = f_{k+2}^U - c_{k+2}\theta_{k+1}^*, \quad \varphi_{k+2} = \phi_{k+2}^U. \tag{31.227}$$

1.3. *Compute*

$$\Lambda_{N+1} = \begin{pmatrix} f_N^U & \phi_N^U h_U(N) \end{pmatrix} Q_{N-1}, \tag{31.228}$$

$$B_{N+1} = \begin{pmatrix} 1 & 0 \\ 0 & h_U(N) \end{pmatrix} Q_{N-1}, \tag{31.229}$$

and determine the numbers $d_B(N+1), g_B(N+1)$ *from the partition*

$$\Lambda_{N+1} = \begin{pmatrix} d_B(N+1) & g_B(N+1) \end{pmatrix}. \tag{31.230}$$

Set

$$x^{(1)}(N) = c_N, \quad y^{(1)}(N) = \theta_N, \tag{31.231}$$

and compute

$$\sigma_{N-1}^{(1)} = d_B(N+1) - x^{(1)}(N)(y^{(1)}(N-1))^*. \tag{31.232}$$

2. *Set*

$$\begin{aligned}
\rho_{N+2} &= \rho_{N+1} = 0, \ \rho_N = 1, \ldots, \rho_{N-m+2} = m-1, \\
\rho_k &= m, \ k = N-m+1, \ldots, 1, \\
\nu'_{N+2} &= \cdots = \nu'_{N-m+2} = 0, \\
\nu'_{N-m+1} &= \cdots = \nu'_2 = 1, \quad \nu'_1 = m, \\
s_{N+2} &= 0, \ s_{N+1} = 1, \ldots, s_{N-m+2} = m, \\
s_{N-m+1} &= \cdots = s_2 = m+1, \\
r_k^U &= r_k + 1, \ k = 2, \ldots, N, \quad r_{N+1}^U = 1.
\end{aligned} \tag{31.233}$$

Set X_{N+2}, z_{N+2} *to be the* $0 \times m, 0 \times 1$ *empty matrices and*

$$h_s(N+2) = 1, \quad Y_{N+2} = 1, \quad g_s(N+1) = g_B(N+1), \tag{31.234}$$

and for $k = N+1, \ldots, 3$ *perform the following:*
Compute the QR *factorizations*

$$V_k^* \begin{pmatrix} x^{(1)}(k-1) \\ X_{k+1} \end{pmatrix} = \begin{pmatrix} X_k \\ 0_{\nu'_k \times m} \end{pmatrix} \tag{31.235}$$

with unitary $(1+\rho_k) \times (1+\rho_k)$ *matrices* V_k *and auxiliary* $\rho_{k-1} \times m$ *matrices* X_k.
Compute the $(1+\rho_k) \times (1+s_k)$ *matrix*

$$Z_k = V_k^* \begin{pmatrix} d_B(k) & g_s(k) \\ X_{k+1}(y^{(1)}(k-2))^* & z_{k+1} \end{pmatrix} \tag{31.236}$$

and determine the matrices $h'_k, h''_k, \Delta_k, \Theta_k$ *of sizes* $\rho_{k-1} \times 1, \rho_{k-1} \times s_k, \nu'_k \times \nu_k, \nu'_k \times s_k$ *from the partition*

$$Z_k = \begin{pmatrix} h'_k & h''_k \\ \Delta_k & \Theta_k \end{pmatrix}. \tag{31.237}$$

The submatrix $\begin{pmatrix} \Delta_k & \Theta_k \end{pmatrix}$ *has orthonormal rows and one can determine the unitary* $(1+s_k) \times (1+s_k)$ *matrix* F_k *from the condition*

$$\begin{pmatrix} \Delta_k & \Theta_k \end{pmatrix} F_k^* = \begin{pmatrix} 0_{\nu'_k \times s_{k-1}} & I_{\nu'_k} \end{pmatrix}. \tag{31.238}$$

Determine the matrices $h_s(k), b_s(k), d_F(k), g_F(k)$ *of sizes* $s_{k-1} \times 1, s_{k-1} \times s_k, \nu'_k \times 1, \nu'_k \times s_k$ *from the partition*

$$F_k = \begin{pmatrix} h_s(k) & b_s(k) \\ d_F(k) & g_F(k) \end{pmatrix}. \tag{31.239}$$

Compute the matrices Y_k *of the size* $r^U_{k-1} \times s_{k-1}$ *and* z_k *of the size* $\rho_{k-1} \times s_{k-1}$ *by the formulas*

$$Y_k = B_k \begin{pmatrix} h^*_s(k) \\ Y_{k+1} b^*_s(k) \end{pmatrix}, \quad z_k = h'_k h^*_s(k) + h''_k b^*_s(k). \tag{31.240}$$

Compute

$$g_s(k-1) = g_B(k-1) Y_k. \tag{31.241}$$

Set

$$\begin{aligned} &g^{(1)}_U(i) = g_s(i+1),\ h^{(1)}_U(i) = h_s(i+2), \quad && i = 1, \dots, N, \\ &b^{(1)}_U(k) = b_s(k+2), && k = 1, \dots, N-1. \end{aligned} \tag{31.242}$$

Proof. By (31.205), we obtain upper triangular generators $g(i)$, $h(i)$ $(i = 1, \dots, N-1)$, $b(k)$ $(k = 1, \dots, N-1)$ of orders $r_k = m + r_k$, $k = 1, \dots, N$ of the matrix A by the formulas

$$g(i) = \begin{pmatrix} g_U(i) & -x(i) \end{pmatrix}, \quad i = 1, \dots, N, \tag{31.243}$$

$$h(i) = \begin{pmatrix} h_U(i) \\ y^*(i) \end{pmatrix}, \ i = 1, \dots, N, \quad b(k) = \begin{pmatrix} b_U(k) & 0 \\ 0 & I_m \end{pmatrix}, \ k = 1, \dots, N-1. \tag{31.244}$$

We apply Theorem 31.4 to the matrix A. For the row vectors β_k used in this theorem we introduce the partitions

$$\beta_k = \begin{pmatrix} f_k & \phi_k \end{pmatrix}, \ \phi_k = \begin{pmatrix} \varphi_k & -c_k \end{pmatrix}, \ k = 2, \dots, N-1, \quad \beta_N = \begin{pmatrix} f_N & \varphi_N \end{pmatrix} \tag{31.245}$$

with numbers f_k, r_k-dimensional rows φ_k and m-dimensional rows c_k.

Using (31.243), (31.244) with $k = 1, 2$ we have

$$g(1)h(1) = \varepsilon_1, \quad g(1)b(1) = \begin{pmatrix} g_U(1)b_U(1) & -x(1) \end{pmatrix}, \quad g(2) = \begin{pmatrix} g_U(2) & -x(2) \end{pmatrix}$$

with the number ε_1 as in (31.211). Inserting this in (31.68) and (31.69) and using (31.245) we obtain the condition (31.212) and the equality

$$\begin{pmatrix} * & * & * \\ f_2 & \varphi_2 & -c_2 \end{pmatrix} = Q^*_1 \begin{pmatrix} \varepsilon_1 & g_U(1)b_U(1) & -x(1) \\ \sigma_1 & g_U(2) & -x(2) \end{pmatrix}. \tag{31.246}$$

Next, using (31.245) and (31.244) we get

$$\phi_{k+1} h(k+1) = \varepsilon_{k+1}, \quad k = 1, \dots, N-1 \tag{31.247}$$

and

$$
\begin{aligned}
&\phi_{k+1}b(k+1) = \begin{pmatrix} \varphi_{k+1}b_U(k+1) & -c_{k+1} \end{pmatrix}, \ k = 1, \dots, N-3, \\
&\phi_{N-1}b(N-1) = \varepsilon_{N-1},
\end{aligned}
\tag{31.248}
$$

with ε_{k+1} as in (31.219).

Thus, for $k = 1, \dots, N-2$ we have the following. Inserting (31.247) in (31.71) we obtain (31.220). Next the condition (31.72) is equivalent to the condition (31.221). Inserting (31.248) and (31.243) in (31.73) we obtain the equalities

$$
\begin{aligned}
&\begin{pmatrix} * & * & * \\ f_{k+2} & \varphi_{k+2} & -c_{k+2} \end{pmatrix} \\
&= Q^*_{k+1} \begin{pmatrix} \Omega_k(1,2) & \varphi_{k+1}b_U(k+1) & -c_{k+1} \\ \Omega_k(2,2) & g_U(k+2) & -x(k+2) \end{pmatrix}, \quad k = 1, \dots, N-3.
\end{aligned}
\tag{31.249}
$$

From the relations (31.208), (31.209) it follows that the formulas (31.215), (31.216), (31.225), (31.226) and (31.231) yield the rows of the perturbation matrices $X^{(1)} = Q^*X$ and $Y^{(1)} = Q^*Y$.

Next, consider the unitary matrix $U^{(1)} = Q^*UQ$. We treat the matrix U in the same way as the matrix A in Theorem 31.4, i.e., as a block matrix with entries of sizes $m_i \times n_j$, $i, j = 1, \dots, N+2$, where the numbers m_i, n_j are defined in (31.78). Relative to this partition, U is a block matrix with the diagonal entries

$$
\begin{aligned}
&d_U(1) \text{ is the } 1 \times 0 \text{ empty matrix}, \\
&d_U(k) = \sigma^U_{k-1} = \sigma_{k-1} + x(k)y^*(k-1), \ k = 2, \dots, N, \\
&d_U(N+1), d_U(N+2) \text{ are the } 0 \times 1, 0 \times 0 \text{ empty matrices},
\end{aligned}
\tag{31.250}
$$

upper quasiseparable generators

$$
\begin{aligned}
&\hat{g}_U(k) = g_U(k), \quad k = 1, \dots, N, && \hat{g}_U(N+1) \text{ is the } 0 \times 0 \text{ empty matrix}, \\
&\hat{h}_U(k) = h_U(k-1), \ k = 2, \dots, N+1, && \hat{h}_U(N+2) \text{ is the } 0 \times 0 \text{ empty matrix}, \\
&\hat{b}_U(k) = b_U(k-1), \ k = 2, \dots, N, && \hat{b}_U(N+1) \text{ is the } r_N \times 0 \text{ empty matrix}
\end{aligned}
\tag{31.251}
$$

and lower quasiseparable generators

$$
\begin{aligned}
&p_U(k) = x(k), \ k = 2, \dots, N, \\
&p_U(N+1), p_U(N+2) \text{ are the } 0 \times m \text{ empty matrices}, \\
&q_U(1) \text{ is the } m \times 0 \text{ empty matrix}, \\
&q_U(k) = y^*(k-1), \ k = 2, \dots, N+1, \\
&a_U(k) = I_m, \ k = 2, \dots, N+1.
\end{aligned}
\tag{31.252}
$$

For the matrices Q^*, Q we use the factorizations (31.83).

We apply Lemma 31.1 to compute (block) upper quasiseparable generators $g_B(i)$ $(i = 1, \dots, N+1)$, $h_B(j)$ $(j = 2, \dots, N+2)$, $b_B(k)$ $(k = 2, \dots, N+1)$ with orders $\tilde{r}_1 = r_1$, $\tilde{r}_k = r_k + 1$, $k = 2, \dots, N$, $\tilde{r}_{N+1} = 1$, of the matrix $U^{(1)} = Q^*UQ$. The matrix $U^{(1)}$ is obtained as a block one with entries of sizes $\tau_i \times \nu_j$, $i, j = 1, \dots, N+2$, where the numbers τ_i, ν_j are defined in (31.86), (31.87). Using (31.9)–(31.11) we obtain that $g_B(1)$ is the $0 \times m$ empty matrix, $\phi_1^U = g_U(1)$ and f_1^U is the 1×0 empty matrix. Using (31.12)–(31.14), (31.84), (31.85) and (31.250)–(31.252) we get

$$\begin{pmatrix} g_B(2) \\ \Gamma_2 \end{pmatrix} = \begin{pmatrix} Q_1^* & 0 \\ 0 & I_m \end{pmatrix} \begin{pmatrix} \phi_1^U h_U(1) & \phi_1^U b_U(1) \\ \sigma_1^U & g_U(2) \\ y^*(1) & 0 \end{pmatrix}, \quad \Gamma_2 = \begin{pmatrix} f_2^U & \phi_2^U \\ \alpha_2 & 0 \end{pmatrix}, \tag{31.253}$$

$$\begin{pmatrix} d_B(k) & g_B(k) \\ * & \Gamma_k \end{pmatrix}$$

$$= \begin{pmatrix} Q_{k-1}^* & 0 \\ 0 & I_m \end{pmatrix} \begin{pmatrix} f_{k-1}^U & \phi_{k-1}^U h_U(k-1) & \phi_{k-1}^U b_U(k-1) \\ x(k)\alpha_{k-1} & \sigma_{k-1}^U & g_U(k) \\ \alpha_{k-1} & y^*(k-1) & 0 \end{pmatrix} \begin{pmatrix} Q_{k-2} & 0 \\ 0 & I_{r_k} \end{pmatrix},$$

$$\Gamma_k = \begin{pmatrix} f_k^U & \phi_k^U \\ \alpha_k & 0 \end{pmatrix}, \quad k = 3, \dots, N, \tag{31.254}$$

$$\begin{pmatrix} d_B(N+1) & g_B(N+1) \end{pmatrix} = \begin{pmatrix} f_N^U & \phi_N^U \end{pmatrix} Q_{N-1}, \tag{31.255}$$

and also

$$\begin{pmatrix} h_B(k) & b_B(k) \end{pmatrix} = \begin{pmatrix} 1 & 0 & 0 \\ 0 & h_U(k-1) & b_U(k-1) \end{pmatrix} \begin{pmatrix} Q_{k-2} & 0 \\ 0 & I_{r_k} \end{pmatrix}, \quad k = 3, \dots, N, \tag{31.256}$$

$$\begin{pmatrix} h_B(N+1) & b_B(N+1) \end{pmatrix} = \begin{pmatrix} 1 & 0 \\ 0 & h_U(N) \end{pmatrix} Q_{N-1}. \tag{31.257}$$

Using (31.17) and (31.18) we get $h_B(N+2) = 1$.

The relations (31.253), (31.254) imply, in particular, that the auxiliary matrices α_k $(k = 2, \dots, N)$ satisfy the relations

$$\alpha_2 = y^*(1), \quad \begin{pmatrix} * & \alpha_k \end{pmatrix} = \begin{pmatrix} \alpha_{k-1} & y^*(k-1) \end{pmatrix} Q_{k-2}, \ k = 3, \dots, N.$$

Comparing this with (31.216), (31.226) we get

$$\alpha_k = \theta_{k-1}^*, \quad k = 2, \dots, N. \tag{31.258}$$

Thus, using(31.253) we obtain (31.213), (31.214), and inserting (31.258) in (31.254) we obtain (31.222), (31.224).

Next we show that the auxiliary variables f_k, φ_k $(k = 2, \dots, N)$ may be determined via relations (31.217), (31.227). Indeed, combining (31.246) and (31.210), (31.211), we get

$$\begin{pmatrix} * & * \\ f_2 & \varphi_2 \end{pmatrix} = Q_1^* \begin{pmatrix} g_U(1)h_U(1) - x(1)y^*(1) & g_U(1)b_U(1) \\ \sigma_1^U - x(1)y^*(1) & g_U(2) \end{pmatrix},$$

whence, using (31.213), (31.214) and (31.215), we obtain (31.217). Set $\theta_1 = y(1)$ and assume that for some k with $1 \le k \le N-2$ the relations

$$f_{k+1} = f^U_{k+1} - c_{k+1}\theta^*_k, \quad \varphi_{k+1} = \phi^U_{k+1} \tag{31.259}$$

hold. By (31.220) and (31.249), we have

$$\begin{pmatrix} \sigma^{(1)}_k & * & * \\ 0 & f_{k+2} & \varphi_{k+2} \end{pmatrix} = Q^*_{k+1} \begin{pmatrix} f_{k+1} & \varepsilon_{k+1} & \varphi_{k+1}b_U(k+1) \\ 0 & \sigma_{k+1} & g_U(k+2) \end{pmatrix} \begin{pmatrix} Q_k & 0 \\ 0 & I_{r_{k+2}} \end{pmatrix}. \tag{31.260}$$

Using (31.218), (31.219) and (31.259) we get

$$\begin{pmatrix} f_{k+1} & \varepsilon_{k+1} \\ 0 & \sigma_{k+1} \end{pmatrix} = \begin{pmatrix} f^U_{k+1} - c_{k+1}\theta^*_k & \phi^U_{k+1}h_U(k+1) - c_{k+1}y^*(k+1) \\ x(k+2)\theta^*_k - x(k+2)\theta^*_k & \sigma^U_{k+1} - x(k+2)y^*(k+1) \end{pmatrix}. \tag{31.261}$$

Thus, combining (31.260) and (31.261) and using (31.222), (31.224) and (31.225), (31.226) we obtain (31.227).

Using (31.256), (31.257) and setting

$$B_k = \begin{pmatrix} h_B(k) & b_B(k) \end{pmatrix}, \quad k = 3, \dots, N+1, \tag{31.262}$$

we obtain (31.223), (31.229).

Next, we apply Theorem 7.5 to determine upper quasiseparable generators $g_s(i)$ $(i = 1, \dots, N-1)$, $h_s(j)(j = 2, \dots, N)$, $b_s(k)$ $(k = 2, \dots, N-1)$ of the matrix $U^{(1)}$ with orders not greater than $m+1$. Since the matrix $A^{(1)}$ is block upper triangular, using (31.207) we obtain lower quasiseparable generators of the matrix $U^{(1)}$ by the formulas

$$\begin{aligned} &p_B(i) = x^{(1)}(i-1),\ i = 2, \dots, N+1, \\ &p_B(N+2) \text{ is the } 0 \times m \text{ empty matrix,} \\ &a_B(k) = I_m,\ k = 2, \dots, N+1, \end{aligned} \tag{31.263}$$

$$\begin{aligned} &q_B(1), q_B(2) \text{ are the } m \times 0 \text{ empty matrices,} \\ &q_B(j) = (y^{(1)}(j-2))^*,\ j = 3, \dots, N+1. \end{aligned} \tag{31.264}$$

These generators have the orders $r^L_k = m,\ k = 1, \dots, N+1$. Inserting $m_k = \tau_k, n_k = \nu_k$ and $r^L_k = m$ in (7.22) and using (31.86), (31.87), we obtain formulas (31.233) for the sizes of matrices and orders of generators used in Theorem 7.5. Since $\tau_{N+2} = 0,\ \nu_{N+2} = 1,\ \rho_{N+1} = 0$ using the formulas (7.28)–(7.34) with $k = N+2$, we get that $V_{N+2}, X_{N+2}, Z_{N+2}$ are the $0 \times 0, 0 \times m, 0 \times 1$ empty matrices and $F_{N+2} = 1$ and, using $h_B(N+2) = 1$, the formulas (31.234) and that z_{N+2} is the 0×1 empty matrix. Next, for $k = N+1, \dots, 3$, applying (7.28)–(7.34) and using (31.263), (31.264) and (31.262) we obtain (31.235)–(31.241).

Finally, by (31.86), (31.87) the elements $U^{(1)}_B(i,j)$ of the block representation of the matrix $U^{(1)}$ are connected with the elements $U^{(1)}(i,j)$ of $U^{(1)}$, treated as an ordinary scalar matrix, via the relations

$$U^{(1)}(i,j) = U_B^{(1)}(i+1, j+2), \quad 1 \le i, j \le N.$$

Hence, the formulas (31.242) for upper triangular generators follow. □

Those steps of the algorithm in the above theorem which are performed about N times have the following complexity. Step 1.2 requires

$$2r_{k+1} + 2m - 2 + 126r_{k+2} + 2(2 + 2r_{k+2})(2 + r_{k+2}) + 12m + m$$

arithmetical operations, while in Step 2 the number of the arithmetical operations is less than

$$\mu(\rho_k+1,m)+\rho_k(2m-1)+(1+\rho_k)(2\rho_k+1)(1+s_k)+r^U_{k-1}(2r^U_k+1)(1+s_{k-1})$$
$$+\rho_{k-1}(2\nu_k-1+1)s_{k-1}+\rho_{k-1}(2s_k-1)s_{k-1}+(2r^U_{k-1}-1)s_{k-1},$$

where $\mu(p,q)$ denotes the number of operations needed for the QR factorization of a $p \times q$ matrix.

§31.4.2 Double shift

Based on the general Theorem 31.5 obtained for the double shift case and Theorem 7.5 for quasiseparable generators of unitary matrices, we obtain the following algorithm.

Theorem 31.9. *Let A be a real matrix from the class $\mathcal{H}_{N,m}$ with subdiagonal entries σ_k $(k = 1, \ldots, N-1)$, perturbation matrices $X = \mathrm{col}(x(i))_{i=1}^N$, $Y = \mathrm{col}(y(i))_{i=1}^N$, upper triangular generators $g_U(i), h_U(i)$ $(i = 1, \ldots, N)$, $b_U(k)$ $(k = 1, \ldots, N-1)$ of orders r_k $(k = 1, \ldots, N)$ of the matrix U, and let s, t be real numbers.*

Then the 3×3 orthogonal matrices Q_k, $k = 1, \ldots, N-2$ and the 2×2 orthogonal matrix Q_{N-1} such that the orthogonal matrix

$$Q = \tilde{Q}_1\tilde{Q}_2 \cdots \tilde{Q}_{N-2}\tilde{Q}_{N-1}, \tag{31.265}$$

with

$$\tilde{Q}_i = I_{i-1} \oplus Q_i \oplus I_{N-i-2},\ i = 1, \ldots, N-2, \quad \tilde{Q}_{N-1} = I_{N-2} \oplus Q_{N-1}, \tag{31.266}$$

satisfies the condition (31.7), and the diagonal entries $d^{(1)}(k)$ $(k = 1, \ldots, N)$, subdiagonal entries $\sigma_k^{(1)}$ $(k = 1, \ldots, N-1)$, of the upper Hessenberg matrix $A^{(1)} = Q^*AQ$, the perturbation matrices $X^{(1)} = \mathrm{col}(x^{(1)}(i))_{i=1}^N$, $Y^{(1)} = \mathrm{col}(y^{(1)}(i))_{i=1}^N$, upper triangular generators $g_U^{(1)}(i), h_U^{(1)}(i)$ $(i = 1, \ldots, N)$, $b_U^{(1)}(k)$ $(k = 1, \ldots, N-1)$ of the unitary matrix $U^{(1)}$ from (31.207) with orders not greater than $m+1$ are determined as follows:

1.1. *Compute*

$$\sigma_1^U = \sigma_1 + x(2)y^*(1), \quad \sigma_2^U = \sigma_2 + x(3)y^*(2). \tag{31.267}$$

Set

$$\begin{gathered} a_{11} = g_U(1)h_U(1) - x(1)y^*(1), \quad a_{12} = g_U(1)b_U(1)h_U(2) - x(1)y^*(2), \\ a_{21} = \sigma_1, \quad a_{22} = g_U(2)h_U(2) - x(2)y^*(2), \quad a_{32} = \sigma_2 \end{gathered} \tag{31.268}$$

and determine the 3×3 *orthogonal matrix* Q_1 *from the condition*

$$Q_1^* \begin{pmatrix} a_{11}^2 + a_{12}a_{21} - sa_{21} + t & a_{21}(a_{11} + a_{22} - s) & a_{32}a_{21} \end{pmatrix}^T = \begin{pmatrix} * & 0 & 0 \end{pmatrix}^T. \tag{31.269}$$

Compute

$$\Lambda_3 = Q_1^* \begin{pmatrix} g_U(1)h_U(1) & g_U(1)b_U(1)h_U(2) & g_U(1)b_U(1)b_U(2) \\ \sigma_1^U & g_U(2)h_U(2) & g_U(2)b_U(2) \\ x(3)y^*(1) & \sigma_2^U & g_U(3) \end{pmatrix} \tag{31.270}$$

and determine the matrices $g_B(3), \beta_3^U, f_3^U, \phi_3^U$ *of sizes* $1 \times (r_3+2), 2 \times (r_3+2), 2 \times 2, 2 \times r_3$ *from the partitions*

$$\Lambda_3 = \begin{pmatrix} g_B(3) \\ \beta_3^U \end{pmatrix}, \quad \beta_3^U = \begin{pmatrix} f_3^U & \phi_3^U \end{pmatrix}. \tag{31.271}$$

Compute

$$\begin{pmatrix} x^{(1)}(1) \\ c_3 \end{pmatrix} = Q_1^* \begin{pmatrix} x(1) \\ x(2) \\ x(3) \end{pmatrix}, \tag{31.272}$$

$$\begin{pmatrix} y^{(1)}(1) \\ \theta_3 \end{pmatrix} = Q_1^* \begin{pmatrix} y(1) \\ y(2) \\ y(3) \end{pmatrix}, \quad \theta_2 = \begin{pmatrix} y(1) \\ y(2) \end{pmatrix}, \tag{31.273}$$

with the m-dimensional rows $x^{(1)}(1), y^{(1)}(1)$ *and* $2 \times m$ *matrices* c_2, θ_2.

Compute

$$f_3 = f_3^U - c_3 \begin{pmatrix} y^*(1) & y^*(2) \end{pmatrix}, \quad \varphi_2 = \phi_2^U. \tag{31.274}$$

1.2. *For* $k = 1, \dots, N-3$, *perform the following: Compute*

$$\sigma_{k+2}^U = \sigma_{k+2} + x(k+3)y^*(k+2), \tag{31.275}$$

$$\varepsilon_{k+2} = -c_{k+2}y^*(k+2) + \varphi_{k+2}h_U(k+2). \tag{31.276}$$

Compute the 3×3 *matrix* Ω_k *by the formula*

$$\Omega_k = \begin{pmatrix} f_{k+2} & \varepsilon_{k+2} \\ 0 & \sigma_{k+2} \end{pmatrix} Q_k. \tag{31.277}$$

Determine the 3×3 *orthogonal matrix* Q_{k+1} *and the number* $\sigma_k^{(1)}$ *such that*

$$Q_{k+1}^* \Omega_k(:,1) = \begin{pmatrix} \sigma_k^{(1)} & 0 & 0 \end{pmatrix}^T. \tag{31.278}$$

Compute

$$\Lambda_{k+3} = Q_{k+1}^* \begin{pmatrix} f_{k+2}^U & \phi_{k+2}^U h_U(k+2) & \phi_{k+2}^U b_U(k+2) \\ x(k+3)\theta_{k+1}^* & \sigma_{k+2}^U & g_U(k+3) \end{pmatrix} \begin{pmatrix} Q_k & 0 \\ 0 & I_{r_{k+3}} \end{pmatrix}, \tag{31.279}$$

$$B_{k+3} = \begin{pmatrix} I_2 & 0 & 0 \\ 0 & h_U(k+2) & b_U(k+2) \end{pmatrix} \begin{pmatrix} Q_k & 0 \\ 0 & I_{r_{k+3}} \end{pmatrix}, \tag{31.280}$$

and determine the matrices $d_B(k+3), g_B(k+3), \beta_{k+3}^U, f_{k+3}^U, \phi_{k+3}^U$ *of sizes* $1 \times 1, 1 \times (r_{k+3}+2), 2 \times (r_{k+3}+2), 2 \times 2, 2 \times r_{k+3}$ *from the partitions*

$$\Lambda_{k+3} = \begin{pmatrix} d_B(k+3) & g_B(k+3) \\ * & \beta_{k+3}^U \end{pmatrix}, \quad \beta_{k+3}^U = \begin{pmatrix} f_{k+3}^U & \phi_{k+3}^U \end{pmatrix}. \tag{31.281}$$

Compute

$$\begin{pmatrix} x^{(1)}(k+1) \\ c_{k+2} \end{pmatrix} = Q_{k+1}^* \begin{pmatrix} c_{k+1} \\ x(k+3) \end{pmatrix}, \tag{31.282}$$

$$\begin{pmatrix} y^{(1)}(k+1) \\ \theta_{k+3} \end{pmatrix} = Q_{k+1}^* \begin{pmatrix} \theta_{k+2} \\ y(k+2) \end{pmatrix}, \tag{31.283}$$

with the m-dimensional rows $x^{(1)}(k+1), y^{(k+1)}(1)$ *and* $2 \times m$ *matrices* c_{k+2}, θ_{k+2}.

Compute

$$f_{k+3} = f_{k+3}^U - c_{k+3}\theta_{k+2}^*, \quad \varphi_{k+3} = \phi_{k+3}^U. \tag{31.284}$$

1.3. *Compute*

$$\varepsilon_N = -c_N y^*(N) + \varphi_N h_U(N) \tag{31.285}$$

and

$$\Omega_{N-2} = \begin{pmatrix} f_N & \varepsilon_N \end{pmatrix} Q_{N-2}. \tag{31.286}$$

Determine the 2×2 *orthogonal matrix* Q_{N-1} *and the number* $\sigma_{N-2}^{(1)}$ *such that*

$$Q_{N-1}^* \Omega_{N-2}(:,1) = \begin{pmatrix} \sigma_{N-2}^{(1)} & 0 \end{pmatrix}^T. \tag{31.287}$$

Compute

$$\Lambda_{N+1} = Q_{N-1}^* \begin{pmatrix} f_N^U & \phi_N^U h_U(N) \end{pmatrix} Q_{N-2}, \tag{31.288}$$

$$B_{N+1} = \begin{pmatrix} I_2 & 0 \\ 0 & h_U(N) \end{pmatrix} Q_{N-2}, \tag{31.289}$$

and determine the matrices $d_B(N+1), g_B(N+1), \beta^U_{N+1}$ *of sizes* $1\times 1, 1\times 2, 1\times 2$ *from the partition*

$$\Lambda_{N+1} = \begin{pmatrix} d_B(N+1) & g_B(N+1) \\ * & \beta^U_{N+1} \end{pmatrix}. \tag{31.290}$$

Compute

$$\Lambda_{N+2} = \beta^U_{N+1} Q_{N-1}, \tag{31.291}$$
$$B_{N+2} = Q_{N-1}, \tag{31.292}$$

and determine the numbers $d_B(N+2), g_B(N+2)$ *from the partition*

$$\Lambda_{N+2} = \begin{pmatrix} d_B(N+2) & g_B(N+2) \end{pmatrix}. \tag{31.293}$$

Compute

$$\begin{pmatrix} x^{(1)}(N-1) \\ x^{(1)}(N) \end{pmatrix} = Q^*_{N-1} c_N, \quad \begin{pmatrix} y^{(1)}(N-1) \\ y^{(1)}(N) \end{pmatrix} = Q^*_{N-1}\theta_N. \tag{31.294}$$

2. *Set*

$$\begin{aligned} &\rho_{N+3} = \rho_{N+2} = 0,\ \rho_{N+1} = 1, \dots, \rho_{N-m+3} = m-1, \\ &\rho_k = m,\ k = N-m+2, \dots, 1, \\ &\nu'_{N+3} = \dots = \nu'_{N-m+3} = 0, \\ &\nu'_{N-m+2} = \dots = \nu'_2 = 1, \quad \nu'_1 = m, \\ &s_{N+3} = 0,\ s_{N+2} = 1, \dots, s_{N-m+3} = m, \\ &s_{N-m+2} = \dots = s_2 = m+1, \\ &r^U_k = r_k + 2,\ k = 3, \dots, N, \quad r^U_{N+1} = 2,\ r^U_{N+2} = 1. \end{aligned} \tag{31.295}$$

Set X_{N+3}, z_{N+3} *to be the* $0\times m, 0\times 1$ *empty matrices and*

$$h_s(N+3) = 1, \quad Y_{N+3} = 1, \quad g_s(N+2) = g_B(N+2) \tag{31.296}$$

and for $k = N+2, \dots, 4$ *perform the following:*

Compute the QR factorizations

$$V^*_k \begin{pmatrix} x^{(1)}(k-2) \\ X_{k+1} \end{pmatrix} = \begin{pmatrix} X_k \\ 0_{\nu'_k \times m} \end{pmatrix}, \tag{31.297}$$

with unitary $(1+\rho_k)\times(1+\rho_k)$ *matrices* V_k *and auxiliary* $\rho_{k-1}\times m$ *matrices* X_k.

Compute the $(1+\rho_k)\times(1+s_k)$ *matrix*

$$Z_k = V^*_k \begin{pmatrix} d_B(k) & g_s(k) \\ X_{k+1}(y^{(1)}(k-2))^* & z_{k+1} \end{pmatrix} \tag{31.298}$$

and determine the matrices $h'_k, h''_k, \Delta_k, \Theta_k$ *of sizes* $\rho_{k-1}\times 1, \rho_{k-1}\times s_k, \nu'_k\times 1, \nu'_k\times s_k$ *from the partition*

$$Z_k = \begin{pmatrix} h'_k & h''_k \\ \Delta_k & \Theta_k \end{pmatrix}. \tag{31.299}$$

The submatrix $\begin{pmatrix} \Delta_k & \Theta_k \end{pmatrix}$ *has orthonormal rows and one can determine the* $(1+s_k)\times(1+s_k)$ *unitary matrix* F_k *from the condition*

$$\begin{pmatrix} \Delta_k & \Theta_k \end{pmatrix} F_k^* = \begin{pmatrix} 0_{\nu'_k\times s_{k-1}} & I_{\nu'_k} \end{pmatrix}. \tag{31.300}$$

Determine the matrices $h_s(k), b_s(k), d_F(k), g_F(k)$ *of sizes* $s_{k-1}\times 1, s_{k-1}\times s_k, \nu'_k\times 1, \nu'_k\times s_k$ *from the partition*

$$F_k = \begin{pmatrix} h_s(k) & b_s(k) \\ d_F(k) & g_F(k) \end{pmatrix}. \tag{31.301}$$

Compute the matrices Y_k *of the size* $r^U_{k-1}\times s_{k-1}$ *and* z_k *of the size* $\rho_{k-1}\times s_{k-1}$ *by the formulas*

$$Y_k = B_k \begin{pmatrix} h_s^*(k) \\ Y_{k+1}b_s^*(k) \end{pmatrix}, \quad z_k = h'_k h_s^*(k) + h''_k b_s^*(k). \tag{31.302}$$

Compute

$$g_s(k-1) = g_B(k-1)Y_k. \tag{31.303}$$

Set

$$\begin{aligned} &g_U^{(1)}(i) = g_s(i+2),\ h_U^{(1)}(i) = h_s(i+3),\ i = 1,\dots,N, \\ &b_U^{(1)}(k) = b_s(k+3), \qquad\qquad\qquad k = 1,\dots,N-1. \end{aligned} \tag{31.304}$$

Proof. As in the proof of Theorem 31.8, we use the formulas (31.243), (31.244) to determine upper triangular generators of the matrix A.

We apply Theorem 31.5 to A. For the matrices β_k used in this theorem we introduce the partitions

$$\beta_k = \begin{pmatrix} f_k & \phi_k \end{pmatrix},\ \phi_k = \begin{pmatrix} \varphi_k & -c_k \end{pmatrix},\ k = 2,\dots,N-1,\ \beta_N = \begin{pmatrix} f_N & \varphi_N \end{pmatrix}, \tag{31.305}$$

with matrices f_k, φ_k, c_k of sizes $2\times 2, 2\times r_k, 2\times m$, respectively.

Using (31.243), (31.244) with $k = 1,2,3$, we have

$$\begin{aligned} g(1)h(1) &= g_U(1)h_U(1) - x(1)y^*(1), \\ g(1)b(1) &= \begin{pmatrix} g_U(1)b_U(1) & -x(1) \end{pmatrix}, \\ g(1)b(1)h(2) &= g_U(1)b_U(1)h_U(2) - x(1)y^*(2), \\ g(1)b(1)b(2) &= \begin{pmatrix} g_U(1)b_U(1)b_U(2) & -x(1) \end{pmatrix}, \\ g(2)h(2) &= g_U(2)h_U(2) - x(2)y^*(2), \\ g(2)b(2) &= \begin{pmatrix} g_U(2)b_U(2) & -x(2) \end{pmatrix}, \\ g(3) &= \begin{pmatrix} g_U(3) & -x(3) \end{pmatrix}. \end{aligned}$$

Inserting this in (31.101), (31.102) and in (31.103), and using (31.305), we obtain the condition (31.269) and the equality

$$\begin{pmatrix} * & * & * \\ f_3 & \varphi_3 & -c_3 \end{pmatrix} = Q_1^* \begin{pmatrix} a_{11} & a_{12} & g_U(1)b_U(1)b_U(2) & -x(1) \\ \sigma_1 & a_{22} & g_U(2)b_U(2) & -x(2) \\ 0 & \sigma_2 & g_U(3) & -x(3) \end{pmatrix}. \tag{31.306}$$

Next, using (31.305) and (31.244) we get

$$\phi_{k+2}h(k+2) = \varepsilon_{k+2}, \qquad k = 1, \dots, N-2, \tag{31.307}$$

and

$$\phi_{k+2}b(k+2) = \begin{pmatrix} \varphi_{k+2}b_U(k+2) & -c_{k+2} \end{pmatrix}, \quad k = 1, \dots, N-3, \tag{31.308}$$

with ε_{k+2} as in (31.276), (31.285).

For $k = 1, \dots, N-3$ we have the following. Inserting (31.307) in (31.105), we obtain (31.277). Next, the condition (31.106) is equivalent to condition (31.278). Inserting (31.308) and (31.243) in (31.107) and using the partitions (31.108), (31.109) and (31.305) we obtain the equalities

$$\begin{pmatrix} * & * & * \\ f_{k+3} & \varphi_{k+3} & -c_{k+3} \end{pmatrix} = Q_{k+1}^* \begin{pmatrix} \Omega_k(1:2,2:3) & \varphi_{k+2}b_U(k+2) & -c_{k+2} \\ \Omega_k(3,2:3) & g_U(k+3) & -x(k+3) \end{pmatrix}, \\ k = 1, \dots, N-4. \tag{31.309}$$

Inserting (31.307) with $k = N-2$ in (31.111), we obtain the equality (31.286). The condition (31.112) is equivalent to the condition (31.287).

From the relations (31.265), (31.266) it follows that the formulas (31.272), (31.273), (31.282), (31.283) and (31.294) yield the rows of the perturbation matrices $X^{(1)} = Q^*X$ and $Y^{(1)} = Q^*Y$.

Next, consider the unitary matrix $U^{(1)} = Q^*UQ$. We treat the matrix U in the same way as the matrix A in Theorem 31.5, i.e., as a block matrix with entries of sizes $m_i \times n_j$, $i, j = 1, \dots, N+3$, where the numbers m_i, n_j are defined in (31.119). Relative to this partition, the matrix U is a block matrix with the following diagonal entries:

$$\begin{aligned} & d_U(1) \text{ is the } 1 \times 0 \text{ empty matrix,} \\ & d_U(k) = \sigma_{k-1}^U = \sigma_{k-1} + x(k)y^*(k-1), \ k = 2, \dots, N, \\ & d_U(N+1), d_U(N+2), d_U(N+3) \text{ are} \\ & \qquad \text{the } 0 \times 1,\ 0 \times 0,\ 0 \times 0 \text{ empty matrices,} \end{aligned} \tag{31.310}$$

upper quasiseparable generators

$$\begin{aligned}
&\hat{g}_U(k) = g_U(k),\ k = 1,\dots,N,\\
&\hat{g}_U(N+1), \hat{g}_U(N+2) \text{ are the } 0\times 0 \text{ empty matrices},\\
&\hat{h}_U(k) = h_U(k-1),\ k = 2,\dots,N+1,\\
&\hat{h}_U(N+2), \hat{h}_U(N+3) \text{ are the } 0\times 0 \text{ empty matrices},\\
&\hat{b}_U(k) = b_U(k-1),\ k = 2,\dots,N,\\
&\hat{b}_U(N+1), \hat{b}_U(N+2) \text{ are the } r_N\times 0,\ 0\times 0 \text{ empty matrices},
\end{aligned} \tag{31.311}$$

and lower quasiseparable generators

$$\begin{aligned}
&p_U(k) = x(k),\ k = 2,\dots,N,\\
&p_U(N+1), p_U(N+2), p_U(N+3) \text{ are the } 0\times m \text{ empty matrices},\\
&q_U(1), q_U(N+2) \text{ are the } m\times 0 \text{ empty matrices},\\
&q_U(k) = y^*(k-1),\ k = 2,\dots,N+1,\\
&a_U(k) = I_m,\ k = 2,\dots,N+2.
\end{aligned} \tag{31.312}$$

For the matrices Q^*, Q we use the factorizations (31.123).

We apply Lemma 31.1 to compute (block) upper quasiseparable generators $g_B(i)$ $(i = 1,\dots,N+2)$, $h_B(j)$ $(j = 2,\dots,N+3)$, $b_B(k)$ $(k = 2,\dots,N+2)$ with orders

$$\tilde{r}_1 = r_1, \quad \tilde{r}_2 = r_2 + 1, \quad \tilde{r}_k = r_k + 2, \quad k = 3,\dots,N, \quad \tilde{r}_{N+1} = 2, \quad \tilde{r}_{N+2} = 1$$

of the matrix $U^{(1)} = Q^*UQ$. The matrix $U^{(1)}$ is obtained as a block one with entries of sizes $\tau_i \times \nu_j$, $i, j = 1,\dots,N+2$, where the numbers τ_i, ν_j are defined in (31.126), (31.127). Using (31.9)–(31.11) we obtain that $g_B(1)$ is the $0\times m$ empty matrix, $\phi_1^U = g_U(1)$ and f_1^U is the 1×0 empty matrix. Using (31.12)–(31.14), (31.124), (31.125) and (31.310)–(31.312) with $k = 2$ we get

$$f_2^U = \begin{pmatrix} g_U(1)h_U(1) \\ \sigma_1^U \end{pmatrix}, \quad \phi_2^U = \begin{pmatrix} g_U(1)b_U(1) \\ g_U(2) \end{pmatrix}, \quad \alpha_2 = y^*(1).$$

Next, using (31.12)–(31.14), (31.124), (31.125) and (31.310)–(31.312) with $k = 3,\dots,N+2$ we get

$$\begin{pmatrix} g_B(3) \\ \Gamma_3 \end{pmatrix} = \begin{pmatrix} Q_1^* & 0 \\ 0 & I_m \end{pmatrix} \begin{pmatrix} g_U(1)h_U(1) & g_U(1)b_U(1)h_U(2) & g_U(1)b_U(1)b_U(2) \\ \sigma_1^U & g_U(2)h_U(2) & g_U(3) \\ x(3)y^*(1) & \sigma_2^U & g_U(3) \\ y^*(1) & y^*(2) & 0 \end{pmatrix},$$

$$\Gamma_3 = \begin{pmatrix} f_3^U & \phi_3^U \\ \alpha_3 & 0 \end{pmatrix}, \tag{31.313}$$

$$\begin{pmatrix} d_B(k) & g_B(k) \\ * & \Gamma_k \end{pmatrix}$$

$$= \begin{pmatrix} Q^*_{k-2} & 0 \\ 0 & I_m \end{pmatrix} \begin{pmatrix} f^U_{k-1} & \phi^U_{k-1} h_U(k-1) & \phi^U_{k-1} b_U(k-1) \\ x(k)\alpha_{k-1} & \sigma^U_{k-1} & g_U(k) \\ \alpha_{k-1} & y^*(k-1) & 0 \end{pmatrix} \begin{pmatrix} Q_{k-3} & 0 \\ 0 & I_{r_k} \end{pmatrix},$$

$$\Gamma_k = \begin{pmatrix} f^U_k & \phi^U_k \\ \alpha_k & 0 \end{pmatrix}, \quad k = 4, \dots, N, \tag{31.314}$$

$$\begin{pmatrix} d_B(N+1) & g_B(N+1) \\ * & \beta_{N+1} \end{pmatrix}$$

$$= Q^*_{N-1} \begin{pmatrix} f^U_N & \phi^U_N h_U(N) \end{pmatrix} Q_{N-2}, \quad \beta_{N+1} = f^U_{N+1}, \tag{31.315}$$

$$\begin{pmatrix} d_B(N+2) & g_B(N+2) \end{pmatrix} = f^U_{N+1} Q_{N-1}, \tag{31.316}$$

and also

$$\begin{pmatrix} h_B(k) & b_B(k) \end{pmatrix} = \begin{pmatrix} I_2 & 0 & 0 \\ 0 & h_U(k-1) & b_U(k-1) \end{pmatrix} \begin{pmatrix} Q_{k-3} & 0 \\ 0 & I_{r_k} \end{pmatrix}, \; k = 4, \dots, N, \tag{31.317}$$

$$\begin{pmatrix} h_B(N+1) & b_B(N+1) \end{pmatrix} = \begin{pmatrix} I_2 & 0 \\ 0 & h_U(N) \end{pmatrix} Q_{N-2}, \tag{31.318}$$

$$\begin{pmatrix} h_B(N+2) & b_B(N+2) \end{pmatrix} = Q_{N-1}. \tag{31.319}$$

Using (31.17), (31.18), we get $h_B(N+3) = 1$.

The relations (31.313), (31.314) imply, in particular, that the auxiliary matrices α_k ($k = 3, \dots, N$) satisfy the relations

$$\alpha_3 = \begin{pmatrix} y^*(1) & y^*(2) \end{pmatrix}, \quad \begin{pmatrix} * & \alpha_k \end{pmatrix} = \begin{pmatrix} \alpha_{k-1} & y^*(k-1) \end{pmatrix} Q_{k-3}, \; k = 4, \dots, N.$$

Comparing this with (31.273), (31.283), we get

$$\alpha_k = \theta^*_{k-1}, \quad k = 3, \dots, N. \tag{31.320}$$

Thus, using(31.313) we obtain (31.270), (31.271), and inserting (31.320) in (31.314) we obtain (31.279), (31.281).

Next we show that the auxiliary variables f_k, φ_k ($k = 2, \dots, N$) may be determined via relations (31.274), (31.284). Indeed, combining (31.306) and (31.267), (31.268) we get

$$\begin{pmatrix} * & * \\ f_3 & \varphi_3 \end{pmatrix}$$

$$= Q^*_1 \begin{pmatrix} g_U(1)h_U(1) - x(1)y^*(1) & g_U(1)b_U(1)h_U(2) - x(1)y^*(2) & g_U(1)b_U(1)b_U(2) \\ \sigma^U_1 - x(2)y^*(1) & g_U(2)h_U(2) - x(2)y^*(2) & g_U(2)b_U(2) \\ x(3)y^*(1) - x(3)y^*(1) & \sigma^U_2 - x(3)y^*(2) & g_U(3) \end{pmatrix},$$

and hence using (31.270), (31.271) and (31.272) we obtain (31.274). Set $\theta_2 = \begin{pmatrix} y(1) \\ y(2) \end{pmatrix}$ and assume that for some k with $1 \le k \le N-2$ the relations

$$f_{k+2} = f^U_{k+2} - c_{k+2}\theta^*_{k+1}, \quad \varphi_{k+2} = \phi^U_{k+2} \tag{31.321}$$

hold. By (31.277) and (31.309), we have

$$\begin{pmatrix} \sigma^{(1)}_k & * & * \\ 0_{2\times 1} & f_{k+3} & \varphi_{k+3} \end{pmatrix} = Q^*_{k+1} \begin{pmatrix} f_{k+2} & \varepsilon_{k+2} & \varphi_{k+2} b_U(k+2) \\ 0_{1\times 2} & \sigma_{k+2} & g_U(k+3) \end{pmatrix} \begin{pmatrix} Q_k & 0 \\ 0 & I_{r_{k+3}} \end{pmatrix}. \tag{31.322}$$

Using (31.275), (31.276) we get

$$\begin{pmatrix} f_{k+2} & \varepsilon_{k+2} \\ 0 & \sigma_{k+2} \end{pmatrix} = \begin{pmatrix} f^U_{k+2} - c_{k+2}\theta^*_{k+1} & \phi^U_{k+2} h_U(k+2) - c_{k+2} y^*(k+2) \\ x(k+3)\theta^*_{k+1} - x(k+3)\theta^*_{k+1} & \sigma^U_{k+2} - x(k+3) y^*(k+2) \end{pmatrix}. \tag{31.323}$$

Thus, combining (31.322) and (31.323) and using (31.279), (31.281) and (31.282), (31.283), we obtain (31.284).

Using (31.317), (31.318) (31.319) and setting

$$B_k = \begin{pmatrix} h_B(k) & b_B(k) \end{pmatrix}, \quad k = 4, \dots, N+2, \tag{31.324}$$

we obtain (31.280), (31.289), (31.292).

Next, we apply Theorem 7.5 to determine upper quasiseparable generators $g_s(i)$ $(i = 1, \dots, N-1)$, $h_s(j) (j = 2, \dots, N)$, $b_s(k)$ $(k = 2, \dots, N-1)$ of the matrix $U^{(1)}$ with orders not greater than $m+1$. Since the matrix $A^{(1)}$ is block upper triangular, using (31.207) we obtain lower quasiseparable generators of the matrix $U^{(1)}$ by the formulas

$$\begin{gathered} p_B(i) = x^{(1)}(i-2), \ i = 3, \dots, N+2, \\ p_B(2), p_B(N+3) \text{ are the } 0 \times m \text{ empty matrices}, \\ a_B(k) = I_m, \ k = 2, \dots, N+2, \end{gathered} \tag{31.325}$$

$$\begin{gathered} q_B(1), q_B(2), q_B(3) \text{ are the } m \times 0 \text{ empty matrices}, \\ q_B(j) = (y^{(1)}(j-3))^*, \ j = 4, \dots, N+2. \end{gathered} \tag{31.326}$$

These generators have the orders $r^L_k = m$, $k = 1, \dots, N+1$.

Inserting $m_k = \tau_k, n_k = \nu_k$ and $r^L_k = m$ in (7.22) and using (31.126), (31.127), we obtain the formulas (31.295) for sizes of matrices and orders of generators used in Theorem 7.5. Since $\tau_{N+3} = 0$, $\nu_{N+3} = 1$, $\rho_{N+2} = 0$, using the formulas (7.28)–(7.34) with $k = N+3$ we get that $V_{N+3}, X_{N+3}, Z_{N+3}$ are the $0\times 0, 0\times m, 0\times 1$ empty matrices and $F_{N+3} = 1$ and, using $h_B(N+3) = 1$, the formulas (31.296) and that z_{N+3} is the 0×1 empty matrix. Next, for $k = N+2, \dots, 4$ applying (7.28)–(7.34) and using (31.325), (31.326) and (31.324) we obtain (31.297)–(31.303).

Finally, by (31.126), (31.127), the elements $U_B^{(1)}(i,j)$ of the block representation of the matrix $U^{(1)}$ are connected with the elements $U^{(1)}(i,j)$ of $U^{(1)}$, treated as an ordinary scalar matrix via the relations

$$U^{(1)}(i,j) = U_B^{(1)}(i+2,j+3), \quad 1 \leq i,j \leq N.$$

Hence the formulas (31.304) for upper triangular generators follow. □

We will compute the complexity of the algorithm in the last theorem without Steps 1.1 and 1.3 which are not performed about N times, but just once. The complexity of Step 1.2 is less than

$$1+2m-1+4m+4r_{k+2}-2+45+15+15(3+r_{k+3})+4r_{k+2}r_{k+3}+6(r_{k+3}+3)^2+30m$$

arithmetical operations, while the complexity of Step 2 is less than

$$\mu(1+\rho_k,m)+2(1+\rho_k)^2(1+s_k)+\rho_k(2m-1)$$

before the computation of F_k and

$$r_{k-1}^U(2s_k+2r_{k-1}^U-1)s_{k-1}+\rho_{k-1}\cdot 2\nu_k s_{k-1}+\rho_{k-1}(2s_k-1)s_{k-1}$$

afterwards.

§31.5 Comments

The implicit QR iteration algorithm for upper Hessenberg matrices which are perturbations of arbitrary rank of Hermitian matrices was developed in [22]. In that paper the backward stability of the proposed algorithm was proved. The general algorithm, as well as the algorithm for upper Hessenberg matrices which are perturbations of arbitrary rank of unitary matrices, appear here for the first time. The case of rank one perturbations of unitary matrices with applications to the polynomial root-finder problem will be treated in detail in the next part.

§31.5 Comments

Part VIII

QR Iterations for Companion Matrices

Introduction to Part VIII

In this part we study algorithms for computing the roots of polynomials based on the fact that the roots are eigenvalues of the corresponding companion matrices. It turns out that companion matrices belong to the class of upper Hessenberg matrices that are rank one perturbations of unitary matrices. This class is invariant under QR iterations and the corresponding unitary matrices have quasiseparable representations with orders not greater than two. Hence, the QR iteration method for companion matrices is reduced to the updating of generators in the iteration process. This allows one to obtain algorithms with the complexity $O(N)$ per step and $O(N^2)$ for the computation of all the set of eigenvalues, in contrast to $O(N^2)$ and $O(N^3)$ when the structure is not used.

In the first chapter of Part VIII we study in detail the aforementioned class of upper Hessenberg matrices and the corresponding class of unitary matrices. The second chapter is devoted to the explicit methods; here we apply results of Chapter 29 to the corresponding upper Hessenberg matrices. In the third chapter, using results of Chapter 31, we derive implicit algorithms. In the next section we present a method in which factorizations of unitary matrices as products of elementary unitary matrices are used. The last section is devoted to an alternative method, which is based on QR factorizations of the considering upper Hessenberg matrices.

Chapter 32

Companion and Unitary Matrices

In this chapter we formulate the main polynomial root finding problem in terms of the eigenvalue problem for the companion matrix. We will easily see that the companion matrix belongs to the class of upper Hessenberg matrices which are rank one perturbations of unitary matrices. We will study this class in detail here.

§32.1 Roots of polynomials and eigenvalues of companion matrices

Consider the polynomial equation

$$z^n + p_{n-1}z^{n-1} + p_{n-2}z^{n-2} + \cdots + p_1 z + p_0 = 0.$$

A complex number z is a solution of this equation if and only if z is an eigenvalue of the companion matrix

$$C = \begin{pmatrix} 0 & 0 & 0 & \dots & 0 & -p_0 \\ 1 & 0 & 0 & \dots & 0 & -p_1 \\ 0 & 1 & 0 & \dots & 0 & -p_2 \\ 0 & 0 & 1 & \dots & 0 & -p_3 \\ \vdots & \vdots & \vdots & \ddots & \vdots & \vdots \\ 0 & 0 & 0 & \dots & 1 & -p_{n-1} \end{pmatrix}. \tag{32.1}$$

The matrix C may be represented in the form

$$C = U - pq^T,$$

where

$$U = \begin{pmatrix} 0 & 0 & \dots & 0 & 1 \\ 1 & 0 & \dots & 0 & 0 \\ 0 & 1 & 0 & \dots & 0 \\ \vdots & \vdots & \ddots & \vdots & \vdots \\ 0 & 0 & \dots & 1 & 0 \end{pmatrix} \tag{32.2}$$

is the circulant matrix and

$$p = \begin{pmatrix} 1+p_0 \\ p_1 \\ p_2 \\ \vdots \\ p_{n-1} \end{pmatrix}, \quad q = \begin{pmatrix} 0 \\ 0 \\ 0 \\ \vdots \\ 1 \end{pmatrix}. \tag{32.3}$$

This means that C is a rank one perturbation of a unitary matrix. Next we consider such matrices in detail.

§32.2 The classes $\mathcal{H}_N$ and $\mathcal{U}_N$

§32.2.1 Definitions

We denote by $\mathcal{H}_N$ the class of upper Hessenberg matrices $A \in \mathbb{C}^{N\times N}$ which are rank one perturbations of unitary matrices, i.e.,

$$A = U - pq^T, \tag{32.4}$$

where $U \in \mathbb{C}^{N\times N}$ is unitary and $p, q \in \mathbb{C}^N$. The vectors $p = (p(i))_{i=1}^N$, $q = (q(i))_{i=1}^N$ are called *the perturbation vectors* for the matrix A.

Since the matrix A is upper Hessenberg, the entries below the first subdiagonal of the matrix U have the form

$$U(i,j) = p(i)q(j), \quad 1 \leq j \leq i-2, \; 3 \leq i \leq N. \tag{32.5}$$

We define *the class* $\mathcal{U}_N$ to be the set of unitary matrices with the elements in the part $i - j \geq 2$ of the form (32.5). The numbers $p(i)$ $(i = 3,\dots,N)$, $q(j)$ $(j = 1,\dots,N-2)$ are called *lower generators* of the matrix U.

In the opposite direction, let U be a unitary matrix from the class $\mathcal{U}_N$ with lower generators $p(i)$ $(i = 3,\dots,N)$, $q(j)$ $(j = 1,\dots,N-2)$. Take arbitrary numbers $p(1), p(2)$ and $q(N-1), q(N)$, and set $p = (p(i))_{i=1}^N$, $q = (q(i))_{i=1}^N$. Then the matrix $A = U - pq^T$ belongs to the class $\mathcal{H}_N$.

As it was shown above, the companion matrix C from (32.1) belongs to the class $\mathcal{H}_N$. The corresponding matrix U to C from the class $\mathcal{U}_N$ is the circulant matrix.

Notice that the class $\mathcal{H}_N$ is a particular case of a more general class considered in Section §31.4, more precisely $\mathcal{H}_N = \mathcal{H}_{N,1}$.

§32.2.2 Lower quasiseparable generators and orders of upper generators

For a matrix from the class $\mathcal{U}_N$, lower quasiseparable generators may be determined as follows.

Lemma 32.1. *Let U be a matrix from the class $\mathcal{U}_N$, with lower generators $p(i)$ ($i = 3,\dots,N$), $q(j)$ ($j = 1,\dots,N-2$) and subdiagonal entries σ_j ($j = 1,\dots,N-1$), i.e., with*

$$U(j+1,j) = \sigma_j, \quad j = 1,\dots,N-1. \tag{32.6}$$

Then a set of lower quasiseparable generators $p_U(i)$ ($i = 2,\dots,N$), $q_U(j)$ $j = 1,\dots,N$, $a_U(k)$ ($k = 2,\dots,N-1$) of U is obtained by the formulas

$$\begin{gathered} p_U(2) = \sigma_1, \qquad p_U(i) = \begin{pmatrix} p(i) & 1 \end{pmatrix}, \\ i = 3,\dots,N-1, \quad p_U(N) = 1; \end{gathered} \tag{32.7}$$

$$\begin{gathered} q_U(1) = 1, \qquad q_U(j) = \begin{pmatrix} q(j) \\ \sigma_j - p(j+1)q(j) \end{pmatrix}, \\ j = 2,\dots,N-2, \quad q_U(N-1) = \sigma_{N-1}; \end{gathered} \tag{32.8}$$

$$\begin{gathered} a_U(2) = \begin{pmatrix} q(1) \\ 0 \end{pmatrix}, \qquad a_U(k) = \begin{pmatrix} 1 & 0 \\ 0 & 0 \end{pmatrix}, \\ k = 3,\dots,N-2, \quad a_U(N-1) = \begin{pmatrix} p(N) & 0 \end{pmatrix}. \end{gathered} \tag{32.9}$$

The orders of these lower quasiseparable generators are

$$r_1 = 1, \quad r_2 = r_3 = \cdots = r_{N-2} = 2, \quad r_{N-1} = 1. \tag{32.10}$$

Proof. One should check that the relations

$$U(i,j) = p_U(i)(a_U)^{>}_{ij} q_U(j), \quad 1 \le j < i \le N, \tag{32.11}$$

hold.

For $i = j+1$, $1 \le j \le N-1$, we have

$$\begin{aligned} & p_U(2)q_U(1) = \sigma_1, \\ & p_U(j+1)q_U(j) = \begin{pmatrix} p(j+1) & 1 \end{pmatrix} \begin{pmatrix} q(j) \\ \sigma_j - p(j+1)q(j) \end{pmatrix} = \sigma_j, \ j = 2,\dots,N-2, \\ & p_U(N)q_U(N-1) = \sigma_{N-1}. \end{aligned}$$

Comparing this with (32.6), we obtain (32.11) with $i = j+1$.

Assume that $i > j+1$. For $j \ne 1$, $i \ne N$, we have

$$(a_U)^{>}_{ij} = \begin{pmatrix} 1 & 0 \\ 0 & 0 \end{pmatrix}.$$

Hence,

$$\begin{aligned} & (a_U)^{>}_{i1} = \begin{pmatrix} q(1) \\ 0 \end{pmatrix}, \qquad N \ge i \ge 3, \\ & (a_U)^{>}_{Nj} = \begin{pmatrix} p(N) & 0 \end{pmatrix}, \quad N-2 \ge j \ge 2, \\ & (a_U)^{>}_{N1} = p(N)q(1). \end{aligned}$$

Thus we obtain the equalities

$$p_U(i)(a_U)^{>}_{i1}q_U(1) = \begin{pmatrix} p(i) & 1 \end{pmatrix} \begin{pmatrix} q(1) \\ 0 \end{pmatrix} = p(i)q(j), \quad 3 \le i \le N-1,$$

$$p_U(N)(a_U)^{>}_{N1}q_U(1) = p(N)q(1),$$

$$p_U(i)(a_U)^{>}_{i1}q_U(j) = \begin{pmatrix} p(i) & 1 \end{pmatrix} \begin{pmatrix} 1 & 0 \\ 0 & 0 \end{pmatrix} \begin{pmatrix} q(j) \\ * \end{pmatrix} = p(i)q(j), \quad j \le 2 < i \le N-1,$$

$$p_U(N)(a_U)^{>}_{Nj}q_U(j) = \begin{pmatrix} p(N) & 0 \end{pmatrix} \begin{pmatrix} q(j) \\ * \end{pmatrix} = p(N)q(j), \; 2 \le j \le N-2.$$

Comparing this with (32.5), we obtain (32.11) with $i > j+1$. □

Applying Corollary 7.2 to the matrix U, we obtain the existence of upper quasiseparable generators of the matrix U with orders r_k $(k = 1, \dots, N-1)$ defined in (32.10).

We can also treat U as a block $(N+1) \times (N+1)$ matrix with entries of sizes $m_i \times n_j$, $i, j = 1, \dots, N+1$, where

$$\begin{aligned} m_1 = m_2 = \cdots = m_N = 1, \qquad m_{N+1} = 0, \\ n_1 = 0, \quad n_2 = n_3 = \cdots = n_N = n_{N+1} = 1. \end{aligned} \tag{32.12}$$

Applying Corollary 7.2, we obtain the existence of upper quasiseparable generators of the matrix U relative to the partition (32.12), i.e., upper triangular generators of U, with orders

$$s_1 = 1, \quad s_2 = s_3 = \cdots = s_{N-1} = 2, \quad s_N = 1. \tag{32.13}$$

§32.2.3 The factorization

Another kind of representations we use for matrices from the class $\mathcal{U}_N$ are factorizations as products of elementary unitary matrices. Such a factorization is obtained via Theorem 7.5. Using this theorem, we obtain also upper triangular generators of a unitary matrix with orders defined in (32.13).

Theorem 32.2. *Let U be a matrix from the class $\mathcal{U}_N$, with lower generators $p(i)$ $(i = 3, \dots, N)$, $q(j)$ $(j = 1, \dots, N-2)$, subdiagonal entries σ_k $(k = 1, \dots, N-1)$ and upper triangular generators $g(i), h(i)$ $(i = 1, \dots, N)$, $b(k)$ $(k = 1, \dots, N-1)$ of orders r^U_k $(k = 1, \dots, N)$, not necessarily minimal.*

The matrix U admits the factorization

$$U = V \cdot F, \tag{32.14}$$

where V is a unitary matrix represented in the form

$$V = \tilde{V}_{N-1} \cdots \tilde{V}_3, \tag{32.15}$$

with

$$\tilde{V}_k = \text{diag}\{I_{k-1}, V_k, I_{N-k-1}\}, \quad k = 3, \dots, N-1, \tag{32.16}$$

where V_k $(k = 3, \dots, N-1)$ *are unitary* 2×2 *matrices. and* F *is a unitary matrix represented in the form*

$$F = \tilde{F}_1 \tilde{F}_2 \cdots \tilde{F}_{N-2}, \tag{32.17}$$

with

$$\tilde{F}_k = \text{diag}\{I_{k-1}, F_k, I_{N-k-2}\}, \quad k = 1, \dots, N-2, \tag{32.18}$$

where F_k $(k = 1, \dots, N-2)$ *are unitary* 3×3 *matrices. Furthermore, the matrices* V_k $(k = 3, \dots, N)$ *and* F_k $(k = 1, \dots, N-2)$*, and upper triangular generators* $\tilde{g}_u(i), \tilde{h}_u(i)$ $(i = 1, \dots, N)$*,* $\tilde{b}(k)$ $(k = 1, \dots, N-1)$ *with orders*

$$s_1 = 1, \quad s_2 = \cdots = s_{N-1} = 2, \quad s_N = 1$$

are obtained using the following algorithm.

1. *Set*

$$X_N = p(N) \tag{32.19}$$

and for $k = N-1, \dots, 3$ *determine the unitary* 2×2 *matrix* V_k *and the number* X_k *such that*

$$V_k^* \begin{pmatrix} p(k) \\ X_{k+1} \end{pmatrix} = \begin{pmatrix} X_k \\ 0 \end{pmatrix}. \tag{32.20}$$

2.1. *Set*

$$h_s(N+1) = 1, \qquad g_s(N) = g(N)h(N), \tag{32.21}$$

$$h_s(N) = \begin{pmatrix} 1 \\ 0 \end{pmatrix}, \qquad b_s(N) = \begin{pmatrix} 0 \\ 1 \end{pmatrix}, \tag{32.22}$$

$$Y_N = \begin{pmatrix} h(N-1) & b(N-1)h(N) \end{pmatrix}, \qquad z_N = \begin{pmatrix} \sigma_{N-1} & g_s(N) \end{pmatrix}, \tag{32.23}$$

$$g_s(N-1) = g(N-1)Y_N. \tag{32.24}$$

2.2. *For* $k = N-1, \dots, 3$*, perform the following; Compute the* 2×3 *matrix*

$$Z_k = V_k^* \begin{pmatrix} \sigma_{k-1} & g_s(k) \\ X_{k+1}q(k-1) & z_{k+1} \end{pmatrix} \tag{32.25}$$

and determine the matrices $h'_k, h''_k, \Delta_k, \Theta_k$ *of sizes* $1 \times 1, 1 \times 2, 1 \times 1, 1 \times 2$ *from the partition*

$$Z_k = \begin{pmatrix} h'_k & h''_k \\ \Delta_k & \Theta_k \end{pmatrix}. \tag{32.26}$$

The three-dimensional row $\begin{pmatrix} \Delta_k & \Theta_k \end{pmatrix}$ *is of unit norm and one can determine the unitary* 3×3 *matrix* Φ_k *from the condition*

$$\begin{pmatrix} \Delta_k & \Theta_k \end{pmatrix} \Phi_k^* = \begin{pmatrix} 0_{1\times 2} & 1 \end{pmatrix}. \tag{32.27}$$

Determine the matrices $h_s(k), b_s(k), d_F(k), g_F(k)$ *of sizes* $2\times 1, 2\times 2, 1\times 1, 1\times 2$ *from the partition*

$$\Phi_k = \begin{pmatrix} h_s(k) & b_s(k) \\ d_F(k) & g_F(k) \end{pmatrix}. \tag{32.28}$$

Compute

$$\begin{aligned} Y_k &= h(k-1)h_s^*(k) + b(k-1)Y_{k+1}b_s^*(k), & (32.29)\\ z_k &= h'_k h_s^*(k) + h''_k b_s^*(k), & (32.30)\\ g_s(k-1) &= g(k-1)Y_k. & (32.31) \end{aligned}$$

2.3. *The* 2×3 *matrix*

$$Z_2 = \begin{pmatrix} \sigma_1 & g_s(2) \\ X_3 q(1) & z_3 \end{pmatrix} \tag{32.32}$$

has orthonormal rows and we determine the unitary 3×3 *matrix* Φ_2 *from the condition*

$$Z_2\Phi_2^* = \begin{pmatrix} 0_{1\times 2} & I_2 \end{pmatrix}. \tag{32.33}$$

Determine the matrices $h_s(2), b_s(2), d_F(2), g_F(2)$ *of sizes* $1\times 1, 1\times 2, 2\times 1, 2\times 2$ *from the partition*

$$\Phi_2 = \begin{pmatrix} h_s(2) & b_s(2) \\ d_F(2) & g_F(2) \end{pmatrix}. \tag{32.34}$$

Compute

$$g_s(1) = g(1)(h(1)h_s^*(2) + b(1)Y_3 b_s^*(2)). \tag{32.35}$$

3. *Set*

$$\begin{aligned} \tilde{g}(k) &= g_s(k), \quad k = 1,\dots,N, \\ \tilde{h}(k-1) &= h_s(k), \quad k = 2,\dots,N+1, \\ \tilde{b}(k-1) &= b_s(k), \quad k = 2,\dots,N. \end{aligned} \tag{32.36}$$

4. *Compute*

$$F_1 = \begin{pmatrix} g_s(1) & 0 \\ 0 & I_2 \end{pmatrix} \Phi_2 \tag{32.37}$$

and set

$$F_k = \Phi_{k+1}, \quad k = 2,\dots,N-2. \tag{32.38}$$

Proof. With respect to the partition (32.12), the matrix U has lower quasiseparable generators

$$\begin{aligned} &p_U(k) = p(k),\ k = 3,\dots,N, \\ &p_U(2), p_U(N+1) \text{ are the } 1\times 0 \text{ and } 0\times 0 \text{ empty matrices}, \\ &q_U(k) = q(k-1),\ k = 2,\dots,N-1, \\ &q_U(1), q_U(N) \text{ are the } 0\times 0 \text{ and } 0\times 1 \text{ empty matrices}, \\ &a_U(k) = 1, \quad k = 3,\dots,N-1, \\ &a_U(2), a_U(N) \text{ are the } 1\times 0 \text{ and } 0\times 1 \text{ empty matrices} \end{aligned} \tag{32.39}$$

of orders

$$r_1^L = 0, \quad r_k^L = 1,\ k = 2, \dots, N-1, \quad r_N^L = 0, \tag{32.40}$$

diagonal entries

$$d_U(k) = \sigma_{k-1},\ k = 2, \dots, N,$$
$$d_U(1), d_U(N+1) \text{ are the } 1 \times 0, \text{ and } 0 \times 1 \text{ empty matrices}$$

and upper quasiseparable generators

$$\begin{aligned} g_U(i) &= g(i), & i &= 1, \dots, N, \\ h_U(j) &= h(j-1), & j &= 2, \dots, N+1, \\ b_U(k) &= b(k-1), & k &= 2, \dots, N. \end{aligned} \tag{32.41}$$

We apply Theorem 7.5 to the matrix U. Inserting (32.12) and (32.40) in (7.22) we obtain the formulas

$$\begin{aligned} &\rho_{N+1} = \rho_N = 0, && \rho_{N-1} = \dots = \rho_2 = 1, \quad \rho_1 = 0, \\ &\nu_{N+1} = \nu_N = 0, && \nu_{N-1} = \dots = \nu_3 = 1, \quad \nu_2 = 2, \ \nu_1 = 1, \\ &s_{N+1} = 0, \ s_N = 1, && s_{N-1} = \dots = s_2 = 2, \quad s_1 = 1. \end{aligned} \tag{32.42}$$

Using (7.28) and (32.39) we get that V_{N+1}, X_{N+1} are 0×0 empty matrices, $V_N = 1$, and the formulas (32.19), (32.20) for determining the numbers X_k ($k = N, \dots, 3$) and the matrices V_k ($k = N-1, \dots, 3$) and the equality $V_2 = I_2$ also, X_2 is the 0×0 empty matrix.

We use the notations Φ_k instead of F_k used in Theorem 7.5. Using (7.29)–(7.34) with $k = N+1$ we get

$$Z_{N+1} = d_U(N+1) = \begin{pmatrix} h'_{N+1} & h''_{N+1} \end{pmatrix} = \begin{pmatrix} \Delta_{N+1} & \Theta_{N+1} \end{pmatrix}$$

with 0×1 empty matrices $d_U(N+1), h'_{N+1}, \Delta_{N+1}$ and 0×0 empty matrices h''_{N+1}, Θ_{N+1}, and hence

$$\Phi_{N+1} = 1 = h_s(N+1) = 1, \quad Y_{N+1} = h(N).$$

Therefore, $g_s(N) = g(N)h(N)$ and z_{N+1} is the 0×1 empty matrix. Using (7.29), (7.30) with $k = N$ we get that

$$h'_N = \sigma_{N-1}, \ h''_N = g_s(N)$$

and that $\begin{pmatrix} \Delta_N & \Theta_N \end{pmatrix}$ is the 0×2 empty matrix. The latter implies $\Phi_N = I_2$ and therefore using (7.32)–(7.34) with $k = N$ we obtain the relations (32.22)–(32.24).

For $k = N-1, \dots, 3$ using (7.29)–(7.34) and (32.41), we obtain the relations (32.25)–(32.31).

For $k = 2$ using (7.29) with $V_2 = I_2$ we obtain the formula (32.32). Using (7.30) and $\rho_1 = 0$ we see that h'_2, h''_2 are $0 \times 1, 0 \times 2$ empty matrices, $Z_2 =$

$\begin{pmatrix} \Delta_2 & \Theta_2 \end{pmatrix}$ and moreover the matrix Z_2 has orthonormal rows and we determine the 3×3 unitary matrix Φ_2 from the condition (32.33). Next, using (7.32), (7.33) and (7.34) we obtain the formula (32.35).

Using (7.35) we get $\Phi_1 = g_s(1)$ and therefore using (32.37), (32.38) and (7.23)–(7.27) we obtain the factorization (32.14)–(32.18). □

§32.2.4 Upper triangular generators via factorization

Here for a matrix from the class $\mathcal{U}_N$ with the given factorization (32.14) we derive formulas for upper triangular generators and subdiagonal entries.

Theorem 32.3. *Let U be a matrix from the class $\mathcal{U}_N$ represented as a product*

$$U = V \cdot F, \tag{32.43}$$

where V is a unitary matrix represented in the form

$$V = \tilde{V}_{N-1} \cdots \tilde{V}_3, \tag{32.44}$$

with

$$\tilde{V}_k = \operatorname{diag}\{I_{k-1}, V_k, I_{N-k-1}\}, \quad k = 3, \dots, N-1, \tag{32.45}$$

where V_k $(k = 3, \dots, N-1)$ are unitary 2×2 matrices, and F is a unitary matrix represented in the form

$$F = \tilde{F}_1 \tilde{F}_2 \cdots \tilde{F}_{N-2}, \tag{32.46}$$

with

$$\tilde{F}_k = \operatorname{diag}\{I_{k-1}, F_k, I_{N-k-2}\}, \quad k = 1, \dots, N-2, \tag{32.47}$$

where F_k $(k = 1, \dots, N-2)$ are unitary 3×3 matrices.

Then a set of upper triangular generators $g(i), h(j)$ $(j = 1, \dots, N)$, $b(k)$ $(k = 1, \dots, N-1)$ with orders r_k $(k = 1, \dots, N)$, where

$$r_k = 1, \quad r_2 = \cdots = r_{N-2} = 2, \quad r_{N-1} = 1,$$

and the subdiagonal entries $\sigma_i = U(i+1, i)$ $(i = 1, \dots, N-1)$ of the matrix U may be may be obtained as follows:

1. *Set $g(1) = 1$ and determine the numbers $h(1), \sigma_1$ and the two-dimensional rows $b(1), g(2), \beta_2$ from the partition*

$$F_1 = \begin{pmatrix} h(1) & b(1) \\ \sigma_1 & g(2) \\ * & \beta_2 \end{pmatrix}. \tag{32.48}$$

2. *For $k = 2, \dots, N-2$ perform the following. Determine the two-dimensional column $h(k)$ and the 2×2 matrix $b(k)$ from the partition*

$$F_k = \begin{pmatrix} h(k) & b(k) \\ * & * \end{pmatrix}. \tag{32.49}$$

Compute the matrix

$$W_k = V_{k+1}\begin{pmatrix} \beta_k & 0 \\ 0 & 1 \end{pmatrix} F_k \tag{32.50}$$

and determine the number σ_k *and two-dimensional rows* $g(k+1), \beta_{k+1}$ *from the partition*

$$W_k = \begin{pmatrix} \sigma_k & g(k+1) \\ * & \beta_{k+1} \end{pmatrix}. \tag{32.51}$$

3. *Determine the numbers* $\sigma_{N-1}, g(N)$ *from the partition*

$$\beta_{N-1} = \begin{pmatrix} \sigma_{N-1} & g(N) \end{pmatrix} \tag{32.52}$$

and set

$$h(N-1) = \begin{pmatrix} 1 \\ 0 \end{pmatrix}, \ b(N-1) = \begin{pmatrix} 0 \\ 1 \end{pmatrix}, \quad h(N) = 1. \tag{32.53}$$

Proof. We treat U as an $(N+1) \times (N+1)$ block matrix with entries of sizes $m_i \times n_j$, $i, j = 1, \dots, N$, with the numbers m_i, n_j defined in (32.12). We define the integers ρ_k, s_k, ν_k $(k = 1, \dots, N+1)$ by

$$\begin{array}{lll} \rho_1 = 0, & \rho_2 = \rho_2 = \cdots = \rho_{N-1} = 1, & \rho_N = \rho_{N+1} = 0, \\ s_1 = 1, & s_2 = s_3 = \cdots = s_{N-1} = 2, & s_N = 1, \ s_{N+1} = 0, \\ \nu_1 = 1, \ \nu_2 = 2, & \nu_3 = \cdots = \nu_{N-1} = 1, & \nu_N = \nu_{N+1} = 0. \end{array} \tag{32.54}$$

One can easily see that the parameters defined in (32.54) satisfy the condition (7.51). Next, we set

$$V_1 = 1, \ V_2 = I_2, \ V_N = 1, \quad \Phi_1 = 1, \qquad \Phi_k = F_{k-1}, \ k = 2, \dots, N-1,$$
$$\Phi_N = I_2, \ \Phi_{N+1} = 1$$

and take for V_{N+1} the 0×0 empty matrix and

$$\tilde{V}_1 = \tilde{V}_N = \tilde{V}_{N+1} = \tilde{\Phi}_1 = \tilde{\Phi}_N = \tilde{\Phi}_{N+1} = I_N.$$

Using (32.44)–(32.47) we obtain the representations of the matrices V and F in the form

$$V = \tilde{V}_{N+1}\tilde{V}_N \cdots \tilde{V}_2\tilde{V}_1$$

with

$$\tilde{V}_1 = \mathrm{diag}\{V_1, I_{\phi_1}\}, \qquad \tilde{V}_k = \mathrm{diag}\{I_{k-1}, V_k, I_{\phi_k}\}, \quad k = 2, \dots, N,$$
$$\tilde{V}_{N+1} = \mathrm{diag}\{I_N, V_{N+1}\}$$

and

$$\tilde{\Phi}_1 = \mathrm{diag}\{\Phi_1, I_{\phi_1}\}, \qquad \tilde{\Phi}_k = \mathrm{diag}\{I_{k-2}, \Phi_k, I_{\phi_k}\}, \quad k = 2, \dots, N,$$
$$\tilde{\Phi}_{N+1} = \mathrm{diag}\{I_{N-1}, \phi_{N+1}\},$$

where $\phi_k = \sum_{i=k+1}^{N} \nu_i$.

Now we are in the position to apply Theorem 7.8 in order to determine upper quasiseparable generators $g_B(i)$ $(i = 1,\dots,N)$, $h_B(j)$ $(j = 2,\dots,N+1)$, $b_B(k)$ $(k = 2,\dots,N)$ and diagonal entries $d_B(k)$ $(k = 1,\dots,N+1)$ of the matrix U, treated as a block one. More precisely, using (7.58), (7.59) we have

$$\Phi_k = \begin{pmatrix} h_B(k) & b_B(k) \\ * & * \end{pmatrix}, \quad k = 2,\dots,N-1, \quad \Phi_{N+1} = \begin{pmatrix} h_B(N+1) \\ * \end{pmatrix} \tag{32.55}$$

with the matrices $h_B(k), b_B(k)$ of sizes $s_{k-1}\times n_k, s_{k-1}\times s_k$, respectively. Inserting $n_1 = \rho_1 = 0$ and $V_1 = \Phi_1 = 1$ in (7.60), (7.61) we get that $g_B(1) = 1$ and β_1 is the 0×1 matrix. Next, using (7.62), (7.63) with $k = 2$ and the fact that $V_2 = I_2$, we get

$$\begin{pmatrix} d_B(2) & g_B(2) \\ * & \beta_2 \end{pmatrix} = \begin{pmatrix} 0_{2\times 1} & I_2 \end{pmatrix} \Phi_2. \tag{32.56}$$

Using (7.62), (7.63) with $k = 3,\dots,N-1$ we have

$$\begin{pmatrix} d_B(k) & g_B(k) \\ * & \beta_k \end{pmatrix} = V_k \begin{pmatrix} \beta_{k-1} & 0 \\ 0 & I_{\nu_k} \end{pmatrix} \Phi_k, \quad k = 3,\dots,N-1. \tag{32.57}$$

Finally, (7.62), (7.63) with $k = N$ and $V_N = 1$, $\Phi_N = I_2$, $\nu_N = 0$ implies

$$\begin{pmatrix} d_B(N) & g_B(N) \end{pmatrix} = \beta_{N-1}. \tag{32.58}$$

By (32.12), we have

$$d_B(k) = \sigma_{k-1}, \qquad k = 2,\dots,N, \qquad g_B(i) = g(i), \qquad i = 1,\dots,N, \tag{32.59}$$

and

$$h_B(k) = h(k-1),\ k = 2,\dots,N+1, \quad b_B(k) = b(k-1),\ k = 2,\dots,N. \tag{32.60}$$

Combining (32.55) with $k = 2$ and (32.56), we get

$$\begin{pmatrix} h_B(2) & b_B(2) \\ d_B(2) & g_B(2) \\ * & \beta_2 \end{pmatrix} = \begin{pmatrix} 1 & 0 \\ 0 & I_2 \end{pmatrix} \Phi_2 = \Phi_2.$$

Hence, using (32.59), (32.60) and $\Phi_2 = F_1$ we obtain (32.48). Using the formulas $F_k = \Phi_{k+1}$ $(k = 2,\dots,N-1)$ and the relations (32.55) and (32.60), we obtain (32.49). Inserting (32.60) and $\Phi_k = F_{k-1}$ in (32.57), we obtain (32.50), (32.51). Combining (32.58) and (32.59), we obtain (32.52). Inserting $\Phi_N = I_2$, $\Phi_{N+1} = 1$ in (32.55) and using (32.60), we obtain (32.53). □

§32.3 The unitary completion for matrices from the class $\mathcal{U}_N$

§32.3.1 Computation of upper generators

Let U be a matrix from the class $\mathcal{U}_N$ with given diagonal and subdiagonal entries and lower generators. Under some conditions one can determine upper quasiseparable generators of U from its lower triangular part. This problem may be treated as a unitary completion problem for matrices with quasiseparable representations and hence we may apply here the unitary completion method presented in Section §12.3. Using these methods we obtain the following algorithm to compute upper quasiseparable generators of the matrix U.

Algorithm 32.4. Let U be a matrix from the class $\mathcal{U}_N$ with lower generators $p(i)$ $(i = 3, \dots, N)$, $q(j)$ $(j = 1, \dots, N-2)$, subdiagonal entries σ_j $(j = 1, \dots, N-1)$, and diagonal entries $d(k)$ $(k = 1, \dots, N)$. Then a set of upper quasiseparable generators $g(i)$ $(i = 1, \dots, N-1)$, $h(j)$ $(j = 2, \dots, N)$, $b(k)$ $(k = 2, \dots, N-1)$ of the matrix U can be determined as follows.

1. Set $\beta_N = p(N)$ and for $k = N-1, \dots, 3$ determine complex Givens matrices V_k (see Section §7.1) and numbers β_k such that

$$V_k^* \begin{pmatrix} p(k) \\ \beta_{k+1} \end{pmatrix} = \begin{pmatrix} \beta_k \\ 0 \end{pmatrix}. \tag{32.61}$$

Compute

$$X_2 = \sqrt{|\sigma_1|^2 + |\beta_3|^2 |q(1)|^2}. \tag{32.62}$$

2. Set $\alpha_1 = q(1)$ and for $k = 2, \dots, N-2$ determine complex Givens matrices F_k and numbers α_k such that

$$\begin{pmatrix} \alpha_{k-1} & q(k) \end{pmatrix} F_k^* = \begin{pmatrix} 0 & \alpha_k \end{pmatrix}. \tag{32.63}$$

Compute

$$Y_{N-1} = \sqrt{|p(N)|^2 |\alpha_{N-2}|^2 + |\sigma_{N-1}|^2}. \tag{32.64}$$

If $Y_{N-1} \neq 0$, compute

$$F_{N-1} = \begin{pmatrix} p(N)\alpha_{N-2} & \sigma_{N-1} \end{pmatrix} / Y_{N-1}; \tag{32.65}$$

otherwise, set

$$F_{N-1} = \begin{pmatrix} 0 & 1 \end{pmatrix}. \tag{32.66}$$

3. Define the matrices B_k, $k = 1, \dots, N-1$ via

$$B_1 = \begin{pmatrix} \sigma_1 \\ \beta_3 q(1) \end{pmatrix}, \tag{32.67}$$

$$B_k = \begin{pmatrix} p(k+1)\alpha_{k-1} & \sigma_k \\ \beta_{k+2}\alpha_{k-1} & \beta_{k+2} q(k) \end{pmatrix}, \quad k = 2, \dots, N-2, \tag{32.68}$$

$$B_{N-1} = \begin{pmatrix} p(N)\alpha_{N-2} & \sigma_{N-1} \end{pmatrix}. \tag{32.69}$$

4. If all the matrices B_k satisfy the condition

$$\|B_k\| < 1, \quad k = 1, \dots, N-1, \tag{32.70}$$

perform the following.

4.1. Determine the generators $g(i)$ by the formulas

$$\begin{aligned} g(1) &= -d(1), \quad g(2) = -\begin{pmatrix} \sigma_1 & d(2) \end{pmatrix}, \\ g(i) &= -\begin{pmatrix} \lambda_i & d(i) \end{pmatrix}, \quad i = 3, \dots, N-2, \\ g(N-1) &= -\begin{pmatrix} \lambda_{N-1} & d(N-1) \end{pmatrix} F^*_{N-1}, \end{aligned} \tag{32.71}$$

where

$$\lambda_2 = \sigma_1, \quad \lambda_i = \sigma_{i-1} y^*_{i-1} + p(i)(\alpha_{i-1} - q(i-1) y^*_{i-1}),\ i = 3, \dots, N-1 \tag{32.72}$$

and $y_i = F_i(2,2)$.

4.2. Determine the generators $h(j)$ by the formulas

$$\begin{aligned} h(2) &= (1 - |X_2|^2)^{-1}(q^*(1)\gamma_2 + \sigma_1^* d(2)), \\ h(j) &= (I - B^*_{j-1} B_{j-1})^{-1} \begin{pmatrix} \alpha^*_{j-2}(\gamma_j + p^*(j) d(j)) \\ q^*(j-1)\gamma_j + \sigma^*_{j-1} d(j) \end{pmatrix}, \quad j = 3, \dots, N-1, \\ h(N) &= (1 - |Y_{N-1}|^2)^{-1} Y_{N-1} d(N), \end{aligned} \tag{32.73}$$

where

$$\begin{aligned} \gamma_j &= |\beta_{j+2}|^2 q(j) + p^*(j+1)\sigma_j, \quad j = 2, \dots, N-2, \\ \gamma_{N-1} &= p^*(N)\sigma_{N-1}. \end{aligned} \tag{32.74}$$

4.3. Determine generators $b(k)$ by the formulas

$$\begin{aligned} b(2) &= \begin{pmatrix} 1 & -h(2) \end{pmatrix}, \\ b(k) &= \begin{pmatrix} w^*_{k-1} & -h(k) \end{pmatrix}, \ k = 3, \dots, N-2, \\ b(N-1) &= \begin{pmatrix} w^*_{N-2} & -h(N-1) \end{pmatrix} F^*_{N-1}, \end{aligned} \tag{32.75}$$

where $w_k = F_k(2,:)$.

To justify this algorithm we proceed as follows. By Lemma 32.1, the lower quasiseparable generators $p_U(i)$ $(i = 2, \dots, N-1)$, $q_U(j)$ $(j = 1, \dots, N-1)$, $a_U(k)$ $(k = 2, \dots, N-1)$ of the unitary matrix U are obtained by the formulas (32.7)–(32.9). The orders of these generators are given in (32.10). We apply Theorem 12.4 to compute upper quasiseparable generators of U.

First, inserting (32.10) and $m_k = 1$ in (12.28) we conclude that the numbers ρ_k, τ_k from (12.28) coincide with the orders r_k:

$$\begin{aligned} r_1 &= \rho_1 = \tau_1 = 1, & r_i &= \rho_i = \tau_i = 2, & i &= 2, \dots, N-2, \\ r_{N-1} &= \rho_{N-1} = \tau_{N-1} = 1, & r_0 &= \rho_0 = \tau_0 = 0, & r_N &= \rho_N = \tau_N = 0. \end{aligned}$$

We use the formulas (12.29) with truncated QR factorizations (see Section §7.1) with lower quasiseparable generators $p_U(k), a_U(k)$. We obtain the relations

$$p_U(N) = \hat{V}_N X_N, \quad \begin{pmatrix} p_U(k) \\ X_{k+1} a_U(k) \end{pmatrix} = \hat{V}_k X_k, \; k = N-1, \dots, 2, \tag{32.76}$$

with matrices $\hat{V}_k, X_k, \; k = N, \dots, 2$ of sizes $(1 + r_k) \times r_{k-1}$ and $r_{k-1} \times r_{k-1}$ such that $\hat{V}_k^* \hat{V}_k = I_{r_{k-1}}$.

Next, we use the formulas (12.30) with truncated L′Q factorizations (see Section §7.1) with lower quasiseparable generators $q_U(k), a_U(k)$. We obtain the relations

$$q_U(1) = Y_1 \hat{F}_1, \quad \begin{pmatrix} a_U(k) Y_{k-1} & q_U(k) \end{pmatrix} = Y_k \hat{F}_k, \; k = 2, \dots, N-1, \tag{32.77}$$

with matrices $Y_k, F_k, \; k = 1, \dots, N-1$ of sizes $r_k \times r_k$ and $r_k \times (r_{k-1} + 1)$ such that $\hat{F}_k \hat{F}_k^* = I_{r_k}$.

Assume that the conditions

$$\|X_{k+1} Y_k\| < 1, \quad k = 1, \dots, N-1 \tag{32.78}$$

hold. By Theorem 12.4, one can apply the formulas (12.36)–(12.38) to compute a set of upper quasiseparable generators $g_U(i)$ $(i = 1, \dots, N-1)$, $h_U(j)$ $(j = 2, \dots, N)$, $b_U(k)$ $(k = 2, \dots, N-1)$ of the matrix U. We get

$$g_U(1) = -d(1) \hat{F}_1^*, \quad g_U(i) = -\begin{pmatrix} p_U(i) Y_{i-1} & d(i) \end{pmatrix} \hat{F}_i^*, \quad i = 2, \dots, N-1, \tag{32.79}$$

$$h_U(j) = (I - Y_{j-1}^* X_j^* X_j Y_{j-1})^{-1} Y_{j-1}^* (p_U^*(j) d(j) + a_U^*(j) X_{j+1}^* X_{j+1} q_U(j)), \quad j = 2, \dots, N-1, \tag{32.80}$$

$$h_U(N) = (I - Y_{N-1}^* X_N^* X_N Y_{N-1})^{-1} Y_{N-1}^* p_U^*(N) d(N),$$

$$b_U(k) = \begin{pmatrix} I_{r_{k-1}} & -h_U(k) \end{pmatrix} \hat{F}_k^*, \; k = 2, \dots, N-1. \tag{32.81}$$

To justify Step 1, notice the following. Using the formulas (32.76) and (32.7) with $k = N$, we get $\hat{V}_N X_N = 1$, and hence one can take $\hat{V}_N = 1, \; X_N = 1$. Using the formulas (32.76) and (32.7), (32.9) with $k = N-1$ we have

$$\begin{pmatrix} p_U(N-1) \\ a_U(N-1) \end{pmatrix} = \begin{pmatrix} p(N-1) & 1 \\ p(N) & 0 \end{pmatrix} = V_{N-1} X_{N-1}, \tag{32.82}$$

where $V_{N-1} = \hat{V}_{N-1}$ is a unitary 2×2 matrix determined from the condition (32.61). Moreover, one can easily check that the 2×2 upper triangular matrix X_{N-1} has the form

$$X_{N-1} = \begin{pmatrix} \beta_{N-1} & * \\ 0 & * \end{pmatrix}. \tag{32.83}$$

Next, for $k = N-2,\dots,3$ using (32.76) we have

$$\begin{pmatrix} p_U(k) \\ X_{k+1}a_U(k) \end{pmatrix} = \hat{V}_k X_k, \tag{32.84}$$

with a 3×2 matrix V_k such that $\hat{V}_k^*\hat{V}_k = I_2$ and 2×2 matrices X_{k+1}, X_k. Assume that the matrix X_{k+1} has the form

$$X_{k+1} = \begin{pmatrix} \beta_{k+1} & * \\ 0 & * \end{pmatrix}. \tag{32.85}$$

From (32.83) it follows that for $k = N-2$ the representation (32.85) holds. Inserting (32.7), (32.9) and (32.85) in (32.84), we get

$$\begin{pmatrix} p(k) & 1 \\ \beta_{k+1} & 0 \\ 0 & 0 \end{pmatrix} = \hat{V}_k X_k.$$

It follows that the matrix $\hat{V}_k$ has the form $\hat{V}_k = \begin{pmatrix} V_k \\ 0_{1\times 2} \end{pmatrix}$, with the unitary 2×2 matrix V_k determined from the condition (32.61), and the 2×2 upper triangular matrix X_k has the form

$$X_k = \begin{pmatrix} \beta_k & * \\ 0 & * \end{pmatrix}.$$

Moreover,

$$V_k X_k = \begin{pmatrix} p(k) & 1 \\ \beta_{k+1} & 0 \end{pmatrix}. \tag{32.86}$$

Finally, using (32.76) and (32.7), (32.9) with $k=2$ we have

$$\begin{pmatrix} p_U(2) \\ X_3 a_U(2) \end{pmatrix} = \begin{pmatrix} \sigma_1 \\ \beta_3 q(1) \\ 0 \end{pmatrix} = \begin{pmatrix} V_2 \\ 0 \end{pmatrix} X_2,$$

where V_2 is a two-dimensional unit column and X_2 is a number. It follows that

$$|X_2|^2 = |\sigma_1|^2 + |\beta_3|^2 |q(1)|^2,$$

and therefore one can compute the number X_2 by the formula (32.62).

To justify Step 2, notice the following. Using the formulas (32.77) and (32.8) with $k=1$ we get $Y_1\hat{F}_1 = 1$ and hence one can take $\hat{F}_1 = 1,\ Y_1 = 1$. Using the formulas (32.77) and (32.8), (32.9) with $k=2$, we have

$$\begin{pmatrix} a_U(2) & q_U(2) \end{pmatrix} = \begin{pmatrix} q(1) & q(2) \\ 0 & \sigma_2 - p(3)q(2) \end{pmatrix} = Y_2 F_2, \tag{32.87}$$

where $F_2 = \hat{F}_2$ is a unitary 2×2 matrix determined from the condition (32.63). Moreover one can easily check that the 2×2 matrix Y_2 has the form

$$Y_2 = \begin{pmatrix} 0 & \alpha_2 \\ * & (\sigma_2 - p(3)q(2))y_2^* \end{pmatrix}, \tag{32.88}$$

where $y_2 = F_2(2,2)$. Next, for $k = 3, \dots, N-2$ using (32.77) we have

$$\begin{pmatrix} a_U(k)Y_{k-1} & q_U(k) \end{pmatrix} = Y_k\hat{F}_k, \tag{32.89}$$

with a 2×3 matrix F_k such that $\hat{F}_k\hat{F}_k^* = I_2$ and 2×2 matrices Y_{k-1}, Y_k. Assume that the matrix Y_{k-1} has the form

$$Y_{k-1} = \begin{pmatrix} 0 & \alpha_{k-1} \\ * & * \end{pmatrix}. \tag{32.90}$$

From (32.88) it follows that for $k = 2$ the representation (32.90) holds. Inserting (32.8), (32.9) and (32.90) in (32.89), we get

$$\begin{pmatrix} 0 & \alpha_{k-1} & q(k) \\ 0 & 0 & \sigma_k - p(k+1)q(k) \end{pmatrix} = Y_k\hat{F}_k.$$

Hence, the matrix $\hat{F}_k$ has the form

$$\hat{F}_k = \begin{pmatrix} 0_{2\times 1} & F_k \end{pmatrix} \tag{32.91}$$

with a unitary 2×2 matrix F_k determined from the condition (32.63) and

$$Y_k = \begin{pmatrix} 0 & \alpha_k \\ * & (\sigma_k - p(k+1)q(k))y_k^* \end{pmatrix}, \tag{32.92}$$

where $y_k = F_k(2,2)$. Moreover,

$$Y_kF_k = \begin{pmatrix} \alpha_{k-1} & q(k) \\ 0 & \sigma_k - p(k+1)q(k) \end{pmatrix}. \tag{32.93}$$

Finally, using (32.77) and (32.8), (32.9) with $k = N-1$, we have

$$\begin{pmatrix} a_U(N-1)Y_{N-2} & q_U(N-1) \end{pmatrix} = \begin{pmatrix} 0 & p(N)\alpha_{N-2} & \sigma_{N-1} \end{pmatrix} = Y_{N-1}\hat{F}_{N-1},$$

with

$$\hat{F}_{N-1} = \begin{pmatrix} 0 & F_{N-1} \end{pmatrix}, \tag{32.94}$$

where F_{N-1} is a two-dimensional unit row and Y_{N-1} is a number. It follows that

$$|Y_{N-1}|^2 = |p(N)|^2|\alpha_{N-2}|^2 + |\sigma_{N-1}|^2,$$

and therefore one can compute the number Y_{N-1} by the formula (32.64). It is clear that the vector F_{N-1} is determined by the formulas (32.65) and (32.66).

Next, consider the matrices $X_{k+1}Y_k$ $(k = 1, \dots, N-1)$ which are used under the conditions (32.78), and the formulas (32.80). Since $Y_1 = 1$, we have

$$X_2 Y_1 = X_2. \tag{32.95}$$

Further, since $X_N = 1$ we have

$$X_N Y_{N-1} = Y_{N-1}. \tag{32.96}$$

For $k = 2, \dots, N-2$ we consider the matrices $V_{k+1}X_{k+1}Y_kF_k$. Using (32.86) and (32.93) we get

$$V_{k+1}X_{k+1}Y_kF_k = \begin{pmatrix} p(k+1) & 1 \\ \beta_{k+2} & 0 \end{pmatrix} \begin{pmatrix} \alpha_{k-1} & q(k) \\ 0 & \sigma_k - p(k+1)q(k) \end{pmatrix} = B_k, \tag{32.97}$$

where the matrices B_k are defined in (32.68). We also define the matrices B_1, B_{N-1} by the formulas (32.67), (32.69). From the equalities (32.95)–(32.97) and (32.62), (32.64) and the fact that the matrices V_{k+1}, F_k $(k = 2, \dots, N-2)$ are unitary it follows that the inequalities (32.78) are equivalent to the conditions (32.70).

Next we derive formulas for upper quasiseparable generators of the matrix U in terms of diagonal and subdiagonal entries and lower generators of this matrix. Using (32.7) and the equalities $Y_1 = 1$ and (32.92) we get

$$\begin{aligned} p_U(2)Y_1 &= \sigma_1, \\ p_U(i)Y_{i-1} &= \begin{pmatrix} p(i) & 1 \end{pmatrix} \begin{pmatrix} 0 & \alpha_{i-1} \\ * & (\sigma_{i-1} - p(i)q(i-1))y_{i-1}^* \end{pmatrix} = \begin{pmatrix} * & \lambda_i \end{pmatrix}, \end{aligned}$$

with λ_i defined in (32.72) and $y_i = F_i(2,2)$. Inserting this in (32.79) and using the equalities $\hat{F}_1 = F_1 = 1$, $\hat{F}_2 = F_2$ and (32.91), (32.94) we get

$$\begin{aligned} g_U(1) = -d(1), \quad g_U(2) &= -\begin{pmatrix} \sigma_1 & d(2) \end{pmatrix} F_2^*, \\ g_U(i) &= -\begin{pmatrix} \lambda_i & d(i) \end{pmatrix} F_i^*, \quad i = 3, \dots, N-2, \\ g_U(N-1) &= -\begin{pmatrix} \lambda_{N-1} & d(N-1) \end{pmatrix} F_{N-1}^*, \end{aligned} \tag{32.98}$$

with the numbers λ_i from (32.72).

Consider the formulas (32.80). For the matrices $X_{j+1}^* X_{j+1}$, $j = 2, \dots, N-2$, using (32.86) we get

$$\begin{aligned} X_{j+1}^* X_{j+1} &= X_{j+1}^* V_{j+1}^* V_{j+1} X_{j+1} \\ &= \begin{pmatrix} p^*(j+1) & \beta_{j+2}^* \\ 1 & 0 \end{pmatrix} \begin{pmatrix} p(j+1) & 1 \\ \beta_{j+2} & 0 \end{pmatrix}, \quad j = 2, \dots, N-2. \end{aligned}$$

Using (32.61) we have

$$|\beta_{j+2}|^2 + |p(j+1)|^2 = |\beta_{j+1}|^2 \tag{32.99}$$

and therefore we obtain the equalities

$$X_{j+1}^* X_{j+1} = \begin{pmatrix} |\beta_{j+1}|^2 & p^*(j+1) \\ p(j+1) & 1 \end{pmatrix}, \quad j = 2, \dots, N-2. \tag{32.100}$$

For the matrices $(I - Y_{j-1}^* X_j^* X_j Y_{j-1})^{-1} Y_{j-1}^*$, $j = 3, \dots, N-1$, we obtain the following representations. Using (32.97) we get

$$\begin{aligned} I - Y_{j-1}^* X_j^* X_j Y_{j-1} &= F_{j-1}(I - F_{j-1}^* Y_{j-1}^* X_j^* V_j^* V_j X_j Y_{j-1} F_{j-1}) F_{j-1}^* \\ &= F_{j-1}(I - B_{j-1}^* B_{j-1}) F_{j-1}^*, \quad j = 3, \dots, N-1. \end{aligned} \tag{32.101}$$

Consequently,

$$(I - Y_{j-1}^* X_j^* X_j Y_{j-1})^{-1} Y_{j-1}^* = F_{j-1}(I - B_{j-1}^* B_{j-1})^{-1} F_{j-1}^* Y_{j-1}^*, \quad j = 3, \dots, N-1$$

and using (32.93) we get

$$(I - Y_{j-1}^* X_j^* X_j Y_{j-1})^{-1} Y_{j-1}^* = F_{j-1}(I - B_{j-1}^* B_{j-1})^{-1} \Phi_j, \quad j = 3, \dots, N-1, \tag{32.102}$$

where

$$\Phi_j = \begin{pmatrix} \alpha_{j-2}^* & 0 \\ q^*(j-1) & \sigma_{j-1}^* - q^*(j-1) p^*(j) \end{pmatrix}. \tag{32.103}$$

Next, using (32.7)–(32.9) and (32.100) with $j = 2$ we get

$$\begin{aligned} & p_U^*(2) d(2) + a_U^*(2) X_3^* X_3 q_U(2) \\ &= \sigma_1^* d(2) + \begin{pmatrix} q^*(1) & 0 \end{pmatrix} \begin{pmatrix} |\beta_3|^2 & p^*(3) \\ p(3) & 1 \end{pmatrix} \begin{pmatrix} q(2) \\ \sigma_2 - p(3) q(2) \end{pmatrix} \\ &= \sigma_1^* d(2) + q^*(1)(|\beta_3|^2 q(2) + p^*(3)\sigma_2 - |p(3)|^2 q(2)), \end{aligned}$$

whence, using (32.99), we get

$$p_U^*(2) d(2) + a_U^*(2) X_3^* X_3 q_U(2) = \sigma_1^* d(2) + q^*(1) \gamma_2, \tag{32.104}$$

with γ_2 defined in (32.74). Inserting (32.104) in the first equality in (32.80) we get

$$h_U(2) = (1 - |X_2|^2)^{-1} (q^*(1) \gamma_2 + \sigma_1^* d(2)). \tag{32.105}$$

For $j = 3, \dots, N-2$ using (32.7)–(32.9) and (32.100) we get

$$\begin{aligned} & p_U^*(j) d(j) + a_U^*(j) X_{j+1}^* X_{j+1} q_U(j) \\ &= \begin{pmatrix} p^*(j) \\ 1 \end{pmatrix} d(j) + \begin{pmatrix} 1 & 0 \\ 0 & 0 \end{pmatrix} \begin{pmatrix} |\beta_{j+1}|^2 & p^*(j+1) \\ * & * \end{pmatrix} \begin{pmatrix} q(j) \\ \sigma_j - p(j+1) q(j) \end{pmatrix}, \end{aligned}$$

whence, using (32.99), we get

$$p_U^*(j) d(j) + a_U^*(j) X_{j+1}^* X_{j+1} q_U(j) = \begin{pmatrix} p^*(j) d(j) + \gamma_j \\ d(j) \end{pmatrix}, \tag{32.106}$$

with γ_j defined in (32.74). Using (32.7)–(32.9) and (32.100) with $j = N-1$ and the equality $X_N = 1$ we get

$$p_U^*(N-1)d(N-1) + a_U^*(N-1)X_N^*X_N q_U(N-1) = \begin{pmatrix} p^*(N-1) \\ 1 \end{pmatrix} d(N-1) + \begin{pmatrix} p^*(N) \\ 0 \end{pmatrix} \sigma(N-1)$$

which means

$$p_U^*(N-1)d(N-1) + a_U^*(N-1)X_N^*X_N q_U(N-1) = \begin{pmatrix} p^*(N-1)d(N-1) + \gamma_{N-1} \\ d(N-1) \end{pmatrix}, \tag{32.107}$$

with $\gamma_{N-1} = p^*(N)\sigma_{N-1}$. Now by virtue of (32.106), (32.107) and (32.103) we obtain

$$\Phi_j(p_U^*(j)d(j) + a_U^*(j)X_{j+1}^*X_{j+1}q_U(j)) = \begin{pmatrix} \alpha_{j-2}^*(p^*(j)d(j) + \gamma_j) \\ q^*(j-1)\gamma_j + \sigma_{j-1}^*d(j) \end{pmatrix}, \quad j = 3, \dots, N-1. \tag{32.108}$$

Thus, combining (32.102), (32.108) and (32.80), we get

$$F_{j-1}^* h_U(j) = (I - B_{j-1}^* B_{j-1})^{-1} \begin{pmatrix} \alpha_{j-2}^*(p^*(j)d(j) + \gamma_j) \\ q^*(j-1)\gamma_j + \sigma_{j-1}^*d(j) \end{pmatrix}, \quad j = 3, \dots, N-1. \tag{32.109}$$

Inserting $p_U^*(N) = X_N = 1$ in the last equality in (32.80) we get

$$h_U(N) = (I - |Y_{N-1}|^2)^{-1} Y_{N-1}^* d(N). \tag{32.110}$$

Consider the formulas (32.81). Using the equalities $\hat{F}_2 = F_2$ and (32.91), (32.94) we get

$$b_U(2) = \begin{pmatrix} 1 & -h_U(2) \end{pmatrix} F_2^*, \quad b_U(k) = \begin{pmatrix} e & -h_U(k) \end{pmatrix} F_k^*, \; k = 3, \dots, N-1,$$

with $e = \begin{pmatrix} 0 & 1 \end{pmatrix}^T$, and using (32.105), (32.109) we get

$$b_U(2) = \begin{pmatrix} 1 & -h(2) \end{pmatrix} F_2^*, \tag{32.111}$$

$$F_{k-1}^* b_U(k) = F_{k-1}^* \begin{pmatrix} e & -h_U(k) \end{pmatrix} F_k^* = \begin{pmatrix} w_{k-1}^* & -h(k) \end{pmatrix} F_k^*, \quad k = 3, \dots, N-1, \tag{32.112}$$

with $w_{k-1} = F_{k-1}(2,:)$ and $h(k)$ defined in (32.73).

Set $S_1 = 1$, $S_k = F_k^*$ $(k = 2, \dots, N-2)$, $S_{N-1} = 1$. From (32.98), (32.105), (32.109), (32.110), (32.111), (32.112) it follows that the relations

$$g_U(k) = g(k)S_k, \qquad k = 1, \dots, N-1, \tag{32.113}$$

$$S_{k-1}b_U(k) = b(k)S_k, \qquad k = 2, \dots, N-1, \tag{32.114}$$

$$h(k) = S_{k-1}h_U(k), \tag{32.115}$$

with the elements $g(i), h(j), b(k)$ defined by the formulas (32.79), (32.73), (32.75), hold. By Lemma 5.23, this means that $g(i)$ $(i = 1, \dots, N-1)$, $h(j)$ $(j = 2, \dots, N)$, $b(k)$ $(k = 2, \dots, N-1)$ are upper quasiseparable generators of the matrix U.

§32.3.2 The diagonal correction

By Theorem 12.5, for a matrix from the class $\mathcal{U}_N$ with fixed subdiagonal entries and lower generators, the diagonal entries satisfy the equalities

$$|d(k) + a_k| = b_k, \quad k = 1, \dots, N, \tag{32.116}$$

with

$$\begin{aligned} a_1 &= 0, \\ a_k &= \frac{g_k^*(I - \Delta_k)^{-1} f_k}{1 + f_k^*(I - \Delta_k)^{-1} f_k}, \quad k = 2, \dots, N-1, \\ a_N &= 0, \end{aligned} \tag{32.117}$$

$$\begin{aligned} b_1^2 &= 1 - h_1^* h_1, \\ b_k^2 &= \frac{1 - g_k^*(I - \Delta_k)^{-1} g_k - h_k^* h_k}{1 + f_k^*(I - \Delta_k)^{-1} f_k} + \frac{|g_k^*(I - \Delta_k)^{-1} f_k|^2}{(1 + f_k^*(I - \Delta_k)^{-1} f_k)^2}, \quad k = 2, \dots, N-1, \\ b_N^2 &= 1 - f_N^* f_N, \end{aligned} \tag{32.118}$$

where

$$\begin{aligned} &f_k = Y_{k-1}^* p_U^*(k), \ k = 2, \dots, N, \quad h_k = X_{k+1} q_U(k), \ k = 1, \dots, N-1, \\ &g_k = Y_{k-1}^* a_U^*(k) X_{k+1}^* h_k, \quad \Delta_k = Y_{k-1}^* X_k^* X_k Y_{k-1}, \quad k = 2, \dots, N-1. \end{aligned} \tag{32.119}$$

Here lower quasiseparable generators $p_U(i)$ $(i = 2, \dots, N-1)$, $q_U(j)$ $(j = 1, \dots, N-1)$, $b_U(k)$ $(k = 2, \dots, N-1)$ are defined in Lemma 32.1. Moreover, from Theorem 12.5 and Theorem 12.4 it follows that if the conditions (32.116) hold, then a matrix U defined via lower generators, diagonal and subdiagonal entries and upper quasiseparable generators obtained in Algorithm 32.4 is unitary. It is clear that U belongs to the class $\mathcal{U}_N$.

Next we derive expressions for the numbers a_k, b_k from formulas (32.116) in terms of lower generators and subdiagonal entries of the matrix U. We check that the following relations hold:

$$a_1 = 0, \quad a_k = \frac{\hat{g}_k^*(I - B_{k-1}^* B_{k-1})^{-1} \hat{f}_k}{1 + \hat{f}_k^*(I - B_{k-1}^* B_{k-1})^{-1} \hat{f}_k}, \ k = 2, \dots, N-1, \quad a_N = 0, \tag{32.120}$$

$$\begin{aligned} b_1^2 &= 1 - c_1, \\ b_k^2 &= \frac{1 - \hat{g}_k^*(I - B_{k-1}^* B_{k-1})^{-1}\hat{g}_k - c_k}{1 + \hat{f}_k^*(I - B_{k-1}^* B_{k-1})^{-1}\hat{f}_k} \\ &\quad + \frac{|\hat{g}_k^*(I - B_{k-1}^* B_{k-1})^{-1}\hat{f}_k|^2}{(1 + \hat{f}_k^*(I - B_{k-1}^* B_{k-1})^{-1}\hat{f}_k)^2}, \\ k &= 2, \dots, N-1, \\ b_N^2 &= 1 - \hat{f}_N^* \hat{f}_N, \end{aligned} \tag{32.121}$$

where

$$\hat{f}_2 = \sigma_1^*, \quad \hat{f}_k = \begin{pmatrix} \alpha_{k-2}^* p^*(k) \\ \sigma_{k-1}^* \end{pmatrix}, \ k = 3, \dots, N, \tag{32.122}$$

$$c_k = |\sigma_k|^2 + |q(k)|^2 |\beta_{k+2}|^2, \ k = 1, \dots, N-2, \quad c_{N-1} = |\sigma_{N-1}|^2 \tag{32.123}$$

and

$$\hat{g}_k = \begin{pmatrix} \alpha_{k-2}^* \\ q^*(k-1) \end{pmatrix} \gamma_k, \quad k = 2, \dots, N-1 \tag{32.124}$$

with

$$\gamma_k = |\beta_{k+2}|^2 q(k) + p^*(k+1)\sigma_k, \ k = 2, \dots, N-2, \quad \gamma_{N-1} = p^*(N)\sigma_{N-1}. \tag{32.125}$$

Using (32.95) and (32.62), (32.67) we get $\Delta_2 = |X_2|^2 = \|B_1\|^2$, whence

$$(I - \Delta_2)^{-1} = (1 - B_1^* B_1)^{-1}. \tag{32.126}$$

Using (32.101), we get

$$(I - \Delta_j)^{-1} = F_{j-1}(I - B_{j-1}^* B_{j-1})^{-1} F_{j-1}^*, \quad j = 3, \dots, N-1. \tag{32.127}$$

Set

$$\begin{aligned} \hat{f}_2 = f_2, \ \hat{f}_k &= F_{k-1}^* f_k, \quad k = 3, \dots, N, \\ \hat{g}_k &= F_{k-1}^* g_k, \quad k = 2, \dots, N, \\ c_k &= h_k^* h_k, \qquad k = 1, \dots, N-1. \end{aligned} \tag{32.128}$$

Using (32.93) and (32.7), we obtain (32.122). Using (32.93) and (32.9), we have

$$\begin{aligned} \hat{g}_k = &\begin{pmatrix} \alpha_{k-2}^* & 0 \\ q^*(k-1) & \sigma_{k-1}^* - q^*(k-1)p^*(k) \end{pmatrix} \\ &\times \begin{pmatrix} 1 & 0 \\ 0 & 0 \end{pmatrix} X_{k+1}^* X_{k+1} \begin{pmatrix} q(k) \\ \sigma_k - p(k+1)q(k) \end{pmatrix} \end{aligned}$$

and therefore (32.99) and (32.100) yield (32.124), (32.125).

Using (32.8) with $k = 1$ we have $h_1 = X_2$, and hence $|h_1|^2 = |X_2|^2$. From here using (32.96) we obtain (32.123) with $k = 1$. Using (32.86) and (32.8) we get

$$V_{k+1}h_k = V_{k+1}X_{k+1}q_U(k) = \begin{pmatrix} p(k+1) & 1 \\ \beta_{k+2} & 0 \end{pmatrix} \begin{pmatrix} q(k) \\ \sigma_k - p(k+1)q(k) \end{pmatrix}$$
$$= \begin{pmatrix} \sigma_k \\ \beta_{k+2}q(k) \end{pmatrix},$$
$$k = 2, \dots, N-2$$

which implies (32.123) with $k = 2, \dots, N-2$. Using (32.8) with $k = N-1$ and the equality $X_N = 1$ we have $h_{N-1} = \sigma_{N-1}$, and therefore $c_{N-1} = |\sigma_{N-1}|^2$.

Finally, inserting (32.126), (32.127) and (32.128) in (32.117), (32.118) we obtain (32.120), (32.121).

The matrices from the class $\mathcal{U}_N$ we consider are obtained in the iteration process. Because of rounding errors, a progressive violation of the condition (32.116) may happen during the iterative process and because of this the obtained matrix U will not be unitary. To overcome this drawback we replace the diagonal entries $d(i)$ with suitably corrected values $\hat{d}(i)$, leaving the other elements σ_i, $p(i)$, $q(i)$ of the structure of the lower triangular part of the matrix U unchanged. In this case the corrected values $\hat{d}(i)$ have to be chosen so that the relations

$$|\hat{d}(i) + a_i| = b_i, \quad i = 1, \dots, N, \tag{32.129}$$

hold. Here the values b_i, a_i are defined via the elements σ_i, $p(i)$, $q(i)$ by the formulas (32.120)–(32.125). Thus, for every $i = 1, \dots, N$ we choose the value $\hat{d}(i)$ such that (32.129) holds and the value $|\hat{d}(i) - d(i)|$ is minimal. One can easily check that the desired values $\hat{d}(i)$ are given by the formulas

$$\hat{d}(i) = b_i \frac{d(i) + a_i}{|d(i) + a_i|} - a_i, \quad i = 1, \dots, N. \tag{32.130}$$

Notice that if the elements σ_i, $p(i)$, $q(i)$ are fixed, then the corrected values $\hat{d}(i)$ are defined uniquely. If the computations were made without rounding errors, then the values $d(i)$ and $\hat{d}(i)$ would coincide and the matrix U would be unitary. Replacing $d(i)$ with $\hat{d}(i)$ we obtain a new matrix $\hat{U}$ which is unitary, and hence the new step of the iteration algorithm will continue starting with this unitary matrix.

§32.4 Comments

The quasiseparable and factorization representations of matrices from the class $\mathcal{U}_N$ were introduced and used in [4]. The results on unitary completion of matrices from the class $\mathcal{U}_N$ were derived in [5].

Chapter 33

Explicit Methods

Here we apply the results on the explicit QR method from Chapter 29 to upper Hessenberg matrices from the class $\mathcal{H}_N$.

§33.1 General description

Let A be a matrix from the class $\mathcal{H}_N$, represented in the form

$$A = U - pq^T, \tag{33.1}$$

where $U \in \mathbb{C}^{N\times N}$ is unitary and $p, q \in \mathbb{C}^N$. The matrix A is completely defined by the following parameters:

1. the diagonal entries $d(k)$ $(k = 1, \dots, N)$ of A;
2. the subdiagonal entries σ_k $(k = 1, \dots, N-1)$ of A;
3. the perturbation vectors $p = (p(i))_{i=1}^N$, $q = (q(i))_{i=1}^N$;
4. the upper quasiseparable generators $g(i)$ $(i = 1, \dots, N-1)$, $h(j)$ $(j = 2, \dots, N)$, $b(k)$ $(k = 2, \dots, N-1)$ of U.

As it was mentioned in Subsection §32.2.2, one can always get a set of upper quasiseparable generators of the matrix U with orders

$$s_1 = 1, \quad s_2 = s_3 = \cdots = s_{N-2} = 2, \quad s_{N-1} = 1. \tag{33.2}$$

For the companion matrix C from (32.1), the diagonal and subdiagonal entries are

$$d(k) = 0,\ k = 1, \dots, N, \quad \sigma_k = 1,\ k = 1, \dots, N-1,$$

and the perturbation vectors are defined in (32.3) and upper quasiseparable generators of the unitary matrix U from (32.2) are

$$g(1)=1, \qquad g(i)=\begin{bmatrix} 0 & 0 \end{bmatrix},\ i=2,\dots,N-2, \quad g(N-1)=1,$$

$$h(2)=0, \qquad h(j)=\begin{bmatrix} 0 \\ 0 \end{bmatrix}, \quad j=3,\dots,N-1, \qquad h(N)=1,$$

$$b(2)=\begin{pmatrix} 1 & 0 \end{pmatrix}, \quad b(k)=I_2, \qquad k=3,\dots,N-2, \quad b(N-1)=\begin{pmatrix} 1 \\ 0 \end{pmatrix}.$$

We consider a QR iteration step to be applied to the matrix A. Let α be a complex number and define the matrix A_1 by the rule

$$\begin{cases} A-\alpha I = QR, \\ \qquad A_1 = \alpha I + RQ, \end{cases}$$

where Q is a unitary matrix and R is an upper triangular matrix. It is clear that the matrix A_1 has the form (33.1):

$$A_1 = U_1 - p_1 q_1^T, \tag{33.3}$$

with the unitary matrix $U_1 = Q^*UQ$ and the vectors $p_1 = Q^*p,\ q_1 = Q^Tq$. We will see also that the matrix Q in the algorithm is obtained in the upper Hessenberg form and therefore the matrix A_1 is upper Hessenberg. Consequently, A_1 belongs to the class $\mathcal{H}_N$.

We derive here algorithms to compute diagonal and subdiagonal entries of the matrix A_1, the perturbation vectors p_1, q_1 and upper quasiseparable generators of the matrix U_1.

§33.2 The QR iteration step

Theorem 33.1. *Let A be a matrix from the class $\mathcal{H}_N$ with diagonal and subdiagonal entries $d(k)$ $(k=1,\dots,N)$ and σ_k $(k=1,\dots,N-1)$, perturbation vectors $p=(p(i))_{i=1}^N$, $q=(q(i))_{i=1}^N$ and upper quasiseparable generators $g(i)$ $(i=1,\dots,N-1)$, $h(j)$ $(j=2,\dots,N)$, $b(k)$ $(k=2,\dots,N-1)$ of the corresponding matrix U with orders r_k $(k=1,\dots,N-1)$, and let α be a number.*

Define the matrix A_1 by the rule

$$\begin{cases} A-\alpha I = QR, \\ \qquad A_1 = \alpha I + RQ, \end{cases}$$

where Q is a unitary matrix and R is an upper triangular matrix.

Then diagonal and subdiagonal entries $d^{(1)}(k)$ $(k=1,\dots,N)$ and $\sigma_k^{(1)}$ $(k=1,\dots,N-1)$ and the perturbation vectors $p_1=(p^{(1)}(i))_{i=1}^N$, $q_1=(q^{(1)}(i))_{i=1}^N$

of the matrix A_1*, and upper quasiseparable generators* $g^{(1)}(i)$ $(i = 1, \dots, N-1)$, $h^{(1)}(j)$ $(j = 2, \dots, N)$, $b^{(1)}(k)$ $(k = 2, \dots, N-1)$ *of the corresponding matrix* U_1 *with orders* s_k $(k = 1, \dots, N-1)$ *defined in* (33.2) *are obtained via the following algorithm.*

1. *Determine diagonal and subdiagonal entries* $d^{(1)}(k)$ $(k = 1, \dots, N)$ *and* $\sigma_k^{(1)}$ $(k = 1, \dots, N-1)$ *and the perturbation vectors* $p_1 = (p^{(1)}(i))_{i=1}^N$, $q_1 = (q^{(1)}(i))_{i=1}^N$ *of the matrix* A_1 *as follows.*

1.1. *Determine a complex Givens rotation matrix* Q_1 *and a number* $d_R(1)$ *from the condition*

$$Q_1^* \begin{pmatrix} d(1) - \alpha \\ \sigma_1 \end{pmatrix} = \begin{pmatrix} d_R(1) \\ 0 \end{pmatrix}. \tag{33.4}$$

Determine the numbers $d_Q(1), g_Q(1), q_Q(1), \beta_1$ *from the partition*

$$Q_1 = \begin{pmatrix} d_Q(1) & g_Q(1) \\ q_Q(1) & \beta_1 \end{pmatrix}. \tag{33.5}$$

Compute

$$y_1' = g_Q^*(1)g(1), \quad y_1'' = g_Q^*(1)p(1), \quad y_1''' = \beta_1^*, \tag{33.6}$$

$$p^{(1)}(1) = d_Q^*(1)p(1) + q_Q^*(1)p(2), \tag{33.7}$$

$$q^{(1)}(1) = d_Q(1)q(1) + q_Q(1)q(2), \quad \varphi_2 = g_Q(1)q(1), \tag{33.8}$$

$$\begin{aligned} d^{(1)}(1) = {} & d_R(1)d_Q(1) + (d_Q^*(1)g(1)h(2) - d_Q^*(1)p(1)q(2))q_Q(1) \\ & + d(2)|q_Q(1)|^2 + \alpha(1 - |q_Q(1)|^2). \end{aligned} \tag{33.9}$$

1.2. *For* $k = 2, \dots, N-1$ *perform the following. Compute*

$$y_k = y_{k-1}'h(k) - y_{k-1}''q(k) + y_{k-1}'''(d(k) - \alpha). \tag{33.10}$$

Determine a complex Givens rotation matrix Q_k *and a number* $d_R(k)$ *from the condition*

$$Q_k^* \begin{pmatrix} y_k \\ \sigma_k \end{pmatrix} = \begin{pmatrix} d_R(k) \\ 0 \end{pmatrix}. \tag{33.11}$$

Determine the numbers $h_Q(k), b_Q(k), q_Q(k), \beta_k$ *from the partition*

$$Q_k = \begin{pmatrix} h_Q(k) & b_Q(k) \\ q_Q(k) & \beta_k \end{pmatrix}. \tag{33.12}$$

Compute

$$z_k' = y_{k-1}'b(k) + y_{k-1}'''g(k), \quad z_k'' = y_{k-1}'' + y_{k-1}'''p(k), \tag{33.13}$$

$$p^{(1)}(k) = h_Q^*(k)z_k'' + q_Q^*(k)p(k+1), \tag{33.14}$$

$$\begin{aligned} q^{(1)}(k) &= h_Q(k)\varphi_k + d_Q(k)q(k) + q_Q(k)q(k+1), \\ \varphi_{k+1} &= b_Q(k)\varphi_k + g_Q(k)q(k), \end{aligned} \tag{33.15}$$

$$y'_k = b^*_Q(k)z'_k, \quad y''_k = b^*_Q(k)z''_k, \quad y'''_k = \beta^*_k, \tag{33.16}$$
$$d_Q(k) = \beta_{k-1}h_Q(k), \quad g_Q(k) = \beta_{k-1}b_Q(k), \tag{33.17}$$
$$d^{(1)}(k) = d_R(k)d_Q(k) + d^*_Q(k)(z'_k h(k+1) - z''_k q(k+1))q_Q(k) + d(k)|q_Q(k)|^2 + \alpha(1 - |q_Q(k)|^2), \tag{33.18}$$
$$\sigma^{(1)}_{k-1} = d_R(k)q_Q(k-1). \tag{33.19}$$

1.3. *Compute*

$$d^{(1)}(N) = (y'_{N-1}h(N) - y''_{N-1}q(N))\beta_{N-1} + d(N)|\beta_{N-1}|^2 + \alpha(1 - |\beta_{N-1}|^2), \tag{33.20}$$

$$\sigma^{(1)}_{N-1} = (y'_{N-1}h(N) - y''_{N-1}q(N) + \beta^*_{N-1}(d(N) - \alpha))q_Q(N-1), \tag{33.21}$$
$$p^{(1)}(N) = y''_{N-1} + \beta^*_{N-1}p(N), \quad q^{(1)}(N) = \varphi_N + \beta_{N-1}q(N). \tag{33.22}$$

2. *Determine diagonal entries $d^{(1)}_U(k)$ $(k = 1, \dots, N)$ and subdiagonal entries $\sigma^{U_1}_k$ $(k = 1, \dots, N-1)$ of the matrix U_1 by the formulas*

$$\begin{aligned} d^{(1)}_U(k) &= d^{(1)}(k) + p^{(1)}(k)q^{(1)}(k), \quad k = 1, \dots, N, \\ \sigma^{U_1}_k &= \sigma^{(1)}_k + p^{(1)}(k+1)q^{(1)}(k), \quad k = 1, \dots, N-1. \end{aligned} \tag{33.23}$$

Using Steps 1–3 of Algorithm 32.4 determine the matrices B_k, $k = 1, \dots, N-1$.

If the conditions (32.70) on the matrices B_k, $k = 1, \dots, N-1$ are satisfied, perform the correction of diagonal entries $d^{(1)}_U(k)$ $(k = 1, \dots, N)$ by the formulas (32.120)–(32.125), (32.130) and compute upper quasiseparable generators $g^{(1)}(i)$ $(i = 1, \dots, N-1)$, $h^{(1)}(j)$ $(j = 2, \dots, N)$, $b^{(1)}(k)$ $(k = 2, \dots, N-1)$ of the matrix $U^{(1)}$ by the formulas (32.71)–(32.75).

3. *If some of the conditions (32.70) are not satisfied, compute upper quasiseparable generators $g^{(1)}(i)$ $(i = 1, \dots, N-1)$, $h^{(1)}(j)$ $(j = 2, \dots, N)$, $b^{(1)}(k)$ $(k = 2, \dots, N-1)$ of the matrix $U^{(1)}$ with orders s_k $(k = 1, \dots, N-1)$ defined in (33.2) as follows.*

3.1. *Set*

$$d_W(k) = d_R(k) + \alpha d_Q(k) + p^{(1)}(k)q(k), \ k = 1, \dots, N \tag{33.24}$$

and

$$\tilde{h}(k) = d(k) + p(k)q(k), \quad k = 2, \dots, N. \tag{33.25}$$

Set

$$\lambda_1 = g^*_Q(1)g(1), \quad \theta_1 = q(1)g_Q(1), \quad \delta_1 = \alpha|g_Q(1)|^2, \tag{33.26}$$

and for $k = 2, \dots, N-1$ compute

$$\lambda_k = b^*_Q(k)\lambda_{k-1}b(k) + g^*_Q(k)g(k), \tag{33.27}$$
$$\theta_k = \theta_{k-1} + q(k)g_Q(k), \quad \delta_k = \delta_{k-1}|b_Q(k)|^2 + \alpha|g_Q(k)|^2. \tag{33.28}$$

Set

$$\begin{aligned}
g'_1 &= d_W(1)g_Q(1), \quad g''_1 = 0_{1\times r_1},\\
g'_k &= d_W(k)g_Q(k) + (p^{(1)}(k)\theta_k + h^*_Q(k)\delta_k)b_Q(k),\\
g''_k &= h^*_Q(k)\lambda_{k-1}b(k),\\
w''_k &= h(k)d_Q(k) + b(k)h(k+1)q_Q(k),\\
w'''_k &= \tilde{h}(k)d_Q(k) + g(k)h(k+1)q_Q(k),\\
k &= 2,\dots,N-1,\\
w''_N &= \beta_{N-1},\ w'''_N = \tilde{h}(N)\beta_{N-1}.
\end{aligned} \tag{33.29}$$

3.2. *Set*

$$\begin{aligned}
&\rho_N = 0,\ \rho_{N-1} = 1, \quad \rho_{N-2} = \cdots = \rho_2 = 2, \quad \rho_1 = 1,\\
&\nu_N = \nu_{N-1} = 0, \quad \nu_{N-2} = \cdots = \nu_3 = 1, \quad \nu_2 = \nu_1 = 2,\\
&s_N = 0,\ s_{N-1} = 1, \quad s_{N-2} = \cdots = s_2 = 2, \quad s_1 = 1.
\end{aligned} \tag{33.30}$$

Set $\mu_N = p^{(1)}(N)$ *and for* $k = N-1,\dots,3$ *determine complex Givens rotation matrices* V_k *and numbers* μ_k *such that*

$$V_k^* \begin{pmatrix} p(k) \\ \mu_{k+1} \end{pmatrix} = \begin{pmatrix} \mu_k \\ 0 \end{pmatrix}. \tag{33.31}$$

Determine a complex Givens matrix V_2 *and a number* μ_2 *such that*

$$V_2^* \begin{pmatrix} \sigma_1^{U_1} \\ \mu_3 q^{(1)}(1) \end{pmatrix} = \begin{pmatrix} \mu_2 \\ 0 \end{pmatrix}. \tag{33.32}$$

Set $h^{(1)}(N) = 1$ *and*

$$h^{(1)}(N-1) = \begin{pmatrix} 1 \\ 0 \end{pmatrix}, \quad b^{(1)}(N-1) = \begin{pmatrix} 0 \\ 1 \end{pmatrix}, \tag{33.33}$$

and compute

$$g^{(1)}(N-1) = g'_{N-1} + g''_{N-1}w''_N + w'''_N, \tag{33.34}$$

$$z_{N-1} = V_{N-1}^* \begin{pmatrix} d_U^{(1)}(N-1) & g^{(1)}(N-1) \\ \sigma_{N-1}^{U_1} & d_U^{(1)}(N) \end{pmatrix}, \tag{33.35}$$

$$\begin{aligned}
\gamma'_{N-1} &= \begin{pmatrix} h_Q(N-1) & b_Q(N-1) \end{pmatrix},\\
\gamma''_{N-1} &= \begin{pmatrix} w''_{N-1} & h(N-1)g_Q(N-1) + b(N-1)w''_N \end{pmatrix},\\
\gamma'''_{N-1} &= \begin{pmatrix} w'''_{N-1} & \tilde{h}(N-1)g_Q(N-1) + g(N-1)w''_N \end{pmatrix},
\end{aligned} \tag{33.36}$$

$$g^{(1)}(N-2) = g'_{N-2}\gamma'_{N-1} + g''_{N-2}\gamma''_{N-1} + \gamma'''_{N-1}. \tag{33.37}$$

For $k = N-2, \dots, 2$ perform the following. Determine the numbers $p_V(k+1), d_V(k+1), a_V(k+1), q_V(k+1)$ from the partition

$$V_{k+1} = \begin{pmatrix} p_V(k+1) & d_V(k+1) \\ a_V(k+1) & q_V(k+1) \end{pmatrix}. \tag{33.38}$$

Compute the vector

$$X'_{k+1} = \begin{pmatrix} \mu_{k+1} q^{(1)}(k) + p_V^*(k+1)\sigma_k^{U_1} \\ d_V^*(k+1)\sigma_k^{U_1} \end{pmatrix}. \tag{33.39}$$

Compute the 3×3 matrix

$$Z_k = \begin{pmatrix} V_k^* & 0 \\ 0 & 1 \end{pmatrix} \begin{pmatrix} d_U^{(1)}(k) & g^{(1)}(k) \\ X'_{k+1} & z_{k+1} \end{pmatrix} \tag{33.40}$$

and determine the matrices $h'_k, h''_k, \Delta_k, \Theta_k$ of sizes $\rho_{k-1} \times 1, \rho_{k-1} \times s_k, \nu_k \times 1, \nu_k \times s_k$ from the partition

$$Z_k = \begin{pmatrix} h'_k & h''_k \\ \Delta_k & \Theta_k \end{pmatrix}. \tag{33.41}$$

The submatrix $\begin{pmatrix} \Delta_k & \Theta_k \end{pmatrix}$ has orthonormal rows and one can determine the $(1+s_k) \times (1+s_k)$ unitary matrix F_k from the condition

$$\begin{pmatrix} \Delta_k & \Theta_k \end{pmatrix} F_k^* = \begin{pmatrix} 0_{\nu_k \times s_{k-1}} & I_{\nu_k} \end{pmatrix}. \tag{33.42}$$

Determine the matrices $h^{(1)}(k), b^{(1)}(k), d_F(k), g_F(k)$ of sizes $s_{k-1} \times n_k, s_{k-1} \times s_k, \nu_k \times n_k, \nu_k \times s_k$ from the partition

$$F_k = \begin{pmatrix} h^{(1)}(k) & b^{(1)}(k) \\ d_F(k) & g_F(k) \end{pmatrix}. \tag{33.43}$$

Compute the $\rho_{k-1} \times s_{k-1}$ matrix z_k by the formula

$$z_k = h'_k(h^{(1)}(k))^* + h''_k(b^{(1)}(k))^*. \tag{33.44}$$

Compute

$$\begin{aligned} \gamma'_k &= h_Q(k)(h^{(1)}(k))^* + b_Q(k)\gamma'_{k+1}(b^{(1)}(k))^*, \\ \gamma''_k &= w''_k(h^{(1)}(k))^* + (h(k)g_Q(k)\gamma'_{k+1} + b(k)\gamma''_{k+1})(b^{(1)}(k))^*, \\ \gamma'''_k &= w'''_k(h^{(1)}(k))^* + (\tilde{h}(k)g_Q(k)\gamma'_{k+1} + g(k)\gamma''_{k+1})(b^{(1)}(k))^*, \end{aligned} \tag{33.45}$$

$$g^{(1)}(k-1) = g'_{k-1}\gamma'_k + g''_{k-1}\gamma''_k + \gamma'''_k. \tag{33.46}$$

Proof. We apply Theorem 29.4 to the matrix A. Lower quasiseparable generators of A are the numbers

$$\begin{aligned} p_A(i) &= \sigma_{i-1}, \ i = 2, \dots, N; \\ q_A(j) &= 1, \quad j = 1, \dots, N-1; \\ a_A(k) &= 0, \quad k = 2, \dots, N-1. \end{aligned} \tag{33.47}$$

The orders of these lower generators equal to one: $r_k^L = 1,\ k = 1, \dots, N-1$. Using the formula (33.1) we obtain the formulas for upper quasiseparable generators of the matrix A:

$$\begin{aligned} g_A(i) &= \begin{pmatrix} g(i) & p(i) \end{pmatrix},\ i = 1, \dots, N-1; \\ h_A(j) &= \begin{pmatrix} h(j) \\ -q(j) \end{pmatrix}, \quad j = 2, \dots, N; \\ b_A(k) &= \begin{pmatrix} b(k) & 0 \\ 0 & 1 \end{pmatrix}, \quad k = 2, \dots, N-1. \end{aligned} \tag{33.48}$$

In accordance with Steps 1 and 2 of the algorithm from Theorem 29.4, we apply Theorem 29.1 to compute the diagonal and subdiagonal entries of the matrix A_1 and the perturbation vectors p_1, q_1. Using the first formula in (29.6) and the formula (29.8) with $p_A(k) = \sigma_{k-1},\ a_A(k) = 0$, we get $C_k = \sigma_{k-1}$ $(k = N, \dots, 2)$, $V_k = I_2$ $(k = N-1, \dots, 2)$. From here, using the second equality in (29.6) and (29.9), we get

$$p_V(N) = 1, \quad p_V(k) = q_V(k) = 1,\ d_V(k) = a_V(k) = 0,\ k = 2, \dots, N-1 \tag{33.49}$$

and $d_V(N)$ to the 1×0 empty matrix. Hence, using the third equality in (29.6) and the equalities (29.10), (29.12) with $C_k = \sigma_{k-1}$, $q_A(k) = 1$ and $g_A(k), h_A(k), b_A(k)$ from (33.48), we obtain the formulas for quasiseparable generators $h_R(k), b_R(k)$ of the matrix R and auxiliary variables Θ_k, Δ_k:

$$h_R(k) = \begin{pmatrix} h(k) & -q(k) & d(k) - \alpha \end{pmatrix}^T, \quad k = 2, \dots, N, \tag{33.50}$$

$$b_R(k) = \begin{pmatrix} b(k) & 0 & 0 \\ 0 & 1 & 0 \\ g(k) & p(k) & 0 \end{pmatrix}, \qquad k = 2, \dots, N-1, \tag{33.51}$$

$$\Delta_1 = \begin{pmatrix} d(1) - \alpha \\ \sigma_1 \end{pmatrix}, \qquad \Theta_1 = \begin{pmatrix} g(1) & p(1) & 0 \\ 0 & 0 & 1 \end{pmatrix}, \tag{33.52}$$

$$\Delta_k = \sigma_k, \qquad \Theta_k = \begin{pmatrix} 0 & 0 & 1 \end{pmatrix}. \tag{33.53}$$

From the equalities (29.13), (29.16), (29.19) we obtain the following. We use the partitions of the row vectors D_k used in (29.13), (29.16), (29.19) in the form

$$D_k = \begin{pmatrix} y'_k & y''_k & y'''_k \end{pmatrix}, \quad k = 1, \dots, N-1 \tag{33.54}$$

with a row vector y'_k and numbers y''_k, y'''_k.

Using (29.13) and the equalities (33.52) we obtain the condition (33.4) to determine a complex Givens rotation matrix Q_1 and a number $d_R(1)$. Using (29.14), (29.15) and (33.5), (33.52), (33.54) we obtain the equality

$$g_R(1) = \begin{pmatrix} d_Q^*(1) g(1) & d_Q^*(1) p(1) & q_Q^*(1) \end{pmatrix} \tag{33.55}$$

and the formulas (33.6).

Next, for $k = 2, \dots, N-1$ we set

$$y_k = D_{k-1}h_R(k), \quad k = 2, \dots, N-1. \tag{33.56}$$

Using (33.50) and (33.54) we obtain (33.10). Using (33.51), (33.54) we obtain the equalities

$$D_{k-1}b_R(k) = \begin{pmatrix} z'_k & z''_k & 0 \end{pmatrix}, \quad k = 2, \dots, N-1, \tag{33.57}$$

with the row vectors z'_k and the number z''_k defined via (33.13). Inserting (33.56) and (33.53) in (29.16), we obtain the condition (33.11) to determine a complex Givens rotation matrix Q_k and a number $d_R(k)$. Using (29.17), (29.18) and (33.12), (33.53), (33.57), (33.54) we obtain the equalities

$$g_R(k) = \begin{pmatrix} h_Q^*(k)z'_k & h_Q^*(k)z''_k & q_Q^*(k) \end{pmatrix}, \quad k = 2, \dots, N-1 \tag{33.58}$$

and the formulas (33.16).

Using (29.19) and (33.50) with $k = N$ we obtain the equality

$$d_R(N) = y'_{N-1}h(N) - y''_{N-1}q(N) + y'''_{N-1}(d(N) - \alpha) \tag{33.59}$$

and also $h_U(N) = 1$ and that $d_U(N)$ is the 0×1 empty matrix.

Next we use the formulas (29.20)–(29.25) to determine quasiseparable generators of the matrix Q. From the partition (29.20) we obtain the partition (33.5) to determine generators $d_Q(1)$, $g_Q(1), q_Q(1)$ of Q and the auxiliary variable β_1. Using (29.17), (29.21) and (33.49) we get

$$p_Q(k) = 1, \quad a_Q(k) = 0, \qquad k = 2, \dots, N-1, \tag{33.60}$$

and

$$Q_k(1,:) = \begin{bmatrix} h_Q(k) & b_Q(k) \end{bmatrix}, \quad k = 2, \dots, N-1. \tag{33.61}$$

Using (29.22) with $V_k = I$ and (29.23) we obtain the equalities

$$\begin{pmatrix} d_Q(k) & g_Q(k) \\ q_Q(k) & \beta_k \end{pmatrix} = \begin{pmatrix} \beta_{k-1} & 0 \\ 0 & 1 \end{pmatrix} Q_k, \quad k = 2, \dots, N-1. \tag{33.62}$$

Combining (33.61) and (33.62) we obtain the partitions (33.12) and the formulas (33.17). Using (29.24), (29.25) with $p_V(N) = h_U(N) = 1$ and empty matrices $d_V(N), d_U(N)$, we get

$$p_Q(N) = h_Q(N) = 1, \quad d_Q(N) = \beta_{N-1}. \tag{33.63}$$

Now using Step 4 of the algorithm from Theorem 29.1 we derive the formulas the for diagonal and subdiagonal entries of the matrix $A^{(1)}$. Using the formulas (29.26), (29.27) with $p_Q(N) = 1$ we get

$$\gamma_N = h_R(N), \quad p^{(1)}(N) = d_R(N), \tag{33.64}$$

and

$$d^{(1)}(N) = d_R(N)d_Q(N) + \alpha. \tag{33.65}$$

Using (29.28) and the equality $q^{(1)}(1) = q_Q(1)$ we get

$$\tilde{q}^{(1)}(k) = q_Q(k), \quad k = 1, \dots, N-1. \tag{33.66}$$

Using (29.29) with $p_Q(k) = 1$, $a_Q(k) = 0$ and (29.30) we have

$$\begin{pmatrix} \gamma_k & \tilde{h}(k) \\ \tilde{p}^{(1)}(k) & d''(k) \end{pmatrix} = \begin{pmatrix} h_R(k) & b_R(k) \\ d_R(k) & g_R(k) \end{pmatrix} \begin{pmatrix} 1 & 0 \\ 0 & \gamma_{k+1} \end{pmatrix} \begin{pmatrix} 1 & d_Q(k) \\ 0 & q_Q(k) \end{pmatrix}, \quad k = 2, \dots, N-1. \tag{33.67}$$

From (33.64) and (§33.2) we obtain

$$\gamma_k = h_R(k), \quad \tilde{p}^{(1)}(k) = d_R(k), \quad k = 2, \dots, N. \tag{33.68}$$

Inserting $\gamma_{k+1} = h_R(k+1)$ in (§33.2) we obtain

$$d''(k) = d_R(k)d_Q(k) + g_R(k)h_R(k+1)q_Q(k), \quad k = 2, \dots, N-1. \tag{33.69}$$

Inserting (33.69) in (29.31) and using (29.33) with $\gamma_2 = h_R(2)$ we obtain

$$d^{(1)}(k) = d_R(k)d_Q(k) + g_R(k)h_R(k+1)q_Q(k) + \alpha, \quad k = 1, \dots, N-1. \tag{33.70}$$

Inserting (33.55), (33.58) and (33.50) in (33.70) we obtain the equalities (33.9), (33.18). Inserting in (33.65) the expression (33.59) with $y'''_{N-1} = \beta^*_{N-1}$ and using also $d_Q(N) = \beta_{N-1}$ we obtain (33.20).

For the subdiagonal entries $\sigma_k^{(1)}$ we have

$$\sigma_k^{(1)} = \tilde{p}^{(1)}(k+1)\tilde{q}^{(1)}(k), \quad k = 1, \dots, N-1.$$

From here using (33.66) with $k = 1, \dots, N-2$ and (33.68) we obtain (33.19). Using (33.64), (33.59) and (33.66) with $k = N-1$ we obtain (33.21).

Next we derive the formulas (33.7), (33.8), (33.14), (33.15) and (33.22) for the perturbation vectors $p_1 = Q^*p$, $q_1 = Q^T q$. At first we apply Algorithm 13.1 to the product $p_1 = Q^*p$. Here the matrix Q^* has quasiseparable generators $h^*_Q(i)$ $(i = 2, \dots, N)$, $g^*_Q(j)$ $(j = 1, \dots, N-1)$, $b^*_Q(k)$ $(k = 2, \dots, N-1)$, $q^*_Q(i)$ $(i = 1, \dots, N-1)$, $p^*_Q(j) = 1$ $(j = 2, \dots, N)$, $a^*_Q(k) = 0$ $(k = 2, \dots, N-1)$, $d^*_Q(k)$ $(k = 1, \dots, N)$. Algorithm 13.1 yields

$$p_1 = p_L + p_D + p_U,$$

with

$$p_L(1) = 0, \quad p_L(i) = h^*_Q(i)\chi_i, \ i = 2, \dots, N,$$

where

$$\chi_2 = g_Q^*(1)p(1), \quad \chi_i = b^*(i-1)\chi_{i-1} + g_Q^*(i-1)p(i-1),\ i = 3, \dots, N, \quad (33.71)$$
$$p_D(i) = d_Q^*(i)p(i), \quad i = 1, \dots, N,$$

and

$$p_U(i) = q_Q^*(i)\eta_i, \quad i = 1, \dots, N-1, \quad x_U(N) = 0,$$

where $\eta_i = p(i+1)$, $i = 1, \dots, N-1$. From here we obtain the relation (33.7), the equalities

$$p^{(1)}(i) = h_Q^*(i)\chi_i + d_Q^*(i)p(i) + q_Q^*(i)p(i+1), \quad i = 2, \dots, N-1 \qquad (33.72)$$

and the equality

$$p^{(1)}(N) = h_Q^*(N)\chi_N + d_Q^*(N)p(N). \qquad (33.73)$$

Inserting (33.17) in (33.71), (33.72) we obtain the recursions

$$\chi_2 = g_Q^*(1)p(1), \quad \chi_{i+1} = b_Q^*(i)(\chi_i + \beta_{i-1}^* p(i)),\ i = 2, \dots, N-1 \qquad (33.74)$$

and the equalities

$$p^{(1)}(i) = h_Q^*(i)(\chi_i + \beta_{i-1}^* p(i)) + q_Q^*(i)p(i+1). \qquad (33.75)$$

Inserting $h_Q(N) = 1$, $d_Q(N) = \beta_{N-1}$ in (33.73) we obtain the equality

$$p^{(1)}(N) = \chi_N + \beta_{N-1}^* p(N). \qquad (33.76)$$

Notice that from the second equality in (33.6), (33.13) and (33.16) it follows that

$$y_1'' = g_Q^*(1)p(1), \quad z_k'' = y_{k-1}'' + \beta_{k-1}^* p(k), \quad y_k'' = b_Q^*(k)z_k'', \quad k = 2, \dots, N-1.$$

Comparing this with (33.74), we get

$$y_k'' = \chi_{k+1},\ k = 1, \dots, N-1, \quad z_k'' = \chi_k + \beta_{k-1}^* p(k),\ k = 2, \dots, N-1,$$

and hence, using the formulas (33.75), (33.76), we obtain the formulas (33.14) and the first equality in (33.22).

Proceeding in the same way as in the proof of the relations (33.7), (33.72), (33.73) one derives the formulas (33.8), (33.15) and (33.22) for the vector $q^{(1)} = Q^T q$.

On Step 2 we use Algorithm 32.4 together with the correction formulas (32.120)–(32.125), (32.130).

If the conditions of Algorithm 32.4 are not satisfied, we proceed as in Steps 3 and 4 of the algorithm from Theorem 29.4. At first we obtain upper quasiseparable generators of the matrix U_1 using Step 3.

Applying Step 3.1 of the algorithm from Theorem 29.4 we obtain the following. Using the formulas (29.51), (29.52) we obtain the formulas (33.24) and

$$\begin{aligned} p_W(i) &= \begin{pmatrix} p^{(1)}(i) & h_Q^*(i) \end{pmatrix}, && i = 2, \dots, N, \\ q_W(j) &= \begin{pmatrix} q(j) \\ \alpha g_Q^*(j) \end{pmatrix}, && j = 1, \dots, N-1, \\ a_W(k) &= \begin{pmatrix} 1 & 0 \\ 0 & b_Q^*(k) \end{pmatrix}, && k = 2, \dots, N-1. \end{aligned} \tag{33.77}$$

Using (29.53), (29.54) we obtain the first equality in (33.26), the relations (33.27) and the formulas

$$\tilde{g}(1) = 0_{1 \times r_1}, \quad \tilde{g}(k) = h_Q^*(k)\lambda_{k-1}b(k). \tag{33.78}$$

Using (29.55), (29.56) and (33.49) we obtain (33.25). Thus using (29.57) and (33.49) we get

$$\begin{aligned} g_W(1) &= \begin{pmatrix} g(1) & 1 \end{pmatrix}, && i = 2, \dots, N-1, \\ g_W(i) &= \begin{pmatrix} g(i) + \tilde{g}(i) & 1 \end{pmatrix}, && i = 2, \dots, N-1, \\ h_W(j) &= \begin{pmatrix} h(j) \\ \tilde{h}(j) \end{pmatrix}, && j = 2, \dots, N, \\ b_W(k) &= \begin{pmatrix} b(k) & 0 \\ g(k) & 0 \end{pmatrix}, && k = 2, \dots, N-1. \end{aligned} \tag{33.79}$$

Applying Step 3.2 of the algorithm from Theorem 29.4 we obtain the following. For the variables $\hat{\beta}_k$, which are two-dimensional columns, we use the partitions

$$\hat{\beta}_k = \begin{pmatrix} \theta_k \\ \delta_k \end{pmatrix}, \quad k = 1, \dots, N-1,$$

with numbers θ_k, δ_k. Using (29.58), (29.59) and (33.79) we obtain the second and the third formula in (33.26), the formulas (33.28) and the equalities

$$\hat{g}(1) = 0, \quad \hat{g}(k) = (p^{(1)}(k)\theta_k + h_Q^*(k)\delta_k)b_Q(k), \; k = 2, \dots, N-1. \tag{33.80}$$

Using (29.60), (29.61) and (33.60), (33.63) we obtain the equalities

$$\begin{gathered} \hat{h}(N) = h_W(N)\beta_{N-1}, \quad \hat{h}(k) = h_W(k)d_Q(k) + b_W(k)h_W(k+1)q_Q(k), \\ k = N-1, \dots, 2. \end{gathered} \tag{33.81}$$

Thus, using (29.62) and (33.78), (33.79), (33.80), (33.81) we obtain upper quasiseparable generators of the matrix U_1 via relations

$$\hat{g}_B(i) = \begin{pmatrix} g'_i & g''_i & 1 \end{pmatrix}, \qquad i = 1, \dots, N-1,$$
$$\hat{h}_B(j) = \begin{pmatrix} h_Q(j) \\ w''_j \\ w'''_j \end{pmatrix}, \qquad j = 2, \dots, N, \tag{33.82}$$
$$\hat{b}_B(k) = \begin{pmatrix} b_Q(k) & 0 & 0 \\ h(k)g_Q(k) & b(k) & 0 \\ \tilde{h}(k)g_Q(k) & g(k) & 0 \end{pmatrix}, k = 2, \dots, N-1$$

with $g', g''_k, g'''_k, h''_k, h'''_k$ defined in (33.29).

Next we justify Step 3.2 of the algorithm. By Lemma 32.1, a set of lower quasiseparable generators $p_U^{(1)}(i)$ $i = 2, \dots, N$, $q_U^{(1)}(j)$ $j = 1, \dots, N$, $a_U^{(1)}(k)$ ($k = 2, \dots, N-1$) of the matrix U_1 may be determined by the formulas

$$p_U^{(1)}(2) = \sigma_1^{(U_1)}, \qquad p_U^{(1)}(i) = \begin{pmatrix} p^{(1)}(i) & 1 \end{pmatrix}, \quad i = 3, \dots, N-1, \qquad p_U^{(1)}(N) = 1, \tag{33.83}$$

$$q_U^{(1)}(1) = 1, \qquad q_U^{(1)}(j) = \begin{pmatrix} q^{(1)}(j) \\ \sigma_j^{U_1} \end{pmatrix}, \quad j = 2, \dots, N-2, \qquad q_U^{(1)}(N-1) = \sigma_{N-1}^{U_1}, \tag{33.84}$$

$$a_U^{(1)}(2) = \begin{pmatrix} q^{(1)}(1) \\ 0 \end{pmatrix}, \qquad a_U^{(1)}(k) = \begin{pmatrix} 1 & 0 \\ 0 & 0 \end{pmatrix}, \quad k = 3, \dots, N-2, \qquad a_U^{(1)}(N-1) = \begin{pmatrix} p^{(1)}(N) & 0 \end{pmatrix}. \tag{33.85}$$

The orders of these lower generators are

$$r_1^L = 1, \quad r_2^L = r_3^L = \cdots = r_{N-2}^L = 2, \quad r_{N-1}^L = 1. \tag{33.86}$$

We apply Theorem 7.5 to compute a set of upper quasiseparable generators of the matrix U_1 with the same orders as in (33.86) (or in (33.2)).

Inserting (33.86) in (7.22) we obtain the equalities (33.30) for the sizes ρ_k, ν_k and orders of upper quasiseparable generators s_k.

Using (7.28) and (33.83) with $k = N$, we get $p_U^{(1)}(N) = 1 = V_N X_N^U$, and therefore $V_N = 1$, $X_N^U = 1$. Hence, using (7.28) with $k = N-1$ and (33.83), (33.85) we get

$$\begin{pmatrix} p^{(1)}(N-1) & 1 \\ p^{(1)}(N) & 0 \end{pmatrix} = V_{N-1} X_{N-1}^U$$

with a unitary 2×2 matrix V_{N-1} and a 2×2 upper triangular matrix X_{N-1}^U. Take the matrix V_{N-1} such that the condition (33.31) holds. From here using (33.38) we get

$$X_{N-1}^U = V_{N-1}^* \begin{pmatrix} p^{(1)}(N-1) & 1 \\ p^{(1)}(N) & 0 \end{pmatrix} = \begin{pmatrix} \mu_{N-1} & p_V^*(N-1) \\ 0 & d_V^*(N-1) \end{pmatrix}. \tag{33.87}$$

For $k = N-2, \dots, 3$, using (7.28), we get

$$\begin{pmatrix} p_U(k) \\ X_{k+1}^U a_U^{(1)}(k) \end{pmatrix} = V_k^U \begin{pmatrix} X_k^U \\ 0_{1\times 2} \end{pmatrix}, \tag{33.88}$$

with unitary 3×3 matrices V_k^U and upper triangular 2×2 matrices X_k^U. Assume that the matrix X_{k+1}^U has the form

$$X_{k+1}^U = \begin{pmatrix} \mu_{k+1} & p_V^*(k+1) \\ 0 & d_V^*(k+1) \end{pmatrix}. \tag{33.89}$$

Then from (33.87) it follows that for $k = N-2$ the representation (33.89) holds. Inserting (33.83), (33.85) and (33.89) in (7.28) we get

$$\begin{pmatrix} p(k) & 1 \\ \mu_{k+1} & 0 \\ 0 & 0 \end{pmatrix} = V_k^U \begin{pmatrix} X_k^U \\ 0_{1\times 2} \end{pmatrix}.$$

It follows that the matrix V_k^U has the form

$$V_k^U = \begin{pmatrix} V_k & 0 \\ 0 & 1 \end{pmatrix}, \tag{33.90}$$

with the unitary 2×2 matrix V_k determined from the condition (33.31). Moreover, the relation

$$X_k^U = V_k^* \begin{pmatrix} p^{(1)}(k) & 1 \\ \mu_{k+1} & 0 \end{pmatrix} = \begin{pmatrix} \mu_k & p_V^*(k) \\ 0 & d_V^*(k) \end{pmatrix}$$

holds. Finally, using (7.28) and (33.83), (33.85) with $k = 2$ we have

$$\begin{pmatrix} p_U^{(1)}(2) \\ X_3^U a_U^{(1)}(2) \end{pmatrix} = \begin{pmatrix} \sigma_1^{U_1} \\ \mu_3 q^{(1)}(1) \\ 0 \end{pmatrix} = V_2^U X_2^U,$$

where

$$V_2^U = \begin{pmatrix} V_2 & 0 \\ 0 & 1 \end{pmatrix}, \tag{33.91}$$

with the unitary 2×2 matrix V_2 determined from the condition (33.32).

Next we apply the formulas (7.29)–(7.34). For the auxiliary Y_k from Theorem 7.5 used in this theorem we introduce the partitions

$$Y_k = \begin{pmatrix} \gamma_k' \\ \gamma_k'' \\ \gamma_k''' \end{pmatrix}, \quad k = N, \dots, 2, \tag{33.92}$$

with the matrices $\gamma_k', \gamma_k'', \gamma_k'''$ of sizes $1 \times s_{k-1}, r_{k-1} \times s_{k-1}, 1 \times s_{k-1}$, respectively. For $k = N$, using $V_N = 1$, $X_N^U = 1$, we obtain the formulas

$$Z_N = d_U^{(1)}(N),\ F_N = 1 = h^{(1)}(N),\ z_N = d_U^{(1)}(N),\ Y_N = \hat{h}_B(N).$$

Moreover, inserting $Y_N = \hat{h}_B(N)$ in (7.34) and using (33.82) and $h_Q(N) = 1$, we obtain (33.34).

For $k = N-1$, using (7.29) and the equality $X_N^U q_U^{(1)}(N-1) = \sigma_{N-1}^{U_1}$, we obtain the formula

$$Z_{N-1} = V_{N-1}^* \begin{pmatrix} d_U^{(1)}(N-1) & g^{(1)}(N-1) \\ \sigma_{N-1}^{U_1} & d_U^{(1)}(N) \end{pmatrix}.$$

Moreover, since $\nu_{N-1} = 0$ we get $F_{N-1} = I_2$, which yields (33.33) and the representation $Z_{N-1} = \begin{pmatrix} h'_{N-1} & h''_{N-1} \end{pmatrix}$. Furthermore, using (7.33) we obtain (33.35) and the equality

$$Y_{N-1} = \begin{pmatrix} \hat{h}_B(N-1) & \hat{b}_B(N-1)Y_N \end{pmatrix}.$$

From here using $Y_N = \hat{h}_B(N)$, the representation (33.92) and (33.82) we obtain (33.36). Moreover, using $Y_N = \hat{h}_B(N)$, the representation (33.92) and (33.82) together with (7.34) we obtain (33.37).

For $k = N-2, \dots, 2$ we obtain the following. Setting $X'_k = X_{k+1}^U q_U^{(1)}(k)$ and using (33.89) and (33.84) we obtain (33.39). Inserting $X_{k+1}^U q_U^{(1)}(k) = X'_{k+1}$ in (7.29) and using (33.90), we obtain (33.40). The formulas (33.41), (33.42), (33.43) follow directly from the relations (7.30), (7.31), (7.32). Using (7.33) we obtain the formula (33.44) and the formula

$$Y_k = \hat{h}_B(k)(h_U^{(1)}(k))^* + \hat{b}_B(k)Y_{k+1}(b_U^{(1)}(k))^*.$$

From here using (33.92) and (33.82) we obtain (33.45). Moreover, using (33.92) and (33.82) together with (7.34) we obtain (33.46). □

We will compute only the complexity of the part which is performed about N times of the algorithm for one step with the explicit method QR as in items 1.2., 2., 3.1. and 3.2. of Theorem 33.1. To this end denote by r the maximal order of the upper quasiseparable generators of the matrix U, i.e., $r = \max_{k=1,\dots,N-1}(r_k)$.

The complexity of Step 1.2 is less than $2r^2 + 6r + 37$ arithmetical operations, the complexity of Step 2 is 40 arithmetical operations, of Step 3.1 is 45 arithmetical operations and of the part of Step 3.2 which is performed about N times is less than $89 + \eta(1,3)$, since $s_k \leq 2$, $k = 1, \dots, N-1$. Here $\eta(j,k)$ denotes the complexity for finding a unitary $k \times k$ matrix as in (33.42) for j orthonormal rows and k columns.

The overall complexity is less than

$$2r^2 + 6r + 211 + \eta(1,3).$$

§33.3 Comments

The invariance of the class $\mathcal{H}_N$ under QR iteration was established by D. Bini, F. Daddi and L. Gemignani in the paper [3]. The explicit method presented in this chapter was developed in [4, 5].

Chapter 34

Implicit Methods with Compression

Here we apply the method presented in Section §31.1 to matrices from the class $\mathcal{H}_N$. As above, for a matrix A from the class $\mathcal{H}_N$ represented in the form

$$A = U - pq^T,$$

where $U \in \mathbb{C}^{N \times N}$ is unitary and $p, q \in \mathbb{C}^N$, the new iterate $A^{(1)} = Q^* A Q$ has the form

$$A^{(1)} = U^{(1)} - p^{(1)} (q^{(1)})^T,$$

with the unitary matrix $U^{(1)} = Q^* U Q$ and the vectors $p^{(1)} = Q^* p$, $q^{(1)} = Q^T q$. But here we use upper triangular generators instead of upper quasiseparable ones.

A matrix A from the class $\mathcal{H}_N$ is completely defined by the following parameters:

1. the subdiagonal entries σ_k $(k = 1, \dots, N-1)$ of A;
2. the perturbation vectors $p = (p(i))_{i=1}^N$, $q = (q(i))_{i=1}^N$;
3. the upper triangular generators $g(i), h(i)$ $(i = 1, \dots, N)$, $b(k)$ $(k = 1, \dots, N-1)$ of the matrix U.

The task of an algorithm is to update these parameters.

For the companion matrix C from (32.1) the subdiagonal entries are $\sigma_k = 1$, $k = 1, \dots, N-1$, the perturbation vectors are defined in (32.3) and upper triangular generators of the unitary matrix U from (32.2) are

$$\begin{aligned} g(1) = 1,\ g(i) &= 0, \quad i = 2, \dots, N, \\ h(j) &= 0, \quad j = 1, \dots, N-1, \quad h(N) = 1, \\ b(k) &= 1, \quad k = 1, \dots, N-1. \end{aligned} \tag{34.1}$$

The results of this chapter follow from more general ones obtained in Section §31.4.

§34.1 The single shift case

We start with the QR iterations with single shift.

Theorem 34.1. *Let $A = U - pq^T$ be a matrix from the class $\mathcal{H}_N$, with subdiagonal entries σ_k $(k = 1, \dots, N-1)$, perturbation vectors $p = \mathrm{col}(p(i))_{i=1}^N$, $q = \mathrm{col}(q(i))_{i=1}^N$, upper triangular generators $g(i), h(i)$ $(i = 1, \dots, N)$, $b(k)$ $(k = 1, \dots, N-1)$ of orders r_k $(k = 1, \dots, N)$ of the matrix U, and let α be a number.*

Then the Givens rotation matrices Q_k, $k = 1, \dots, N-1$, such that the unitary matrix

$$Q = \tilde{Q}_1\tilde{Q}_2 \cdots \tilde{Q}_{N-2}\tilde{Q}_{N-1}, \tag{34.2}$$

with

$$\tilde{Q}_i = I_{i-1} \oplus Q_i \oplus I_{N-i-1},\ i = 1, \dots, N-1, \tag{34.3}$$

satisfies the condition (31.4), as well as subdiagonal entries $\sigma_k^{(1)}$ $(k = 1, \dots, N-1)$, of the upper Hessenberg matrix $A^{(1)} = Q^*AQ$, perturbation vectors $p^{(1)} = \mathrm{col}(p^{(1)}(i))_{i=1}^N$, $q^{(1)} = \mathrm{col}(q^{(1)}(i))_{i=1}^N$ and upper triangular generators $g^{(1)}(i)$, $h^{(1)}(i)$ $(i = 1, \dots, N)$, $b^{(1)}(k)$ $(k = 1, \dots, N)$ of the unitary matrix $U^{(1)} = Q^*UQ$ are determined as follows:

1.1. *Compute*

$$\sigma_1^U = \sigma_1 + p(2)q(1), \tag{34.4}$$

$$\varepsilon_1 = g(1)h(1) - p(1)q(1), \quad \theta_1 = q(1). \tag{34.5}$$

Determine the complex Givens transformation matrix Q_1 from the condition

$$Q_1^* \begin{pmatrix} \varepsilon_1 - \alpha \\ \sigma_1 \end{pmatrix} = \begin{pmatrix} * \\ 0 \end{pmatrix}. \tag{34.6}$$

Compute

$$\Lambda_2 = Q_1^* \begin{pmatrix} g(1)h(1) & g(1)b(1) \\ \sigma_1^U & g(2) \end{pmatrix} \tag{34.7}$$

and determine the matrices $g_B(2), \beta_2^U, f_2^U, \phi_2^U$ of sizes $1 \times (r_2+1), 1 \times (r_2+1), 1 \times 1, 1 \times r_2$ from the partitions

$$\Lambda_2 = \begin{pmatrix} g_B(2) \\ \beta_2^U \end{pmatrix}, \quad \beta_2^U = \begin{pmatrix} f_2^U & \phi_2^U \end{pmatrix}. \tag{34.8}$$

Compute

$$\begin{pmatrix} p^{(1)}(1) \\ c_2 \end{pmatrix} = Q_1^* \begin{pmatrix} p(1) \\ p(2) \end{pmatrix}, \quad \begin{pmatrix} q^{(1)}(1) \\ \theta_2 \end{pmatrix} = Q_1^T \begin{pmatrix} q(1) \\ q(2) \end{pmatrix}, \tag{34.9}$$

with the numbers $p^{(1)}(1), c_2, q^{(1)}(1), \theta_2$.

Compute

$$f_2 = f_2^U - c_2q(1), \quad \varphi_2 = \phi_2^U. \tag{34.10}$$

1.2. *For* $k = 1, \dots, N-2$ *perform the following. Compute*

$$\begin{aligned} \sigma^U_{k+1} &= \sigma_{k+1} + p(k+2)q(k+1), \\ \varepsilon_{k+1} &= \varphi_{k+1}h(k+1) - c_{k+1}q(k+1). \end{aligned} \tag{34.11}$$

Compute the 2×2 *matrix* Ω_k *by the formula*

$$\Omega_k = \begin{pmatrix} f_{k+1} & \varepsilon_{k+1} \\ 0 & \sigma_{k+1} \end{pmatrix} Q_k \tag{34.12}$$

and determine the complex Givens rotation matrix Q_{k+1} *and the number* $\sigma^{(1)}_k$ *such that*

$$Q^*_{k+1}\Omega_k(:,1) = \begin{pmatrix} \sigma^{(1)}_k \\ 0 \end{pmatrix}. \tag{34.13}$$

Compute

$$\Lambda_{k+2} = Q^*_{k+1} \begin{pmatrix} f^U_{k+1} & \phi^U_{k+1}h(k+1) & \phi^U_{k+1}b(k+1) \\ p(k+2)\theta_k & \sigma^U_{k+1} & g(k+2) \end{pmatrix} \begin{pmatrix} Q_k & 0 \\ 0 & I_{r_{k+2}} \end{pmatrix}, \tag{34.14}$$

$$B_{k+2} = \begin{pmatrix} 1 & 0 & 0 \\ 0 & h(k+1) & b(k+1) \end{pmatrix} \begin{pmatrix} Q_k & 0 \\ 0 & I_{r_{k+2}} \end{pmatrix}, \tag{34.15}$$

and determine the matrices $d_B(k+2), g_B(k+2), \beta^U_{k+2}, f^U_{k+2}, \phi^U_{k+2}$ *of sizes* $1 \times 1, 1 \times (r_{k+2}+1), 1 \times (r_{k+2}+1), 1 \times 1, 1 \times r_{k+2}$ *from the partitions*

$$\Lambda_{k+2} = \begin{pmatrix} d_B(k+2) & g_B(k+2) \\ * & \beta^U_{k+2} \end{pmatrix}, \quad \beta^U_{k+2} = \begin{pmatrix} f^U_{k+2} & \phi^U_{k+2} \end{pmatrix}. \tag{34.16}$$

Compute

$$\begin{pmatrix} p^{(1)}(k+1) \\ c_{k+2} \end{pmatrix} = Q^*_{k+1} \begin{pmatrix} c_{k+1} \\ p(k+2) \end{pmatrix}, \quad \begin{pmatrix} q^{(1)}(k+1) \\ \theta_{k+2} \end{pmatrix} = Q^T_{k+1} \begin{pmatrix} \theta_{k+1} \\ q(k+2) \end{pmatrix}, \tag{34.17}$$

with the numbers $p^{(1)}(k+1), c_{k+2}, q^{(k+1)}(1), \theta_{k+2}$.

Compute

$$f_{k+2} = f^U_{k+2} - c_{k+2}\theta_{k+1}, \quad \varphi_{k+2} = \phi^U_{k+2}. \tag{34.18}$$

1.3. *Compute*

$$\Lambda_{N+1} = \begin{pmatrix} f^U_N & \phi^U_N h(N) \end{pmatrix} Q_{N-1}, \quad B_{N+1} = \begin{pmatrix} 1 & 0 \\ 0 & h(N) \end{pmatrix} Q_{N-1} \tag{34.19}$$

and determine the numbers $d_B(N+1), g_B(N+1)$ *from the partition*

$$\Lambda_{N+1} = \begin{pmatrix} d_B(N+1) & g_B(N+1) \end{pmatrix}. \tag{34.20}$$

Set

$$p^{(1)}(N)=c_N,\quad q^{(1)}(N)=\theta_N, \tag{34.21}$$

and compute

$$\sigma^{(1)}_{N-1}=d_B(N+1)-p^{(1)}(N)q^{(1)}(N-1). \tag{34.22}$$

2. *Set*

$$\begin{gathered}\rho_{N+2}=\rho_{N+1}=0,\ \rho_k=1,\ k=N,\dots,1,\ \nu'_{N+2}=\nu'_{N+1}=0,\ \nu'_N=\dots=\nu'_1=1,\\ s_{N+2}=0,\ s_{N+1}=1,\ s_N=\dots=s_2=2,\ r^U_k=r_k+1,\ k=2,\dots,N,\ r^U_{N+1}=1.\end{gathered} \tag{34.23}$$

Set X_{N+2}, z_{N+2} to be the 0×1 empty matrices and

$$h_s(N+2)=1,\quad Y_{N+2}=1,\quad g_s(N+1)=g_B(N+1), \tag{34.24}$$

and for $k=N+1,\dots,3$ perform the following.

Determine the unitary $(1+\rho_k)\times(1+\rho_k)$ matrix V_k and the number X_k such that

$$V_k^*\begin{pmatrix} p^{(1)}(k-1)\\ X_{k+1}\end{pmatrix}=\begin{pmatrix} X_k\\ 0_{\nu'_k\times 1}\end{pmatrix}. \tag{34.25}$$

Compute the $(1+\rho_k)\times(1+s_k)$ matrix

$$Z_k=V_k^*\begin{pmatrix} d_B(k) & g_s(k)\\ X_{k+1}q^{(1)}(k-2) & z_{k+1}\end{pmatrix} \tag{34.26}$$

and determine the matrices $h'_k, h''_k, \Delta_k, \Theta_k$ of sizes $\rho_{k-1}\times 1, \rho_{k-1}\times s_k, \nu'_k\times 1, \nu'_k\times s_k$ from the partition

$$Z_k=\begin{pmatrix} h'_k & h''_k\\ \Delta_k & \Theta_k\end{pmatrix}. \tag{34.27}$$

The submatrix $\begin{pmatrix}\Delta_k & \Theta_k\end{pmatrix}$ has orthonormal rows and one can determine the unitary $(1+s_k)\times(1+s_k)$ matrix F_k from the condition

$$\begin{pmatrix}\Delta_k & \Theta_k\end{pmatrix}F_k^*=\begin{pmatrix}0_{\nu'_k\times s_{k-1}} & I_{\nu'_k}\end{pmatrix}. \tag{34.28}$$

Determine the matrices $h_s(k), b_s(k), d_F(k), g_F(k)$ of sizes $s_{k-1}\times 1, s_{k-1}\times s_k, \nu'_k\times 1, \nu'_k\times s_k$ from the partition

$$F_k=\begin{pmatrix} h_s(k) & b_s(k)\\ d_F(k) & g_F(k)\end{pmatrix}. \tag{34.29}$$

Compute the matrices Y_k of the size $r^U_{k-1}\times s_{k-1}$ and z_k of the size $\rho_{k-1}\times s_{k-1}$ by the formulas

$$Y_k=B_k\begin{pmatrix} h_s^*(k)\\ Y_{k+1}b_s^*(k)\end{pmatrix},\quad z_k=h'_kh_s^*(k)+h''_kb_s^*(k). \tag{34.30}$$

Compute

$$g_s(k-1) = g_B(k-1)Y_k. \tag{34.31}$$

Set

$$\begin{aligned} g^{(1)}(i) &= g_s(i+1),\ h^{(1)}(i) = h_s(i+2), \quad && i = 1,\dots,N, \\ b^{(1)}(k) &= b_s(k+2), && k = 1,\dots,N-1. \end{aligned} \tag{34.32}$$

This result follows directly from Theorem 31.8 by taking $m = 1$.

We will compute only the complexity of the part which is performed about N times of the algorithm for the one-step single-shift case, with the implicit method with compression as in items 1.2 and 2 of Theorem 34.1. To this end, denote by r the maximal order of the upper quasiseparable generators of the matrix U, i.e., $r = \max_{k=1,\dots,N-1}(r_k)$. Notice that after the first iteration we get $r_k \leq 2$ and hence one can take $r = 2$.

The complexity of Step 1.2 is less than $95+\eta(2,1)$ arithmetical operations and the complexity of Step 2 is less than $109+\eta(2,1)+\eta(1,3)$ arithmetical operations, since $s_k \leq 2$, $k = 1,\dots,N$. Here $\eta(j,k)$ denotes the complexity of finding a unitary $k \times k$ matrix as in (34.13) or (34.28) for j orthonormal rows and k columns.

The overall complexity is less than

$$204 + 2\eta(2,1) + \eta(1,3).$$

§34.2 The double shift case

The corresponding result for the double shift case is the following.

Theorem 34.2. *Let $A = U - pq^T$ be a real matrix from the class $\mathcal{H}_N$, with subdiagonal entries σ_k ($k = 1,\dots,N-1$), perturbation vectors $p = \operatorname{col}(p(i))_{i=1}^N$, $q = \operatorname{col}(q(i))_{i=1}^N$, upper triangular generators $g(i), h(i)$ ($i = 1,\dots,N$), $b(k)$ ($k = 1,\dots,N-1$) of orders r_k ($k = 1,\dots,N$) of the orthogonal matrix U, and let s,t be real numbers.*

Then the orthogonal 3×3 matrices Q_k, $k = 1,\dots,N-2$, and the orthogonal 2×2 matrix Q_{N-1} such that the orthogonal matrix

$$Q = \tilde{Q}_1\tilde{Q}_2\cdots\tilde{Q}_{N-2}\tilde{Q}_{N-1}, \tag{34.33}$$

with

$$\tilde{Q}_i = I_{i-1}\oplus Q_i \oplus I_{N-i-2},\ i = 1,\dots,N-2, \quad \tilde{Q}_{N-1} = I_{N-2}\oplus Q_{N-1}, \tag{34.34}$$

satisfies the condition (31.7), and subdiagonal entries $\sigma_k^{(1)}$ ($k = 1,\dots,N-1$) of the upper Hessenberg matrix $A^{(1)} = Q^*AQ$, perturbation vectors $p^{(1)} = \operatorname{col}(p^{(1)}(i))_{i=1}^N$, $q^{(1)} = \operatorname{col}(q^{(1)}(i))_{i=1}^N$, upper triangular generators $g^{(1)}(i), h^{(1)}(i)$ ($i = 1,\dots,N$),

$b^{(1)}(k)$ $(k = 1, \dots, N-1)$ *with orders not greater than two of the orthogonal matrix* $U^{(1)} = Q^*UQ$ *are determined as follows.*

1.1. *Set*

$$a_{11} = g(1)h(1) - p(1)q(1),\ a_{12} = g(1)b(1)h(2) - p(1)q(2),$$
$$a_{21} = \sigma_1,\ a_{22} = g(2)h(2) - p(2)q(2),\ a_{32} = \sigma_2 \tag{34.35}$$

and determine the orthogonal 3×3 *matrix* Q_1 *from the condition*

$$Q_1^* \begin{pmatrix} a_{11}^2 + a_{12}a_{21} - sa_{21} + t & a_{21}(a_{11} + a_{22} - s) & a_{32}a_{21} \end{pmatrix}^T = \begin{pmatrix} * & 0 & 0 \end{pmatrix}^T. \tag{34.36}$$

Compute

$$\Lambda_3 = Q_1^* \begin{pmatrix} g(1)h(1) & g(1)b(1)h(2) & g(1)b(1)b(2) \\ \sigma_1^U & g(2)h(2) & g(2)b(2) \\ p(3)q(1) & \sigma_2^U & g(3) \end{pmatrix} \tag{34.37}$$

and determine the matrices $g_B(3), \beta_3^U, f_3^U, \phi_3^U$ *of sizes* $1 \times (r_3+2), 2 \times (r_3+2), 2 \times 2, 2 \times r_3$ *from the partitions*

$$\Lambda_3 = \begin{pmatrix} g_B(3) \\ \beta_3^U \end{pmatrix}, \quad \beta_3^U = \begin{pmatrix} f_3^U & \phi_3^U \end{pmatrix}. \tag{34.38}$$

Compute

$$\begin{pmatrix} p^{(1)}(1) \\ c_3 \end{pmatrix} = Q_1^* \begin{pmatrix} p(1) \\ p(2) \\ p(3) \end{pmatrix}, \quad \begin{pmatrix} q^{(1)}(1) \\ \theta_3 \end{pmatrix} = Q_1^T \begin{pmatrix} q(1) \\ q(2) \\ q(3) \end{pmatrix}, \quad \theta_2 = \begin{pmatrix} q(1) \\ q(2) \end{pmatrix}, \tag{34.39}$$

with the numbers $p^{(1)}(1), q^{(1)}(1)$ *and two-dimensional columns* c_2, θ_2.

Compute

$$f_3 = f_3^U - c_3 \begin{pmatrix} q(1) & q(2) \end{pmatrix}, \quad \varphi_2 = \phi_2^U. \tag{34.40}$$

1.2. *For* $k = 1, \dots, N-3$, *perform the following. Compute*

$$\sigma_{k+2}^U = \sigma_{k+2} + p(k+3)q(k+2), \tag{34.41}$$
$$\varepsilon_{k+2} = -c_{k+2}y^*(k+2) + \varphi_{k+2}h(k+2). \tag{34.42}$$

Compute the 3×3 *matrix* Ω_k *by the formula*

$$\Omega_k = \begin{pmatrix} f_{k+2} & \varepsilon_{k+2} \\ 0 & \sigma_{k+2} \end{pmatrix} Q_k. \tag{34.43}$$

Determine the orthogonal 3×3 *matrix* Q_{k+1} *and the number* $\sigma_k^{(1)}$ *such that*

$$Q_{k+1}^* \Omega_k(:,1) = \begin{pmatrix} \sigma_k^{(1)} & 0 & 0 \end{pmatrix}^T. \tag{34.44}$$

Compute

$$\Lambda_{k+3} = Q^*_{k+1} \begin{pmatrix} f^U_{k+2} & \phi^U_{k+2}h(k+2) & \phi^U_{k+2}b(k+2) \\ p(k+3)\theta^T_{k+1} & \sigma^U_{k+2} & g(k+3) \end{pmatrix} \begin{pmatrix} Q_k & 0 \\ 0 & I_{r_{k+3}} \end{pmatrix}, \tag{34.45}$$

$$B_{k+3} = \begin{pmatrix} I_2 & 0 & 0 \\ 0 & h(k+2) & b(k+2) \end{pmatrix} \begin{pmatrix} Q_k & 0 \\ 0 & I_{r_{k+3}} \end{pmatrix}, \tag{34.46}$$

and determine the matrices $d_B(k+3), g_B(k+3), \beta^U_{k+3}, f^U_{k+3}, \phi^U_{k+3}$ *of sizes* $1\times 1, 1\times (r_{k+3}+2), 2\times(r_{k+3}+2), 2\times 2, 2\times r_{k+3}$ *from the partitions*

$$\Lambda_{k+3} = \begin{pmatrix} d_B(k+3) & g_B(k+3) \\ * & \beta^U_{k+3} \end{pmatrix}, \quad \beta^U_{k+3} = \begin{pmatrix} f^U_{k+3} & \phi^U_{k+3} \end{pmatrix}. \tag{34.47}$$

Compute

$$\begin{pmatrix} p^{(1)}(k+1) \\ c_{k+2} \end{pmatrix} = Q^*_{k+1} \begin{pmatrix} c_{k+1} \\ p(k+3) \end{pmatrix}, \quad \begin{pmatrix} q^{(1)}(k+1) \\ \theta_{k+3} \end{pmatrix} = Q^T_{k+1} \begin{pmatrix} \theta_{k+2} \\ q(k+2) \end{pmatrix} \tag{34.48}$$

with the numbers $p^{(1)}(k+1), q^{(k+1)}(1)$ *and two-dimensional columns* c_{k+2}, θ_{k+2}.

Compute

$$f_{k+3} = f^U_{k+3} - c_{k+3}\theta^T_{k+2}, \quad \varphi_{k+3} = \phi^U_{k+3}. \tag{34.49}$$

1.3. *Compute*

$$\varepsilon_N = -c_N q(N) + \varphi_N h(N), \quad \Omega_{N-2} = \begin{pmatrix} f_N & \varepsilon_N \end{pmatrix} Q_{N-2}. \tag{34.50}$$

Determine the orthogonal 2×2 *matrix* Q_{N-1} *and the number* $\sigma^{(1)}_{N-2}$ *such that*

$$Q^*_{N-1}\Omega_{N-2}(:,1) = \begin{pmatrix} \sigma^{(1)}_{N-2} & 0 \end{pmatrix}^T. \tag{34.51}$$

Compute

$$\Lambda_{N+1} = Q^*_{N-1} \begin{pmatrix} f^U_N & \phi^U_N h(N) \end{pmatrix} Q_{N-2}, \tag{34.52}$$

$$B_{N+1} = \begin{pmatrix} I_2 & 0 \\ 0 & h(N) \end{pmatrix} Q_{N-2} \tag{34.53}$$

and determine the matrices $d_B(N+1), g_B(N+1), \beta^U_{N+1}$ *of sizes* $1\times 1, 1\times 2, 1\times 2$ *from the partition*

$$\Lambda_{N+1} = \begin{pmatrix} d_B(N+1) & g_B(N+1) \\ * & \beta^U_{N+1} \end{pmatrix}. \tag{34.54}$$

Compute

$$\Lambda_{N+2} = \beta^U_{N+1} Q_{N-1}, \quad B_{N+2} = Q_{N-1}, \tag{34.55}$$

and determine the numbers $d_B(N+2), g_B(N+2)$ *from the partition*

$$\Lambda_{N+2} = \begin{pmatrix} d_B(N+2) & g_B(N+2) \end{pmatrix}. \tag{34.56}$$

Compute

$$\begin{pmatrix} p^{(1)}(N-1) \\ p^{(1)}(N) \end{pmatrix} = Q^*_{N-1}c_N, \quad \begin{pmatrix} q^{(1)}(N-1) \\ q^{(1)}(N) \end{pmatrix} = Q^T_{N-1}\theta_N. \tag{34.57}$$

2. *Set*

$$\begin{aligned} &\rho_{N+3} = \rho_{N+2} = 0, && \rho_k = 1,\ k = N+1, \dots, 1, \\ &\nu'_{N+3} = \nu'_{N+2} = 0, && \nu'_{N+1} = \dots = \nu'_1 = 1, \\ &s_{N+3} = 0,\ s_{N+2} = 1, && s_{N+1} = \dots = s_2 = 2, \\ &r^U_k = r_k + 2,\ k = 3, \dots, N, && r^U_{N+1} = 2,\ r^U_{N+2} = 1. \end{aligned} \tag{34.58}$$

Set X_{N+3}, z_{N+3} *to be the* 0×1 *empty matrices and*

$$h_s(N+3) = 1, \quad Y_{N+3} = 1, \quad g_s(N+2) = g_B(N+2), \tag{34.59}$$

and for $k = N+2, \dots, 4$ *perform the following.*
Compute the QR *factorizations*

$$V_k^* \begin{pmatrix} p^{(1)}(k-2) \\ X_{k+1} \end{pmatrix} = \begin{pmatrix} X_k \\ 0_{\nu'_k \times 1} \end{pmatrix} \tag{34.60}$$

with unitary $(1+\rho_k) \times (1+\rho_k)$ *matrices* V_k *and auxiliary* $\rho_{k-1} \times 1$ *matrices* X_k.
Compute the $(1+\rho_k) \times (1+s_k)$ *matrix*

$$Z_k = V_k^* \begin{pmatrix} d_B(k) & g_s(k) \\ X_{k+1}q^{(1)}(k-2) & z_{k+1} \end{pmatrix} \tag{34.61}$$

and determine the matrices $h'_k, h''_k, \Delta_k, \Theta_k$ *of sizes* $\rho_{k-1}\times 1, \rho_{k-1}\times s_k, \nu'_k \times 1, \nu'_k \times s_k$ *from the partition*

$$Z_k = \begin{pmatrix} h'_k & h''_k \\ \Delta_k & \Theta_k \end{pmatrix}. \tag{34.62}$$

The submatrix $\begin{pmatrix} \Delta_k & \Theta_k \end{pmatrix}$ *has orthonormal rows and one can determine the unitary* $(\nu_k + s_k) \times (\nu_k + s_k)$ *matrix* F_k *from the condition*

$$\begin{pmatrix} \Delta_k & \Theta_k \end{pmatrix} F_k^* = \begin{pmatrix} 0_{\nu'_k \times s_{k-1}} & I_{\nu'_k} \end{pmatrix}. \tag{34.63}$$

Determine the matrices $h_s(k), b_s(k), d_F(k), g_F(k)$ *of sizes* $s_{k-1} \times 1, s_{k-1} \times s_k, \nu'_k \times 1, \nu'_k \times s_k$ *from the partition*

$$F_k = \begin{pmatrix} h_s(k) & b_s(k) \\ d_F(k) & g_F(k) \end{pmatrix}. \tag{34.64}$$

Compute the matrices Y_k *of size* $r^U_{k-1} \times s_{k-1}$ *and* z_k *of size* $\rho_{k-1} \times s_{k-1}$ *by the formulas*

$$Y_k = B_k \begin{pmatrix} h^*_s(k) \\ Y_{k+1} b^*_s(k) \end{pmatrix}, \qquad z_k = h'_k h^*_s(k) + h''_k b^*_s(k). \tag{34.65}$$

Compute

$$g_s(k-1) = g_B(k-1) Y_k. \tag{34.66}$$

Set

$$\begin{aligned} &g^{(1)}_U(i) = g_s(i+2),\ h^{(1)}_U(i) = h_s(i+3),\ i = 1, \dots, N, \\ &b^{(1)}_U(k) = b_s(k+3), \qquad\qquad\qquad\qquad k = 1, \dots, N-1. \end{aligned} \tag{34.67}$$

This result follows directly from Theorem 31.9 by taking $m = 1$.

We will compute only the complexity of the part which is performed about N times of the algorithm for the one-step double-shift case with the implicit method with compression as in items 1.2 and 2 of Theorem 34.2. To this end denote by r the maximal order of the upper quasiseparable generators of the matrix U, i.e., $r = \max_{k=1,\dots,N-1}(r_k)$. Notice that after the first iteration we get $r_k \leq 2$ and hence one can take $r = 2$.

The complexity of Step 1.2 is less than $\eta(3,1) + 265$ arithmetical operations and the complexity of Step 2 is less than $\eta(3,1) + \eta(1,3) + 147$ arithmetical operations, since $s_k \leq 2$, $k = 2, 3, \dots, N+3$. Here $\eta(j,k)$ denotes the complexity for finding a unitary $j \times k$ matrix as in (34.44) or (34.63) for j orthonormal rows and k columns.

The overall complexity is less than

$$412 + 2\eta(2,1) + \eta(1,3).$$

§34.3 Comments

The results presented in this chapter was obtained in the paper [6] but the proofs here are different. That paper contains also results of numerical tests with the developed algorithms.

Chapter 35

The Factorization Based Implicit Method

Here we use another representation of a matrix from the class $\mathcal{H}_N$, which may be better for ensuring the stability of the iteration process. As data we use, instead of generators, the elementary unitary matrices whose product forms a corresponding unitary matrix from the class $\mathcal{U}_N$. Notice that in this case also it is necessary to compute upper triangular generators in the algorithm.

Let A be a matrix from the class $\mathcal{H}_N$ represented as above in the form

$$A = U - pq^T, \tag{35.1}$$

with a matrix U from the class $\mathcal{U}_N$. By Theorem 32.2, U admits the factorization

$$U = V \cdot F,$$

with a unitary lower Hessenberg matrix V and a unitary matrix F with all entries below the second subdiagonal equal to zero. First we show that the matrix V may be chosen in such a way that the vector V^*p has all entries except for the two first ones equal to zero.

Definiton 35.1. Let $p = (p(i))_{i=1}^N$ be a vector and let V_i, $i = 1, \dots, N-1$, be unitary 2×2 matrices. We say that the vector p is reduced by the matrices V_i ($i = 1, \dots, N-1$), if there exist complex numbers β_i, $i = 2, \dots, N$, such that

$$\beta_N = p(N), \quad V_i^* \begin{pmatrix} p(i) \\ \beta_{i+1} \end{pmatrix} = \begin{pmatrix} \beta_i \\ 0 \end{pmatrix}, \ i = N-1, \dots, 1. \tag{35.2}$$

Define the unitary matrix V by

$$V = \tilde{V}_{N-1} \cdots \tilde{V}_1, \tag{35.3}$$

with

$$\tilde{V}_k = I_{k-1} \oplus V_k \oplus I_{N-k-1}, \quad k = 1, \dots, N-1. \tag{35.4}$$

We say also that the vector p is reduced by the matrix V.

One can easily see that the matrix V in (35.3), (35.4) is a lower Hessenberg matrix. Moreover, if the vector p is reduced by the matrices V_i $(i = 1, \dots, N-1)$, then

$$V^* p = \begin{pmatrix} \beta_1 & 0 & \dots & 0 \end{pmatrix}^T. \tag{35.5}$$

Lemma 35.2. *Let U be a unitary matrix from the class $\mathcal{U}_N$, with lower generators $p(i)$ $(i = 3, \dots, N)$, $q(j)$ $(j = 1, \dots, N-2)$, and let $p(1), p(2)$ be arbitrary numbers. Assume the vector $p = (p(i))_{i=1}^N$ is reduced by the unitary 2×2 matrices V_i, $i = 1, \dots, N-1$, and define the unitary matrix V by the relations (35.3), (35.4).*

Then for the unitary matrix $F = V^* U$ all the entries $F(i,j)$ below the second subdiagonal are zeros, i.e.,

$$F(i,j) = 0, \quad i > j+2. \tag{35.6}$$

Proof. Set $q(N-1), q(N)$ to be arbitrary numbers and next set $p = \mathrm{col}(p(i))_{i=1}^N$, $q = \mathrm{col}(q(i))_{i=1}^N$. Define the matrix A by the formula (35.1). It is clear that V is an upper Hessenberg matrix. Furthermore, the relations (35.3), (35.4) and (35.2) imply that the vector $V^* p$ contains zeros in all positions greater than 3. Hence, using the fact that A and V are upper Hessenberg matrices, we conclude that the matrix

$$F = V^* U = V^* A + V^* p q^T$$

satisfies the relations (35.6). □

From here we obtain a factorization similar to the one obtained in Theorem 32.2, but suitable for algorithms presented in this section.

Corollary 35.3. *Let U be a unitary matrix from the class $\mathcal{U}_N$, with lower generators $p(i)$ $(i = 3, \dots, N)$, $q(j)$ $(j = 1, \dots, N-2)$ and let $p(1), p(2)$ be arbitrary numbers.*

The matrix U admits the factorization

$$U = V \cdot F, \tag{35.7}$$

where V is a unitary matrix represented in the form

$$V = \tilde{V}_{N-1} \cdots \tilde{V}_1, \tag{35.8}$$

with

$$\tilde{V}_k = \mathrm{diag}\{I_{k-1}, V_k, I_{N-k-1}\}, \quad k = 1, \dots, N-1, \tag{35.9}$$

where V_k $(k = 1, \dots, N-1)$ are unitary 2×2 matrices such that the vector $p = (p(i))_{i=1}^N$ is reduced by the matrices V_k, and F is a unitary matrix represented in the form

$$F = \tilde{F}_1 \tilde{F}_2 \cdots \tilde{F}_{N-2}, \tag{35.10}$$

with

$$\tilde{F}_k = \mathrm{diag}\{I_{k-1}, F_k, I_{N-k-2}\}, \quad k = 1, \dots, N-2, \tag{35.11}$$

where F_k $(k = 1, \dots, N-2)$ *are unitary* 3×3 *matrices. Moreover, one can determine the matrices* F_k, $k = N-2, \dots, 2$, *as in Theorem 32.2, and the matrix* F_1 *by the formula*

$$F_1 = \begin{pmatrix} V_1^* & 0 \\ 0 & 1 \end{pmatrix} \begin{pmatrix} g_s(1) & 0 \\ 0 & V_2^* \end{pmatrix} \Phi_2, \tag{35.12}$$

with the matrix Φ_2 *defined in* (32.33) *and the number* $g_s(1)$ *defined in* (32.35).

Proof. Using Theorem 32.2, we obtain the factorization (32.14)–(32.18). Moreover, (32.19), (32.20) means (35.2) with $k = N-1, \dots, 3$. Determine the unitary 2×2 matrices V_2, V_1 via (35.2) with $k = 2, 1$. Using (32.14), (32.15), (32.17) and (32.37) we get

$$U = (\tilde{V}_{N-1} \cdots \tilde{V}_1)(\tilde{V}_1^* \tilde{V}_2^* \tilde{\Phi}_2)\tilde{F}_2 \cdots \tilde{F}_{N-2},$$

where $\tilde{\Phi}_2 = \operatorname{diag}\{\Phi_2, I_{N-3}\}$. From here using (35.12) we obtain the desired factorization of the matrix U. □

Thus a matrix $A \in \mathcal{H}_N$ is completely specified by the following parameters:

1. the unitary matrices V_k, $k = 1, \dots, N-1$, defining the matrix V;
2. the unitary matrices F_k, $k = 1, \dots, N-2$, defining the matrix F;
3. the perturbation vectors p and q such that the vector p is reduced by the matrices V_k $(k = 1, \dots, N-1)$.

These parameters also called *the generating elements* of the matrix A.

§35.1 The initialization

To initialize the iteration process with the companion matrix C from (32.1) we need to first determine its generating elements. The perturbation vectors of the matrix C are defined in (32.3). We determine the generating elements V_i $(i = 1, \dots, N)$ via (35.2), we get

$$\beta_N = p_{N-1}, \quad V_i^* \begin{pmatrix} p_{i-1} \\ \beta_{i+1} \end{pmatrix} = \begin{pmatrix} \beta_i \\ 0 \end{pmatrix}, \ i = N-1, \dots, 2, \quad V_1^* \begin{pmatrix} p_0 + 1 \\ \beta_2 \end{pmatrix} = \begin{pmatrix} \beta_1 \\ 0 \end{pmatrix}.$$

The generating elements F_k $(k = 1, \dots, N-2)$ are obtained as follows. The unitary matrix U in (32.2) has subdiagonal entries $\sigma_k = 1$, $k = 1, \dots, N-1$, lower generators $q(1) = \cdots = q(N-1)$, and upper triangular generators defined in (34.1). Using Theorem 32.2 with these generators we get

$$\begin{aligned} &g_s(N) = 0, && g_s(k) = \begin{pmatrix} 0 & 0 \end{pmatrix}, && k = N-1, \dots, 2, \quad g_s(1) = Y_3 b_s^*(2), \\ &Y_N = \begin{pmatrix} 0 & 1 \end{pmatrix}, && Y_k = Y_{k+1} b_s^*(k), && k = N-1, \dots, 3, \\ &z_N = \begin{pmatrix} 1 & 0 \end{pmatrix}, && Z_k = V_k^* \begin{pmatrix} 1 & 0_{1\times 2} \\ 0 & z_{k+1} \end{pmatrix}, && k = N-1, \dots, 3, \quad Z_2 = \begin{pmatrix} 1 & 0_{1\times 2} \\ 0 & z_3 \end{pmatrix}. \end{aligned}$$

Using (32.25)–(32.30) with $k = N-1$ we get

$$\Phi_{N-1}^* = \begin{pmatrix} 0 & V_{N-1} \\ 1 & 0 \end{pmatrix}, \quad Y_{N-1} = \begin{pmatrix} 1 & 0 \end{pmatrix}, \quad z_{N-1} = \begin{pmatrix} 0 & 1 \end{pmatrix}.$$

Next, using (32.25)–(32.30) we prove by induction that

$$Y_k = \begin{pmatrix} 1 & 0 \end{pmatrix}, \; z_k = \begin{pmatrix} 0 & 1 \end{pmatrix}, \quad k = N-1, \dots, 3,$$

and also that

$$\Phi_k^* = \begin{pmatrix} 1 & 0 & 0 \\ 0 & 0 & 1 \\ 0 & 1 & 0 \end{pmatrix} \begin{pmatrix} 0 & V_k \\ 1 & 0 \end{pmatrix}, \quad k = N-2, \dots, 3.$$

Using (32.32)–(32.35) we get

$$\Phi_2 = \begin{pmatrix} 0 & 1 & 0 \\ 1 & 0 & 0 \\ 0 & 0 & 1 \end{pmatrix}, \quad g_s(1) = 1.$$

Thus, using Corollary 35.3 we get

$$F_1 = \begin{pmatrix} V_1^* & 0 \\ 0 & 1 \end{pmatrix} \begin{pmatrix} 1 & 0 \\ 0 & V_2^* \end{pmatrix} \begin{pmatrix} 0 & 1 & 0 \\ 1 & 0 & 0 \\ 0 & 0 & 1 \end{pmatrix},$$

$$F_k = \begin{pmatrix} 0 & 1 \\ V_{k+1}^* & 0 \end{pmatrix} \begin{pmatrix} 1 & 0 & 0 \\ 0 & 0 & 1 \\ 0 & 1 & 0 \end{pmatrix}, \quad k = 2, \dots, N-3,$$

$$F_{N-2} = \begin{pmatrix} 0 & 1 \\ V_{N-1}^* & 0 \end{pmatrix}.$$

§35.2 The single shift case

Next we derive algorithms to update generating elements of a matrix A from the class $\mathcal{H}_N$ in the iteration step.

Theorem 35.4. *Let*

$$A = U - pq^T \tag{35.13}$$

be a matrix from the class $\mathcal{H}_N$, with generating elements V_k $(k = 1, \dots, N-1)$, F_k $(k = 1, \dots, N-2)$, $p = (p(i))_{i=1}^N$, $q = (q(i))_{i=1}^N$, and let α be a number. Then the Givens rotation matrices Q_k, $k = 1, \dots, N-1$ such that the unitary matrix

$$Q = \tilde{Q}_1 \tilde{Q}_2 \cdots \tilde{Q}_{N-2} \tilde{Q}_{N-1}, \tag{35.14}$$

with

$$\tilde{Q}_i = I_{i-1} \oplus Q_i \oplus I_{N-i-1},\ i = 1, \dots, N-1,$$

satisfies the condition (31.4), *and the generating elements* $V_k^{(1)}$ $(k = 2, \dots, N-1)$, $F_k^{(1)}$ $(k = 1, \dots, N-2)$ *and* $p^{(1)} = \mathrm{col}(p^{(1)}(i))_{i=1}^N$, $q^{(1)} = \mathrm{col}(q^{(1)}(i))_{i=1}^N$ *of the matrix* $A^{(1)}$ *are determined using the following algorithm.*

1. *The first step of the algorithm is to compute upper triangular generators* $g(i)$, $h(i)$ $(i = 1, \dots, N)$, $b(k)$ $(k = 1, \dots, N-1)$ *with orders*

$$r_k = 1, \quad r_2 = \dots = r_{N-2} = 2, \quad r_{N-1} = 1,$$

and subdiagonal entries σ_k $(k = 1, \dots, N-1)$ *of the matrix* U.

1.1. *Set* $g(1) = 1$. *Compute the matrix*

$$Y_1 = \begin{pmatrix} 1 & 0 \\ 0 & V_2 \end{pmatrix} \begin{pmatrix} V_1 & 0 \\ 0 & 1 \end{pmatrix} F_1$$

and determine the numbers $h(1), \sigma_1$ *and the two-dimensional rows* $b(1), g(2), \beta_1$ *from the partition*

$$Y_1 = \begin{pmatrix} h(1) & b(1) \\ \sigma_1 & g(2) \\ * & \beta_1 \end{pmatrix}.$$

1.2. *For* $k = 2, \dots, N-2$ *perform the following. Determine the two-dimensional column* $h(k)$ *and the* 2×2 *matrix* $b(k)$ *from the partition*

$$F_k = \begin{pmatrix} h(k) & b(k) \\ * & * \end{pmatrix}.$$

Compute the matrix

$$Y_k = V_{k+1} \begin{pmatrix} \beta_{k-1} & 0 \\ 0 & 1 \end{pmatrix} F_k$$

and determine the number σ_k *and the two-dimensional rows* $g(k+1), \beta_k$ *from the partition*

$$Y_k = \begin{pmatrix} \sigma_k & g(k+1) \\ * & \beta_k \end{pmatrix}.$$

1.3. *Determine the numbers* $\sigma_{N-1}, g(N)$ *from the partition*

$$\beta_{N-1} = \begin{pmatrix} \sigma_{N-1} & g(N) \end{pmatrix}$$

and set

$$h(N-1) = \begin{pmatrix} 1 \\ 0 \end{pmatrix}, \ b(N-1) = \begin{pmatrix} 0 \\ 1 \end{pmatrix}, \quad h(N) = 1.$$

2. *The second step of the algorithm is to compute the Givens rotation matrices* Q_k, $k = 1, \dots, N-1$, *and the updated perturbation vectors* $p^{(1)}, q^{(1)}$.

2.1. *Determine the complex Givens transformation matrix* Q_1 *from the condition*

$$Q_1^* \begin{pmatrix} g(1)h(1) - p(1)q(1) - \alpha \\ \sigma_1 - p(2)q(1) \end{pmatrix} = \begin{pmatrix} * \\ 0 \end{pmatrix}. \tag{35.15}$$

Compute

$$\beta_2^U = Q_1^*(2,:) \begin{pmatrix} g(1)h(1) & g(1)b(1) \\ \sigma_1 & g(2) \end{pmatrix} \tag{35.16}$$

and determine the number f_2^U *and the* r_2*-dimensional row* ϕ_2^U *from the partition*

$$\beta_2^U = \begin{pmatrix} f_2^U & \phi_2^U \end{pmatrix}. \tag{35.17}$$

Compute

$$\begin{pmatrix} p^{(1)}(1) \\ c_2 \end{pmatrix} = Q_1^* \begin{pmatrix} p(1) \\ p(2) \end{pmatrix}, \quad \begin{pmatrix} q^{(1)}(1) \\ \theta_2 \end{pmatrix} = Q_1^T \begin{pmatrix} q(1) \\ q(2) \end{pmatrix}, \quad \theta_1 = q(1) \tag{35.18}$$

with the numbers $p^{(1)}(1), c_2, q^{(1)}(1), \theta_2$.

Compute

$$f_2 = f_2^U - c_2 q(1), \quad \varphi_2 = \phi_2^U. \tag{35.19}$$

2.2. *For* $k = 1, \dots, N-2$ *perform the following.*

Compute

$$\varepsilon_{k+1} = \varphi_{k+1} h(k+1) - c_{k+1} q(k+1), \quad \tilde{\sigma}_{k+1} = \sigma_{k+1} - p(k+2)q(k+1) \tag{35.20}$$

and the 2×2 *matrix* Ω_k *by the formula*

$$\Omega_k = \begin{pmatrix} f_{k+1} & \varepsilon_{k+1} \\ 0 & \tilde{\sigma}_{k+1} \end{pmatrix} Q_k. \tag{35.21}$$

Determine the complex Givens rotation matrix Q_{k+1} *such that*

$$Q_{k+1}^* \Omega_k(:,1) = \begin{pmatrix} * \\ 0 \end{pmatrix}. \tag{35.22}$$

Compute

$$\beta_{k+2}^U = Q_{k+1}^*(2,:) \begin{pmatrix} f_{k+1}^U & \phi_{k+1}^U h(k+1) & \phi_{k+1}^U b(k+1) \\ p(k+2)\theta_k & \sigma_{k+1} & g(k+2) \end{pmatrix} \begin{pmatrix} Q_k(;,2) & 0 \\ 0 & I_{r_{k+2}} \end{pmatrix}, \tag{35.23}$$

and determine the number f_{k+2}^U *and the* r_{k+2}*-dimensional row* ϕ_{k+2}^U *from the partition*

$$\beta_{k+2}^U = \begin{pmatrix} f_{k+2}^U & \phi_{k+2}^U \end{pmatrix}. \tag{35.24}$$

Compute

$$\begin{pmatrix} p^{(1)}(k+1) \\ c_{k+2} \end{pmatrix} = Q_{k+1}^* \begin{pmatrix} c_{k+1} \\ p(k+2) \end{pmatrix}, \quad \begin{pmatrix} q^{(1)}(k+1) \\ \theta_{k+2} \end{pmatrix} = Q_{k+1}^T \begin{pmatrix} \theta_{k+1} \\ q(k+2) \end{pmatrix}, \tag{35.25}$$

with the numbers $p^{(1)}(k+1), c_{k+2}, q^{(k+1)}(1), \theta_{k+2}$.

Compute

$$f_{k+2} = f^U_{k+2} - c_{k+2}\theta^*_{k+1}, \quad \varphi_{k+2} = \phi^U_{k+2}. \tag{35.26}$$

2.3. *Set*

$$p^{(1)}(N) = c_N, \quad q^{(1)}(N) = \theta_N. \tag{35.27}$$

3. *The third step of the algorithm is to compute the generating elements* $V_k^{(1)}$ $(k = 1, \dots, N-1)$, $F_k^{(1)}$ $(k = 1, \dots, N-2)$ *of the matrix* $A^{(1)}$.

3.1. *Set* $V_1^{(\tau)} = V_1$ *and for* $k = 1, \dots, N-2$ *perform the following.*
Compute the 3×3 *matrix*

$$\hat{V}_k = \begin{pmatrix} Q_k^* & 0 \\ 0 & 1 \end{pmatrix} \begin{pmatrix} 1 & 0 \\ 0 & V_{k+1} \end{pmatrix} \begin{pmatrix} V_k^{(\tau)} & 0 \\ 0 & 1 \end{pmatrix}. \tag{35.28}$$

Determine the complex Givens rotation matrix Z_{k+1} *from the condition*

$$\hat{V}_k(1, 2:3) Z_{k+1} = \begin{pmatrix} * & 0 \end{pmatrix} \tag{35.29}$$

and compute the matrix

$$W_k = \hat{V}_k \begin{pmatrix} 1 & 0 \\ 0 & Z_{k+1} \end{pmatrix}. \tag{35.30}$$

The unitary 3×3 *matrix* W_k *has the zero entry in the position* $(1,3)$; *compute the factorization*

$$W_k = \begin{pmatrix} 1 & 0 \\ 0 & V_{k+1}^{(\tau)} \end{pmatrix} \begin{pmatrix} V_k^{(1)} & 0 \\ 0 & 1 \end{pmatrix}. \tag{35.31}$$

with unitary 2×2 *matrices* $V_{k+2}^{(\tau)}, V_{k+1}^{(1)}$.

3.2. *Compute the matrix*

$$F_1^{(\tau)} = \begin{pmatrix} 1 & 0 \\ 0 & Z_2^* \end{pmatrix} F_1 \tag{35.32}$$

and for $k = 1, \dots, N-3$ *perform the following.*
Compute the 4×4 *matrix*

$$D_k = \begin{pmatrix} I_2 & 0 \\ 0 & Z_{k+2}^* \end{pmatrix} \begin{pmatrix} F_k^{(\tau)} & 0 \\ 0 & 1 \end{pmatrix} \begin{pmatrix} 1 & 0 \\ 0 & F_{k+1} \end{pmatrix} \begin{pmatrix} Q_k & 0 \\ 0 & I_2 \end{pmatrix}. \tag{35.33}$$

The unitary matrix D_k *has a zero entry in the position* $(4,1)$; *compute the factorization*

$$D_k = \begin{pmatrix} F_k^{(1)} & 0 \\ 0 & 1 \end{pmatrix} \begin{pmatrix} 1 & 0 \\ 0 & F_{k+1}^{(\tau)} \end{pmatrix} \tag{35.34}$$

with unitary 3×3 *matrices* $F_k^{(1)}, F_{k+1}^{(\tau)}$.

3.3. *Compute*

$$V_{N-1}^{(1)} = Q_{N-1}^* V_{N-1}^{(\tau)}, \tag{35.35}$$

$$F_{N-2}^{(1)} = F_{N-2}^{(\tau)} \begin{pmatrix} Q_{N-2} & 0 \\ 0 & 1 \end{pmatrix} \begin{pmatrix} 1 & 0 \\ 0 & Q_{N-1} \end{pmatrix}. \tag{35.36}$$

Thus, all generating elements $V_k^{(1)}$ $(k = 2, \dots, N-1)$, $F_k^{(1)}$ $(k = 1, \dots, N-2)$, $p^{(1)} = (p^{(1)}(i))_{i=1}^N$, $q^{(1)} = (q^{(1)}(i))_{i=1}^N$ of the matrix $A^{(1)}$ are computed.

Proof. We rewrite the factorization (35.7) in the form

$$U = (\tilde{V}_{N-1} \cdots \tilde{V}_3)(\tilde{F}_1' \tilde{F}_2 \cdots \tilde{F}_{N-2}),$$

with $\tilde{F}_1' = \tilde{V}_2 \tilde{V}_1 \tilde{F}_1$. Hence, using Theorem 32.3 we obtain Step 1 of the algorithm.

Stage 2 of the algorithm follows from Theorem 34.1.

Now we will justify Stage 3 of the algorithm. We have

$$U^{(1)} = Q^* U Q = (Q^* V)(FQ).$$

Define the unitary matrix Z by

$$Z = \tilde{Z}_2 \cdots \tilde{Z}_{N-1}, \tag{35.37}$$

where

$$\tilde{Z}_i = I_{i-1} \oplus Z_i \oplus I_{N-i-1}, \quad i = 2, \dots, N-1. \tag{35.38}$$

We get

$$U^{(1)} = V^{(1)} \cdot F^{(1)},$$

with $V^{(1)} = Q^* V Z$, $F^{(1)} = Z^* F Q$. We need to prove that the factorizations

$$V^{(1)} = \tilde{V}_{N-1}^{(1)} \cdots \tilde{V}_1^{(1)}, \tag{35.39}$$

where

$$\tilde{V}_i^{(1)} = I_{i-1} \oplus V_i^{(1)} \oplus I_{N-i-1}$$

and

$$F^{(1)} = \tilde{F}_1^{(1)} \cdots \tilde{F}_{N-2}^{(1)}, \tag{35.40}$$

where

$$\tilde{F}_i^{(1)} = I_{i-1} \oplus F_i^{(1)} \oplus I_{N-i-2}$$

hold and that the vector $p^{(1)}$ is reduced by the matrix $V^{(1)}$.

Set

$$\Psi_1 = V, \quad \Psi_{k+1} = \tilde{Q}_k^* \Psi_k \tilde{Z}_{k+1}, \ k = 1, \dots, N-2.$$

We prove by induction that

$$\Psi_k = \tilde{V}_{N-1} \cdots \tilde{V}_{k+1} \tilde{V}_k^{(\tau)} \tilde{V}_{k-1}^{(1)} \cdots \tilde{V}_1^{(1)}, \quad k = 1, \dots, N-1, \tag{35.41}$$

where

$$\tilde{V}_k^{(\tau)} = I_{k-1} \oplus V_k^{(\tau)} \oplus I_{N-k-1}. \tag{35.42}$$

For $k = 1$ the equality (35.41) follows from (35.8), (35.9) and $V_1^{(\tau)} = V_1$. Suppose that for some k with $1 \le k \le N-2$ the relation (35.41) holds. Hence, the matrix $\Psi'_{k+1} = \tilde{Q}_k^* \Psi_k$ has the form

$$\Psi'_{k+1} = (\tilde{V}_{N-1} \cdots \tilde{V}_{k+2})(\tilde{Q}_k^* \tilde{V}_{k+1} \tilde{V}_k^{(\tau)}) \tilde{V}_{k-1}^{(1)} \cdots \tilde{V}_1^{(1)}.$$

We have

$$\tilde{Q}_k^* \tilde{V}_{k+1} \tilde{V}_k^{(\tau)} = I_{k-1} \oplus \hat{V}_k \oplus I_{N-k-2},$$

with the unitary 3×3 matrix $\hat{V}_k$ from (35.28). Next we determine a unitary 2×2 matrix Z_{k+1} from the condition (35.29) and compute the matrix W_k by the formula (35.30). The unitary 3×3 matrix W_k has a zero entry in the position $(1,3)$. Hence this matrix may be factorized in the form (35.31) with unitary 2×2 matrices $V_{k+1}^{(\tau)}, V_k^{(1)}$. Hence it follows that the matrix $\Psi_{k+1} = \Psi'_{k+1} \tilde{Z}_k$ has the form

$$\begin{aligned}\Psi_{k+1} &= (\tilde{V}_{N-1} \cdots \tilde{V}_{k+2})(\tilde{Q}_k^* \tilde{V}_{k+1} \tilde{V}_k^{(\tau)} \tilde{Z}_{k+1})(\tilde{V}_{k-1}^{(1)} \cdots \tilde{V}_1^{(1)}) \\ &= (\tilde{V}_{N-1} \cdots \tilde{V}_{k+2})(\tilde{V}_{k+1}^{(\tau)} \tilde{V}_k^{(1)})(\tilde{V}_{k-1}^{(1)} \cdots \tilde{V}_1^{(1)}),\end{aligned}$$

which completes the proof of (35.41). Now using (35.41) with $k = N-1$ and (35.35), we obtain the factorization (35.39) for the matrix $V^{(1)} = \tilde{Q}_{N-1}^* \Psi_{N-1}$.

Next, using (35.5), (35.37), (35.38) and the formula $p^{(1)} = Q^* p$, we get

$$(V^{(1)})^* p^{(1)} = Z^* V^* Q p^{(1)} = Z^* V^* p = Z^* \begin{pmatrix} \beta_1 \\ 0_{(N-1)\times 1} \end{pmatrix} = \begin{pmatrix} \beta_1 \\ 0_{(N-1)\times 1} \end{pmatrix}.$$

It follows that the vector $p^{(1)}$ is reduced by the matrix $V^{(1)}$.

Applying Lemma 35.2 to the matrix $U^{(1)}$ we obtain the equalities

$$F^{(1)}(j+3:N, j) = 0, \quad j = 1, \ldots, N-3. \tag{35.43}$$

Set

$$\Phi_1 = \tilde{Z}_2^* F, \quad \Phi_{k+1} = \tilde{Z}_{k+2}^* \Phi_k \tilde{Q}_k, \ k = 1, \ldots, N-3,$$

which means that

$$\Phi_{k+1} = \tilde{Z}_{k+2}^* \cdots \tilde{Z}_2^* F \tilde{Q}_1 \cdots \tilde{Q}_k, \quad k = 0, 1, \ldots, N-3. \tag{35.44}$$

We prove by induction that

$$\Phi_k = \tilde{F}_1^{(1)} \cdots \tilde{F}_{k-1}^{(1)} \tilde{F}_k^{(\tau)} \tilde{F}_{k+1} \cdots \tilde{F}_{N-2}, \quad k = 1, \ldots, N-2, \tag{35.45}$$

where

$$\tilde{F}_i^{(\tau)} = I_{i-1} \oplus F_i^{(\tau)} \oplus I_{N-i-2}.$$

For $k = 1$, (35.45) follows from (35.10), (35.11) and (35.32). Suppose that for some k with $1 \le k \le N-3$ the relation (35.45) holds. Hence, the matrix $\Phi_{k+1} = \tilde{Z}^*_{k+2}\Phi_k\tilde{Q}_k$ has the form

$$\Phi_{k+1} = (\tilde{F}_1^{(1)} \cdots \tilde{F}_{k-1}^{(1)})(\tilde{Z}^*_{k+2}\tilde{F}_k^{(\tau)}\tilde{F}_{k+1}\tilde{Q}_k)(\tilde{F}_{k+2} \cdots \tilde{F}_{N-2}). \tag{35.46}$$

We have

$$\tilde{Z}^*_{k+2}\tilde{F}_k^{(\tau)}\tilde{F}_{k+1}\tilde{Q}_k = I_{k-1} \oplus D_k \oplus I_{N-k-3}, \tag{35.47}$$

with the unitary 4×4 matrix D_k from (35.33). Using the relations $F^{(1)} = Z^*FQ$ and (35.14), (35.37), (35.44) we get

$$\Phi_{k+1} = (\tilde{Z}_{k+3} \cdots \tilde{Z}_{N-1})F^{(1)}(\tilde{Q}^*_{N-1} \cdots \tilde{Q}^*_{k+1}).$$

From here using (35.43) we obtain $\Phi_{k+1}(k+3 : N, k) = 0$, which implies

$$\Phi_{k+1}(k+3, k) = 0. \tag{35.48}$$

By (35.46) and (35.47), we have $D_k(4,1) = \Phi_{k+1}(k+3,k)$ and using (35.48) we conclude that the unitary 4×4 matrix D_k has a zero entry in the $(4,1)$ position. Hence, the matrix D_k admits the factorization (35.34) with unitary 3×3 matrices $F_k^{(1)}, F_{k+1}^{(\tau)}$. It follows that

$$\Phi_{k+1} = (\tilde{F}_1^{(1)} \cdots \tilde{F}_{k-1}^{(1)})(\tilde{F}_k^{(1)}\tilde{F}_{k+1}^{(\tau)})(\tilde{F}_{k+2} \cdots \tilde{F}_{N-2}),$$

which completes the proof of (35.45). Now using (35.45) with $k = N-2$ and (35.36) we obtain the factorization (35.40) for the matrix $F^{(1)} = \Phi_{N-2}\tilde{Q}_{N-2}\tilde{Q}_{N-1}$. □

We will compute only the complexity of the part which is performed about N times of the algorithm for the one-step single-shift case with the implicit method based on factorization, as in items 1.2, 2.2 and 3.2 of Theorem 35.4. To this end note that the maximal order of the upper triangular generators of the matrix is 2.

The complexity of Step 1.2 is 35 arithmetical operations, the complexity of Step 2.2 is less than $122 + \eta(2,2)$ arithmetical operations, and the complexity of Step 3.2 is $\eta(2,2) + \mu(3,3) + \mu(4,4) + 336$ operations, but most of the latter are multiplications by 1 or 0 and the corresponding additions, so that only $\eta(2,2) + \mu(3,3) + \mu(4,4) + 32$ arithmetical operations count for this step.

The overall complexity of the algorithm is

$$189 + 2\eta(2,2) + \mu(3,3) + \mu(4,4)$$

arithmetical operations. Here $\eta(j,k)$ denotes the complexity of finding a orthonormal $k \times k$ matrix as in (35.29) and $\mu(j,k)$ denotes the complexity of finding a factorization like in (35.31).

§35.3 The double shift case

For the double shift case we obtain the following algorithm.

Theorem 35.5. *Let $A = U - pq^T$ be a matrix from the class $\mathcal{H}_N$, with generating elements V_k $(k = 2,\dots,N-1)$, F_k $(k = 1,\dots,N-2)$, $p = (p(i))_{i=1}^N$, $q = (q(i))_{i=1}^N$, and let s,t be real numbers.*

Then the orthogonal 3×3 matrices Q_k, $k = 1,\dots,N-2$ and the orthogonal 2×2 matrix Q_{N-1} such that the orthogonal matrix

$$Q = \tilde{Q}_1\tilde{Q}_2\cdots\tilde{Q}_{N-2}\tilde{Q}_{N-1}, \tag{35.49}$$

with

$$\tilde{Q}_i = I_{i-1}\oplus Q_i\oplus I_{N-i-2},\ i = 1,\dots,N-2,\quad \tilde{Q}_{N-1} = I_{N-2}\oplus Q_{N-1},$$

satisfies the condition (31.7), and the generating elements $V_k^{(1)}$ $(k = 2,\dots,N-1)$, $F_k^{(1)}$ $(k = 1,\dots,N-2)$ and $p^{(1)} = \mathrm{col}(p^{(1)}(i))_{i=1}^N$, $q^{(1)} = \mathrm{col}(q^{(1)}(i))_{i=1}^N$ of the matrix $A^{(1)}$ are determined using the following algorithm.

1. *Using Step 1 of the algorithm from Theorem 35.4 compute upper triangular generators $g(i), h(i)$ $(i = 1,\dots,N)$, $b(k)$ $(k = 1,\dots,N-1)$ and subdiagonal entries σ_k $(k = 1,\dots, N-1)$ of the matrix U.*

2. *The second step of the algorithm is to compute the orthogonal matrices Q_k, $k = 1,\dots,N-1$ and the updated perturbation vectors $p^{(1)}, q^{(1)}$.*

2.1. *Set*

$$\begin{aligned} &a_{11} = g(1)h(1) - p(1)q(1),\ a_{12} = g(1)b(1)h(2) - p(1)q(2),\\ &a_{21} = \sigma_1 - p(2)q(1),\ a_{22} = g(2)h(2) - p(2)q(2),\ a_{32} = \sigma_2 - p(3)q(2), \end{aligned} \tag{35.50}$$

and determine the orthogonal 3×3 matrix Q_1 from the condition

$$Q_1^*\begin{pmatrix} a_{11}^2 + a_{12}a_{21} - sa_{21} + t & a_{21}(a_{11}+a_{22}-s) & a_{32}a_{21}\end{pmatrix}^T = \begin{pmatrix} * & 0 & 0\end{pmatrix}^T. \tag{35.51}$$

Compute

$$\beta_3^U = Q_1^*(2:3,:)\begin{pmatrix} g(1)h(1) & g(1)b(1)h(2) & g(1)b(1)b(2)\\ \sigma_1^U & g(2)h(2) & g(2)b(2)\\ p(3)q(1) & \sigma_2^U & g(3)\end{pmatrix} \tag{35.52}$$

and determine the matrices f_3^U, ϕ_3^U of sizes $2\times 2, 2\times r_3$ from the partition

$$\beta_3^U = \begin{pmatrix} f_3^U & \phi_3^U \end{pmatrix}. \tag{35.53}$$

Compute

$$\begin{pmatrix} p^{(1)}(1)\\ c_3\end{pmatrix} = Q_1^*\begin{pmatrix} p(1)\\ p(2)\\ p(3)\end{pmatrix},\quad \begin{pmatrix} q^{(1)}(1)\\ \theta_3\end{pmatrix} = Q_1^T\begin{pmatrix} q(1)\\ q(2)\\ q(3)\end{pmatrix},\quad \theta_2 = \begin{pmatrix} q(1)\\ q(2)\end{pmatrix}, \tag{35.54}$$

with the numbers $p^{(1)}(1), q^{(1)}(1)$ and two-dimensional columns c_2, θ_2.

Compute

$$f_3 = f_3^U - c_3 \begin{pmatrix} q(1) & q(2) \end{pmatrix}, \quad \varphi_2 = \phi_2^U. \tag{35.55}$$

2.2. *For $k = 1, \dots, N-3$ perform the following.*
Compute

$$\begin{aligned} \varepsilon_{k+2} &= \varphi_{k+2} h(k+2) - c_{k+2} q(k+2), \\ \tilde{\sigma}_{k+2} &= \sigma_{k+2} - p(k+3) q(k+2), \end{aligned} \tag{35.56}$$

and the 3×3 matrix Ω_k by the formula

$$\Omega_k = \begin{pmatrix} f_{k+2} & \varepsilon_{k+2} \\ 0 & \tilde{\sigma}_{k+2} \end{pmatrix} Q_k. \tag{35.57}$$

Determine the orthogonal 3×3 matrix Q_{k+1} such that

$$Q_{k+1}^* \Omega_k(:,1) = \begin{pmatrix} * & 0 & 0 \end{pmatrix}^T \tag{35.58}$$

Compute

$$\begin{aligned} \beta_{k+3}^U = Q_{k+1}^*(:,2:3) &\begin{pmatrix} f_{k+2}^U & \phi_{k+2}^U h(k+2) & \phi_{k+2}^U b(k+2) \\ p(k+3)\theta_{k+1}^T & \sigma_{k+2} & g(k+3) \end{pmatrix} \\ &\times \begin{pmatrix} Q_k(:,2:3) & 0 \\ 0 & I_{r_{k+3}} \end{pmatrix}, \end{aligned} \tag{35.59}$$

and determine the matrices f_{k+3}^U, ϕ_{k+3}^U of sizes $2 \times 2, 2 \times r_{k+3}$ from the partition

$$\beta_{k+3}^U = \begin{pmatrix} f_{k+3}^U & \phi_{k+3}^U \end{pmatrix}. \tag{35.60}$$

Compute

$$\begin{pmatrix} p^{(1)}(k+1) \\ c_{k+2} \end{pmatrix} = Q_{k+1}^* \begin{pmatrix} c_{k+1} \\ p(k+3) \end{pmatrix}, \quad \begin{pmatrix} q^{(1)}(k+1) \\ \theta_{k+3} \end{pmatrix} = Q_{k+1}^T \begin{pmatrix} \theta_{k+2} \\ q(k+2) \end{pmatrix}, \tag{35.61}$$

with the numbers $p^{(1)}(k+1), q^{(k+1)}(1)$ and two-dimensional columns c_{k+2}, θ_{k+2}.
Compute

$$f_{k+3} = f_{k+3}^U - c_{k+3}\theta_{k+2}^T, \quad \varphi_{k+3} = \phi_{k+3}^U. \tag{35.62}$$

2.3. *Compute*

$$\varepsilon_N = -c_N q(N) + \varphi_N h(N), \tag{35.63}$$

and the 2×3 matrix Ω_{N-2} by the formula

$$\Omega_{N-2} = \begin{pmatrix} f_N & \varepsilon_N \end{pmatrix} Q_{N-2}. \tag{35.64}$$

Determine the orthogonal 2×2 matrix Q_{N-1} such that

$$Q_{N-1}^* \Omega_{N-2}(:,1) = \begin{pmatrix} * & 0 \end{pmatrix}^T \tag{35.65}$$

Compute

$$\begin{pmatrix} p^{(1)}(N-1) \\ p^{(1)}(N) \end{pmatrix} = Q_{N-1}^* c_N, \quad \begin{pmatrix} q^{(1)}(N-1) \\ q^{(1)}(N) \end{pmatrix} = Q_{N-1}^T \theta_N. \tag{35.66}$$

3. *The third step of the algorithm is to compute the generating elements* $V_k^{(1)}$ $(k = 1, \dots, N-1)$, $F_k^{(1)}$ $(k = 1, \dots, N-2)$ *of the matrix* $A^{(1)}$.

3.1. *Compute the* 3×3 *matrix*

$$V_1^{(\tau)} = \begin{pmatrix} V_1 & 0 \\ 0 & 1 \end{pmatrix} \begin{pmatrix} 1 & 0 \\ 0 & V_2 \end{pmatrix}, \tag{35.67}$$

and for $k = 1, \dots, N-3$ *perform the following.*

Compute the 4×4 *matrix*

$$\hat{V}_k = \begin{pmatrix} Q_k^* & 0 \\ 0 & 1 \end{pmatrix} \begin{pmatrix} I_2 & 0 \\ 0 & V_{k+2} \end{pmatrix} \begin{pmatrix} V_k^{(\tau)} & 0 \\ 0 & I_2 \end{pmatrix} \tag{35.68}$$

and determine the orthogonal 3×3 *matrix* Z_{k+1} *from the condition*

$$\hat{V}_k(1, 2:4) Z_{k+1} = \begin{pmatrix} * & 0 & 0 \end{pmatrix}. \tag{35.69}$$

Compute the matrix

$$W_k = \hat{V}_k \begin{pmatrix} 1 & 0 \\ 0 & Z_{k+1} \end{pmatrix}. \tag{35.70}$$

The orthogonal 4×4 *matrix* W_k *has zero entries in the positions* $(1,3)$ *and* $(1,4)$. *Compute the factorization*

$$W_k = \begin{pmatrix} 1 & 0 \\ 0 & V_{k+1}^{(\tau)} \end{pmatrix} \begin{pmatrix} V_k^{(1)} & 0 \\ 0 & 1 \end{pmatrix} \tag{35.71}$$

with the orthogonal 2×2 *matrix* $V_k^{(1)}$ *and orthogonal* 3×3 *matrix* $V_{k+1}^{(\tau)}$.

3.2. *Compute the* 4×4 *matrix*

$$F_1^{(\tau)} = \begin{pmatrix} 1 & 0 \\ 0 & Z_2^* \end{pmatrix} \begin{pmatrix} F_1 & 0 \\ 0 & 1 \end{pmatrix} \begin{pmatrix} 1 & 0 \\ 0 & F_2 \end{pmatrix} \tag{35.72}$$

and for $k = 1, \dots, N-4$ *perform the following.*

Compute the 5×5 *matrix*

$$D_k = \begin{pmatrix} I_2 & 0 \\ 0 & Z_{k+2}^* \end{pmatrix} \begin{pmatrix} F_k^{(\tau)} & 0 \\ 0 & 1 \end{pmatrix} \begin{pmatrix} I_2 & 0 \\ 0 & F_{k+2} \end{pmatrix} \begin{pmatrix} Q_k & 0 \\ 0 & I_2 \end{pmatrix}. \tag{35.73}$$

The orthogonal 5×5 *matrix* D_k *has zero entries in the positions* $(4,1)$ *and* $(5,1)$. *Compute the factorization*

$$D_k = \begin{pmatrix} F_k^{(1)} & 0 \\ 0 & I_2 \end{pmatrix} \begin{pmatrix} 1 & 0 \\ 0 & F_{k+1}^{(\tau)} \end{pmatrix} \tag{35.74}$$

with the orthogonal 3×3 *matrix* $F_k^{(1)}$ *and orthogonal* 4×4 *matrix* $F_{k+1}^{(\tau)}$.

3.3. *Compute the* 3×3 *matrix*

$$\hat{V}_{N-2} = Q^*_{N-2} V^{(\tau)}_{N-2} \tag{35.75}$$

and determine the orthogonal 2×2 *matrix* Z_{N-1} *from the condition*

$$\hat{V}_{N-2}(1, 2:3) Z_{N-1} = \begin{pmatrix} * & 0 \end{pmatrix}. \tag{35.76}$$

Compute the 3×3 *matrix*

$$W_{N-2} = \hat{V}_{N-2} \begin{pmatrix} 1 & 0 \\ 0 & Z_{N-1} \end{pmatrix} \tag{35.77}$$

and the 4×4 *matrix*

$$D_{N-3} = \begin{pmatrix} I_2 & 0 \\ 0 & Z^*_{N-1} \end{pmatrix} \begin{pmatrix} F^{(\tau)}_{N-3} & 0 \\ 0 & 1 \end{pmatrix} \begin{pmatrix} Q_{N-3} & 0 \\ 0 & 1 \end{pmatrix}. \tag{35.78}$$

The orthogonal matrix W_{N-3} *has a zero entry in the position* $(1,3)$ *and the orthogonal matrix* D_{N-3} *has a zero entry in the position* $(4,1)$*. Compute the factorization*

$$W_{N-2} = \begin{pmatrix} 1 & 0 \\ 0 & V^{(\tau)}_{N-1} \end{pmatrix} \begin{pmatrix} V^{(1)}_{N-2} & 0 \\ 0 & 1 \end{pmatrix}, \tag{35.79}$$

with orthogonal 2×2 *matrices* $V^{(\tau)}_{N-1}, V^{(1)}_{N-2}$ *and the factorization*

$$D_{N-3} = \begin{pmatrix} F^{(1)}_{N-3} & 0 \\ 0 & 1 \end{pmatrix} \begin{pmatrix} 1 & 0 \\ 0 & F^{(\tau)}_{N-2} \end{pmatrix} \tag{35.80}$$

with orthogonal 3×3 *matrices* $F^{(1)}_{N-3}, F^{(\tau)}_{N-2}$.

3.4. *Compute*

$$V^{(1)}_{N-1} = Q^*_{N-1} V^{(\tau)}_{N-1}, \tag{35.81}$$

$$F^{(1)}_{N-2} = F^{(\tau)}_{N-2} Q_{N-2} \begin{pmatrix} 1 & 0 \\ 0 & Q_{N-1} \end{pmatrix}. \tag{35.82}$$

Thus all generating elements $V^{(1)}_k$ $(k = 2, \ldots, N-1)$, $F^{(1)}_k$ $(k = 1, \ldots, N-2)$, $p^{(1)} = (p^{(1)}(i))_{i=1}^N$, $q^{(1)} = (q^{(1)}(i))_{i=1}^N$ *of the matrix* $A^{(1)}$ *are computed.*

Proof. Step 1 of the algorithm is the same as in Theorem 35.4.

Step 2 follows from Theorem 34.2.

Now we will justify Step 3 of the algorithm. We have $U^{(1)} = Q^*UQ = (Q^*V)(FQ)$. Define the unitary matrix Z by

$$Z = \tilde{Z}_2 \cdots \tilde{Z}_{N-1}, \tag{35.83}$$

where

$$\tilde{Z}_i = I_{i-1} \oplus Z_i \oplus I_{N-i-1},\ i = 2, \ldots, N-2, \quad \tilde{Z}_{N-1} = I_{N-2} \oplus Z_{N-1}. \tag{35.84}$$

We get

$$U^{(1)} = V^{(1)} \cdot F^{(1)}$$

with $V^{(1)} = Q^* V Z$, $U^{(1)} = Z^* F Q$. We need to prove that the factorizations (35.39) and (35.40) hold and that the vector $p^{(1)}$ is reduced by the matrix $V^{(1)}$.

Set

$$\Psi_1 = V, \quad \Psi_{k+1} = \tilde{Q}_k^* \Psi_k \tilde{Z}_{k+1}, \; k = 1, \dots, N-2.$$

We prove by induction that

$$\Psi_k = \tilde{V}_{N-1} \cdots \tilde{V}_{k+2} \tilde{V}_k^{(\tau)} \tilde{V}_{k-1}^{(1)} \cdots \tilde{V}_1^{(1)}, \quad k = 1, \dots, N-2, \tag{35.85}$$

with

$$\tilde{V}_k^{(\tau)} = I_{k-1} \oplus V_k^{(\tau)} \oplus I_{N-k-2}.$$

For $k = 1$, (35.85) follows from the relations (35.8), (35.9) and (35.67). Suppose that for some k with $1 \leq k \leq N-3$ the relation (35.85) holds. Hence, the matrix $\Psi'_{k+1} = \tilde{Q}_k^* \Psi_k$ has the form

$$\Psi'_{k+1} = (\tilde{V}_{N-1} \cdots \tilde{V}_{k+3})(\tilde{Q}_k^* \tilde{V}_{k+2} \tilde{V}_k^{(\tau)}) \tilde{V}_{k-1}^{(1)} \cdots \tilde{V}_1^{(1)}.$$

We have

$$\tilde{Q}_k^* \tilde{V}_{k+2} \tilde{V}_k^{(\tau)} = I_{k-1} \oplus \hat{V}_k \oplus I_{N-k-2},$$

with the orthogonal 4×4 matrix $\hat{V}_k$ from (35.68). Next we determine an orthogonal 3×3 matrix Z_{k+1} from the condition (35.69) and compute the matrix W_k by the formula (35.70). The orthogonal 4×4 matrix W_k has zero entries in the positions $(1,3)$ and $(1,4)$. Hence, this matrix may be factorized in the form (35.71), with an orthogonal 2×2 matrix $V_k^{(1)}$ and an orthogonal 3×3 matrix $V_{k+1}^{(\tau)}$. It follows that the matrix $\Psi_{k+1} = \Psi'_{k+1} \tilde{Z}_{k+1}$ has the form

$$\begin{aligned} \Psi_{k+1} &= (\tilde{V}_{N-1} \cdots \tilde{V}_{k+3})(\tilde{Q}_k^* \tilde{V}_{k+2} \tilde{V}_k^{(\tau)} \tilde{Z}_{k+1})(\tilde{V}_{k-1}^{(1)} \cdots \tilde{V}_1^{(1)}) \\ &= (\tilde{V}_{N-1} \cdots \tilde{V}_{k+3})(\tilde{V}_{k+1}^{(\tau)} \tilde{V}_k^{(1)})(\tilde{V}_{k-1}^{(1)} \cdots \tilde{V}_1^{(1)}), \end{aligned}$$

which completes the proof of (35.85). Now using (35.85) with $k = N-2$ we obtain the representation of the matrix $\Psi'_{N-1} = \tilde{Q}_{N-2}^* \Psi_{N-2}$ in the form

$$\Psi'_{N-1} = (\tilde{Q}_{N-2}^* \tilde{V}_{N-2}^{(\tau)}) \tilde{V}_{N-3}^{(1)} \cdots \tilde{V}_1^{(1)}.$$

Next we determine an orthogonal 2×2 matrix Z_{N-1} from the condition (35.76) and compute the matrix W_{N-2} by the formula (35.70). The orthogonal 3×3 matrix W_{N-2} has zero entries in the positions $(1,3)$ and $(1,4)$. Hence, this matrix may be factorized in the form (35.79) with orthogonal 2×2 matrices $V_{N-1}^{(\tau)}, V_{N-2}^{(1)}$. It follows that the matrix $\Psi_{N-1} = \Psi'_{N-1} \tilde{Z}_{N-1}$ has the form

$$\begin{aligned} \Psi_{N-1} &= (\tilde{Q}_{N-2}^* \tilde{V}_{N-2}^{(\tau)} \tilde{Z}_{N-1})(\tilde{V}_{k-1}^{(1)} \cdots \tilde{V}_1^{(1)}) \\ &= (\tilde{V}_{N-1}^{(\tau)} \tilde{V}_{N-2}^{(1)})(\tilde{V}_{N-3}^{(1)} \cdots \tilde{V}_1^{(1)}) \end{aligned}$$

with $\tilde{V}_{N-1}^{(\tau)} = I_{N-2} \oplus V_{N-1}^{(\tau)}$. From here using (35.81) we obtain the factorization (35.39) for the matrix $V^{(1)} = \tilde{Q}_{N-1}^{*}\Psi_{N-1}$.

In the same way as in the proof of Theorem 35.4 we prove that the vector $p^{(1)}$ is reduced by the matrix $V^{(1)}$.

Applying Lemma 35.2 to the matrix $U^{(1)}$ we obtain the equalities (35.43). Set

$$\Phi_1 = \tilde{Z}_2^{*}F, \quad \Phi_{k+1} = \tilde{Z}_{k+2}^{*}\Phi_k\tilde{Q}_k,\ k = 1,\dots,N-3,$$

which means that

$$\Phi_{k+1} = \tilde{Z}_{k+2}^{*}\cdots\tilde{Z}_2^{*}F\tilde{Q}_1\cdots\tilde{Q}_k, \quad k = 0,1,\dots,N-3. \tag{35.86}$$

We prove by induction that

$$\Phi_k = \tilde{F}_1^{(1)}\cdots\tilde{F}_{k-1}^{(1)}\tilde{F}_k^{(\tau)}\tilde{F}_{k+2}\cdots\tilde{F}_{N-2}, \quad k = 1,\dots,N-3, \tag{35.87}$$

where

$$\tilde{F}_i^{(\tau)} = I_{i-1} \oplus F_i^{(\tau)} \oplus I_{N-i-3}.$$

For $k = 1$, (35.87) follows from (35.10), (35.11) and (35.72). Suppose that for some k with $1 \le k \le N-4$ the relation (35.87) holds. Hence the matrix $\Phi_{k+1} = \tilde{Z}_{k+2}^{*}\Phi_k\tilde{Q}_k$ has the form

$$\Phi_{k+1} = (\tilde{F}_1^{(1)}\cdots\tilde{F}_{k-1}^{(1)})(\tilde{Z}_{k+2}^{*}\tilde{F}_k^{(\tau)}\tilde{F}_{k+2}\tilde{Q}_k)(\tilde{F}_{k+3}\cdots\tilde{F}_{N-2}). \tag{35.88}$$

We have

$$\tilde{Z}_{k+2}^{*}\tilde{F}_k^{(\tau)}\tilde{F}_{k+2}\tilde{Q}_k = I_{k-1} \oplus D_k \oplus I_{N-k-4}, \tag{35.89}$$

with the orthogonal 5×5 matrix D_k from (35.73). Using the relations $F^{(1)} = Z^{*}FQ$ and (35.49), (35.83), (35.86) we get

$$\Phi_{k+1} = (\tilde{Z}_{k+3}\cdots\tilde{Z}_{N-1})F^{(1)}(\tilde{Q}_{N-1}\cdots\tilde{Q}_{k+1}).$$

From here using (35.43) we obtain $\Phi_{k+1}(k+3:N,k) = 0$, which implies

$$\Phi_{k+1}(k+3:k+4,k) = 0. \tag{35.90}$$

By (35.88) and (35.89), we have $D_k(4:5,1) = \Phi_{k+1}(k+3:k+4,k)$ and using (35.90) we conclude that the unitary 5×5 matrix D_k has zero entries in the $(4,1)$ and $(5,1)$ positions. Hence the matrix D_k admits the factorization (35.74) with an orthogonal 3×3 matrix $F_k^{(1)}$ and an orthogonal 4×4 matrix $F_{k+1}^{(\tau)}$. It follows that

$$\Phi_{k+1} = (\tilde{F}_1^{(1)}\cdots\tilde{F}_{k-1}^{(1)})(\tilde{F}_k^{(1)}\tilde{F}_{k+1}^{(\tau)})(\tilde{F}_{k+3}\cdots\tilde{F}_{N-2}),$$

which completes the proof of (35.87). Now using (35.87) with $k = N-3$ we obtain the representation of the matrix $\Phi_{N-2} = \tilde{Z}_{N-1}^{*}\Phi_{N-3}\tilde{Q}_{N-3}$ in the form

$$\Phi_{N-2} = (\tilde{F}_1^{(1)}\cdots\tilde{F}_{N-4}^{(1)})(\tilde{Z}_{N-1}^{*}\tilde{F}_{N-3}^{(\tau)}\tilde{Q}_{N-3}).$$

We have

$$\tilde{Z}^*_{N-1}\tilde{F}^{(\tau)}_{N-3}\tilde{Q}_{N-3} = I_{N-4} \oplus D_{N-3}$$

with the orthogonal 4×4 matrix D_{N-3} from (35.78). From the relations

$$\Phi_{N-2} = F^{(1)}(\tilde{Q}_{N-2}\tilde{Q}_{N-1})^*$$

and (35.43) it follows that

$$\Phi_{N-2}(N, N-3) = 0. \tag{35.91}$$

We have $D_{N-3}(4,1) = \Phi_{N-2}(N, N-3)$ and using (35.91) we conclude that the unitary 4×4 matrix D_{N-3} has a zero entry in the $(4,1)$ position. Hence the matrix D_{N-3} admits the factorization (35.80) with orthogonal 3×3 matrices $F^{(1)}_{N-3}, F^{(\tau)}_{N-2}$. It follows that the matrix Φ_{N-2} has the form

$$\Phi_{N-2} = (\tilde{F}^{(1)}_1 \cdots \tilde{F}^{(1)}_{N-4})(\tilde{F}^{(1)}_{N-3}\tilde{F}^{(\tau)}_{N-2}),$$

with $\tilde{F}^{(\tau)}_{N-2} = I_{N-3} \oplus F^{(\tau)}_{N-2}$. From here using (35.82) we obtain the factorization (35.40) for the matrix $F^{(1)} = \Phi_{N-2}\tilde{Q}_{N-2}\tilde{Q}_{N-1}$. □

We will compute only the complexity of the part which is performed about N times of the algorithm for the one-step double-shift case with the implicit method based on factorization, as in item 1.2 of Theorem 35.4, which is reused by item 1 of Theorem 35.5 and in items 2.2, 3.1 and 3.2 of Theorem 35.5. To this end note that the maximal order of the upper triangular generators of the matrix is 2.

The complexity of Step 1.2 is 35 arithmetical operations again, the complexity of Step 2.2 is less than $\eta(1,3) + 233$ arithmetical operations, the complexity of Step 3.1 is $\eta(1,3) + \mu(4,4)$ plus 196 arithmetical operations (all but 24 of which are multiplications with 0 or 1 and the corresponding additions), and finally the complexity of Step 3.2 is $\mu(5,5)$ plus 675 arithmetical operations (all but 32 of which are multiplications with 0 or 1 and the corresponding additions). Here $\eta(j,k)$ denotes the complexity of finding an orthonormal $k \times k$ matrix as in (35.69) and $\mu(j,k)$ denotes the complexity of finding a factorization like in (35.74).

The overall complexity is less than

$$324 + 2\eta(1,3) + \mu(4,4) + \mu(5,5).$$

§35.4 Comments

The algorithms of this chapter were developed in [2]. Concerning implicit algorithms for companion matrices based on representation as products of Givens matrices see [36] and the literature cited therein.

Chapter 36

Implicit Algorithms Based on the QR Representation

In this chapter we discuss an alternative method which is based on the using of the QR factorization $A = \Theta R$ of an upper Hessenberg matrix A from the class $\mathcal{H}_N$, with a unitary upper Hessenberg matrix Θ and an upper triangular matrix R. It turns out that the upper triangular matrix R in this factorization is a rank one perturbation of a unitary quasiseparable of order one matrix. This fact lies in the basis of the method presented here.

For a matrix A from the class $\mathcal{H}_N$ represented in the form

$$A = U - wq^T, \tag{36.1}$$

with a unitary matrix $U \in \mathbb{C}^{N \times N}$ and $w, q \in \mathbb{C}^N$, consider the QR factorization

$$A = \Theta R, \tag{36.2}$$

with a unitary upper Hessenberg matrix Θ and an upper triangular matrix R. Since Θ is unitary Hessenberg it admits the factorization

$$\Theta = \tilde{\Theta}_1 \tilde{\Theta}_2 \cdots \tilde{\Theta}_{N-1}, \tag{36.3}$$

where

$$\tilde{\Theta}_i = I_{i-1} \oplus \Theta_i \oplus I_{N-i-1}, \quad i = 1, \ldots, N-1, \tag{36.4}$$

with unitary 2×2 matrices Θ_i.

Using (36.1) and (36.2) we have

$$U - wq^T = \Theta R,$$

which means that

$$W - pq^T = R, \tag{36.5}$$

with the unitary matrix $W = \Theta^* U$ and the vector $p = \Theta^* w \in \mathbb{C}^N$. Since the matrix R is upper triangular the entries below the first subdiagonal of the matrix

W have the form

$$W(i,j) = p(i)q(j), \quad 1 \le j < i \le N. \tag{36.6}$$

We define *the class* $\mathcal{W}_N$ to be the set of unitary matrices with the elements in the strictly lower triangular part $i > j$ of the form (36.6). This means that the numbers $p(i)$ $(i = 2, \dots, N)$, $q(j)$ $(j = 1, \dots, N-1)$ are lower semiseparable generators of order one of the matrix W. The vectors $p, q \in \mathbb{C}^N$ are called *the perturbation vectors.*

Lemma 36.1. *Let W be a unitary matrix from the class $\mathcal{W}_N$, with lower semiseparable generators $p(i)$ $(i = 2, \dots, N)$, $q(j)$ $(j = 1, \dots, N-1)$, and let $p(1)$ be an arbitrary number. Assume the vector $p = (p(i))_{i=1}^N$ is reduced by unitary 2×2 matrices V_i, $i = 1, \dots, N-1$, and define the unitary matrix V via the relations (35.3), (35.4).*

Then $F = V^*W$ is a unitary upper Hessenberg matrix.

Proof. The matrix F is unitary as a product of unitary matrices. Set $q(N)$ to be an arbitrary number and next set $p = \mathrm{col}(p(i))_{i=1}^N$, $q = \mathrm{col}(q(i))_{i=1}^N$. Define the matrix R by the formula (36.5). It is clear that R is an upper triangular matrix and V is a lower Hessenberg matrix. Hence, since then V^* is also an upper Hessenberg matrix, we conclude that the product V^*R is an upper Hessenberg matrix. Using the equality (35.5) we conclude that

$$F = V^*W = V^*R + (V^*p)q^T$$

is an upper Hessenberg matrix. □

Corollary 36.2. *Let W be a unitary matrix from the class $\mathcal{W}_N$ with lower semiseparable generators $p(i)$ $(i = 2, \dots, N)$, $q(j)$ $(j = 1, \dots, N-1)$, and let $p(1)$ be an arbitrary number.*

Then W admits the factorization

$$W = VF, \tag{36.7}$$

where V is a unitary matrix represented in the form

$$V = \tilde{V}_{N-1} \cdots \tilde{V}_1, \tag{36.8}$$

with

$$\tilde{V}_k = \mathrm{diag}\{I_{k-1}, V_k, I_{N-k-1}\}, \quad k = 1, \dots, N-1, \tag{36.9}$$

where V_k $(k = 1, \dots, N-1)$ are unitary 2×2 matrices such that the vector p is reduced by the matrices V_k and F is a unitary matrix represented in the form

$$F = \tilde{F}_1 \tilde{F}_2 \cdots \tilde{F}_{N-1}, \tag{36.10}$$

with

$$\tilde{F}_k = \mathrm{diag}\{I_{k-1}, F_k, I_{N-k-1}\}, \quad k = 1, \dots, N-1, \tag{36.11}$$

where F_k $(k = 1, \dots, N-1)$ are unitary 2×2 matrices.

Proof. By Lemma 36.1, we obtain the factorization (36.7) such that the vector p is reduced by the matrix V defined via (36.8), (36.9) and the matrix F is upper Hessenberg. We treat the unitary matrix F as a block upper triangular matrix with entries of sizes $\nu_i \times n_j$, $i, j = 1, \dots, N-1$, where

$$\nu_1 = 2,\ \nu_2 = \cdots = \nu_{N-1} = 1, \quad n_1 = \cdots = n_{N-2} = 1,\ n_{N-1} = 2.$$

Using Lemma 7.3, we obtain the factorization (36.10), (36.11), which completes the proof. □

Theorem 36.3. *Let $W = (w_{i,j})$ be a unitary matrix from the class $\mathcal{W}_N$, hence the matrix W admits the representation described in Corollary 36.2 with unitary 2×2 matrices V_i, F_i $(i = 1, \dots, N-1)$.*

Then upper quasiseparable generators $g(i)$ $(i = 1, \dots, N-1)$, $h(j)$ $(j = 2, \dots, N)$, $b(k)$ $(k = 2, \dots, N-1)$ with orders equal to one and the diagonal entries $d(k)$ $(k = 1, \dots, N)$ of the matrix W are obtained by the formulas

$$h(k) = F_k(1,1), \quad b(k) = F_k(1,2), \quad k = 2, \dots, N-1, \quad h(N) = 1, \tag{36.12}$$

and via the following recursion relations: set $\gamma_0 = 1$ and for $k = 1, \dots, N-1$ compute

$$\begin{pmatrix} d(k) & g(k) \\ * & \gamma_k \end{pmatrix} = V_k \begin{pmatrix} \gamma_{k-1} & 0 \\ 0 & 1 \end{pmatrix} F_k, \tag{36.13}$$

and finally set

$$d(N) = \gamma_{N-1}. \tag{36.14}$$

Here $\gamma_k \in \mathbb{C}$ are auxiliary variables.

Proof. Using Corollary 36.2 we obtain the factorization

$$W = (\tilde{V}_N \tilde{V}_{N-1} \cdots \tilde{V}_2)(\tilde{F}'_1 \tilde{F}_2 \cdots \tilde{F}_{N-1} \tilde{F}_N),$$

where $\tilde{F}'_1 = \tilde{V}_1 \tilde{F}_1$ and $\tilde{V}_N = I_{N-1} \oplus V_N$, $\tilde{F}_N = I_{N-1} \oplus F_N$, with $V_N = F_N = 1$. Now the statement of the theorem follows directly from Theorem 7.8. □

Thus every matrix A from the class $\mathcal{H}_N$ is completely defined by the following parameters:

1. the unitary 2×2 matrices Θ_k, $k = 1, \dots, N-1$ from the factorization (36.3), (36.4) of the matrix Θ;
2. the perturbation vectors $p = (p(i))_{i=1}^N$, $q = (q(i))_{i=1}^N$;
3. the unitary 2×2 matrices V_k, F_k $k = 1, 2, \dots, N-1$ from the factorization (36.7) of the matrix W, such that the vector p is reduced by the matrices V_k $(k = 1, \dots, N-1$ or upper quasiseparable generators $g(i)$, $i = 1, \dots, N-1$, $h(j)$, $j = 2, \dots, N$, $b(k)$, $k = 2, \dots, N-1$ of the matrix W and diagonal entries $d(k)$, $k = 1, \dots, N$ of the matrix R.

These parameters are called *the Q-generating elements* of the matrix A.

§36.1 General description

As above, we determine the unitary upper Hessenberg matrix Q and the matrix $A^{(1)} = Q^*AQ$ such that the condition (31.4) or the condition (31.7) holds. Using (36.2) we have

$$A^{(1)} = Q^*\Theta RQ.$$

We determine the unitary matrix S such that $\Theta^{(1)} = Q^*\Theta S$ is a unitary upper Hessenberg and the matrix

$$R^{(1)} = S^*RQ \tag{36.15}$$

is upper triangular. Combining (36.15) and (36.5), we obtain the representation

$$R^{(1)} = W^{(1)} - p^{(1)}(q^{(1)})^T,$$

with the unitary matrix $W^{(1)} = S^*WQ$ and the vectors $p^{(1)} = S^*p$, $q^{(1)} = Q^Tq$.

§36.1.1 The initialization

For the companion matrix C from (32.1) we have the factorization

$$C = \Theta R,$$

where the matrix Θ is represented in the form (36.3), (36.4) with $\Theta_i = \begin{pmatrix} 0 & 1 \\ 1 & 0 \end{pmatrix}$, $i = 1, \dots, N-1$, and

$$R = \begin{pmatrix} 1 & 0 & \dots & 0 & -p_1 \\ 0 & 1 & \dots & 0 & -p_2 \\ 0 & 0 & 1 & \dots & -p_3 \\ \vdots & \vdots & \ddots & \vdots & \vdots \\ 0 & 0 & \dots & 0 & -p_0 \end{pmatrix}.$$

We get the representation (36.5) with $W = I_N$ and

$$p = \begin{pmatrix} p_1 \\ p_2 \\ p_3 \\ \vdots \\ p_0 + 1 \end{pmatrix}, \quad q = \begin{pmatrix} 0 \\ 0 \\ 0 \\ \vdots \\ 1 \end{pmatrix}. \tag{36.16}$$

Next determine the unitary 2×2 matrices V_k, $k = N-1, \dots, 1$ via the relations

$$\beta_N = p_0 + 1, \quad V_i^* \begin{pmatrix} p_i \\ \beta_{i+1} \end{pmatrix} = \begin{pmatrix} \beta_i \\ 0 \end{pmatrix}, \; i = N-1, \dots, 1. \tag{36.17}$$

Thus the companion matrix C has the following Q-generating elements:

1. the unitary 2×2 unitary matrices $\Theta_k = \begin{pmatrix} 0 & 1 \\ 1 & 0 \end{pmatrix}$, $k = 1, \dots, N-1$;
2. the perturbation vectors p, q defined in (36.16);
3. the unitary 2×2 matrices V_k, $k = 1, 2, \dots, N-1$ defined in (36.17) and $F_k = V_k^*$, $k = 1, \dots, N-1$, or upper quasiseparable generators $g(i) = 0$, $i = 1, \dots, N-1$, $h(j) = 0$, $j = 2, \dots, N$, $b(k) = 0$, $k = 2, \dots, N-1$ of the matrix $W = I_N$ and the diagonal entries $d(k) = 1$, $k = 1, \dots, N-1$, $d(N) = -p_0$ of the matrix R.

§36.2 The single shift case

We consider the iteration step with a single shift with using Q-generating elements.

We start with the case where upper quasiseparable generators of the matrix W from (36.5) are used.

Theorem 36.4. *Let A be a matrix from the class $\mathcal{H}_N$ with Q-generating elements Θ_k $(k = 1, \dots, N-1)$, $g(i)$ $(i = 1, \dots, N-1), h(j)$ $(j = 2, \dots, N)$, $b(k)$ $(k = 2, \dots, N-1)$ with orders r_k $(k = 1, \dots, N-1)$, $d(k)$ $(k = 1, \dots, N)$ and $p = (p(i))_{i=1}^N$, $q = (q(i))_{i=1}^N$, and let α be a number. Then the Givens rotation matrices Q_k, $k = 1, \dots, N-1$ such that the unitary matrix*

$$Q = \tilde{Q}_1 \tilde{Q}_2 \cdots \tilde{Q}_{N-2} \tilde{Q}_{N-1}, \tag{36.18}$$

with

$$\tilde{Q}_i = I_{i-1} \oplus Q_i \oplus I_{N-i-1}, \; i = 1, \dots, N-1 \tag{36.19}$$

satisfies the condition (31.4), and the generating elements $\Theta_k^{(1)}$ $(k = 1, \dots, N-1)$; $g^{(1)}(i)$ $(i = 1, \dots, N-1), h^{(1)}(j)$ $(j = 2, \dots, N)$, $b^{(1)}(k)$ $(k = 2, \dots, N-1)$ with orders equal to one, $d^{(1)}(k)$ $(k = 1, \dots, N)$ and $p^{(1)} = (p^{(1)}(i))_{i=1}^N$, $q^{(1)} = (q^{(1)}(i))_{i=1}^N$ of the matrix $A^{(1)}$ are determined using the following algorithm.

1.1. *Compute*

$$d_W(1) = d(1) + p(1)q(1) \tag{36.20}$$

and set

$$f_1^W = d_W(1), \, \phi_1^W = g(1), \tag{36.21}$$

$$f_1 = d(1), \, \varphi_1 = g(1), \qquad c_1 = p(1), \, \theta_1 = q(1). \tag{36.22}$$

Compute

$$\begin{pmatrix} b_1 \\ b_2 \end{pmatrix} = \Theta_1(1:2, 1)d(1). \tag{36.23}$$

Determine the complex Givens rotation matrix Q_1 such that

$$Q_1^* \begin{pmatrix} b_1 - \alpha \\ b_2 \end{pmatrix} = \begin{pmatrix} * \\ 0 \end{pmatrix} \tag{36.24}$$

and compute

$$\Theta_1^{(\tau)} = Q_1^* \Theta_1, \tag{36.25}$$

1.2. *For $k = 1, \dots, N-2$ perform the following.*
Compute

$$\begin{pmatrix} q^{(1)}(k) \\ \theta_{k+1} \end{pmatrix} = Q_k^T \begin{pmatrix} \theta_k \\ q(k+1) \end{pmatrix}, \tag{36.26}$$

with the numbers $q^{(1)}(k), \theta_{k+1}$.
Compute the numbers

$$\begin{aligned} d_W(k+1) &= d(k+1) + p(k+1)q(k+1), \\ \varepsilon_k &= \varphi_k h(k+1) - c_k q(k+1) \end{aligned} \tag{36.27}$$

and the 2×2 matrix Ω_k by the formula

$$\Omega_k = \begin{pmatrix} f_k & \varepsilon_k \\ 0 & d(k+1) \end{pmatrix} Q_k. \tag{36.28}$$

Determine the complex Givens rotation matrix S_k and the number $d_k^{(1)}$ such that

$$S_k^* \Omega_k(:,1) = \begin{pmatrix} d_k^{(1)} \\ 0 \end{pmatrix}. \tag{36.29}$$

Compute

$$\Lambda_{k+1} = S_k^* \begin{pmatrix} f_k^W & \phi_k^W h(k+1) & \phi_k^W b(k+1) \\ p(k+1)\theta_k & d_W(k+1) & g(k+1) \end{pmatrix} \begin{pmatrix} Q_k & 0 \\ 0 & I_{r_{k+1}} \end{pmatrix}, \tag{36.30}$$

$$B_{k+1} = \begin{pmatrix} 1 & 0 & 0 \\ 0 & h(k+1) & b(k+1) \end{pmatrix} \begin{pmatrix} Q_k & 0 \\ 0 & I_{r_{k+1}} \end{pmatrix}, \tag{36.31}$$

and determine the matrices $d_B(k+1), g_B(k+1), \beta_{k+1}^W, f_{k+1}^W, \phi_{k+1}^W$ of sizes $1 \times 1, 1 \times (r_{k+1}+1), 1 \times (r_{k+1}+1), 1 \times 1, 1 \times r_{k+1}$ from the partitions

$$\Lambda_{k+1} = \begin{pmatrix} d_B(k+1) & g_B(k+1) \\ * & \beta_{k+1}^W \end{pmatrix}, \quad \beta_{k+1}^W = \begin{pmatrix} f_{k+1}^W & \phi_{k+1}^W \end{pmatrix}. \tag{36.32}$$

Compute

$$\begin{pmatrix} p^{(1)}(k) \\ c_{k+1} \end{pmatrix} = S_k^* \begin{pmatrix} c_k \\ p(k+1) \end{pmatrix}, \tag{36.33}$$

with the numbers $p^{(1)}(k), c_{k+1}$.
Compute

$$f_{k+1} = f_{k+1}^W - c_{k+1}\theta_{k+1}, \quad \varphi_{k+1} = \phi_{k+1}^W. \tag{36.34}$$

Compute the unitary 3×3 *matrix*

$$\Psi_k = \begin{pmatrix} \Theta_k^{(\tau)} & 0 \\ 0 & 1 \end{pmatrix} \begin{pmatrix} 1 & 0 \\ 0 & \Theta_{k+1} \end{pmatrix} \begin{pmatrix} S_k & 0 \\ 0 & 1 \end{pmatrix}. \tag{36.35}$$

Determine the complex Givens rotation matrix Q_{k+1} *from the condition*

$$Q_{k+1}^* \Psi_k(2:3,1) = \begin{pmatrix} * \\ 0 \end{pmatrix}. \tag{36.36}$$

Compute the unitary 3×3 *matrix*

$$\Phi_k = \begin{pmatrix} 1 & 0 \\ 0 & Q_{k+1}^* \end{pmatrix} \Psi_k. \tag{36.37}$$

Compute the factorization

$$\Phi_k = \begin{pmatrix} \Theta_k^{(1)} & 0 \\ 0 & 1 \end{pmatrix} \begin{pmatrix} 1 & 0 \\ 0 & \Theta_{k+1}^{(\tau)} \end{pmatrix}, \tag{36.38}$$

with unitary 2×2 *matrices* $\Theta_k^{(1)}, \Theta_{k+1}^{(\tau)}$.

1.3. *Compute*

$$\begin{pmatrix} q^{(1)}(N-1) \\ q^{(1)}(N) \end{pmatrix} = Q_{N-1}^T \begin{pmatrix} \theta_{N-1} \\ q(N) \end{pmatrix}, \tag{36.39}$$

with the numbers $q^{(1)}(N-1), q^{(1)}(N)$.

Compute the number

$$\varepsilon_{N-1} = \varphi_{N-1} h(N) - c_{N-1} q(N) \tag{36.40}$$

and the 2×2 *matrix* Ω_{N-1} *by the formula*

$$\Omega_{N-1} = \begin{pmatrix} f_{N-1} & \varepsilon_{N-1} \\ 0 & d(N) \end{pmatrix} Q_{N-1}. \tag{36.41}$$

Determine the complex Givens rotation matrix S_{N-1} *and the number* $d_{N-1}^{(1)}$ *such that*

$$S_{N-1}^* \Omega_{N-1}(:,1) = \begin{pmatrix} d_{N-1}^{(1)} \\ 0 \end{pmatrix}. \tag{36.42}$$

Compute

$$\Lambda_N = S_{N-1}^* \begin{pmatrix} f_{N-1}^U & \phi_{N-1}^U h(N) \\ p(N)\theta_{N-1} & d_W(N) \end{pmatrix} Q_{N-1}, \tag{36.43}$$

$$B_N = \begin{pmatrix} 1 & 0 \\ 0 & h(N) \end{pmatrix} Q_{N-1}, \tag{36.44}$$

and determine the numbers $d_B(N), g_B(N), d_B(N+1)$ *from the partitions*

$$\Lambda_N = \begin{pmatrix} d_B(N) & g_B(N) \\ * & d_B(N+1) \end{pmatrix}. \tag{36.45}$$

Compute

$$\begin{pmatrix} p^{(1)}(N-1) \\ p^{(1)}(N) \end{pmatrix} = S_{N-1}^* \begin{pmatrix} c_{N-1} \\ p(N) \end{pmatrix}, \tag{36.46}$$

with the numbers $p^{(1)}(N-1), p^{(1)}(N)$. *Compute*

$$d^{(1)}(N) = d_B(N+1) + p^{(1)}(N)q^{(1)}(N), \tag{36.47}$$

$$\Theta_{N-1}^{(1)} = \Theta_{N-1}^{(\tau)} S_{N-1}. \tag{36.48}$$

2.1. *Set*

$$X_N = p^{(1)}(N) \tag{36.49}$$

and compute

$$h^{(1)}(N) = 1, \quad Y_N = 1, \quad z_N = d_B(N+1), \quad g^{(1)}(N-1) = g_B(N). \tag{36.50}$$

2.2. *For* $k = N-1, \dots, 2$ *perform the following. Determine the unitary* 2×2 *unitary matrix* V_k *and the number* X_k *such that*

$$V_k^* \begin{pmatrix} p^{(1)}(k) \\ X_{k+1} \end{pmatrix} = \begin{pmatrix} X_k \\ 0 \end{pmatrix}. \tag{36.51}$$

Compute the 2×2 *matrix*

$$Z_k = V_k^* \begin{pmatrix} d_W(k) & g^{(1)}(k) \\ X_{k+1}q^{(1)}(k) & z_{k+1} \end{pmatrix} \tag{36.52}$$

and determine the numbers $h'_k, h''_k, \Delta_k, \Omega_k$ *from the partition*

$$Z_k = \begin{bmatrix} h'_k & h''_k \\ \Delta_k & \Omega_k \end{bmatrix}. \tag{36.53}$$

The two-dimensional row $\begin{pmatrix} \Delta_k & \Omega_k \end{pmatrix}$ *has unit norm and one can determine the unitary* 2×2 *matrix* F_k *from the condition*

$$\begin{pmatrix} \Delta_k & \Omega_k \end{pmatrix} F_k^* = \begin{pmatrix} 0 & 1 \end{pmatrix}. \tag{36.54}$$

Determine the numbers $h^{(1)}(k), b^{(1)}(k), d_F(k), g_F(k)$ *from the partition*

$$F_k = \begin{pmatrix} h^{(1)}(k) & b^{(1)}(k) \\ d_F(k) & g_F(k) \end{pmatrix}. \tag{36.55}$$

Compute

$$Y_k = B_{k+1} \begin{pmatrix} (h^{(1)}(k))^* \\ Y_{k+1}(b^{(1)}(k))^* \end{pmatrix}, \quad z_k = h'_k(h^{(1)}(k))^* + h''_k(b^{(1)}(k))^*, \tag{36.56}$$

$$g^{(1)}(k-1) = g_B(k)Y_k. \tag{36.57}$$

Proof. Using the formulas (36.2), (36.3) and (36.5) we get

$$A(1:2,1) = \Theta_1(1:2,1)d(1).$$

Hence, to satisfy the condition $(Q^*(A-\alpha I))(2,1) = 0$ we need to take the Givens rotation matrix Q_1 as in (36.23), (36.24).

Define the unitary matrix S by

$$S = \tilde{S}_1\tilde{S}_2\cdots\tilde{S}_{N-2}\tilde{S}_{N-1}, \tag{36.58}$$

with

$$\tilde{S}_i = I_{i-1} \oplus S_i \oplus I_{N-i-1},\ i = 1,\dots,N-1. \tag{36.59}$$

We show that $\Theta^{(1)} = Q^*\Theta S$ is a unitary upper Hessenberg matrix represented in the form

$$\Theta^{(1)} = \tilde{\Theta}_1^{(1)}\tilde{\Theta}_2^{(1)}\cdots\tilde{\Theta}_{N-1}^{(1)}, \tag{36.60}$$

where

$$\tilde{\Theta}_i^{(1)} = I_{i-1} \oplus \Theta_i^{(1)} \oplus I_{N-i-1}, \quad i = 1,\dots,N-1, \tag{36.61}$$

and that $R^{(1)} = S^*RQ$ is an upper triangular matrix, and we derive the formulas to compute its upper quasiseparable generators and diagonal entries.

Set

$$\hat{\Theta}^{(k)} = \tilde{Q}_k^*\cdots\tilde{Q}_1^*\Theta\tilde{S}_1\cdots\tilde{S}_{k-1}, \quad k = 1,\dots,N-1. \tag{36.62}$$

To get (36.60) and (36.61), we prove by induction that

$$\hat{\Theta}^{(k)} = \tilde{\Theta}_1^{(1)}\cdots\tilde{\Theta}_{k-1}^{(1)}\tilde{\Theta}_k^{(\tau)}\tilde{\Theta}_{k+1}\cdots\tilde{\Theta}_{N-1}, \quad k = 1,\dots,N-1, \tag{36.63}$$

where

$$\tilde{\Theta}_i^{(\tau)} = I_{i-1} \oplus \Theta_i^{(\tau)} \oplus I_{N-i-1}.$$

Using (36.3) and (36.62) we get

$$\hat{\Theta}^{(1)} = \tilde{Q}_1^*\Theta = (\tilde{Q}_1^*\tilde{\Theta}_1)\tilde{\Theta}_2\cdots\tilde{\Theta}_{N-1},$$

whence, using (36.25), we obtain (36.63) with $k = 1$. Suppose that for some k with $1 \le k \le N-2$ the representation (36.63) holds. Using (36.59) we get

$$\hat{\Theta}^{(k)}\tilde{S}_k = \tilde{\Theta}_1^{(1)}\cdots\tilde{\Theta}_{k-1}^{(1)}(\tilde{\Theta}_k^{(\tau)}\tilde{\Theta}_{k+1}\tilde{S}_k)\tilde{\Theta}_{k+2}\cdots\tilde{\Theta}_{N-1}$$

which means that

$$\hat{\Theta}^{(k)}\tilde{S}_k = \tilde{\Theta}_1^{(1)}\cdots\tilde{\Theta}_{k-1}^{(1)}\tilde{\Psi}_k\tilde{\Theta}_{k+2}\cdots\tilde{\Theta}_{N-1}, \tag{36.64}$$

where $\tilde{\Psi}_k = I_{k-1} \oplus \Psi_k \oplus I_{N-k-2}$, with unitary 3×3 matrix Ψ_k determined via (36.35). We determine the complex Givens rotation matrix Q_{k+1} from the condition (36.36). The matrix Φ_k defined in (36.37) has a zero entry in the (3,1) position and therefore Φ_k admits the factorization (36.38). From (36.37) and (36.38) it follows that $\tilde{Q}^*_{k+1}\tilde{\Psi}_k = \tilde{\Theta}^{(1)}_k\tilde{\Theta}^{(\tau)}_{k+1}$. Inserting this in (36.64) we get

$$\hat{\Theta}^{(k+1)} = \tilde{Q}^*_{k+1}\hat{\Theta}^{(k)}\tilde{S}_k = \tilde{\Theta}^{(1)}_1\cdots\tilde{\Theta}^{(1)}_k\tilde{\Theta}^{(\tau)}_{k+1}\tilde{\Theta}_{k+2}\cdots\tilde{\Theta}_{N-1}, \tag{36.65}$$

which completes the proof of (36.63). Using (36.63) with $k = N-1$ and the equality (36.48) we obtain the representation (36.60) for the matrix $\Theta^{(1)} = \hat{\Theta}^{(N-1)}\tilde{S}_{N-1}$.

Next we consider the matrices $R^{(1)} = S^*RQ$, $W^{(1)} = S^*WQ$. Set

$$m_1 = m_2 = \cdots = m_N = 1,\ m_{N+1} = 0; \quad n_1 = 0, \quad n_2 = \cdots = n_N = n_{N+1} = 1. \tag{36.66}$$

We treat R and W as $(N+1)\times(N+1)$ matrices with entries of sizes $m_i \times m_j$, $i,j = 1,\dots,N+1$ and Q,S as $(N+1)\times(N+1)$ matrices with entries of sizes $m_i \times n_j$, $i,j = 1,\dots,N+1$.

Using (36.18), (36.58) and setting

$$\begin{aligned} &\tilde{C}_1 = I_N, \quad \tilde{C}_i = \tilde{S}^*_{i-1},\ i = 2,\dots,N, \quad \tilde{C}_{N+1} = I_N,\\ &\tilde{T}_1 = I_N, \quad \tilde{T}_i = \tilde{Q}_{i-1},\ i = 2,\dots,N, \quad \tilde{T}_{N+1} = I_N \end{aligned} \tag{36.67}$$

we get

$$S^* = \tilde{C}_{N+1}\cdots\tilde{C}_1, \quad Q = \tilde{T}_1\cdots\tilde{T}_{N+1}. \tag{36.68}$$

We have

$$\tilde{C}_1 = \operatorname{diag}\{C_1, I_{N-1}\}, \quad \tilde{C}_k = \operatorname{diag}\{I_{k-2}, C_k, I_{N-k}\},\ k = 2,\dots,N,$$
$$\tilde{C}_{N+1} = \operatorname{diag}\{I_{N-1}, C_{N+1}\},$$

with

$$C_1 = 1, \quad C_k = S^*_{k-1},\ k = 2,\dots,N, \quad C_{N+1} = 1, \tag{36.69}$$

and

$$\tilde{T}_1 = \operatorname{diag}\{T_1, I_{N-1}\}, \quad \tilde{T}_k = \operatorname{diag}\{I_{k-2}, T_k, I_{N-k}\},\ k = 2,\dots,N,$$
$$\tilde{T}_{N+1} = \operatorname{diag}\{I_{N-1}, T_{N+1}\},$$

with

$$T_1 = 1, \quad T_k = Q_{k-1},\ k = 2,\dots,N, \quad T_{N+1} = 1. \tag{36.70}$$

The matrix R has the following upper quasiseparable generators:

$$\begin{aligned} &g_R(i) = \begin{pmatrix} g(i) & -p(i) \end{pmatrix},\ i = 1,\dots,N-1, \quad g_R(N) \text{ is the } 1\times 0 \text{ empty matrix},\\ &h_R(j) = \begin{pmatrix} h(j)\\ q(j) \end{pmatrix}, \qquad j = 2,\dots,N, \quad h_R(N+1) \text{ is the } 0\times 0 \text{ empty matrix},\\ &b_R(k) = \begin{pmatrix} b(k) & 0\\ 0 & 1 \end{pmatrix}, \quad k = 2,\dots,N-1, \quad b_R(N) \text{ is the } 2\times 0 \text{ empty matrix}, \end{aligned} \tag{36.71}$$

with orders $r^R_k = r_k + 1,\ k = 1, \dots, N-1,\ r^R_N = 0$ and diagonal entries

$$d_R(k) = d(k),\ k = 1, \dots, N, \quad d_R(N+1) \text{ is the } 0 \times 0 \text{ empty matrix.} \tag{36.72}$$

The matrix $W = R + pq^T$ has the following quasiseparable generators:

$$\begin{aligned} &p_W(i) = p(i),\ i = 2, \dots, N, && p_W(N+1) \text{ is the } 0 \times 0 \text{ empty matrix,} \\ &q_W(j) = q(j),\ j = 1, \dots, N-1, && q_W(N) \text{ is the } 0 \times 1 \text{ empty matrix,} \\ &a_W(k) = 1, \quad k = 2, \dots, N-1, && a_W(N) \text{ is the } 0 \times 1 \text{ empty matrix;} \end{aligned} \tag{36.73}$$

$$\begin{aligned} &g_W(i) = g(i),\ i = 1, \dots, N-1, && g_W(N) \text{ is the } 1 \times 0 \text{ empty matrix,} \\ &h_W(j) = h(j),\ j = 2, \dots, N, && h_W(N+1) \text{ is the } 1 \times 0 \text{ empty matrix,} \\ &b_W(k) = b(k),\ k = 2, \dots, N-1, && b_W(N) \text{ is the } 0 \times 0 \text{ empty matrix;} \end{aligned} \tag{36.74}$$

$$\begin{aligned} d_W(k) = d(k) + p(k)q(k),\ k = 1, \dots, N, \\ d_W(N+1) \text{ is the } 0 \times 0 \text{ empty matrix.} \end{aligned} \tag{36.75}$$

Now we apply Lemma 31.1 and Corollary 31.2, with the matrices S^*, Q represented in the form (36.68), (36.67), to the matrices W and R, respectively. The sizes of the blocks and the orders of generators used in Lemma 31.1 and Corollary 31.2 are given by the formulas

$$\tau_1 = \nu_1 = 0, \quad \tau_i = \nu_i = 1,\ i = 2, \dots, N+1;$$
$$r^S_i = r^T_i = 1,\ i = 1, \dots, N, \quad \rho^A_k = 1,\ r^A_k = r_k,\ k = 1, \dots, N-1, \quad \rho^A_N = r^A_N = 0.$$

We apply Lemma 31.1 to compute (block) upper quasiseparable generators $g_B(i)$ $(i = 1, \dots, N)$, $h_B(j)$ $(j = 2, \dots, N+1)$, $b_B(k)$ $(k = 2, \dots, N)$ with orders $\tilde{r}_k = r_k + 1,\ k = 1, \dots, N-1,\ \tilde{r}_N = 1$ of the matrix $U^{(1)} = S^*UQ$. The matrix $U^{(1)}$ is obtained as a block one with entries of sizes $\tau_i \times \nu_j,\ i, j = 1, \dots, N+1$. Using (31.9)–(31.11) with $\tau_1 = \nu_1 = 0$ and the equalities $d_W(1) = d(1),\ q_W(1) = q(1),\ g_W(1) = g(1)$, we get

$$f^W_1 = d_W(1), \quad \phi^W_1 = g(1), \quad \alpha_1 = q(1). \tag{36.76}$$

Using (31.12)–(31.14), (36.69), (36.70) and (36.73)–(36.75) we get

$$\begin{pmatrix} d_B(k) & g_B(k) \\ * & \Gamma_k \end{pmatrix} = \begin{pmatrix} S^*_{k-1} & 0 \\ 0 & 1 \end{pmatrix} \begin{pmatrix} f^W_{k-1} & \phi^W_{k-1}h(k) & \phi^W_{k-1}b(k) \\ p(k)\alpha_{k-1} & d_W(k) & g(k) \\ \alpha_{k-1} & q(k) & 0 \end{pmatrix} \begin{pmatrix} Q_{k-1} & 0 \\ 0 & I_{r_k} \end{pmatrix},$$
$$\Gamma_k = \begin{pmatrix} f^W_k & \phi^W_k \\ \alpha_k & 0 \end{pmatrix}, \quad k = 2, \dots, N-1, \tag{36.77}$$

$$\begin{pmatrix} d_B(N) & g_B(N) \\ * & \Gamma_N \end{pmatrix} = S^*_{N-1} \begin{pmatrix} f^W_{N-1} & \phi^W_{N-1}h(N) \\ p(N)\alpha_{N-1} & d_W(N) \end{pmatrix} Q_{N-1}, \quad \Gamma_N = f^W_N, \tag{36.78}$$

and also

$$\begin{pmatrix} h_B(k) & b_B(k) \end{pmatrix} = \begin{pmatrix} 1 & 0 & 0 \\ 0 & h(k) & b(k) \end{pmatrix} \begin{pmatrix} Q_{k-1} & 0 \\ 0 & I_{r_k} \end{pmatrix}, \quad k = 2, \dots, N-1, \tag{36.79}$$

$$\begin{pmatrix} h_B(N) & b_B(N) \end{pmatrix} = \begin{pmatrix} 1 & 0 \\ 0 & h(N) \end{pmatrix} Q_{N-1}. \tag{36.80}$$

Using (31.17), (31.18) we get $h_B(N+1) = 1,\ d_B(N+1) = f_N^W$.

The relations (36.76), (36.77) imply, in particular, that the auxiliary matrices α_k $(k = 1, \dots, N-1)$ satisfy the relations

$$\alpha_1 = q(1), \quad \begin{pmatrix} * & \alpha_k \end{pmatrix} = \begin{pmatrix} \alpha_{k-1} & q(k) \end{pmatrix} Q_{k-1},\ k = 2, \dots, N-1.$$

Comparing this with (36.22), (36.26), we get

$$\alpha_k = \theta_k, \quad k = 1, \dots, N-1. \tag{36.81}$$

Thus inserting (36.81) in (36.77) and (36.78) and using also the fact that $d_B(N+1) = f_N^W$ we obtain (36.30), (36.32) and (36.43), (36.45).

Applying Lemma 31.2 to the matrix $R^{(1)} = S^*RQ$, treated as an $(N+1) \times (N+1)$ matrix with entries of sizes $n_i \times n_j$, $i, j = 1, \dots, N+1$, we obtain quasiseparable generators $q_F(j)$ $(j = 2, \dots, N)$, $g_F(i)$ $(i = 2, \dots, N)$, $h_F(j)$ $(j = 3, \dots, N+1)$, $b_F(k)$ $(k = 3, \dots, N)$, $d_F(k)$ $(k = 2, \dots, N+1)$ of $R^{(1)}$ by the formulas

$$f_1 = d_R(1), \quad \phi_1 = g_R(1), \tag{36.82}$$

$$\begin{pmatrix} d_F(k) & g_F(k) \\ q_F(k) & \beta_k \end{pmatrix} = S_{k-1}^* \begin{pmatrix} f_{k-1} & \phi_{k-1}h_R(k) & \phi_{k-1}b_R(k) \\ 0 & d_R(k) & g_R(k) \end{pmatrix} \begin{pmatrix} Q_{k-1} & 0 \\ 0 & I_{r_k^R} \end{pmatrix},$$
$$\beta_k = \begin{pmatrix} f_k & \phi_k \end{pmatrix}, \quad k = 2, \dots, N. \tag{36.83}$$

For the (r_k+1)-dimensional rows ϕ_k we use the partitions

$$\phi_k = \begin{pmatrix} \varphi_k & -c_k \end{pmatrix}, \quad k = 1, \dots, N-2, \tag{36.84}$$

with r_k-dimensional rows φ_k and numbers c_k. Using (36.71), (36.72), (36.84) with $k = 1$ and (36.82) we get

$$f_1 = d(1), \quad \varphi_1 = g(1), \quad c_1 = p(1). \tag{36.85}$$

Next, using (36.71) and (36.84) we obtain the relations

$$\phi_k h_R(k+1) = \varepsilon_k, \quad k = 1, \dots, N-1, \tag{36.86}$$

with ε_k as in (36.27), and

$$\phi_k b_R(k+1) = \begin{pmatrix} \varphi_k b(k+1) & -c_k \end{pmatrix}, \quad k = 1, \dots, N-2. \tag{36.87}$$

Inserting (36.86), (36.87) in (36.83) and using (36.28), (36.41) we obtain

$$\begin{pmatrix} d_F(k+1) & g_F(k+1) \\ q_F(k+1) & \beta_{k+1} \end{pmatrix} = S_k^* \begin{pmatrix} \Omega_k(1,1) & \Omega_k(1,2) & \varphi_k b(k+1) & -c_k \\ \Omega_k(2,1) & \Omega_k(2,2) & g(k+1) & -p(k+1) \end{pmatrix},$$
$$\beta_{k+1} = \begin{pmatrix} f_{k+1} & \phi_{k+1} \end{pmatrix}, \quad k = 1, \dots, N-2, \tag{36.88}$$
$$\begin{pmatrix} d_F(N) & g_F(N) \\ q_F(N) & \beta_N \end{pmatrix} = S_{N-1}^* \Omega_{N-1}.$$

Now (36.29) and (36.42) yield

$$d_F(k+1) = d^{(1)}(k), \quad k = 1, \dots, N-1 \tag{36.89}$$

and

$$q_F(k+1) = 0, \qquad k = 1, \dots, N-1. \tag{36.90}$$

The formulas (36.89) and (36.47) mean that $d^{(1)}(k)$, $k = 1, \dots, N$, are diagonal entries of the matrix $R^{(1)}$. The equalities (36.90) imply that $R^{(1)}$ is an upper triangular matrix. Using (36.88) and (36.84) we have also

$$\begin{pmatrix} * & * & * \\ f_{k+1} & \varphi_{k+1} & -c_{k+1} \end{pmatrix} = S_k^* \begin{pmatrix} \Omega_k(1,2) & \varphi_k b(k+1) & -c_k \\ \Omega_k(2,2) & g(k+1) & -p(k+1) \end{pmatrix},$$
$$k = 1, \dots, N-2. \tag{36.91}$$

Using the relations (36.18), (36.19) we obtain the formulas (36.26), (36.39) to compute the coordinates of the perturbation vector $q^{(1)} = Q^T q$. From the relations (36.58), (36.59) and it follows that the formulas (36.33), (36.46) yield the coordinates of the perturbation vector $p^{(1)} = S^* p$.

Using (36.79), (36.80) and setting

$$B_k = \begin{pmatrix} h_B(k) & b_B(k) \end{pmatrix}, \quad k = 2, \dots, N, \tag{36.92}$$

we obtain (36.31), (36.44).

It remains to show that the auxiliary variables f_k, φ_k $(k = 2, \dots, N)$ may be determined via the relations (36.34). Comparing (36.76), (36.20) with (36.85) we get

$$f_1 = f_1^W - c_1\theta_1, \quad \varphi_1 = \phi_1^W.$$

Assume that for some k with $1 \le k \le N-2$ the relations

$$f_k = f_k^W - c_k\theta_k, \quad \varphi_k = \phi_k^W \tag{36.93}$$

hold. By (36.28) and (36.91), we have

$$\begin{pmatrix} d^{(1)}(k) & * & * \\ 0 & f_{k+1} & \varphi_{k+1} \end{pmatrix} = S_k^* \begin{pmatrix} f_k & \varepsilon_k & \varphi_k b(k+1) \\ 0 & d(k+1) & g(k+1) \end{pmatrix} \begin{pmatrix} Q_k & 0 \\ 0 & I_{r_{k+1}} \end{pmatrix}. \tag{36.94}$$

Using (36.27) and (36.93) we get

$$\begin{pmatrix} f_k & \varepsilon_k \\ 0 & d(k+1) \end{pmatrix} = \begin{pmatrix} f_k^W - c_k\theta_k & \phi_k^W h(k+1) - c_k q(k+1) \\ p(k+1)\theta_k - p(k+1)\theta_k & d_W(k+1) - p(k+1)q(k+1) \end{pmatrix}. \tag{36.95}$$

Thus, combining (36.94) and (36.95) and using (36.30), (36.32) and (36.26), (36.33) we obtain (36.34).

We have obtained upper quasiseparable generators and diagonal entries of the matrix $W^{(1)}$, treated as a block one with $n_i \times n_j$ $(i,j = 1,\dots,N+1)$ entries with the sizes n_i defined in (36.66). The matrix $W^{(1)}$, treated as an ordinary scalar matrix, has upper quasiseparable generators $g_B(i+1)$ $(i = 1\dots,N-1)$, $h_B(j+1)$ $(j = 2,\dots,N)$, $b_B(k+1)$ $(k = 2,\dots,N-1)$ and diagonal entries $d_B(k+1)$ $(k = 1,\dots,N)$. Next we apply Theorem 7.5 to determine upper quasiseparable generators $g^{(1)}(i)$ $(i = 1,\dots,N-1)$, $h^{(1)}(j)$ $(j = 2,\dots,N)$, $b^{(1)}(k)$ $(k = 2,\dots,N-1)$ of the matrix $W^{(1)}$ with orders equal to one. The matrix $W^{(1)}$ has lower quasiseparable generators $p^{(1)}(k)$ $(k = 2,\dots,N)$, $q^{(1)}(k)$ $(k = 1,\dots,N-1)$, $a(k) = 1$ $(k = 2,\dots,N-1)$ of orders $r_k^L = 1$, $k = 1,\dots,N$. Using (7.28) we obtain $V_N = 1$ and the formulas (36.49), (36.51) to determine the numbers X_k $(k = N,\dots,2)$ and the matrices V_k $(k = N-1,\dots,2)$.

Using (7.29)–(7.34) with $k = N$ we get

$$Z_N = d_B(N+1) = h'_N, \quad F_N = h^{(1)}(N) = 1, \quad Y_N = h_B(N+1) = 1,$$
$$z_N = d_B(N+1), \quad g^{(1)}(N-1) = g_B(N).$$

For $k = N-1,\dots,2$ applying (7.29)–(7.34) and using (36.92) we obtain the relations (36.52)–(36.57). □

Next we consider an algorithm in which the factorizations from Corollary 36.2 are used. The result is similar to one obtained in Theorem 35.4.

Theorem 36.5. *Let A be a matrix from the class $\mathcal{H}_N$ with Q-generating elements Θ_k $(k = 1,\dots,N-1)$, V_k, F_k $(k = 1,2,\dots,N-1)$ and $p = (p(i))_{i=1}^N$, $q = (q(i))_{i=1}^N$, and let α be a number. Then the Givens rotation matrices Q_k, $k = 1,\dots,N-1$, such that the unitary matrix*

$$Q = \tilde{Q}_1\tilde{Q}_2\cdots\tilde{Q}_{N-2}\tilde{Q}_{N-1}, \tag{36.96}$$

with

$$\tilde{Q}_i = I_{i-1} \oplus Q_i \oplus I_{N-i-1}, \ i = 1,\dots,N-1$$

satisfies the condition (31.4), and Q-generating elements $\Theta_k^{(1)}$ $(k = 1,\dots,N-1)$, $V_k^{(1)}$, $F_k^{(1)}$ $(k = 1,2,\dots,N-1)$ and $p^{(1)} = (p^{(1)}(i))_{i=1}^N$, $q^{(1)} = (q^{(1)}(i))_{i=1}^N$ of the matrix $A^{(1)}$ are determined using the following algorithm.

1. *Using the algorithm from Theorem* 36.3 *compute upper quasiseparable generators $g(i)$ $(i = 1,\dots,N-1)$, $h(j)$ $(j = 2,\dots,N)$, $b(k)$ $(k = 2,\dots,N-1)$ with orders equal to one and the diagonal entries $d(k)$ $(k = 1,\dots,N)$ of the matrix W.*

2. *Using Step* 1 *of the algorithm from Theorem* 36.4, *compute the matrices* Q_k $(k = 1, \dots, N-1)$, *the updated matrices* $\Theta_k^{(1)}$ $(k = 1, \dots, N-1)$, *the matrices* S_k $(k = 1, \dots, N-1)$ *such that the matrix* $R^{(1)} = S^*RQ$ *with* S *defined via* (36.58), (36.59) *is an upper triangular matrix, and the updated perturbation vectors* $p^{(1)}, q^{(1)}$.

3. *Determine the* Q*-generating elements* $V_k^{(1)}, F_k^{(1)}$ $(k = 1, \dots, N-1)$ *of the matrix* $A^{(1)}$ *as follows.*

3.1. *Set* $V_1^{(\tau)} = V_1$, $F_1^{(\tau)} = F_1$.

3.2. *For* $k = 1, \dots, N-2$ *perform the following.*

Compute the unitary 3×3 *matrix*

$$\Gamma_k = \begin{pmatrix} S_k^* & 0 \\ 0 & 1 \end{pmatrix} \begin{pmatrix} 1 & 0 \\ 0 & V_{k+1} \end{pmatrix} \begin{pmatrix} V_k^{(\tau)} & 0 \\ 0 & 1 \end{pmatrix} \tag{36.97}$$

and determine the complex Givens rotation matrix Z_{k+1} *from the condition*

$$\Gamma_k(1, 2:3) Z_{k+1} = \begin{pmatrix} * & 0 \end{pmatrix}. \tag{36.98}$$

Compute the unitary 3×3 *matrices*

$$L_k = \Gamma_k \begin{pmatrix} 1 & 0 \\ 0 & Z_{k+1} \end{pmatrix} \tag{36.99}$$

and

$$D_k = \begin{pmatrix} 1 & 0 \\ 0 & Z_k^* \end{pmatrix} \begin{pmatrix} F_k^{(\tau)} & 0 \\ 0 & 1 \end{pmatrix} \begin{pmatrix} 1 & 0 \\ 0 & F_{k+1} \end{pmatrix} \begin{pmatrix} Q_k & 0 \\ 0 & 1 \end{pmatrix}. \tag{36.100}$$

The unitary matrices L_k *and* M_k *have zero entries in the positions* $(1,3)$ *and* $(3,1)$ *respectively.*

Compute the factorization

$$L_k = \begin{pmatrix} 1 & 0 \\ 0 & V_{k+1}^{(\tau)} \end{pmatrix} \begin{pmatrix} V_k^{(1)} & 0 \\ 0 & 1 \end{pmatrix} \tag{36.101}$$

and the factorization

$$D_k = \begin{pmatrix} F_k^{(1)} & 0 \\ 0 & 1 \end{pmatrix} \begin{pmatrix} 1 & 0 \\ 0 & F_{k+1}^{(\tau)} \end{pmatrix}, \tag{36.102}$$

with unitary 2×2 *matrices* $V_k^{(1)}, V_{k+1}^{(\tau)}, F_k^{(1)}, F_{k+1}^{(\tau)}$.

4.3. *Set*

$$V_{N-1}^{(1)} = S_{N-1}^* V_{N-1}^{(\tau)}, \quad F_{N-1}^{(1)} = F_{N-1}^{(\tau)} Q_{N-1}. \tag{36.103}$$

Proof. Step 1 of the algorithm follows from Theorem 34.1.

Step 2 follows from Theorem 36.4.

Now we will justify Step 3. We have

$$W^{(1)} = S^*WQ = (S^*V)(FQ).$$

Define the unitary matrix Z via relations (35.37), (35.38). We get

$$W^{(1)} = V^{(1)} \cdot F^{(1)},$$

with $V^{(1)} = S^*VZ$, $U^{(1)} = Z^*FQ$. We need to prove that the factorizations

$$V^{(1)} = \tilde{V}^{(1)}_{N-1} \cdots \tilde{V}^{(1)}_1, \tag{36.104}$$

where

$$\tilde{V}^{(1)}_i = I_{i-1} \oplus V^{(1)}_i \oplus I_{N-i-1}, \quad i = 1, \dots, N-1,$$

and

$$F^{(1)} = \tilde{F}^{(1)}_1 \cdots \tilde{F}^{(1)}_{N-1}, \quad i = 1, \dots, N-1, \tag{36.105}$$

where

$$\tilde{F}^{(1)}_i = I_{i-1} \oplus F^{(1)}_i \oplus I_{N-i-1},$$

hold and that the vector $p^{(1)}$ is reduced by the matrix $V^{(1)}$. This is done in the same way as in the proof of Theorem 35.4.

Applying Lemma 36.1 to the matrix $W^{(1)}$ we obtain the equalities

$$F^{(1)}(j+2, j) = 0, \quad j = 1, \dots, N-2. \tag{36.106}$$

Set

$$\Phi_1 = F, \quad \Phi_{k+1} = \tilde{Z}^*_{k+1}\Phi_k\tilde{Q}_k, \ k = 1, \dots, N-2,$$

which means that

$$\Phi_{k+1} = \tilde{Z}^*_{k+1} \cdots \tilde{Z}^*_2 F \tilde{Q}_1 \cdots \tilde{Q}_k, \quad k = 1, \dots, N-3. \tag{36.107}$$

We prove by induction that

$$\Phi_k = \tilde{F}^{(1)}_1 \cdots \tilde{F}^{(1)}_{k-1}\tilde{F}^{(\tau)}_k\tilde{F}_{k+1} \cdots \tilde{F}_{N-1}, \quad k = 1, \dots, N-1, \tag{36.108}$$

where

$$\tilde{F}^{(\tau)}_i = I_{i-1} \oplus F^{(\tau)}_i \oplus I_{N-i-1}.$$

For $k = 1$ the equality (36.108) follows from (36.10), (36.11) and $F^{(\tau)}_1 = F_1$. Suppose that for some k with $1 \le k \le N-2$ the relation (36.108) holds. Hence the matrix $\Phi_{k+1} = \tilde{Z}^*_{k+1}\Phi_k\tilde{Q}_k$ has the form

$$\Phi_{k+1} = (\tilde{F}^{(1)}_1 \cdots \tilde{F}^{(1)}_{k-1})(\tilde{Z}^*_{k+1}\tilde{F}^{(\tau)}_k\tilde{F}_{k+1}\tilde{Q}_k)(\tilde{F}_{k+2} \cdots \tilde{F}_{N-2}). \tag{36.109}$$

We have

$$\tilde{Z}^*_{k+1}\tilde{F}^{(\tau)}_k\tilde{F}_{k+1}\tilde{Q}_k = I_{k-1} \oplus D_k \oplus I_{N-k-2} \tag{36.110}$$

with the unitary 3×3 matrix D_k from (36.100). Also, the relations $F^{(1)} = Z^*FQ$ and (36.96), (35.37), (36.107) yield

$$\Phi_{k+1} = (\tilde{Z}_{k+2} \cdot \tilde{Z}_{N-1})F^{(1)}\tilde{Q}^*_{N-1} \cdot \tilde{Q}^*_{k+1},$$

whence, using (36.106), we get

$$\Phi_{k+1}(k+2, k) = 0. \tag{36.111}$$

By (36.109) and (36.110) we have $D_k(3,1) = \Phi_{k+1}(k+2,k)$ and using (36.111) we conclude that the unitary 3×3 matrix D_k has a zero entry in the $(3,1)$ position. Hence, D_k admits the factorization (36.102) with unitary 2×2 matrices $F^{(1)}_k, F^{(\tau)}_{k+1}$. It follows that

$$\Phi_{k+1} = (\tilde{F}^{(1)}_1 \cdots \tilde{F}^{(1)}_{k-1})(\tilde{F}^{(1)}_k \tilde{F}^{(\tau)}_{k+1})(\tilde{F}_{k+2} \cdots \tilde{F}_{N-1}).$$

which completes the proof of (36.108). Now using (36.108) with $k = N-1$ and (36.103) we obtain the factorization (36.105) for the matrix $F^{(1)} = \Phi_{N-1}\tilde{Q}_{N-1}$. □

§36.3 The double shift case

Here we present two algorithms in a similar way as in the previous subsection but for real matrices using the double shift.

Theorem 36.6. *Let A be a real matrix from the class $\mathcal{H}_N$ with Q-generating elements Θ_k $(k = 1,\dots,N-1)$, $g(i)$ $(i = 1,\dots,N-1)$, $h(j)$ $(j = 2,\dots,N)$, $b(k)$ $(k = 2,\dots,N-1)$ with orders r_k $(k = 1,\dots,N-1)$, $d(k)$ $(k = 1,\dots,N)$ and $p = (p(i))_{i=1}^N$, $q = (q(i))_{i=1}^N$, and let s,t be real numbers. Then the orthogonal 3×3 matrices Q_k, $k = 1,\dots,N-2$ and the orthogonal 2×2 matrix Q_{N-1} such that the orthogonal matrix*

$$Q = \tilde{Q}_1\tilde{Q}_2 \cdots \tilde{Q}_{N-2}\tilde{Q}_{N-1}, \tag{36.112}$$

with

$$\tilde{Q}_i = I_{i-1} \oplus Q_i \oplus I_{N-i-2},\ i = 1,\dots,N-2, \quad \tilde{Q}_{N-1} = I_{N-2} \oplus Q_{N-1} \tag{36.113}$$

satisfies the condition (31.7), and Q-generating elements $\Theta^{(1)}_k$ $(k = 1,\dots,N-1)$ $g^{(1)}(i)$ $(i = 1,\dots,N-1), h^{(1)}(j)$ $(j = 2,\dots,N)$, $b^{(1)}(k)$ $(k = 2,\dots,N-1)$ with orders equal to one, $d^{(1)}(k)$ $(k = 1,\dots,N)$ and $p^{(1)} = (p^{(1)}(i))_{i=1}^N$, $q^{(1)} = (q^{(1)}(i))_{i=1}^N$ of the matrix $A^{(1)}$ are determined using the following algorithm.

1.1. *Compute*

$$d_W(1) = d(1) + p(1)q(1), \quad d_W(2) = d(2) + p(2)q(2), \tag{36.114}$$

and set

$$f_2^W = \begin{pmatrix} d_W(1) & g(1)h(2) \\ p(2)q(1) & d_W(2) \end{pmatrix}, \ \phi_2^W = \begin{pmatrix} g(1)b(2) \\ g(2) \end{pmatrix}, \tag{36.115}$$

$$\varepsilon = g(1)h(2) - p(1)q(2), \tag{36.116}$$

$$f_2 = \begin{pmatrix} d(1) & \varepsilon \\ 0 & d(2) \end{pmatrix}, \ \varphi_2 = \begin{pmatrix} g(1)b(2) \\ g(2) \end{pmatrix}, \quad c_2 = \begin{pmatrix} p(1) \\ p(2) \end{pmatrix}, \ \theta_2 = \begin{pmatrix} q(1) \\ q(2) \end{pmatrix}. \tag{36.117}$$

Determine the numbers $\rho_{11}, \rho_{12}, \rho_{21}, \rho_{22}, \lambda_1, \lambda_2$ *from the partitions*

$$\Theta_1 = \begin{pmatrix} \rho_{11} & \rho_{12} \\ \rho_{21} & \rho_{22} \end{pmatrix}, \quad \Theta_2(:,1) = \begin{pmatrix} \lambda_1 \\ \lambda_2 \end{pmatrix}, \tag{36.118}$$

and compute

$$\begin{aligned} &a_{11} = \rho_{11} d(1), \quad a_{12} = \rho_{11}\varepsilon + \rho_{12}\lambda_1 d(2), \quad a_{21} = \rho_{21} d(1), \\ &a_{22} = \rho_{21}\varepsilon + \rho_{22}\lambda_1 d(2), \quad a_{32} = \lambda_2 d(2). \end{aligned} \tag{36.119}$$

Determine the orthogonal 3×3 *matrix* Q_1 *from the condition*

$$Q_1^* \begin{pmatrix} a_{11}^2 + a_{12}a_{21} - sa_{21} + t & a_{21}(a_{11} + a_{22} - s) & a_{32}a_{21} \end{pmatrix}^T = \begin{pmatrix} * & 0 & 0 \end{pmatrix}^T. \tag{36.120}$$

Compute

$$\Theta_1^{(\tau)} = Q_1^* \begin{pmatrix} \Theta_1 & 0 \\ 0 & 1 \end{pmatrix} \begin{pmatrix} 1 & 0 \\ 0 & \Theta_2 \end{pmatrix}. \tag{36.121}$$

1.2. *For* $k = 1, \ldots, N-3$, *perform the following.*
Compute

$$\begin{pmatrix} q^{(1)}(k) \\ \theta_{k+2} \end{pmatrix} = Q_k^T \begin{pmatrix} \theta_{k+1} \\ q(k+2) \end{pmatrix}, \tag{36.122}$$

with the number $q^{(1)}(k)$ *and two-dimensional column* θ_{k+2}.

Compute the number $d_W(k+2)$ *and the two-dimensional column* ε_{k+1} *via*

$$d_W(k+2) = d(k+2) + p(k+2)q(k+2), \quad \varepsilon_{k+1} = \varphi_{k+1}h(k+2) - c_{k+1}q(k+2), \tag{36.123}$$

and the 3×3 *matrix* Ω_k *with the formula*

$$\Omega_k = \begin{pmatrix} f_{k+1} & \varepsilon_{k+1} \\ 0 & d(k+2) \end{pmatrix} Q_k. \tag{36.124}$$

Determine the orthogonal 3×3 *matrix* S_k *and the number* $d^{(1)}(k)$ *such that*

$$S_k^* \Omega_k(:,1) = \begin{pmatrix} d^{(1)}(k) \\ 0 \\ 0 \end{pmatrix}. \tag{36.125}$$

Compute

$$\Lambda_{k+2} = S_k^* \begin{pmatrix} f_{k+1}^W & \phi_{k+1}^W h(k+2) & \phi_{k+1}^W b(k+2) \\ p(k+2)\theta_{k+1}^T & d_W(k+2) & g(k+2) \end{pmatrix} \begin{pmatrix} Q_k & 0 \\ 0 & I_{r_{k+2}} \end{pmatrix}, \tag{36.126}$$

$$B_{k+2} = \begin{pmatrix} I_2 & 0 & 0 \\ 0 & h(k+2) & b(k+2) \end{pmatrix} \begin{pmatrix} Q_k & 0 \\ 0 & I_{r_{k+2}} \end{pmatrix}, \tag{36.127}$$

and determine the matrices $d_B(k+2), g_B(k+2), \beta_{k+2}^W, f_{k+2}^W, \phi_{k+2}^W$ *of sizes* $1\times 1, 1\times (r_{k+2}+1), 2\times(r_{k+2}+2), 2\times 2, 2\times r_{k+2}$ *from the partitions*

$$\Lambda_{k+2} = \begin{pmatrix} d_B(k+2) & g_B(k+2) \\ * & \beta_{k+1}^W \end{pmatrix}, \quad \beta_{k+2}^W = \begin{pmatrix} f_{k+2}^W & \phi_{k+2}^W \end{pmatrix}. \tag{36.128}$$

Compute

$$\begin{pmatrix} p^{(1)}(k) \\ c_{k+2} \end{pmatrix} = S_k^* \begin{pmatrix} c_{k+1} \\ p(k+2) \end{pmatrix}, \tag{36.129}$$

with the number $p^{(1)}(k)$ *and the two-dimensional vector column* c_{k+2}.

Compute the orthogonal 4×4 *matrix*

$$\Psi_k = \begin{pmatrix} \Theta_k^{(\tau)} & 0 \\ 0 & 1 \end{pmatrix} \begin{pmatrix} I_2 & 0 \\ 0 & \Theta_{k+2} \end{pmatrix} \begin{pmatrix} S_k & 0 \\ 0 & 1 \end{pmatrix}. \tag{36.130}$$

Determine the orthogonal 3×3 *matrix* Q_{k+1} *from the condition*

$$Q_{k+1}^* \Psi_k(2:4,1) = \begin{pmatrix} * \\ 0 \\ 0 \end{pmatrix}. \tag{36.131}$$

Compute the orthogonal 4×4 *matrix*

$$\Phi_k = \begin{pmatrix} 1 & 0 \\ 0 & Q_{k+1}^* \end{pmatrix} \Psi_k. \tag{36.132}$$

Compute the factorization

$$\Phi_k = \begin{pmatrix} \Theta_k^{(1)} & 0 \\ 0 & I_2 \end{pmatrix} \begin{pmatrix} 1 & 0 \\ 0 & \Theta_{k+1}^{(\tau)} \end{pmatrix}, \tag{36.133}$$

with orthogonal 2×2 *matrix* $\Theta_k^{(1)}$ *and orthogonal* 3×3 *matrix* $\Theta_{k+1}^{(\tau)}$.

Compute

$$f_{k+2} = f_{k+2}^W - c_{k+2}\theta_{k+2}^T. \tag{36.134}$$

1.3. *Compute*

$$\begin{pmatrix} q^{(1)}(N-2) \\ \theta_N \end{pmatrix} = Q_{N-2}^T \begin{pmatrix} \theta_{N-1} \\ q(N) \end{pmatrix} \tag{36.135}$$

with the number $q^{(1)}(N-2)$ *and two-dimensional column* θ_N.

Compute the two-dimensional column

$$d_W(N) = d(N) + p(N)q(N), \quad \varepsilon_{N-1} = \varphi_{N-1}h(N) - c_{N-1}q(N), \tag{36.136}$$

and the 3×3 *matrix* Ω_{N-2} *by the formula*

$$\Omega_{N-2} = \begin{pmatrix} f_{N-1} & \varepsilon_{N-1} \\ 0 & d(N) \end{pmatrix} Q_{N-2}. \tag{36.137}$$

Determine the orthogonal 3×3 *matrix* S_{N-2} *and the number* $d^{(1)}(N-2)$ *such that*

$$S^*_{N-2}\Omega_{N-2}(:,1) = \begin{pmatrix} d^{(1)}(N-2) \\ 0 \\ 0 \end{pmatrix}. \tag{36.138}$$

Compute

$$\Lambda_N = S^*_{N-2} \begin{pmatrix} f^W_{N-1} & \phi^W_{N-1}h(N) \\ p(N)\theta^T_{N-1} & d_W(N) \end{pmatrix} Q_{N-2}, \tag{36.139}$$

$$B_N = \begin{pmatrix} I_2 & 0 \\ 0 & h(N) \end{pmatrix} Q_{N-2}, \tag{36.140}$$

and determine the matrices $d_B(N), g_B(N), f^W_N$ *of sizes* $1 \times 1, 1 \times 2, 2 \times 2$ *from the partition*

$$\Lambda_N = \begin{pmatrix} d_B(N) & g_B(N) \\ * & f^W_N \end{pmatrix}. \tag{36.141}$$

Compute

$$\begin{pmatrix} p^{(1)}(N-2) \\ c_N \end{pmatrix} = S^*_{N-2} \begin{pmatrix} c_{N-1} \\ p(N) \end{pmatrix}, \tag{36.142}$$

with the number $p^{(1)}(N-2)$ *and the vector column* c_N.

Compute the orthogonal 3×3 *matrix*

$$\Psi_{N-2} = \Theta^{(\tau)}_{N-2} S_{N-2}. \tag{36.143}$$

Determine the orthogonal 2×2 *matrix* Q_{N-1} *from the condition*

$$Q^*_{N-1}\Psi_{N-2}(2:3,1) = \begin{pmatrix} * \\ 0 \end{pmatrix}. \tag{36.144}$$

Compute the orthogonal 3×3 *matrix*

$$\Phi_{N-2} = \begin{pmatrix} 1 & 0 \\ 0 & Q^*_{N-1} \end{pmatrix} \Psi_{N-2}. \tag{36.145}$$

Compute the factorization

$$\Phi_{N-2} = \begin{pmatrix} \Theta^{(1)}_{N-2} & 0 \\ 0 & 1 \end{pmatrix} \begin{pmatrix} 1 & 0 \\ 0 & \Theta^{(\tau)}_N \end{pmatrix}, \tag{36.146}$$

with orthogonal 2×2 *matrices* $\Theta^{(1)}_{N-2}, \Theta^{(\tau)}_{N-1}$.

Compute

$$f_N = f_N^W - c_N q_N^T. \tag{36.147}$$

Compute

$$\begin{pmatrix} q^{(1)}(N-1) \\ q^{(1)}(N) \end{pmatrix} = Q_{N-1}^T \theta_N, \tag{36.148}$$

with the numbers $q^{(1)}(N-1), q^{(1)}(N)$, *and set*

$$B_{N+1} = Q_{N-1}. \tag{36.149}$$

Compute the 2×2 *matrix* Ω_{N-1} *by the formula*

$$\Omega_{N-1} = f_N Q_{N-1}. \tag{36.150}$$

Determine the complex Givens rotation matrix S_{N-1} *and the number* $d_{N-1}^{(1)}$ *such that*

$$S_{N-1}^* \Omega_{N-1}(:,1) = \begin{pmatrix} d_{N-1}^{(1)} \\ 0 \end{pmatrix}. \tag{36.151}$$

Compute

$$\Lambda_{N+1} = S_{N-1}^* f_N^W Q_{N-1} \tag{36.152}$$

and determine the numbers $d_B(N+1), g_B(N+1), d_B(N+2)$ *from the partition*

$$\Lambda_{N+1} = \begin{pmatrix} d_B(N+1) & g_B(N+1) \\ * & d_B(N+2) \end{pmatrix}. \tag{36.153}$$

Compute

$$\begin{pmatrix} p^{(1)}(N-1) \\ p^{(1)}(N) \end{pmatrix} = S_{N-1}^* c_N, \tag{36.154}$$

with the numbers $p^{(1)}(N-1), p^{(1)}(N)$.

Compute

$$d^{(1)}(N) = d_B(N+2) + p^{(1)}(N) q^{(1)}(N), \tag{36.155}$$

$$\Theta_{N-1}^{(1)} = \Theta_{N-1}^{(\tau)} S_{N-1}. \tag{36.156}$$

2.1. *Set*

$$X_N = p^{(1)}(N) \tag{36.157}$$

and compute

$$h^{(1)}(N) = 1, \quad Y_N = 1, \quad g^{(1)}(N-1) = g_B(N+1). \tag{36.158}$$

2.2. *For* $k = N-1, \dots, 2$ *perform the following. Determine the unitary* 2×2 *matrix* V_k *and the number* X_k *such that*

$$V_k^* \begin{bmatrix} p^{(1)}(k) \\ X_{k+1} \end{bmatrix} = \begin{pmatrix} X_k \\ 0 \end{pmatrix}. \tag{36.159}$$

Compute the 2×2 *matrix*

$$Z_k = V_k^* \begin{pmatrix} d_W(k) & g^{(1)}(k) \\ X_{k+1}q^{(1)}(k) & z_{k+1} \end{pmatrix} \tag{36.160}$$

and determine the numbers $h'_k, h''_k, \Delta_k, \Omega_k$ *from the partition*

$$Z_k = \begin{pmatrix} h'_k & h''_k \\ \Delta_k & \Omega_k \end{pmatrix}. \tag{36.161}$$

The two-dimensional row $\begin{pmatrix} \Delta_k & \Omega_k \end{pmatrix}$ *has unit norm and one can determine the unitary* 2×2 *matrix* F_k *from the condition*

$$\begin{pmatrix} \Delta_k & \Omega_k \end{pmatrix} F_k^* = \begin{pmatrix} 0 & 1 \end{pmatrix}. \tag{36.162}$$

Determine the matrices $h^{(1)}(k), b^{(1)}(k), d_F(k), g_F(k)$ *from the partition*

$$F_k = \begin{pmatrix} h^{(1)}(k) & b^{(1)}(k) \\ d_F(k) & g_F(k) \end{pmatrix}. \tag{36.163}$$

Compute

$$Y_k = B_{k+2} \begin{pmatrix} (h^{(1)}(k))^* \\ Y_{k+1}(b^{(1)}(k))^* \end{pmatrix}, \quad z_k = h'_k(h^{(1)}(k))^* + h''_k(b^{(1)}(k))^*, \tag{36.164}$$

$$g^{(1)}(k-1) = g_B(k+1)Y_k. \tag{36.165}$$

Proof. Using the formulas (36.2), (36.3) and (36.5) we get

$$A(1:3,1:2) = \begin{pmatrix} \Theta_1 & 0 \\ 0 & 1 \end{pmatrix} \begin{pmatrix} 1 & 0 \\ 0 & \Theta_2 \end{pmatrix} \begin{pmatrix} d(1) & g(1)h(2)-p(1)q(2) \\ 0 & d(2) \\ 0 & 0 \end{pmatrix}.$$

Hence, using (36.118) we get

$$A(1:3,1:2) = \begin{pmatrix} \rho_{11} & \rho_{12} & 0 \\ \rho_{21} & \rho_{22} & 0 \\ 0 & 0 & 1 \end{pmatrix} \begin{pmatrix} 1 & 0 \\ 0 & \lambda_1 \\ 0 & \lambda_2 \end{pmatrix} \begin{pmatrix} d(1) & g(1)h(2)-p(1)q(2) \\ 0 & d(2) \\ 0 & 0 \end{pmatrix},$$

which implies the equalities (36.119) for determining the entries a_{11}, a_{12}, a_{21}, a_{22}, a_{32} of the matrix A. In the same way as above we take the orthogonal 3×3 matrix Q_1 from the condition (36.120).

Define the orthogonal matrix S via relation (36.58), with

$$\tilde{S}_i = I_{i-1} \oplus S_i \oplus I_{N-i-1},\ i=1,\dots,N-2, \quad \tilde{S}_{N-2} = I_{N-2} \oplus S_{N-2}. \tag{36.166}$$

We show that $\Theta^{(1)} = Q^*\Theta S$ is an orthogonal upper Hessenberg matrix represented in the form (36.60), (36.61) and that $R^{(1)} = S^*RQ$ is an upper triangular matrix, and derive the formulas to compute its upper quasiseparable generators and diagonal entries.

Define the matrices $\hat{\Theta}^{(k)}$ $(k = 1, \dots, N-1)$ via the relations (36.62). To get (36.60), (36.61) we prove by induction that

$$\hat{\Theta}^{(k)} = \tilde{\Theta}_1^{(1)} \cdots \tilde{\Theta}_{k-1}^{(1)} \tilde{\Theta}_k^{(\tau)} \tilde{\Theta}_{k+2} \cdots \tilde{\Theta}_{N-1}, \quad k = 1, \dots, N-2, \tag{36.167}$$

where

$$\tilde{\Theta}_i^{(\tau)} = I_{i-1} \oplus \Theta_i^{(\tau)} \oplus I_{N-i-2}.$$

Using (36.3) and (36.62) we get

$$\hat{\Theta}^{(1)} = \tilde{Q}_1^* \Theta = (\tilde{Q}_1^* \tilde{\Theta}_1 \tilde{\Theta}_2) \tilde{\Theta}_3 \cdots \tilde{\Theta}_{N-1}.$$

From here using (36.121) we obtain (36.167) with $k = 1$. Suppose that for some k with $1 \le k \le N-3$ the representation (36.167) holds. Using (36.166) we get

$$\hat{\Theta}^{(k)} \tilde{S}_k = \tilde{\Theta}_1^{(1)} \cdots \tilde{\Theta}_{k-1}^{(1)} (\tilde{\Theta}_k^{(\tau)} \tilde{\Theta}_{k+2} \tilde{S}_k) \tilde{\Theta}_{k+3} \cdots \tilde{\Theta}_{N-1},$$

which means that

$$\hat{\Theta}^{(k)} \tilde{S}_k = \tilde{\Theta}_1^{(1)} \cdots \tilde{\Theta}_{k-1}^{(1)} \tilde{\Psi}_k \tilde{\Theta}_{k+3} \cdots \Theta_{N-1}, \tag{36.168}$$

where $\tilde{\Psi}_k = I_{k-1} \oplus \Psi_k \oplus I_{N-k-2}$, with the orthogonal 4×4 matrix Ψ_k determined via (36.130). We determine the orthogonal 3×3 matrix Q_{k+1} from the condition (36.131). The matrix Φ_k defined in (36.132) has zero entries in the positions (3,1), (4,1), and therefore Φ_k admits the factorization (36.133). From (36.132), (36.133) it follows that $\tilde{Q}_{k+1}^* \tilde{\Psi}_k = \tilde{\Theta}_k^{(1)} \tilde{\Theta}_{k+1}^{(\tau)}$. Inserting this in (36.168) we get

$$\hat{\Theta}^{(k+1)} = \tilde{Q}_{k+1}^* \hat{\Theta}^{(k)} \tilde{S}_k = \tilde{\Theta}_1^{(1)} \cdots \tilde{\Theta}_k^{(1)} \tilde{\Theta}_{k+1}^{(\tau)} \tilde{\Theta}_{k+3} \cdots \tilde{\Theta}_{N-1}, \tag{36.169}$$

which completes the proof of (36.167). Using (36.167) with $k = N-2$ we get

$$\hat{\Theta}^{(N-2)} \tilde{S}_{N-2} = \tilde{\Theta}_1^{(1)} \cdots \tilde{\Theta}_{N-3}^{(1)} (\tilde{\Theta}_{N-2}^{(\tau)} \tilde{S}_{N-2}),$$

which means that

$$\hat{\Theta}^{(N-2)} \tilde{S}_{N-2} = \tilde{\Theta}_1^{(1)} \cdots \tilde{\Theta}_{N-3}^{(1)} \tilde{\Psi}_{N-2}, \tag{36.170}$$

where $\tilde{\Psi}_{N-2} = I_{N-3} \oplus \Psi_{N-2}$, with the orthogonal 3×3 matrix Ψ_{N-2} determined via (36.143). We determine the 2×2 orthogonal matrix Q_{N-1} from the condition (36.144). The matrix Φ_{N-2} defined in (36.145) has a zero entry in the position (3,1) and therefore Φ_{N-2} admits the factorization (36.146). From (36.145), (36.146) it follows that $\tilde{Q}_{N-1}^* \tilde{\Psi}_{N-2} = \tilde{\Theta}_{N-2}^{(1)} \tilde{\Theta}_{N-1}^{(\tau)}$. Inserting this in (36.170) we get

$$\hat{\Theta}^{(N-1)} = \tilde{Q}_{N-1}^* \hat{\Theta}^{(N-2)} \tilde{S}_{N-2} = \tilde{\Theta}_1^{(1)} \cdots \tilde{\Theta}_{N-2}^{(1)} \tilde{\Theta}_{N-1}^{(\tau)}.$$

From here using the equality (36.156) we obtain the representation (36.60) for the matrix $\Theta^{(1)} = \hat{\Theta}^{(N-1)}\tilde{S}_{N-1}$.

Next we consider the matrices $R^{(1)} = S^*RQ,\ W^{(1)} = S^*WQ$. Set

$$\begin{aligned} m_1 = m_2 = \cdots = m_N = 1,\ m_{N+1} = m_{N+2} = 0; \\ n_1 = n_2 = 0, \quad n_3 = \cdots = n_{N+2} = 1. \end{aligned} \tag{36.171}$$

We treat R, W as $(N+2)\times(N+2)$ matrices with entries of sizes $m_i \times m_j,\ i,j = 1,\dots,N+2$ and Q, S as $(N+2)\times(N+2)$ matrices with entries of sizes $m_i \times n_j,\ i,j = 1,\dots,N+2$.

Using (36.112), (36.58) and setting

$$\begin{aligned} &\tilde{C}_1 = \tilde{C}_2 = I_N, \quad \tilde{C}_i = \tilde{S}^*_{i-2},\ i = 2,\dots,N+1, \quad \tilde{C}_{N+2} = I_N, \\ &\tilde{T}_1 = \tilde{T}_2 = I_N, \quad \tilde{T}_i = \tilde{Q}_{i-2},\ i = 2,\dots,N+1, \quad \tilde{T}_{N+2} = I_N \end{aligned} \tag{36.172}$$

we get

$$S^* = \tilde{C}_{N+2}\cdots\tilde{C}_1, \quad Q = \tilde{T}_1\cdots\tilde{T}_{N+2}. \tag{36.173}$$

Further, we have

$$\begin{aligned} \tilde{C}_1 &= \mathrm{diag}\{C_1, I_{N-1}\},\ \tilde{C}_2 = \mathrm{diag}\{C_2, I_{N-2}\}, \\ \tilde{C}_k &= \mathrm{diag}\{I_{k-3}, C_k, I_{N-k}\},\ k = 3,\dots,N, \\ \tilde{C}_{N+1} &= \mathrm{diag}\{I_{N-2}, C_{N+1}\},\ \tilde{C}_{N+2} = \mathrm{diag}\{I_{N-1}, C_{N+2}\}, \end{aligned}$$

with

$$C_1 = 1,\ C_2 = I_2, \quad C_k = S^*_{k-2},\ k = 3,\dots,N+1, \quad C_{N+2} = 1, \tag{36.174}$$

and

$$\begin{aligned} \tilde{T}_1 &= \mathrm{diag}\{T_1, I_{N-1}\},\ \tilde{T}_2 = \mathrm{diag}\{T_2, I_{N-2}\}, \\ \tilde{T}_k &= \mathrm{diag}\{I_{k-3}, T_{k-2}, I_{N-k}\},\ k = 3,\dots,N, \\ \tilde{T}_{N+1} &= \mathrm{diag}\{I_{N-2}, T_{N+1}\},\ \tilde{T}_{N+2} = \mathrm{diag}\{I_{N-1}, T_{N+2}\}, \end{aligned}$$

with

$$T_1 = 1,\ T_2 = I_2, \quad T_k = Q_{k-2},\ k = 3,\dots,N+1, \quad T_{N+2} = 1. \tag{36.175}$$

The matrix R has upper quasiseparable generators and diagonal entries defined in (36.71), (36.72) and by setting $g_R(N+1), h_R(N+2), b_R(N+1), d_R(N+2)$ to be 0×0 empty matrices. The orders of these upper generators are $r^R_k = r_k+1,\ k = 1,\dots,N-1,\ r^R_N = r^R_{N+1} = 0$. The matrix $W = R + pq^T$ has quasiseparable generators defined in (36.73)–(36.75) and by setting $p_W(N+2), q_W(N+1), a_W(N+1), g_W(N+1), h_W(N+2), b_W(N+1), d_W(N+2)$ to be 0×0 empty matrices.

Now we apply Lemma 31.1 and Corollary 31.2, with the matrices S^*, Q represented in the form (36.173), (36.172), to the matrices W and R, respectively. The sizes of the blocks and the orders of generators used in Lemma 31.1 and Corollary 31.2 are given by the formulas

$$\tau_1 = \nu_1 = \tau_2 = \nu_2 = 0, \quad \tau_i = \nu_i = 1,\ i = 3, \dots, N+2;$$
$$r_1^S = r_1^T = 1, \quad r_i^S = r_i^T = 2,\ i = 2, \dots, N, \quad r_{N+1}^S = r_{N+1}^T = 1;$$
$$\rho_k^A = 1,\ r_k^A = r_k,\ k = 1, \dots, N-1, \quad \rho_N^A = r_N^A = \rho_{N+1}^A = r_{N+1}^A = 0.$$

We apply Lemma 31.1 to compute (block) upper quasiseparable generators $g_B(i)$ $(i = 1, \dots, N)$, $h_B(j)$ $(j = 2, \dots, N+1)$, $b_B(k)$ $(k = 2, \dots, N)$ with orders $\tilde{r}_k = r_k + 1,\ k = 1, \dots, N-1,\ \tilde{r}_N = 1$ of the matrix $U^{(1)} = S^*UQ$. The matrix $U^{(1)}$ is obtained as a block one with entries of sizes $\tau_i \times \nu_j,\ i, j = 1, \dots, N+1$. Using (31.9)–(31.11) with $\tau_1 = \nu_1 = 0$ and the equalities $d_W(1) = d(1),\ q_W(1) = q(1),\ g_W(1) = g(1)$, we get

$$f_1^W = d_W(1), \quad \phi_1^W = g(1), \quad \alpha_1 = q(1). \tag{36.176}$$

Using (31.12)–(31.14), (36.174), (36.175) and (36.176), (36.73)–(36.75), we get

$$f_2^W = \begin{pmatrix} d_W(1) & g(1)h(2) \\ p(2)q(1) & d_W(2) \end{pmatrix}, \ \phi_2^W = \begin{pmatrix} g(1)b(2) \\ g(2) \end{pmatrix}, \ \alpha_2 = \begin{pmatrix} q(1) & q(2) \end{pmatrix}, \tag{36.177}$$

$$\begin{aligned} &\begin{pmatrix} d_B(k) & g_B(k) \\ * & \Gamma_k \end{pmatrix} \\ &= \begin{pmatrix} S_{k-2}^* & 0 \\ 0 & 1 \end{pmatrix} \begin{pmatrix} f_{k-1}^W & \phi_{k-1}^W h(k) & \phi_{k-1}^W b(k) \\ p(k)\alpha_{k-1} & d_W(k) & g(k) \\ \alpha_{k-1} & q(k) & 0 \end{pmatrix} \begin{pmatrix} Q_{k-2} & 0 \\ 0 & I_{r_k} \end{pmatrix}, \\ &\Gamma_k = \begin{pmatrix} f_k^W & \phi_k^W \\ \alpha_k & 0 \end{pmatrix}, \quad k = 3, \dots, N-1, \end{aligned} \tag{36.178}$$

$$\begin{pmatrix} d_B(N) & g_B(N) \\ * & \Gamma_N \end{pmatrix} = S_{N-2}^* \begin{pmatrix} f_{N-1}^W & \phi_{N-1}^W h(N) \\ p(N)\alpha_{N-1} & d_W(N) \end{pmatrix} Q_{N-2}, \quad \Gamma_N = f_N^W, \tag{36.179}$$

$$\begin{pmatrix} d_B(N+1) & g_B(N+1) \\ * & \Gamma_{N+1} \end{pmatrix} = S_{N-1}^* f_N^W Q_{N-1}, \quad \Gamma_{N+1} = f_{N+1}^W, \tag{36.180}$$

and also

$$\begin{pmatrix} h_B(k) & b_B(k) \end{pmatrix} = \begin{pmatrix} I_2 & 0 & 0 \\ 0 & h(k) & b(k) \end{pmatrix} \begin{pmatrix} Q_{k-2} & 0 \\ 0 & I_{r_k} \end{pmatrix}, \ k = 3, \dots, N-1, \tag{36.181}$$

$$\begin{pmatrix} h_B(N) & b_B(N) \end{pmatrix} = \begin{pmatrix} I_2 & 0 \\ 0 & h(N) \end{pmatrix} Q_{N-2}, \tag{36.182}$$

$$\begin{pmatrix} h_B(N+1) & b_B(N+1) \end{pmatrix} = Q_{N-1}. \tag{36.183}$$

Using (31.17), (31.18) we get $h_B(N+2) = 1,\ d_B(N+2) = f_{N+1}^W$.

The relations (36.177), (36.178) imply, in particular, that the auxiliary variables α_k $(k = 2, \dots, N-1)$ satisfy the relations

$$\alpha_2 = \begin{pmatrix} q(1) & q(2) \end{pmatrix}, \quad \begin{pmatrix} * & \alpha_k \end{pmatrix} = \begin{pmatrix} \alpha_{k-1} & q(k) \end{pmatrix} Q_{k-2}, \; k = 3, \dots, N-1.$$

Comparing this with (36.117), (36.122) we get

$$\alpha_k = \theta_k^T, \quad k = 2, \dots, N-1. \tag{36.184}$$

Thus, inserting (36.184) in (36.178) and (36.179) we obtain (36.126), (36.128) and (36.139), (36.141). Combining (36.180) and $d_B(N+2) = f_{N+1}^W$ we obtain (36.152), (36.153).

Applying Lemma 31.2 to the matrix $R^{(1)} = S^*RQ$, treated as an $(N+2) \times (N+2)$ matrix with entries of sizes $n_i \times n_j$, $i, j = 1, \dots, N+2$, we obtain quasiseparable generators $q_F(j)$ $(j = 2, \dots, N+1)$, $g_F(i)$ $(i = 2, \dots, N+1)$, $h_F(j)$ $(j = 3, \dots, N+2)$, $b_F(k)$ $(k = 3, \dots, N+1)$, $d_F(k)$ $(k = 2, \dots, N+2)$ of $R^{(1)}$ by the formulas

$$f_1 = d_R(1), \quad \phi_1 = g_R(1), \tag{36.185}$$

$$f_2 = \begin{pmatrix} d_R(1) & \phi_1 h_R(2) \\ 0 & d_R(2) \end{pmatrix}, \quad \phi_2 = \begin{pmatrix} g_R(1) b_R(2) \\ g_R(2) \end{pmatrix}, \tag{36.186}$$

$$\begin{pmatrix} d_F(k) & g_F(k) \\ q_F(k) & \beta_k \end{pmatrix} = S_{k-2}^* \begin{pmatrix} f_{k-1} & \phi_{k-1} h_R(k) & \phi_{k-1} b_R(k) \\ 0 & d_R(k) & g_R(k) \end{pmatrix} \begin{pmatrix} Q_{k-2} & 0 \\ 0 & I_{r_k^R} \end{pmatrix},$$
$$\beta_k = \begin{pmatrix} f_k & \phi_k \end{pmatrix}, \quad k = 3, \dots, N+1. \tag{36.187}$$

For the $2 \times (r_k + 1)$ matrices ϕ_k $(k = 1, \dots, N-1)$ we use the partitions

$$\phi_k = \begin{pmatrix} \varphi_k & -c_k \end{pmatrix}, \quad k = 1, \dots, N-2 \tag{36.188}$$

with matrices φ_k, c_k of sizes $2 \times r_k, 2 \times 1$. Using (36.71), (36.72), (36.188) with $k = 1, 2$ and (36.185), (36.186) we get

$$f_2 = \begin{pmatrix} d(1) & \varepsilon \\ 0 & d(2) \end{pmatrix}, \quad \varphi_2 = \begin{pmatrix} g(1)b(2) \\ g(2) \end{pmatrix}, \quad c_2 = \begin{pmatrix} p(1) \\ p(2) \end{pmatrix}, \tag{36.189}$$

with ε as in (36.116). Using (36.71) and (36.188) we obtain the relations

$$\phi_{k+1} h_R(k+2) = \varepsilon_{k+1}, \quad k = 1, \dots, N-1, \tag{36.190}$$

with ε_{k+1} as in (36.123), and

$$\phi_{k+1} b_R(k+2) = \begin{pmatrix} \varphi_{k+1} b(k+2) & -c_{k+1} \end{pmatrix}, \quad k = 1, \dots, N-3. \tag{36.191}$$

Inserting (36.190), (36.191) in (36.187) and using (36.124), (36.137), (36.150) we obtain

$$\begin{pmatrix} d_F(k+2) & g_F(k+2) \\ q_F(k+2) & \beta_{k+2} \end{pmatrix}$$
$$= S_k^* \begin{pmatrix} \Omega_k(1:2,1) & \Omega_k(1:2,2:3) & \varphi_{k+1}b(k+2) & -c_{k+1} \\ \Omega_k(3,1) & \Omega_k(3,2:3) & g(k+2) & -p(k+2) \end{pmatrix},$$
$$\beta_{k+2} = \begin{pmatrix} f_{k+2} & \phi_{k+2} \end{pmatrix}, \quad k = 1,\dots,N-3, \tag{36.192}$$
$$\begin{pmatrix} d_F(N) & g_F(N) \\ q_F(N) & \beta_N \end{pmatrix} = S_{N-2}^*\Omega_{N-2}, \ \beta_N = f_N,$$
$$\begin{pmatrix} d_F(N+1) & g_F(N+1) \\ q_F(N+1) & \beta_{N+1} \end{pmatrix} = S_{N-1}^*\Omega_{N-1}.$$

Hence, using (36.125), (36.138), (36.151) we have

$$d_F(k+2) = d^{(1)}(k), \quad k = 1,\dots,N-1. \tag{36.193}$$

and

$$q_F(k+2) = 0, \qquad k = 1,\dots,N-1. \tag{36.194}$$

The formulas (36.193) and (36.155) mean that $d^{(1)}(k)$, $k = 1,\dots,N$, are the diagonal entries of the matrix $R^{(1)}$. The equalities (36.194) imply that $R^{(1)}$ is an upper triangular matrix. Using (36.192) and (36.188) we have also

$$\begin{pmatrix} * & * & * \\ f_{k+2} & \varphi_{k+2} & -c_{k+2} \end{pmatrix} = S_k^* \begin{pmatrix} \Omega_k(1:2,2:3) & \varphi_{k+1}b(k+2) & -c_{k+1} \\ \Omega_k(3,2:3) & g(k+2) & -p(k+2) \end{pmatrix},$$
$$k = 1,\dots,N-3, \tag{36.195}$$
$$\begin{pmatrix} * \\ f_N \end{pmatrix} = S_{N-2}^*\Omega_{N-2}(:,2:3).$$

Further, the relations (36.112), (36.113) yield the formulas (36.117), (36.122), (36.135), (36.148) for computing the coordinates of the perturbation vector $q^{(1)} = Q^T q$. From the relations (36.58), (36.166) it follows that the formulas (36.117), (36.129), (36.142), (36.154) yield the coordinates of the perturbation vector $p^{(1)} = S^* p$.

Using (36.181), (36.182), (36.183) and setting

$$B_k = \begin{pmatrix} h_B(k) & b_B(k) \end{pmatrix}, \quad k = 3,\dots,N+1 \tag{36.196}$$

we obtain (36.127), (36.140), (36.149).

It remains to show that the auxiliary variables f_k, φ_k $(k = 3,\dots,N-1)$ may be determined via relations (36.134), and also that the relation (36.147) holds. Comparing (36.177), (36.114) with (36.189) we get

$$f_2 = f_2^W - c_2\theta_2^T, \quad \varphi_2 = \phi_2^W.$$

Assume that for some k with $1 \le k \le N-2$ the relations

$$f_{k+1} = f^W_{k+1} - c_{k+1}\theta^T_{k+1}, \quad \varphi_{k+1} = \phi^W_{k+1} \tag{36.197}$$

hold. By (36.124) and (36.195),

$$\begin{pmatrix} d^{(1)}(k) & * & * \\ 0 & f_{k+2} & \varphi_{k+2} \end{pmatrix} = S^*_k \begin{pmatrix} f_{k+1} & \varepsilon_{k+1} & \varphi_{k+1}b(k+2) \\ 0 & d(k+2) & g(k+2) \end{pmatrix} \begin{pmatrix} Q_k & 0 \\ 0 & I_{r_{k+2}} \end{pmatrix},$$
$$k = 1, \dots, N-3, \tag{36.198}$$
$$\begin{pmatrix} d^{(1)}(N-2) & * \\ 0 & f_N \end{pmatrix} = S^*_{N-2} \begin{pmatrix} f_{N-1} & \varepsilon_{N-1} \\ 0 & d(N) \end{pmatrix} Q_{N-2}.$$

Using (36.123) and (36.197) we get

$$\begin{pmatrix} f_{k+1} & \varepsilon_{k+1} \\ 0 & d(k+2) \end{pmatrix}$$
$$= \begin{pmatrix} f^W_{k+1} - c_{k+1}\theta^T_{k+1} & \phi^W_{k+1}h(k+2) - c_{k+1}q(k+2) \\ p(k+2)\theta^T_{k+1} - p(k+2)\theta^T_{k+1} & d_W(k+2) - p(k+2)q(k+2) \end{pmatrix}, \tag{36.199}$$
$$k = 1, \dots, N-2.$$

Thus, combining (36.198) and (36.199) and using (36.126), (36.128) and (36.122), (36.129) we obtain (36.134) and (36.147).

We have obtained upper quasiseparable generators and the diagonal entries of the matrix $W^{(1)}$, treated as a block one with $n_i \times n_j$ $(i, j = 1, \dots, N+2)$ entries, with the sizes n_i defined in (36.171). The matrix $W^{(1)}$, treated as an ordinary scalar matrix, has upper quasiseparable generators $g_B(i+2)$ $(i = 1 \dots, N-1)$, $h_B(j+2)$ $(j = 2, \dots, N)$, $b_B(k+2)$ $(k = 2, \dots, N-1)$ and diagonal entries $d_B(k+2)$ $(k = 1, \dots, N)$. Next we apply Theorem 7.5 to determine upper quasiseparable generators $g^{(1)}(i)$ $(i = 1, \dots, N-1)$, $h^{(1)}(j)$ $(j = 2, \dots, N)$, $b^{(1)}(k)$ $(k = 2, \dots, N-1)$ of $W^{(1)}$ with orders equal to one. The matrix $W^{(1)}$ has lower quasiseparable generators $p^{(1)}(k)$ $(k = 2, \dots, N)$, $q^{(1)}(k)$ $(k = 1, \dots, N-1)$, $a(k) = 1$ $(k = 2, \dots, N-1)$ of orders $r^L_k = 1$, $k = 1, \dots, N$. Using (7.28) we obtain $V_N = 1$ and the formulas (36.157), (36.159) for determining the numbers X_k $(k = N, \dots, 2)$ and the matrices V_k $(k = N-1, \dots, 2)$.

Using (7.29)–(7.33) with $k = N$ we get

$$Z_N = d_B(N+2) = h'_N, \qquad F_N = h^{(1)}(N) = 1, \quad Y_N = h_B(N+2) = 1,$$
$$z_N = d_B(N+2), \qquad g^{(1)}(N-1) = g_B(N+1).$$

For $k = N-1, \dots, 2$, applying (7.29)–(7.34) and using (36.196), we obtain the relations (36.160)–(36.165). □

The analogue of Theorem 36.5 for the double shift case reads

Theorem 36.7. *Let A be a real matrix from the class $\mathcal{H}_N$ with Q-generating elements Θ_k $(k = 1, \dots, N-1)$, V_k, F_k $(k = 1, 2, \dots, N-1)$ and $p = (p(i))_{i=1}^N$, $q = (q(i))_{i=1}^N$, and let s, t be real numbers.*

Then the orthogonal 3×3 matrices Q_k, $k = 1, \dots, N-2$ and the orthogonal 2×2 matrix Q_{N-1} such that the orthogonal matrix

$$Q = \tilde{Q}_1 \tilde{Q}_2 \cdots \tilde{Q}_{N-2} \tilde{Q}_{N-1}, \tag{36.200}$$

with

$$\tilde{Q}_i = I_{i-1} \oplus Q_i \oplus I_{N-i-2},\ i = 1, \dots, N-2, \quad \tilde{Q}_{N-1} = I_{N-2} \oplus Q_{N-1},$$

satisfies the condition (31.7), and Q-generating elements $\Theta_k^{(1)}$ $(k = 1, \dots, N-1)$; $V_k^{(1)}$, $F_k^{(1)}$ $(k = 1, 2, \dots, N-1)$ and $p^{(1)} = (p^{(1)}(i))_{i=1}^N$, $q^{(1)} = (q^{(1)}(i))_{i=1}^N$ of the matrix $A^{(1)}$ are determined using the following algorithm.

1. *Using the algorithm from Theorem 36.3, compute upper quasiseparable generators $g(i)$ $(i = 1, \dots, N-1)$, $h(j)$ $(j = 2, \dots, N)$, $b(k)$ $(k = 2, \dots, N-1)$ and the diagonal entries $d(k)$ $(k = 1, \dots, N)$ of the matrix W.*

2. *Compute the matrices Q_k $(k = 1, \dots, N-1)$, the updated matrices $\Theta_k^{(1)}$ $(k = 1, \dots, N-1)$, the matrices S_k $(k = 1, \dots, N-1)$ such that the matrix $R^{(1)} = S^* R Q$, with S defined via (36.58), (36.166), is an upper triangular matrix, and the updated perturbation vectors $p^{(1)}$, $q^{(1)}$ via Step 1 of the algorithm from Theorem 36.6.*

3.1. *Compute the 3×3 matrices*

$$V_1^{(\tau)} = \begin{pmatrix} 1 & 0 \\ 0 & V_2 \end{pmatrix} \begin{pmatrix} V_1 & 0 \\ 0 & 1 \end{pmatrix} \tag{36.201}$$

and

$$F_1^{(\tau)} = \begin{pmatrix} F_1 & 0 \\ 0 & 1 \end{pmatrix} \begin{pmatrix} 1 & 0 \\ 0 & F_2 \end{pmatrix}. \tag{36.202}$$

3.2. *For $k = 1, \dots, N-3$ one performs the following.*

Compute the orthogonal 4×4 matrix

$$\Gamma_k = \begin{pmatrix} S_k^* & 0 \\ 0 & 1 \end{pmatrix} \begin{pmatrix} I_2 & 0 \\ 0 & V_{k+2} \end{pmatrix} \begin{pmatrix} V_k^{(\tau)} & 0 \\ 0 & 1 \end{pmatrix}$$

and determine the orthogonal 3×3 matrix Z_{k+1} from the condition

$$\Gamma_k(1:2, 2:4) Z_k = \begin{pmatrix} * & 0 & 0 \\ * & * & 0 \end{pmatrix}.$$

Compute the orthogonal 4×4 matrices

$$L_k = \Gamma_k \begin{pmatrix} 1 & 0 \\ 0 & Z_{k+1} \end{pmatrix}$$

and

$$M_k = \begin{pmatrix} 1 & 0 \\ 0 & Z_{k+1}^* \end{pmatrix} \begin{pmatrix} F_k^{(\tau)} & 0 \\ 0 & 1 \end{pmatrix} \begin{pmatrix} I_2 & 0 \\ 0 & F_{k+2} \end{pmatrix} \begin{pmatrix} Q_k(\alpha) & 0 \\ 0 & 1 \end{pmatrix}. \tag{36.203}$$

The orthogonal matrices L_k and M_k contain zero entries in the positions $(1,3)$, $(1,4)$ and $(3,1)$, $(4,1)$, respectively, and we compute the factorizations

$$L_k = \begin{pmatrix} 1 & 0 \\ 0 & V_{k+1}^{(\tau)} \end{pmatrix} \begin{pmatrix} V_k^{(1)} & 0 \\ 0 & I_2 \end{pmatrix}$$

and

$$M_k = \begin{pmatrix} F_k^{(1)} & 0 \\ 0 & I_2 \end{pmatrix} \begin{pmatrix} 1 & 0 \\ 0 & F_{k+1}^{(\tau)} \end{pmatrix}, \tag{36.204}$$

with orthogonal 2×2 matrices $V_k^{(1)}, F_k^{(1)}$ and orthogonal 3×3 matrices $V_{k+1}^{(\tau)}, F_{k+1}^{(\tau)}$.
3.3. *Compute the orthogonal 3×3 matrices*

$$\Gamma_{N-2} = S_{N-2}^* V_{N-2}^{(\tau)}$$

and determine the orthogonal 2×2 matrix Z_{N-1} from the condition

$$\Gamma_{N-2}(1, 2:3) Z_{N-1} = \begin{pmatrix} * & 0 \end{pmatrix}.$$

Compute the 3×3 matrix

$$L_{N-2} = \Gamma_{N-2} \begin{pmatrix} 1 & 0 \\ 0 & Z_{N-1} \end{pmatrix}$$

and the 3×3 matrix

$$M_{N-2} = \begin{pmatrix} 1 & 0 \\ 0 & Z_{N-1}^* \end{pmatrix} F_{N-2}^{(\tau)} Q_{N-2}. \tag{36.205}$$

The orthogonal matrix L_{N-2} has a zero entry in the position $(1,3)$ and the orthogonal matrix M_{N-2} has a zero entry in the position $(3,1)$ and we compute the factorization

$$L_{N-2} = \begin{pmatrix} 1 & 0 \\ 0 & V_{N-1}^{(\tau)} \end{pmatrix} \begin{pmatrix} V_{N-2}^{(1)} & 0 \\ 0 & 1 \end{pmatrix},$$

with orthogonal 2×2 matrices $V_{N-1}^{(\tau)}, V_{N-2}^{(1)}$, and the factorization

$$M_{N-2} = \begin{pmatrix} F_{N-2}^{(1)} & 0 \\ 0 & 1 \end{pmatrix} \begin{pmatrix} 1 & 0 \\ 0 & F_{N-1}^{(\tau)} \end{pmatrix}, \tag{36.206}$$

with orthogonal 2×2 matrices $F_{N-2}^{(1)}, F_{N-1}^{(\tau)}$.

3.4. *Set*

$$V_{N-1}^{(1)} = S_{N-1}^* V_{N-1}^{(\tau)}, \quad F_{N-1}^{(1)} = F_{N-1}^{(\tau)} Q_{N-1}. \tag{36.207}$$

Proof. Step 1 of the algorithm follows from Theorem 34.1. Step 2 follows from Theorem 36.6. Now we will justify Step 3.

We have

$$W^{(1)} = S^*WQ = (S^*V)(FQ).$$

Define the unitary matrix Z via relations (35.83), (35.84). We get

$$W^{(1)} = V^{(1)} \cdot F^{(1)},$$

with $V^{(1)} = S^*VZ$, $U^{(1)} = Z^*FQ$. We need to prove that the factorizations (36.104) and (36.105) hold and that the vector $p^{(1)}$ is reduced by the matrix $V^{(1)}$. In the same way as in the proof of Theorem 35.5 we obtain (36.104) and prove that the vector $p^{(1)}$ is reduced by $V^{(1)}$.

Applying Lemma 36.1 to the matrix $W^{(1)}$ we obtain the equalities (36.106). Set

$$\Phi_1 = F, \quad \Phi_{k+1} = \tilde{Z}^*_{k+1}\Phi_k\tilde{Q}_k, \; k = 1, \dots, N-3;$$

this means that

$$\Phi_{k+1} = \tilde{Z}^*_{k+1} \cdots \tilde{Z}^*_2 F\tilde{Q}_1 \cdots \tilde{Q}_k, \quad k = 1, \dots, N-4. \tag{36.208}$$

We prove by induction that

$$\Phi_k = \tilde{F}^{(1)}_1 \cdots \tilde{F}^{(1)}_{k-1}\tilde{F}^{(\tau)}_k \tilde{F}_{k+2} \cdots \tilde{F}_{N-1}, \quad k = 1, \dots, N-2, \tag{36.209}$$

where

$$\tilde{F}^{(\tau)}_i = I_{i-1} \oplus F^{(\tau)}_i \oplus I_{N-i-2}.$$

For $k = 1$ the equality (36.209) follows from (36.10), (36.11) and (36.202). Suppose that for some k with $1 \le k \le N-3$ the relation (36.209) holds. Hence the matrix $\Phi_{k+1} = \tilde{Z}^*_{k+1}\Phi_k\tilde{Q}_k$ has the form

$$\Phi_{k+1} = (\tilde{F}^{(1)}_1 \cdots \tilde{F}^{(1)}_{k-1})(\tilde{Z}^*_{k+1}\tilde{F}^{(\tau)}_k\tilde{F}_{k+2}\tilde{Q}_k)(\tilde{F}_{k+3} \cdots \tilde{F}_{N-1}). \tag{36.210}$$

We have

$$\tilde{Z}^*_{k+1}\tilde{F}^{(\tau)}_k\tilde{F}_{k+2}\tilde{Q}_k = I_{k-1} \oplus M_k \oplus I_{N-k-3}, \tag{36.211}$$

with the orthogonal 4×4 matrix M_k from (36.203). But also using the relations $F^{(1)} = Z^*FQ$ and (36.200), (35.83), (36.208) we get

$$\Phi_{k+1} = (\tilde{Z}_{k+2} \cdot \tilde{Z}_{N-1})F^{(1)}\tilde{Q}^*_{N-1} \cdot \tilde{Q}^*_{k+1},$$

whence, using (36.106),

$$\Phi_{k+1}(k+2 : k+3, k) = 0. \tag{36.212}$$

By (36.210) and (36.211) we have $M_k(3:4,1) = \Phi_{k+1}(k+2 : k+3, k)$, and using (36.212) we conclude that the orthogonal 4×4 matrix M_k has zero entries in the $(3,1), (4,1)$ positions. Hence the matrix M_k admits the factorization (36.204) with orthogonal 2×2 matrix $F_k^{(1)}$ and orthogonal 3×3 matrix $F_{k+1}^{(\tau)}$. It follows that

$$\Phi_{k+1} = (\tilde{F}_1^{(1)} \cdots \tilde{F}_{k-1}^{(1)})(\tilde{F}_k^{(1)} \tilde{F}_{k+1}^{(\tau)})(\tilde{F}_{k+3} \cdots \tilde{F}_{N-1}),$$

which completes the proof of (36.209). Now using (36.209) with $k = N-2$ we obtain the representation of the matrix $\Phi_{N-1} = \tilde{Z}_{N-1}^* \Phi_{N-2} \tilde{Q}_{N-2}$ in the form

$$\Phi_{N-1} = (\tilde{F}_1^{(1)} \cdots \tilde{F}_{N-3}^{(1)})(\tilde{Z}_{N-1}^* \tilde{F}_{N-2}^{(\tau)} \tilde{Q}_{N-2}).$$

We have

$$\tilde{Z}_{N-1}^* \tilde{F}_{N-2}^{(\tau)} \tilde{Q}_{N-2} = I_{N-3} \oplus M_{N-2},$$

with the orthogonal 3×3 matrix M_{N-2} from (36.205). From the relations $\Phi_{N-1} = F^{(1)} \tilde{Q}_{N-1}^*$ and (36.106) it follows that

$$\Phi_{N-1}(N, N-2) = 0. \tag{36.213}$$

We have $M_{N-2}(3,1) = \Phi_{N-2}(N, N-3)$ and using (35.91) we conclude that the unitary 4×4 matrix D_{N-3} has a zero entry in the $(4,1)$ position. Hence, D_{N-3} admits the factorization (35.80) with orthogonal 2×2 matrices $F_{N-2}^{(1)}, F_{N-1}^{(\tau)}$. It follows that the matrix Φ_{N-1} has the form

$$\Phi_{N-1} = (\tilde{F}_1^{(1)} \cdots \tilde{F}_{N-3}^{(1)})(\tilde{F}_{N-2}^{(1)} \tilde{F}_{N-1}^{(\tau)}),$$

with $\tilde{F}_{N-1}^{(\tau)} = I_{N-2} \oplus F_{N-1}^{(\tau)}$. From here using (36.207) we obtain the factorization (35.40) for the matrix $F^{(1)} = \Phi_{N-1} \tilde{Q}_{N-1}$. □

The complexity of the part which is performed about N times of the algorithm for one step with the implicit method based on QR using a single shift as in items 1.2 and 2.2 of Theorem 36.4 is less than

$$256 + 6r + 4r^2 + 2r^3$$

arithmetical operations, where r is the maximal order of the upper quasiseparable generators plus two. One QR-based step with double shift, as in Theorem 36.6, costs

$$492 + 8\tilde{r} = 8\tilde{r}^2 + 2\tilde{r}^3$$

arithmetical operations, where $\tilde{r}$ is the maximal order of the upper quasiseparable generators plus three.

§36.4 Comments

The idea of the method presented here and a stable algorithm were suggested by S. Chandrasekaran, M. Gu, J. Xia and J. Zhu in the paper [17]. The algorithm presented here is improved essentially thanks to a more accurate usage of the quasiseparable structure. Also, an alternative factorization method is given.

Bibliography

[1] Atkinson, F.V., Discrete and Continuous Boundary Problems, Academic Press, New York, London, 1964.

[2] Bini, D.A., Boito, P., Eidelman, Y., Gemignani, L., Gohberg, I., A fast implicit QR eigenvalue algorithm for companion matrices, *Linear Algebra Appl.* **432**, no. 8, 2006–2031 (2010).

[3] Bini, D.A., Daddi, F., Gemignani, L., On the shifted QR iteration applied to companion matrices. *Electron. Trans. Numer. Anal.* **18** 137–152 (2004).

[4] Bini, D.A., Eidelman, Y., Gemignani, L., Gohberg, I., Fast QR eigenvalue algorithms for Hessenberg matrices which are rank-one perturbations of unitary matrices. *SIAM J. Matrix Anal. Appl.* **29**, no. 2, 566–585 (2007).

[5] Bini, D.A., Eidelman, Y., Gemignani, L., Gohberg, I., The unitary completion and QR iterations for a class of structured matrices. *Math. Comp.* **77**, no. 261, 353–378 (2008).

[6] Boito, P., Eidelman, Y., Gemignani, L., Gohberg, I., Implicit QR with compression. *Indag. Math.*, (N.S.) **23**, no. 4, 733–761 (2012).

[7] Bella, T., Eidelman, Y., Gohberg, I., Koltracht, I., Olshevsky, V., A Björk–Pereyra-type algorithm for Szego-Vandermonde matrices based on properties of unitary Hessenberg matrices, *Linear Algebra Appl.*, **420**, no. 2-3, 634–647 (2007).

[8] Bella, T., Eidelman, Y., Gohberg, I., Koltracht, I., Olshevsky, V., A fast Björk–Pereyra-type algorithm for solving Hessenberg-quasiseparable-Vandermonde systems. *SIAM J. Matrix Anal. Appl.*, **31**, no. 9, 790–815 (2009).

[9] Bella, T., Eidelman, Y., Gohberg, I., Olshevsky, V., Computations with quasiseparable polynomials and matrices. *Theoret. Comput. Sci.*, **409**, no. 2, 158–179 (2008).

[10] Bini, D.A., Gemignani, L., Pan, V., Inverse power and Durand-Kerner iterations for univariate polynomial root-finding, *Comput. Math. Appl.*, **47**, no. 2-3, 447–459 (2004).

[11] Bini, D.A., Gemignani, L., Pan, V., Fast and stable QR eigenvalue algorithms for generalized companion matrices and secular equations, *Numer. Math.* **100**, no. 3, 373–408 (2005).

[12] Bunch, J.R., Nielsen, C.P., Sorensen, D.C., Rank-one modification of the symmetric eigenproblem. *Numer. Math.* **31**: 31–48 (1978).

[13] Bella, T., Olshevsky, V., Zhlobich, P., Signal flow graph approach to inversion of (H, m)-quasiseparable-Vandermonde matrices and new filter structures. *Linear Algebra Appl.*, **432**, no. 8, 2032–2051 (2010).

[14] Bella, T., Olshevsky, V., Zhlobich, P., A quasiseparable approach to five-diagonal CMV and Fiedler matrices. *Linear Algebra Appl.*, **434**, no. 4, 957–976 (2011).

[15] Cuppen, J., A divide and conquer method for symmetric tridiagonal eigenproblem. *Numer. Math.*. **36**, no. 2, 177–195 (1981).

[16] Chandrasekaran, S., Gu, M., Fast and stable eigendecomposition of symmetric banded plus semiseparable matrices. *Linear Algebra Appl.*, **313**, 107–114 (2000).

[17] Chandrasekaran, S., Gu, M., Xia, J., Zhu, J., A fast QR algorithm for companion matrices. In: Recent advances in matrix and operator theory, *Operator Theory: Advances and Applications*, **179**, 111–143, Birkhäuser, Basel, 2008.

[18] Demmel, J.W., Applied Numerical Linear Algebra, SIAM, Philadelphia, 1997.

[19] Delvaux, S., Van Barel, M., A Hessenberg reduction algorithm for rank structured matrices. *SIAM J. Matrix Anal. Appl.*, **29**, no. 3, 895–926 (2008).

[20] Delvaux, S., Van Barel, M., Eigenvalue computation for unitary rank structured matrices. *J. Comput. Appl. Math.*, **213**, no. 1, 268–287 (2008).

[21] Eidelman, Y., Gohberg, I., Gemignani, L., On the fast reduction of a quasiseparable matrix to Hessenberg and tridiagonal forms. *Linear Algebra Appl.*, **420**, no. 1, 86–101 (2007).

[22] Eidelman, Y., Gohberg, I., Gemignani, L., Efficient eigenvalue computation for quasiseparable Hermitian matrices under low rank perturbations. *Numer. Algorithms*, **47**, no. 3, 253–273 (2008).

[23] Eidelman, Y., Gohberg, I., Olshevsky, V., The QR iteration method for Hermitian quasiseparable matrices of an arbitrary order. *Linear Algebra Appl.*, **404**, 305–324 (2005).

[24] Eidelman, Y., Gohberg, I., Olshevsky, V., Eigenstructure of order-one-quasiseparable matrices. Three-term and two-term recurrence relations. *Linear Algebra Appl.*, **405**, 1–40 (2005).

[25] Eidelman, Y., Haimovici, I., Divide and conquer method for eigenstructure of quasiseparable matrices using zeroes of rational matrix functions. In: A panorama of modern operators theory and related topics, *Operator Theory: Advances and Applications*, **218**, 299–328, Birkhäuser/Springer, Basel, 2012.

[26] Fasino, D., Gemignani, L., Direct and inverse eigenvalue problems for diagonal-plus-semiseparable matrices. *Numer. Algorithms*, **34**, no. 2-4, 313–324 (2003).

[27] Gu, M., and Eisenstat, S., A divide-and-conquer algorithm for the symmetric tridiagonal eigenproblem. *SIAM J. Matrix Anal. Appl.*, **16**, no. 1, 172–191 (1995).

[28] Gu, M., Guzzo, R., Chi, X.-B., Cao, X.-Q., A stable divide and conquer algorithm for the unitary eigenproblem. *SIAM J. Matrix Anal. Appl.*, **25**, no. 2, 385–404 (2003).

[29] Golub, G.L, Van Loan, C.F., Matrix Computations, John Hopkins University Press, Baltimore 1989.

[30] Gragg, W., Reichel, L., A divide and conquer method for unitary and orthogonal eigenproblem. *Numer. Math.*, **57**, no. 8, 695–718 (1990).

[31] Lancaster, P., Tismenetsky, M., The Theory of Matrices, Academic Press, New York, London, 1985.

[32] Melman, A., Numerical solution of a secular equation. *Numer. Math.*, **69**, no. 4, 483–493 (1995).

[33] Mastronardi, N., Chandrasekaran, S., Van Huffel, S., Fast and stable reduction of diagonal plus semiseparable matrices to tridiagonal and bidiagonal form. *BIT*, **41**, no. 1, 149–157 (2001).

[34] Mastronardi, N., Van Camp, E., Van Barel, M., Divide and conquer algorithms for computing the eigendecomposition of symmetric diagonal-plus-semiseparable matrices. *Numerical Algorithms*, **9**, no. 4, 379–398 (2005).

[35] Stoer, J., Bulirsch, R., Introduction to Numerical Analysis, Springer-Verlag, New York, London, 1993.

[36] Vandebril, R., Van Barel, M., Mastronardi, N., Matrix Computations and Semiseparable Matrices. Volume II: Eigenvalues and Singular Value Methods, The John Hopkins University Press, Baltimore, 2008.

[37] Wilkinson, J.H., The Algebraic Eigenvalue Problem, Monographs on Numerical Analysis, The Clarendon Press Oxford University Press, New York, 1988.

Index

MIX
Paper from responsible sources
FSC® C105338

If you have any concerns about our products
you can contact us on
ProductSafety@springernature.com

In case Publisher is established outside the EU,
the EU authorized representative is:
Springer Nature Customer Service Center GmbH
Europaplatz 3, 69115 Heidelberg, Germany

Printed by Libri Plureos GmbH
in Hamburg, Germany